Catalysis by Materials with WELL-DEFINED STRUCTURES

Catalysis by Materials with
WELL-DEFINED STRUCTURES

Edited by

ZILI WU & STEVEN H. OVERBURY
Chemical Science Division,
Center for Nanophase Materials Sciences,
Oak Ridge National Laboratory, Oak Ridge, TN, USA

ELSEVIER

Amsterdam • Boston • Heidelberg • London • New York • Oxford
Paris • San Diego • San Francisco • Singapore • Sydney • Tokyo

Elsevier
Radarweg 29, PO Box 211, 1000 AE Amsterdam, Netherlands
The Boulevard, Langford Lane, Kidlington, Oxford OX5 1GB, UK
225 Wyman Street, Waltham, MA 02451, USA

ISBN: 978-0-12-801217-8

British Library Cataloguing in Publication Data
A catalogue record for this book is available from the British Library

Library of Congress Cataloging-in-Publication Data
A catalog record for this book is available from the Library of Congress

For Information on all Elsevier publications visit our website at http://store.elsevier.com/

CONTENTS

CONTRIBUTORS

S. Agarwal
Catalytic Processes and Materials, MESA+ Institute for Nanotechnology, University of Twente, AE Enschede, The Netherlands

Robert A. Bartynski
Department of Physics and Astronomy, and Laboratory for Surface Modification, Rutgers, The State University of New Jersey, Piscataway, NJ, USA

Tian Cao
Hefei National Laboratory for Physical Sciences at the Microscale, Collaborative Innovation Center of Suzhou Nano Science and Technology, CAS Key Laboratory of Materials for Energy Conversion and Department of Chemical Physics, University of Science and Technology of China, China

Yuxiang Chen
Department of Chemistry, Carnegie Mellon University, Pittsburgh, PA, USA

Wenhua Chen
Department of Physics and Astronomy, and Laboratory for Surface Modification, Rutgers, The State University of New Jersey, Piscataway, NJ, USA

A.K. Datye
University of New Mexico, Department of Chemical & Biological Engineering and Center for Microengineered Materials, Albuquerque, NM, USA

Maria Flytzani-Stephanopoulos
Department of Chemical and Biological Engineering, Tufts University, Medford, MA, USA

Pu-Xian Gao
Nanomaterials Science Laboratory, Department of Materials Science and Engineering & Institute of Materials Science, University of Connecticut, Storrs, CT, USA

Yejun Guan
Shanghai Key Laboratory of Green Chemistry and Chemical Processes, Department of Chemistry, East China Normal University, Shanghai, China

Yanbing Guo
Nanomaterials Science Laboratory, Department of Materials Science and Engineering & Institute of Materials Science, University of Connecticut, Storrs, CT, USA

Emiel J.M. Hensen
Laboratory of Inorganic Materials Chemistry, Schuit Institute of Catalysis, Department of Chemical Engineering and Chemistry, Eindhoven University of Technology, Eindhoven, The Netherlands

Weixin Huang
Hefei National Laboratory for Physical Sciences at the Microscale, Collaborative Innovation Center of Suzhou Nano Science and Technology, CAS Key Laboratory of Materials for Energy Conversion and Department of Chemical Physics, University of Science and Technology of China, China

Yu-Xi Huang
Department of Chemistry, Brown University, Providence, Rhode Island, USA

Rongchao Jin
Department of Chemistry, Carnegie Mellon University, Pittsburgh, PA, USA

L. Lefferts
Catalytic Processes and Materials, MESA+ Institute for Nanotechnology, University of Twente, AE Enschede, The Netherlands

Gao Li
Department of Chemistry, Carnegie Mellon University, Pittsburgh, PA, USA

Yuyuan Lin
Northwestern University, Evanston, IL, USA

Amanda K.P. Mann
Analytical Sciences, Merck Research Laboratories, Rahway, NJ, USA;
Chemical Science Division, Oak Ridge National Laboratory, Oak Ridge, TN, USA

Laurence D. Marks
Northwestern University, Evanston, IL, USA

James A. McCarthy
Northwestern University, Evanston, IL, USA

B.L. Mojet
Catalytic Processes and Materials, MESA+ Institute for Nanotechnology, University of Twente, AE Enschede, The Netherlands

Melanie Moses-DeBusk
Energy & Transportation Sciences Division, Oak Ridge National Laboratory, Oak Ridge, TN, USA

Chaitanya K. Narula
Materials Science & Technology Division, Oak Ridge National Laboratory, Oak Ridge, TN, USA

Michael Nolan
Materials Modelling for Devices Group, Tyndall National Institute, University College Cork, Lee Maltings, Dyke Parade, Cork, Ireland

Steven H. Overbury
Chemical Science Division, Oak Ridge National Laboratory, Oak Ridge, TN, USA;
Center for Nanophase Materials Sciences, Oak Ridge National Laboratory, Oak Ridge, TN, USA

Kenneth R. Poeppelmeier
Northwestern University, Evanston, IL, USA

Huifeng Qian
Department of Chemistry, Carnegie Mellon University, Pittsburgh, PA, USA

Zheng Ren
Nanomaterials Science Laboratory, Department of Materials Science and Engineering & Institute of Materials Science, University of Connecticut, Storrs, CT, USA

Weiyu Song
Laboratory of Inorganic Materials Chemistry, Schuit Institute of Catalysis, Department of Chemical Engineering and Chemistry, Eindhoven University of Technology, Eindhoven, The Netherlands

Shouheng Sun
Department of Chemistry, Brown University, Providence, Rhode Island, USA

Hao Wang
Department of Physics and Astronomy, and Laboratory for Surface Modification, Rutgers, The State University of New Jersey, Piscataway, NJ, USA

Zili Wu
Chemical Science Division, Oak Ridge National Laboratory, Oak Ridge, TN, USA; Center for Nanophase Materials Sciences, Oak Ridge National Laboratory, Oak Ridge, TN, USA

Wen Xiao
Nanomaterials Science Laboratory, Department of Materials Science and Engineering & Institute of Materials Science, University of Connecticut, Storrs, CT, USA

Nan Yi
Department of Chemical and Biological Engineering, Tufts University, Medford, MA, USA

Sen Zhang
Department of Chemistry, Brown University, Providence, Rhode Island, USA

Huiyuan Zhu
Department of Chemistry, Brown University, Providence, Rhode Island, USA

structure with atomic resolution. In Chapter 7 Lin et al. describe recent advances in electron microscopy that remarkably have made it possible to achieve direct images of surface atomic structure, including even the imaging of the columns of low Z scatters such as the crucial O atoms on an oxide surface, as well as resolving single atoms on supports. Of great importance for catalysis, it is now possible to monitor the effects of adsorbates and reactive atmosphere as described by Lin et al.

Other approaches, distinct from collections of crystallographically controlled nanoparticles, are also described herein. Catalytic structures of a smaller size scale are considered. Specifically, these include Au clusters where researchers make use of the remarkable ability to create size-specific clusters with precisely defined numbers of atoms. In the Chapter 8, Jin et al. describe the use of "size focusing" in the synthesis of n-atom Au-thiolate clusters, Au_nSR_m, where n can assume specific values from 25 to 333. These model clusters provide a new approach to understand catalysis by Au and the authors review use of these cluster catalysts, both supported and in homogeneous media, for selective oxidation and selective hydrogenation. Achieving the small size limit, in Chapter 9 Narula and DeBusk describe catalysts where the active sites consist of single atoms of Pt or Ir localized in well-characterized sites of Fe or Al oxides supports. Catalysts with well-defined metal, alloy, and core–shell particles can also be created and in Chapter 10, the use of such particles in electrocatalytic reactions is reported.

Beyond working with isolated clusters or particles, another approach is to build aligned structures on monolithic surfaces as a method to explore structure-dependent surface chemistry. In Chapter 11 Chen, Wang, and Bartynski, describe how surface faceting on a monolithic support can alter surface structure and therefore surface chemistry. They compare faceted vs unfaceted surfaces to explore structure sensitivity, with the in situ ability to continuously alter facet size. Another form of monolithically supported catalysts result from synthetic approaches to create arrays of free-standing 1D nanostructures with consistent morphologies assembled onto solid substrates. In Chapter 12 Gao et al. review the latest progress in the fabrication, characterization, and catalytic performance of representative types of nanoarrays. These nanoarrays may be useful for thermochemical, electro, and photocatalysts.

Overall, this book provides a snap view of recent progress in catalysis research making use of model materials with well-defined structures. The knowledge generated from studying these uniform model catalysts sheds light on the structure–catalysis relationship and provides basis for designing better catalytic materials.

Enjoy!

Steven H. Overbury and Zili Wu
Oak Ridge National Laboratory, TN, USA
September 2014

ACKNOWLEDGMENTS

There are a number of people that we wish to thank for all their efforts without whom this book would not have been possible. First and foremost, we need to thank all the authors for their hardwork in completing a great selection of chapters. We would like to thank Dr Kostas Marinakis and Ms Christine McElvenny at Elsevier for their help and patience during the compilation of this book. We are also grateful for the reviewers for their critical comments and suggestions on the chapters. We acknowledge the financial support from Oak Ridge National Laboratory and the U.S. Department of Energy, Office of Science, Basic Energy Sciences, Chemical Sciences, Geosciences, and Biosciences Division.

Zili Wu and Steven H. Overbury
Chemical Science Division
Center for Nanophase Materials Sciences
Oak Ridge National Laboratory
1 Bethel Valley Road, Oak Ridge, TN 37831, USA

Cover art by Yuyuan Lin

CHAPTER 1

Surface Chemistry and Catalytic Properties of Well-Defined Cu_2O Nanocrystals

Weixin Huang*, Tian Cao
Hefei National Laboratory for Physical Sciences at the Microscale, Collaborative Innovation Center of Suzhou Nano Science and Technology, CAS Key Laboratory of Materials for Energy Conversion and Department of Chemical Physics, University of Science and Technology of China, China
*Corresponding author: E-mail: huangwx@ustc.edu.cn

1. INTRODUCTION

Catalysts are the heart of technologies for chemical and materials synthesis, fuel production, power generation and conversion, and environmental remediation. As the resource (energy) shortage and environmental pollution become serious, green chemical processes must be developed; accordingly, the catalyst must be not only active but also highly selective. To achieve this goal, fundamental understanding of structure–activity relation of catalysts is needed for the structural design of highly selective catalysts; meanwhile, the strategy for the controlled synthesis of catalyst with the designed structure is also needed.

Heterogeneous catalysis always occurs on the surface of solid catalyst, thus the surface composition and surface electronic/geometric structures of a solid particle cooperatively determine its catalytic property [1]. As schematically illustrated in Figure 1, the surface composition of a solid particle is determined by the composition and crystal planes exposed on the surface (for oxides), the surface electronic structure of a solid particle is determined by the composition and size, and the surface geometric structure of a solid particle is determined by the crystal planes exposed on the surface. According to the Wulff's rule [2], the crystal planes exposed on the surface of a crystalline are determined by its morphology. Therefore, macroscopically the catalytic property of a solid particle varies with their composition, size, and morphology. For multicomponent catalysts, the interfacial structure between adjacent components is also vital for their catalytic property. Thus fundamental studies of heterogeneous catalysis aim to establish the size/morphology-surface structure–catalytic property relation of solid catalysts in which one of the greatest challenges is the diversity of catalyst particles within a catalyst in their sizes and morphologies. Although we are able to comprehensively characterize the catalyst surface structure, surface adsorbates/intermediates and catalytic property, the inhomogeneity of catalyst particles in their sizes and morphologies makes the unambiguous correlation between the catalyst surface structure and the catalytic property very difficult. With this respect, a model catalyst approach

Catalysis by Materials with Well-Defined Structures
http://dx.doi.org/10.1016/B978-0-12-801217-8.00001-3

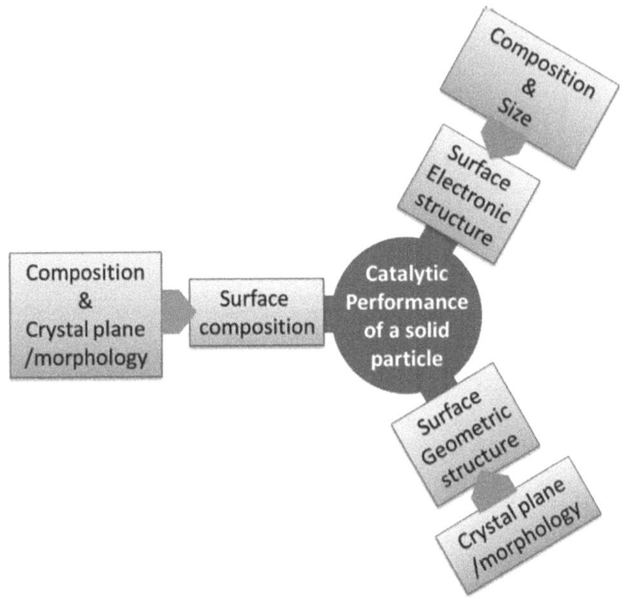

Figure 1 Schematic illustration of microscopic and macroscopic structural parameters of a solid particle that affect its catalytic property.

has been developed to employ the uniform and well-defined surface as the model surface to establish the surface structure–catalytic property relation [3]. Traditional model catalysts are mainly based on single crystals with very low specific surface areas and the studies are usually performed under ultrahigh vacuum conditions. Single crystals-based model catalysts are very useful in addressing the effect of crystal planes exposed on a solid surface on its catalytic property [4–6]; however, the so-called "materials gap" and "pressure gap" exist between single crystals-based model catalysts and corresponding powder catalysts. Powder catalysts consist of supported three-dimensional nanoparticles exhibiting a large specific surface area and the catalytic reactions are carried out at atmospheric or higher pressures, whereas single crystals-based model catalysts are extended two-dimensional surfaces exhibiting a low specific surface area and the reaction pressure usually does not exceed several mbar. Thus the structure–catalytic property relation learned from single crystals-based model catalysts sometimes cannot be simply extended to practical heterogeneous catalytic reactions and meanwhile several issues in practical heterogeneous catalytic reactions cannot be adequately approached using single crystals-based model catalysts [7–11].

Benefiting from the recent great advancement of nanotechnology, nanocrystals of catalytic materials with uniform composition and structure (size and morphology) can be successfully prepared [12–15]. Because of their uniform composition and structure, these nanocrystals constitute a novel type of model catalysts to explore the surface structure–catalytic property relation. The study of structure–catalytic property relation employing nanocrystals-based model catalysts can be carried out under the working conditions of practical heterogeneous catalytic reactions, thus nanocrystals-based model catalysts can bridge the "materials gap" and "pressure gap" between single crystals-based model catalysts and corresponding powder catalysts. Oxides are widely used as catalysts and catalyst supports. Consisting of cations and anions, oxides exhibit more complicated surface composition and structure than metals. Crystal planes exposed on the oxide catalyst surface determine both surface composition and surface structure and thus strongly affect the catalytic property of oxide catalyst. Since the morphology of oxide nanocrystals determines their exposed crystal planes, oxide nanocrystals with uniform morphology provide a nice platform to study the crystal plane effect for oxide catalysts. In recent years, there has been increasing number of examples reporting the morphology-dependent catalytic property of oxide nanocrystals, such as MoO_3 [16,17], CeO_2 [18–36], MgO [37–39], Co_3O_4 [40–46], and Fe_2O_3 [47–51]. Several nice reviews have also been published on this topic [52–57].

Cuprous oxide (Cu_2O) has a *cubic* crystal structure (Space group of $O^4{}_K$-Pn3m) with a unit cell length of 4.27 Å Cu_2O is a nonstoichiometric defect p-type semiconductor (direct band gap: 2.17 eV) [58] and widely applied as an active component of catalysts for CO oxidation, water–gas shift reaction, CO hydrogenation reaction, partial oxidation of propylene, organic synthesis, photocatalysis, and photoelectrocatalysis [59–72]. Meanwhile, uniform Cu_2O nanocrystals that selectively expose one or two types of crystal planes with well-defined surface structures have been successfully synthesized and recently their surface chemistry and catalytic properties have been much explored. In this chapter, the progresses on the surface chemistry and catalytic properties of well-defined Cu_2O nanocrystals are reviewed, with which we aim to establish the influence of surface composition and surface structure of Cu_2O catalysts on their surface chemistry and catalytic properties and exemplify the concept of nanocrystals-based model catalysts. We begin with a brief review of the synthesis of well-defined Cu_2O nanocrystals, and then review the surface composition and surface structure of well-defined Cu_2O nanocrystals, the surface chemistry of well-defined Cu_2O nanocrystals in chemical reactions involving the breaking of Cu(I)–O bonds, the catalytic properties of well-defined Cu_2O nanocrystals, and finally we will give the summary and outlook.

2. SYNTHESIS OF WELL-DEFINED Cu_2O NANOCRYSTALS

The synthesis of Cu_2O crystals with various particle sizes and morphologies has attracted extensive studies and achieved great success. Since McFadyen and Matijević systematically studied the particle size and shape of Cu_2O crystals prepared by reducing copper(II) tartrate complex with glucose as early as in 1973 [73], Cu_2O crystals with novel nanostructures including hierarchical architectures, nanowires, nanocages, multipods, nanospheres, hollow structures, and various polyhedra have been synthesized employing different methods. The comprehensive progresses on the synthesis of Cu_2O crystals have demonstrate that PVP can preferentially adsorb on the Cu_2O {111} crystal planes and thus suppress their growth. Zhang et al. prepared Cu_2O octahedra by reducing a mixed solution of $CuSO_4$, sodium tartrate, and PVP with glucose at 60 °C [89]. The acquired octahedral morphology of Cu_2O crystals was found to be independent on the concentrations of the reactants and their sizes varied with the amount of added PVP.

The reported synthesis methods for uniform Cu_2O rhombic dodecahedra enclosed with 12 {110} crystal planes are not as many as those for uniform Cu_2O cubes and octahedra. Liang et al. successfully synthesized uniform rhombic dodecahedral Cu_2O microcrystals by reducing a mixed solution of $CuSO_4$, NaOH, and anhydrous alcohol using glucose at 60 °C in the presence of appropriate amount of oleic acid (OA) that acting as the capping ligand preferentially adsorbed on Cu_2O {110} crystal planes [90]. Yao et al. prepared Cu_2O rhombic dodecahedra by reducing $Cu(CH_3COO)_2$ in a mixed solution of 1-hexadecylamine and undecane between 160 and 220 °C [91].

In several synthesis methods listed in Table 1, the morphology of Cu_2O polyhedra could be continuously changed from one to another by controlling the amount of added capping ligand, and in between Cu_2O polyhedra selectively exposing two types of crystal planes could be synthesized. In Zhang et al.'s method [85], by increasing the amount of added PVP, the acquired Cu_2O polyhedra varied from uniform cubes to uniform octahedra via uniform 14-facet polyhedra (corner-truncated octahedra) exposing the six {100} crystal planes and eight {111} crystal planes with different ratios. In Liang et al.'s method [90], by increasing the amount of added OA, the acquired Cu_2O polyhedra varied from uniform octahedra to uniform rhombic dodecahedra via uniform 20-facet polyhedra (edge-truncated octahedra) exposing the 12 {110} crystal planes and eight {111} crystal planes. 18-facet Cu_2O polyhedra exposing 6 {100} crystal planes and 12 {110} crystal planes were also synthesized by reducing a mixed solution of $Cu(CH_3COO)_2$ and NaOH at 80 °C using vitamin C [84] and by reducing a mixed solution of $CuSO_4$ and NaOH at 60 °C using ascorbic acid [105].

3. SURFACE COMPOSITION/STRUCTURE OF WELL-DEFINED Cu_2O NANOCRYSTALS

Figure 2 shows scanning electron microscope (SEM), transmission electron microscope (TEM), and high-resolution transmission electron microscope (HRTEM) images of Cu_2O cubes, octahedra, and rhombic dodecahedra that we synthesized. The synthesis of Cu_2O cubes, and octahedra followed Zhang et al.'s recipe [85] and the synthesis of Cu_2O rhombic dodecahedra followed Liang et al.'s recipe [90]. PVP and OA capping ligands were employed respectively to prepare Cu_2O octahedra and rhombic dodecahedra while no capping ligand was used to prepare Cu_2O cubes. All three types of Cu_2O nanocrystals exhibit uniform morphologies. Cu_2O cubes (denoted as *c*-Cu_2O) are with sizes between 400 and 700 nm and selectively expose six {100} crystal planes, Cu_2O octahedra (denoted

Figure 2 SEM, TEM, and HRTEM images of Cu_2O cubes (A1–A5), octahedra (B1–B5), and rhombic dodecahedra (C1–C5). The inserts show the electronic diffraction pattern of corresponding Cu_2O nanocrystals. 0.43 and 0.30 nm respectively correspond to the lattice fringe of {001} and {110} crystal planes of *cubic* Cu_2O.

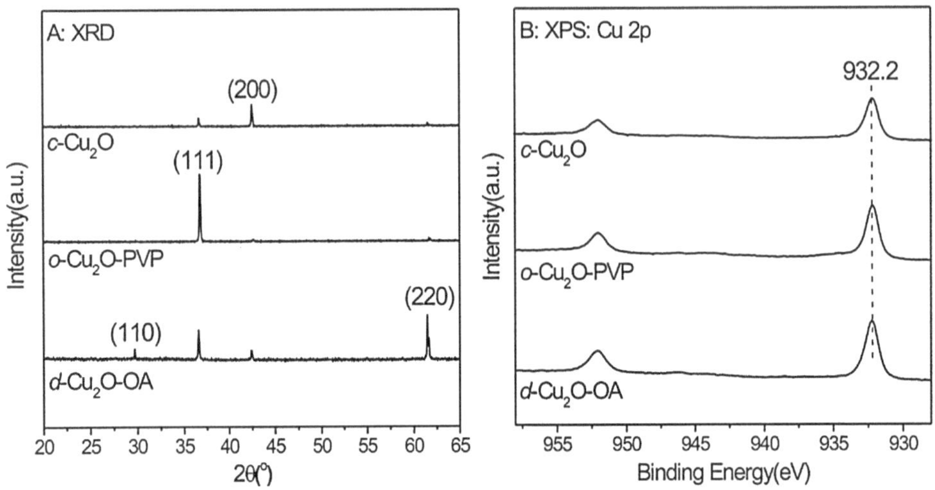

Figure 3 XRD patterns and Cu 2p XPS spectra of Cu_2O cubes (*c*-Cu_2O), octahedra (*o*-Cu_2O-PVP), and rhombic dodecahedra (*d*-Cu_2O-OA).

as *o*-Cu_2O-PVP) are with sizes between 300 and 600 nm and selectively expose eight {111} crystal planes, and Cu_2O rhombic dodecahedra (denoted as *d*-Cu_2O-OA) are with sizes between 600 and 900 nm and selectively expose 12 {110} crystal planes. X-ray diffraction (XRD) patterns (Figure 3(A)) demonstrate that all three types of Cu_2O nanocrystals are well-crystallized with the *cubic* Cu_2O phase structure (JCPDS card no. 78-2076) [113]. The {200} diffraction peak in the XRD pattern of *c*-Cu_2O exhibits a much stronger intensity than the {111} diffraction peaks, in consistence with the fact that they exclusively expose {100} crystal planes. The {111} diffraction peak dominates the XRD pattern of *o*-Cu_2O-PVP, in consistence with the fact that they exclusively expose {111} crystal planes. The {220} diffraction peak in the XRD pattern of *d*-Cu_2O-OA exhibits a much stronger intensity than the {111} and {200} diffraction peaks, in consistence with the fact that they exclusively expose {110} crystal planes. The Cu 2p x-ray photoelectron spectroscopy (XPS) spectra of all three types of Cu_2O nanocrystals (Figure 3(B)) show a single component with a Cu $2p_{3/2}$ binding energy at 932.2 eV corresponding to Cu(I) in Cu_2O, demonstrating that their surface compositions remain as Cu_2O [113].

c-Cu_2O cubes, *o*-Cu_2O-PVP, and *d*-Cu_2O-OA selectively expose Cu_2O (100), (111), and (110) crystal planes, respectively. We have optimized the structures of Cu_2O(100), (111), and (110) surfaces within the spin-polarized density functional theory (DFT) implemented in the DMol package whose results are shown in Figure 4 [113,114]. The structural parameters of optimized Cu_2O(100), (111), and (110) surfaces are summarized in Table 2. The data of bulk Cu_2O are also included in Figure 4 and Table 2 for comparison. In the bulk Cu_2O, coordinated saturated Cu is two-coordinated (Cu_{CSA}) and coordinated saturated O is four-coordinated (O_{CSA}), and the Cu–O bond length is 1.85 Å. Comparing with the bulk Cu_2O,

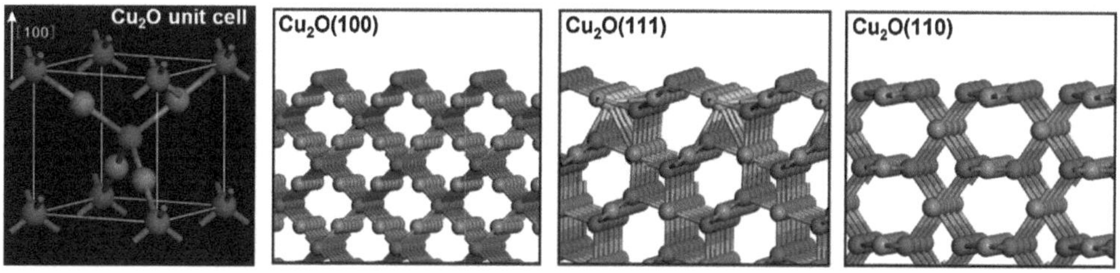

Figure 4 Unit cell of Cu_2O and optimized structures of $Cu_2O(100)$, (111), and (110) surfaces. The red (dark gray in print versions), brick red (light gray in print versions), and green (lighter gray in print versions) balls represent oxygen, coordinated saturated copper, and coordinated unsaturated copper atoms, respectively.

Table 2 Structural parameters of bulk Cu_2O and optimized $Cu_2O(100)$, (111), and (110) surfaces

	Topmost layer	Second layer	d_{Cu-O} (Å)
Bulk Cu_2O	Two-coordinated Cu_{CSA} and four-coordinated O_{CSA}		1.85
$Cu_2O(100)$	Two-coordinated O_{CUS}	Two-coordinated Cu_{CSA}	1.76
$Cu_2O(111)$	Three-coordinated O_{CUS}	Two-coordinated Cu_{CSA} (75%)	1.83
		One-coordinated Cu_{CUS} (25%)	1.91
$Cu_2O(110)$	Three-coordinated O_{CUS} two-coordinated Cu_{CSA}	Two-coordinated Cu_{CSA}	1.82

Cu_2O (100), (111), and (110) surfaces are with different surface compositions and their structures relax more or less. On $Cu_2O(100)$, the topmost layer consists of two-coordinated O (O_{CUS}) and the second layer consists of coordinated saturated Cu (Cu_{CSA}). The bond length between O_{CUS} and Cu_{CSA} is 1.76 Å. On $Cu_2O(111)$, the topmost layer consists of three-coordinated O_{CUS} and the second layer consists of Cu_{CSA} (75%) and one-coordinated unsaturated Cu (Cu_{CUS}) (25%) with a distance along z-direction of 0.23 Å. The O_{CUS} is coordinated to Cu_{CSA} with a bond length of 1.83 Å, and the Cu_{CUS} is coordinated to underneath O_{CSA} with a bond length of 1.91 Å. On $Cu_2O(110)$, the topmost layer consists of three-coordinated O_{CUS} and Cu_{CSA} and the second layer consists of Cu_{CSA}. The bond length between O_{CUS} and Cu_{CSA} is 1.82 Å. Thus *c*-Cu_2O cubes, *o*-Cu_2O-PVP, and *d*-Cu_2O-OA exhibit different surface compositions and structures.

4. SURFACE CHEMISTRY OF WELL-DEFINED CU_2O NANOCRYSTALS

Cu_2O possesses Cu(I) that can either be reduced to Cu(0) or oxidized to Cu(II), thus it exhibits very rich chemical reactivity. For chemical reactions involving solid reactants, the reaction always occurs on the solid surface. Therefore, similar to its catalytic property, the reactivity of a solid particle in chemical reactions is also determined by the surface composition/structure; however, different from catalytic reactions where the catalyst particle keeps intact, the surface chemical bonds of a solid particle get broken, and the solid particle reacts off in chemical reactions.

As-synthesized Cu_2O nanocrystals were observed to undergo the structural evolution with their local chemical environment [115–121] and were employed as the sacrificial template for the synthesis of nanocomposites [122–126]. Interestingly, the reactivity of Cu_2O nanocrystals in all these liquid–solid chemical reactions depends on their morphology. Octahedral Cu_2O nanocrystals were reported to transform into octahedral Cu_2O nanocages and corner-etched Cu_2O octahedra in Fehling's solutions containing glucose and $PdCl_2$ without and with the presence of PVP, respectively [115]. Kuo and Huang reported the morphological transformation from {100} truncated rhombic dodecahedral Cu_2O nanocages to Cu_2O nanoframes with elliptical pores on the {110} faces in an aqueous HCl solution with the addition of ethanol [116]. Sui et al. reported that polyhedral Cu_2O nanoparticles exposed to air at room temperature could undergo oxidative etching and transform to Cu_2O nanoframes and nanocages with single crystal walls [118]. Liu reported the selective deposition of Au nanoparticles on the {111} crystal planes of corner-truncated Cu_2O octahedra exposing {100} and {111} crystal planes during the redox reaction of corner-truncated Cu_2O octahedra and $HAuCl_4$ aqueous solution [125]. Sun et al. reported the selective growth of Cu nanoparticles on the {111} crystal planes of 26-facet Cu_2O polyhedra exposing {100}, {111}, and {110} crystal planes when they were reduced by hydrazine hydrate [126].

We have comparatively investigated the reactivity of *c*-Cu_2O cubes, *o*-Cu_2O-PVP, and *d*-Cu_2O-OA in typical chemical reactions involving Cu_2O [113,114,127–129]. The first reaction the oxidative dissolution reaction of Cu_2O in a weak acetic acid solution following the chemical equation of $Cu_2O+2H^+ \rightarrow Cu^{2+}+H_2O$ [113]. After etched in acetic acid solution ($pH=3.5$) for 150 min, XRD and XPS results confirm that the phase structure and surface composition of resulting etched nanocrystals are pure Cu_2O, but *c*-Cu_2O cubes, *o*-Cu_2O-PVP, and *d*-Cu_2O-OA exhibit different morphological evolutions (Figure 5). Etched *c*-Cu_2O well keep their original morphology and each facet is smooth although some pinholes and adlayers form on the surfaces; however, etched *o*-Cu_2O-PVP and *d*-Cu_2O-OA keep their original morphologies but the original smooth facets are destroyed and stepped layers of square facets with different sizes develop. As demonstrated by HRTEM images, the surfaces of etched *c*-Cu_2O expose the original {100} crystal planes, but for etched Cu_2O octahedra and rhombic dodecahedra, the originally exposed {111} and {110} crystal planes disappear whereas new microfacets exposing the {100} crystal planes form and grow. These results demonstrate that Cu_2O {100} crystal planes are most stable among the investigated Cu_2O {100}, {111}, and {110} crystal planes during the oxidative dissolution reaction of Cu_2O. Employing {110}-truncated octahedral Cu_2O nanocrystals that expose both {111} and {110} crystal planes, the {111} crystal planes were found to be slightly more stable than the {110} crystal planes. Therefore the stability of Cu_2O {100}, {110}, and {111} crystal planes in the oxidative dissolution reaction in a weak acid solution follows the order of $\{100\} >> \{111\} \geq \{110\}$. The stability order established by microscopic results is supported by the amount of Cu(II) formed as the product of the oxidative

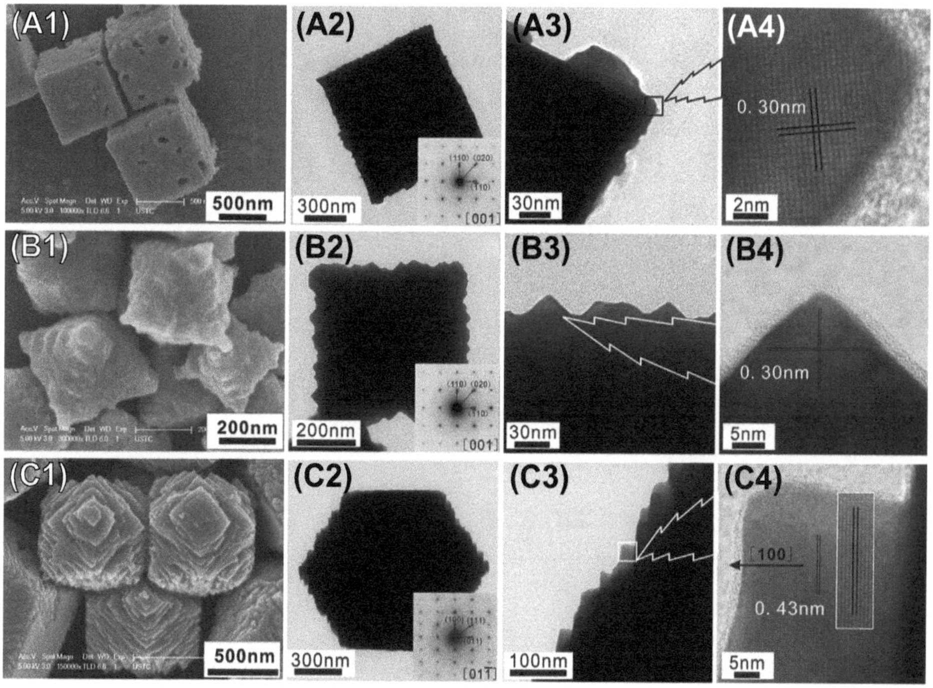

Figure 5 SEM, TEM, and HRTEM images of (A1–A4) Cu_2O cubes, (B1–B4) octahedra, and (C1–C4) rhombic dodecahedra etched in acetic acid solution (pH = 3.5) for 150 min. 0.43 and 0.30 nm respectively correspond to the lattice fringe of {001} and {110} crystal planes of *cubic* Cu_2O. *Adapted from reference [113]. Copyright 2011 American Chemical Society.*

dissolution reaction of Cu_2O that follows the order of *c*-Cu_2O cubes<<*o*-Cu_2O-PVP≤*d*-Cu_2O-OA (Figure 6). The stability order of Cu_2O (100}, (110), and (111) crystal planes can be well rationalized by their surface structures. As listed in Table 2, the Cu–O bond on Cu_2O(100) surface is 1.76 Å, much shorter than that on Cu_2O(111) and (110) surfaces. For the liquid–solid reactions that need to break the surface Cu–O bond of Cu_2O, the Cu_2O(100) crystal plane with the shortest and thus the strongest surface Cu–O bond reasonably exhibit a much higher stability than Cu_2O(111) and (110) crystal planes. In addition to the Cu_{CSA}–O_{CUS} bond of 1.83 Å, the Cu_2O(111) surface exhibit a weak Cu_{CUS}–O_{CSA} bond of 1.91 Å, but the break of this bond only releases the Cu atom but not the O atom and thus does not result in the reaction of Cu_2O(111) surface.

We propose a likely mechanism for the morphological evolution of various Cu_2O nanocrystals in weak acid solution, as schematically illustrated in Figure 7. Since Cu_2O(100) is most stable, the morphological evolution always tends to expose the {100} crystal planes on Cu_2O nanocrystals as much as possible, no matter which crystal planes the original Cu_2O nanocrystals expose. *c*-Cu_2O already exclusively exposing the {100} crystal planes retains their shape during the course of oxidative dissolution reaction (type I mechanism). The

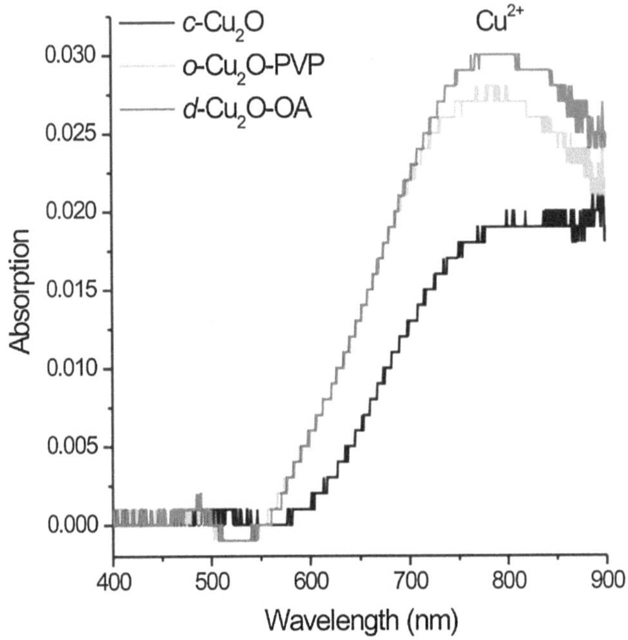

Figure 6 UV–vis spectra of the aqueous solutions after Cu_2O cubes (*c*-Cu_2O), octahedra (*o*-Cu_2O-PVP), and rhombic dodecahedra were etched in acetic acid solution (pH = 3.5) for 150 min. *Reproduced from reference [113]. Copyright 2011 American Chemical Society.*

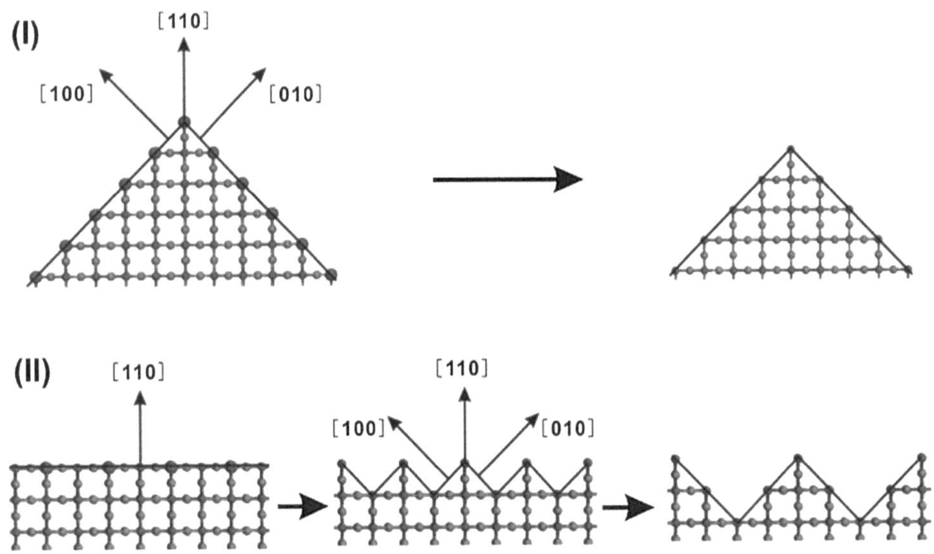

Figure 7 Likely mechanism for the morphological evolution of various Cu_2O nanocrystals during the oxidative dissolution reaction in weak acid solution. *Adapted from reference [113]. Copyright 2011 American Chemical Society.*

morphological evolution of *d*-Cu_2O-OA follows the type II mechanism. The {100} crystal planes will remain as long as they are formed by the likely dissolution of some ions on the Cu_2O {110} crystal planes as indicated by the dashed lines in type II mechanism. This will form square facets exposing the {100} crystal planes on *d*-Cu_2O-OA. The morphological evolution of *o*-Cu_2O-PVP exposing the {111} crystal planes initiates from the edge of octahedron whose normal direction is along [109] and thus also follows the type II mechanism.

The second reaction is the reaction of Cu_2O in aqueous ammonia solution in which three consecutive reactions occurs: the coordinative dissolution reaction of Cu_2O following the equation of $Cu_2O+8NH_3+H_2O \rightarrow 2[Cu(NH_3)_4]^+ + 2OH^-$, the subsequent oxidation reaction of $[Cu(NH_3)_4]^+$ to $[Cu(NH_3)_4]^{2+}$, and the precipitation reaction of $[Cu(NH_3)_4]^{2+}$ with OH^- to form $Cu(OH)_2$ precipitate [127]. The morphology of Cu_2O cubes, octahedra, and rhombic dodecahedra evolves in different ways during this reaction system (Figure 8). The reactivity of *c*-Cu_2O cubes, *o*-Cu_2O-PVP, and *d*-Cu_2O-OA in the coordinative dissolution reaction of Cu_2O to form $[Cu(NH_3)_4]^+$ follows the same order as that in the oxidative dissolution reaction in a weak acid: *c*-Cu_2O cubes $<<$ *o*-Cu_2O-PVP $\leq$ *d*-Cu_2O-OA. This further supports our conclusion that the surface structure of crystal planes exposed on Cu_2O nanocrystals determines their reactivity in chemical reactions. In the subsequent precipitation reaction of $[Cu(NH_3)_4]^{2+}$ with OH^- to form $Cu(OH)_2$ precipitate, the nucleation and growth of $Cu(OH)_2$ nanostructures also depend on the Cu_2O morphology. For *c*-Cu_2O, $Cu(OH)_2$ nanostructures nucleate and grow on the Cu_2O surfaces, forming dense $Cu(OH)_2$ shells (Figure 8(A) and(B)) that prevent the coordinative dissolution reaction of Cu_2O and further enhance the stability of *c*-Cu_2O. For *o*-Cu_2O-PVP, initially no nucleation and grow of $Cu(OH)_2$ nanostructures could be observed on the Cu_2O surfaces and *o*-Cu_2O-PVP well preserves their original morphology and exposes the {111} crystal planes (Figure 8(C)) although corresponding XPS and XRD results demonstrate the occurrence of the reactions, with the reactions proceeding sparsely distributed $Cu(OH)_2$ nanostructures nucleate and grow on the Cu_2O surfaces (Figure 8(D)). For *d*-Cu_2O-OA, some $Cu(OH)_2$ nanostructures nucleate and grow on the Cu_2O surfaces, forming loosely packed $Cu(OH)_2$ shells (Figure 8(E) and (F)). These observations could also be rationalized by the crystal plane-dependent surface composition and stoichiometric of Cu_2O nanocrystals that affect the local concentrations of $[Cu(NH_3)_4]^+$ and OH^- at the liquid–solid interface. The etching of Cu_2O nanocrystals in aqueous ammonia solution initiates from the coordination of Cu(I) in Cu_2O with NH_3 to produce $[Cu(NH_3)_4]^+$, and then oxygen anions bonded with Cu(I) in Cu_2O subsequently hydrolysis to produce OH^-. On $Cu_2O(100)$ surface, O_{CUS} in the topmost layer simultaneously gets released and hydrolysis to produce OH^- as soon as Cu_{CSA} in the second layer initially occurs the coordination reaction. The surface Cu:O stoichiometric on $Cu_2O(100)$ is 1, thus the $[OH^-]$ is high at the liquid–Cu_2O interface, facilitating the occurrence of $Cu(OH)_2$ precipitation reaction. Meanwhile the surface composition of $Cu_2O(100)$ is homogeneous. These microscopic factors cooperatively result in the observed

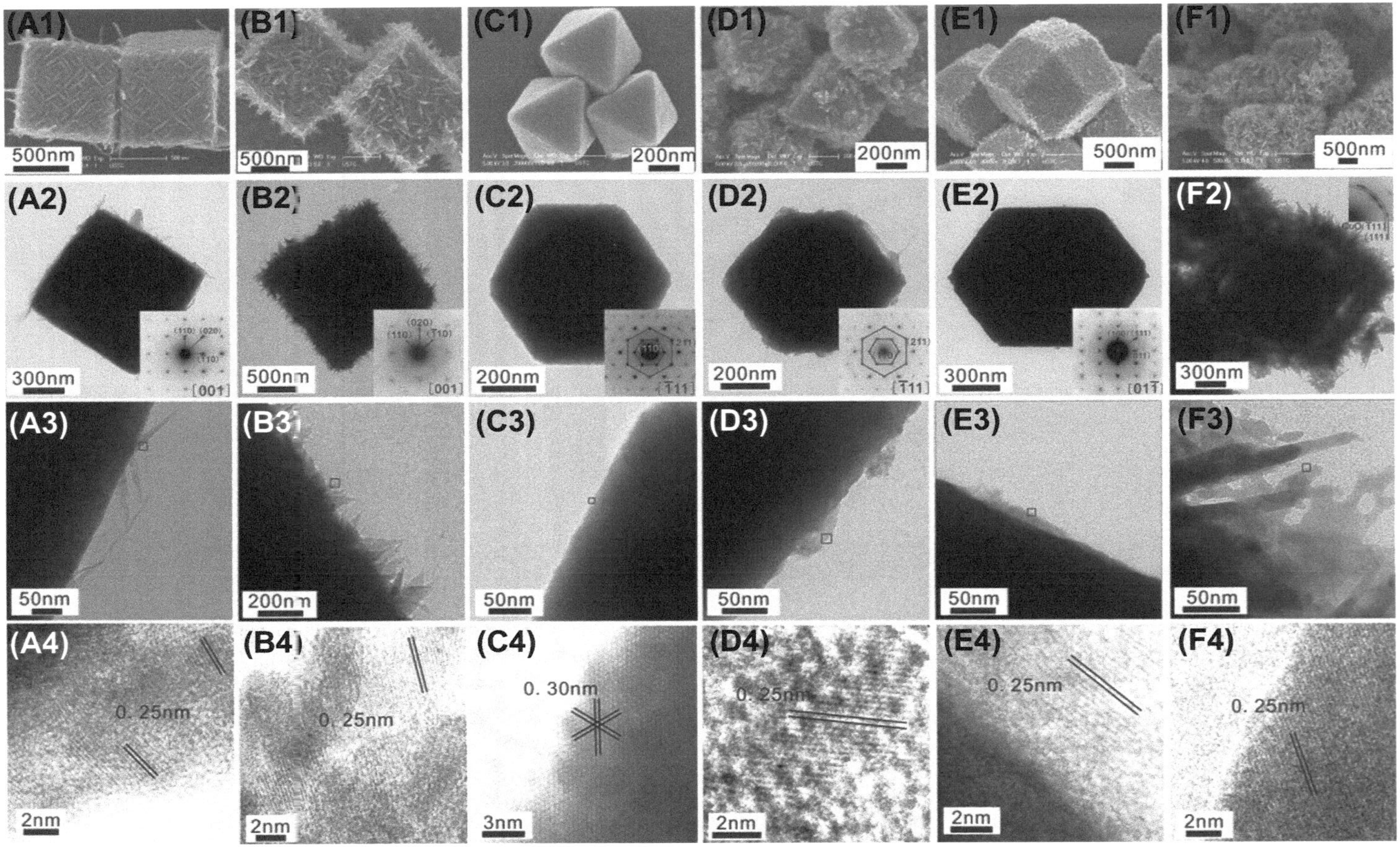

Figure 8 SEM, TEM, and HRTEM images of Cu_2O nanocrystals etched in aqueous ammonia solution with pH = 10.5 for different times: Cu_2O cubes for 10 (A) and 150 min (B), Cu_2O octahedra for 10 (C) and 150 min (D), Cu_2O rhombic dodecahedra for 10 (E) and 150 min (F). The HRTEM image was acquired from the region indicated by the square in the corresponding TEM image. 0.30 nm corresponds to the lattice fringe of {110} crystal planes of *cubic* Cu_2O and 0.25 nm corresponds to the lattice fringe of (111) crystal planes of $Cu(OH)_2$. *Adapted from reference [127]. Copyright 2011 American Chemical Society.*

homogeneous nucleation and growth of $Cu(OH)_2$ precipitate on the surfaces of *c*-Cu_2O. On $Cu_2O(111)$ surface, due to its weak Cu–O bond, Cu_{CUS} in the second layer is more reactive than Cu_{CSA} and preferentially undergoes the coordinative reaction, but the formation of $[Cu(NH_3)_4]^+$ from Cu_{CUS} only reduces the coordination number of O_{CSA} in the third layer of $Cu_2O(111)$ to which it is bonded and does not release any free O anions, thus the local OH^- concentration at the liquid–Cu_2O interface does not increase with the formation of $[Cu(NH_3)_4]^+$ from Cu_{CUS} and the formed $[Cu(NH_3)_4]^+$ can migrate away from the liquid–Cu_2O interface and undergo the subsequent oxidation and precipitation reaction in the aqueous ammonia solution. Therefore no $Cu(OH)_2$ nanostructures nucleates on the Cu_2O surfaces during the initial etching reaction of *o*-Cu_2O-PVP in aqueous ammonia solution. On $Cu_2O(110)$, three-coordinated O_{CUS} is bonded to one Cu_{CSA} in the topmost layer and the other Cu_{CSA} in the second layers and gets released only when both Cu_{CSA} cations simultaneously undergo the coordination reaction. Meanwhile the surface Cu:O stoichiometric on $Cu_2O(110)$ is larger than that on $Cu_2O(100)$. Therefore, the formed $[Cu(NH_3)_4]^+$ can both migrate away from the liquid–Cu_2O interface and undergo the subsequent precipitation reaction on the Cu_2O surface, resulting in the loose nucleation and growth of $Cu(OH)_2$ precipitates on the surfaces of *d*-Cu_2O-OA.

We have also comparatively studied the reactivity of *c*-Cu_2O and *o*-Cu_2O-PVP in the redox reaction between Cu_2O and Ce(IV) [128]. In this reaction system, CeO_2 eventually forms as one of the products. Under the same reaction condition, the homogeneous nucleation and growth of CeO_2 occur on the surface of *c*-Cu_2O, developing smooth CeO_2 shell and eventually forming cubic Cu_2O(core)/CeO_2(shell) nanostructures; however, heterogeneous nucleation and growth of CeO_2 occur on the surface of *o*-Cu_2O-PVP, developing rough CeO_2 shell and eventually forming hollow octahedral nanocages. These observations provide further experimental evidence that the crystal plane exposed on a solid particle in liquid–solid chemical reactions involving the formation of solid products not only determines its reactivity but also strongly affects the nucleation and growth of solid products.

The reducibility of an oxide is closely related with its catalytic property in oxidation reactions, we have comparatively studied the reducibility of *c*-Cu_2O and *o*-Cu_2O-PVP by H_2 and CO that are the gas–solid chemical reactions [114]. As shown in Figure 9(A) and (B), reduced either by H_2 or CO, the reduction temperature of *o*-Cu_2O-PVP is lower than that of *c*-Cu_2O by more than 120 °C. This demonstrates that $Cu_2O(111)$ surface is much more facile to be reduced than $Cu_2O(100)$ surface. Two factors determine the reactivity of a solid particle in the gas–solid chemical reactions: the surface reactivity of the solid particle and the adsorption/activation of gas molecules on the solid particle surface. On one hand, the surface Cu–O bond of $Cu_2O(100)$ is stronger than that of $Cu_2O(111)$ and thus $Cu_2O(100)$ surface is more stable than $Cu_2O(111)$ surface; on the other hand, due to the presence of coordination unsaturated Cu_{CUS} on $Cu_2O(111)$ surface, the adsorption/activation of both CO and H_2 are much stronger on $Cu_2O(111)$ surface than on $Cu_2O(100)$ surface. As shown by DFT calculation results (Figure 9(C)–(F)), CO preferentially chemisorbs atop on

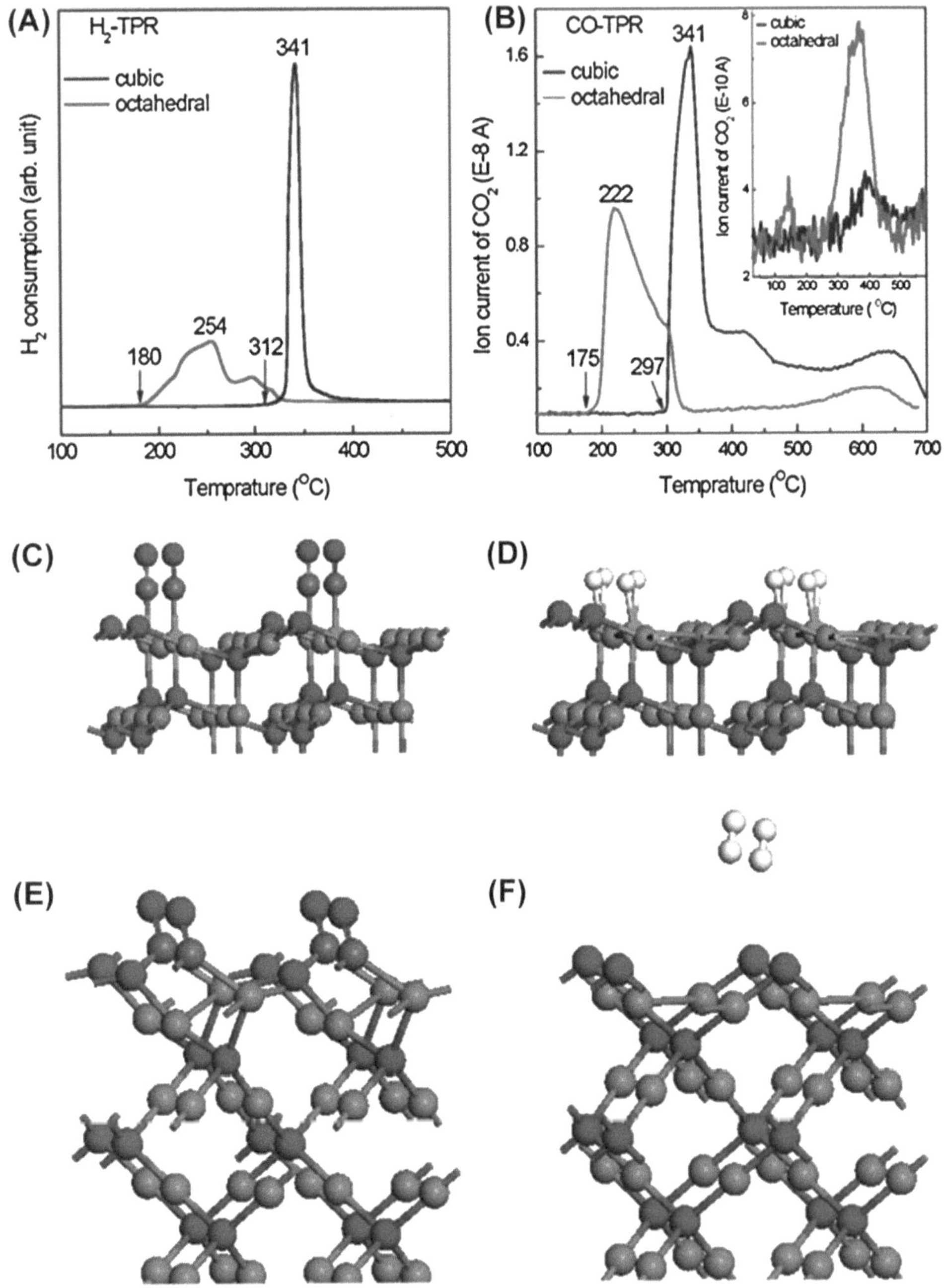

Figure 9 (A) H_2-TPR and (B) CO-TPR spectra of Cu_2O cubes and octahedra, and the optimized most stable adsorption configurations of (C) CO and (D) H_2 on $Cu_2O(111)$ surface, (E) CO and (F) H_2 on $Cu_2O(100)$ surface. The inset shows the CO_2-TPRS spectra of Cu_2O octahedra and cubes in Ar. The red (darkest gray in print versions), brick red (dark gray in print versions), green (lighter gray in print versions), gray, and white balls represent oxygen, Cu_{CSA}, Cu_{CUS}, carbon, and hydrogen atoms, respectively. *Adapted from reference [114]. Copyright 2010 American Chemical Society.*

the Cu_{CUS} site of $Cu_2O(111)$ with an adsorption energy of 1.52 eV and bridgely on the Cu_{CSA}–O_{CUS} bridge site of $Cu_2O(100)$ with an adsorption energy of 1.12 eV; H_2 chemisorbs on the Cu_{CUS} top site of $Cu_2O(111)$ with an adsorption energy of 0.41 eV but only weakly physisorbs on $Cu_2O(100)$ surface. The H–H bond length of $H_2(a)$ is 0.80 Å on $Cu_2O(111)$ but 0.74 Å on $Cu_2O(100)$, suggesting an enhanced activation of H_2 upon the adsorption on $Cu_2O(111)$. Thus the high stability of $Cu_2O(100)$ surface over $Cu_2O(111)$ surface and the strong ability to adsorb/activate H_2 and CO of $Cu_2O(111)$ surface over $Cu_2O(100)$ surface cooperatively result in the experimental observation that *o*-Cu_2O-PVP are much more facilely reduced than *c*-Cu_2O.

A common phenomenon occurring during the heterogeneous catalytic reaction catalyzed by a polycrystalline particle exposing various types of crystal planes is the coupling between adjacent crystal planes with different catalytic activities by the migration of surface species. This has been directly observed employing model catalyst approaches [7–9] and points to an important concept that the reaction kinetics of a polycrystalline particle cannot be regarded as a simple superposition of the contributions from each individual crystal plane exposed on its surface. We have comparatively studied the reducibility of the same type of crystal plane exposed on Cu_2O nanocrystals with different morphologies [129]. As shown in Figure 10, in both H_2-TPR (temperature-programmed reduction) and CO-TPR profiles, the reduction of {100} truncated octahedra capped with PVP (Cu_2O-14f-PVP) exposing the {111} and {100} crystal planes initiates at similar temperatures to that of *o*-Cu_2O-PVP exposing the {111} crystal planes, which can be attributed to the reduction of the {111} crystal plane on Cu_2O-14f-PVP; however, the reduction feature of the {100} crystal plane on cubes (Cu_2O-6f) can be hardly observed in the reduction of

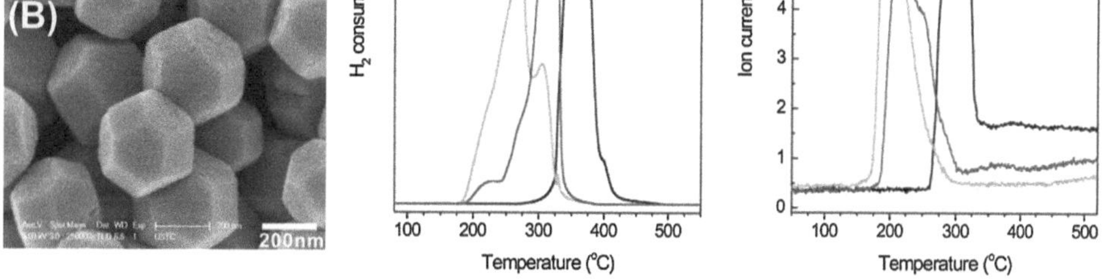

Figure 10 SEM images of (A) Cu_2O cubes (Cu_2O-6f) and (B) {100} truncated octahedra capped with PVP (Cu_2O-14f-PVP), and their H_2-TPR (C) and CO-TPR profiles (D). The H_2-TPR and CO-TPR profiles of octahedra capped with PVP (*o*-Cu_2O-PVP) were also included for comparisons. *Adapted from Ref. [129] with permission from the Royal Society of Chemistry.*

Cu_2O-14f-PVP, and the reduction of Cu_2O-14f-PVP completes at a much lower temperature than that of Cu_2O-6f. Therefore, the reduction kinetics of the {100} crystal plane on Cu_2O-14f-PVP differs much from that on Cu_2O-6f, which can be attributed to the occurrence of coupling between adjacent {111} and {100} crystal planes on Cu_2O-14f-PVP during the course of reduction reaction. The {111} crystal plane on a Cu_2O-14f-PVP nanocrystal can chemisorb and activate H_2 or CO and thus be reduced at relatively low temperatures, meanwhile, the activated species on the {111} crystal plane can migrate to the adjacent {100} crystal plane on the Cu_2O-14f-PVP nanocrystal and results in its reduction at temperatures where the {100} crystal plane on Cu_2O-6f cannot be reduced. Therefore, due to the coupling between adjacent {111} and {100} crystal planes in the reduction reactions, the reduction kinetics of Cu_2O-14f-PVP exposing the {111} and {100} crystal planes is obviously not a simple superposition of the contributions from the individual {111} and {100} crystal planes. The coupling was also observed to occur between adjacent {111} and {110} crystal planes exposed on {110} truncated octahedra capped with OA (Cu_2O-20f-OA) during the reduction reactions in which the activated H_2 or CO species formed on the more reactive {111} crystal plane can migrate to the adjacent {110} crystal plane on the Cu_2O-20f-OA nanocrystal and promote its reduction at temperatures where the {110} crystal plane on *d*-Cu_2O-OA cannot be reduced [129]. Therefore, employing various types of Cu_2O nanocrystals with uniform and well-defined shape as model catalysts, we nicely validate the concept of the coupling between adjacent crystal planes exposed on a polycrystalline particle during the heterogeneous catalytic reaction.

Therefore different types of crystal planes exposed on Cu_2O nanocrystals exhibit different surface compositions/structures that determine their surface reactivities in the chemical reactions. Among three low-index Cu_2O (100), (110), and (111) surfaces, the (100) surface exhibits the shortest Cu–O bond and thus is the least reactive and most stable in chemical reactions, and the $Cu_2O(111)$ surface exhibits the coordination unsaturated Cu(I) sites and thus is most active in adsorbing/activating gas-phase reactants. Thus *c*-Cu_2O selectively exposing the {100} crystal planes is most stable in chemical reactions while *o*-Cu_2O-PVP selectively exposing the {111} crystal planes is very facile to be reduced by H_2 and CO. The reactivity of crystal planes exposed on Cu_2O nanocrystal affects the local concentration of products at the liquid–Cu_2O interface and the subsequent nucleation and growth of solid products in chemical reactions. These results establish the crystal plane-surface composition/structure–surface reactivity relation of Cu_2O nanocrystals at the molecular level that guides the compositional, morphological, and structural evolutions of Cu_2O nanocrystals in chemical reactions.

5. CATALYTIC PROPERTY OF WELL-DEFINED Cu_2O NANOCRYSTALS

The catalytic property of various types of well-defined Cu_2O nanocrystals has also been much explored. Cu_2O is a visible light-responsive semiconductor and the photocatalytic activity of Cu_2O nanocrystals in photocatalytic degradation of dye molecules has been studied [77,94,130–134]. Zheng et al. studied the photocatalytic performance of Cu_2O microcrystals exposing {100} and {110} crystal planes in photodegradation of methyl orange under visible light irradiation [130]. Cu_2O microcrystals exhibited considerable photocatalytic activity in the first runs, which was ascribed to the charge separation among crystal planes, but Cu_2O {100} and {110} crystal planes were observed to transform into nanosheets mainly exposing {111} crystal planes, and the finally formed Cu_2O nanosheets presented stable photocatalytic activity. However, Zhang et al. reported that 26-facet and 18-facet Cu_2O polyhedra with dominant {110} crystal planes exhibited a higher adsorption and photocatalytic activity than Cu_2O octahedra with dominant {111} crystal planes and cubes with {100} crystal planes in photodegradation of methyl orange under visible light irradiation [94]. Cao et al. observed that hollow octahedral Cu_2O particles exhibited a higher photocatalytic activity for photodegradation of *p*-nitrophenol aqueous solution under visible light illumination than other Cu_2O particles with different morphologies (hollow spheres, solid octahedra, and solid spheres) [131]. Huang et al. synthesized Cu_2O nanocrystals with systematic shape evolution from cubic to face-raised cubic, edge- and corner-truncated octahedral, all-corner-truncated rhombic dodecahedral, {100}-truncated rhombic dodecahedral, and rhombic dodecahedral and found that the rhombic dodecahedra selectively exposing {110} crystal planes exhibit an exceptionally good photocatalytic activity toward the fast and complete photodegradation of methyl orange under xenon lamp illumination [77]. Thus among the Cu_2O {111}, {100}, and {110} crystal planes, the Cu_2O {110} crystal planes is most photocatalytically active in photodegradation of methyl orange under visible light illumination, which could be attributed to their highest number density of surface copper atoms.

Cu_2O nanocrystals were also examined as the catalyst for organic synthesis and strong morphology-dependent catalytic activity was observed [88,135,136]. Xu et al. reported that the activity of Cu_2O {111} crystal planes in catalyzing the N-arylation reaction of iodobenzene with imidazole at 30 °C was as two times as that of Cu_2O {100} crystal planes [88]. Li et al. studied the catalytic activity of cubic, rhombic dodecahedral, and {100}-truncated dodecahedral Cu_2O nanocrystals in a ligand-free arylation of phenylacetylene with arylboronic acids and found that {100}-truncated dodecahedral Cu_2O nanocrystals exhibited the highest activity and stability [135]. Chanda et al. employed Cu_2O cubes, octahedra, and rhombic dodecahedra as catalysts for the multicomponent synthesis of 1,2,3-triazoles in ethanol by 1,3-dipolar cycloaddition reactions and found that all these nanocrystals could serve as efficient catalysts but Cu_2O rhombic dodecahedra bounded by {110} crystal planes were the best catalysts for a broad range of reagents used [136].

The reported studies of Cu_2O nanocrystals in gas-phase heterogeneous catalytic reactions were limited to CO oxidation [106,107,137,138]. Because gas-phase heterogeneous catalytic reactions usually operate at high temperatures, the surfaces of Cu_2O nanocrystals facilely undergo reaction-dependent restructuring although their morphology does not

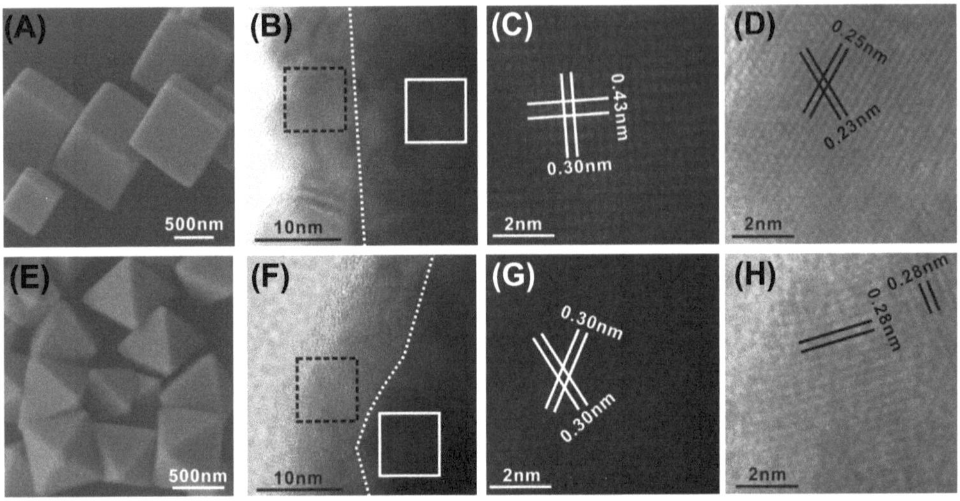

Figure 11 (A and B) SEM and HRTEM images of Cu_2O cubes after the first run of catalytic activity evaluation up to 240 °C (CuO/*c*-Cu_2O) in CO oxidation with excess O_2. (C) Zoom-in image of the inner region of CuO/*c*-Cu_2O indicated by the solid white square in (B). The lattice fringe arises from Cu_2O. (D) Zoom-in image of the surface region of CuO/*c*-Cu_2O indicated by the dash black square in (B). The lattice fringe arises from CuO. (E & F) SEM and HRTEM images of Cu_2O cubes after the first run of catalytic activity evaluation up to 240 °C (CuO/*o*-Cu_2O) in CO oxidation with excess O_2. (G) Zoom-in image of the inner region of CuO/*o*-Cu_2O indicated by the solid white square in (F). The lattice fringe arises from Cu_2O. (H) Zoom-in image of the surface region of CuO/*o*-Cu_2O indicated by the dash black square in (F). The lattice fringe arises from CuO. The dash white line in (B) and (F) indicates the CuO-Cu_2O boundary. *Adapted from reference [137]. Copyright 2011 WILEY-VCH Verlag GmbH & Co.*

change. Leng et al. [107] observed an enhanced catalytic activity of 50-facet Cu_2O nanocrystals in CO oxidation with excess O_2 that was attributed to the presence of (311) high-index crystal planes; however, such a correlation is questionable because the structural characterization results demonstrated that the surface of 50-facet Cu_2O nanocrystals was oxidized to CuO during CO oxidation. We have also studied the catalytic property of Cu_2O cubes and octahedra in CO oxidation with excess O_2 (1% CO in dry air) [137]. As shown in Figure 11, after the first run of catalytic activity evaluation up to 240 °C, Cu_2O cubes and octahedra well preserve their original morphologies, but the surfaces are oxidized to CuO thin film. This is also verified by Cu 2p XPS results. The surface oxidation meanwhile removes the PVP capping ligands on Cu_2O octahedra. We observed that the CuO thin film formed on Cu_2O octahedra are much more active in catalyzing CO oxidation and exhibit a lower apparent activation energy than that formed on Cu_2O cubes (Figure 12). These experimental observations demonstrate that under the same CO oxidation reaction condition, cubic and octahedral Cu_2O nanocrystals undergo surface restructuring to form CuO thin films with different surface structures catalyzing CO oxidation via different mechanisms. DFT calculation results demonstrate that CuO thin films grown on $Cu_2O(111)$ and (100) substrates are with different surface compositions and structures

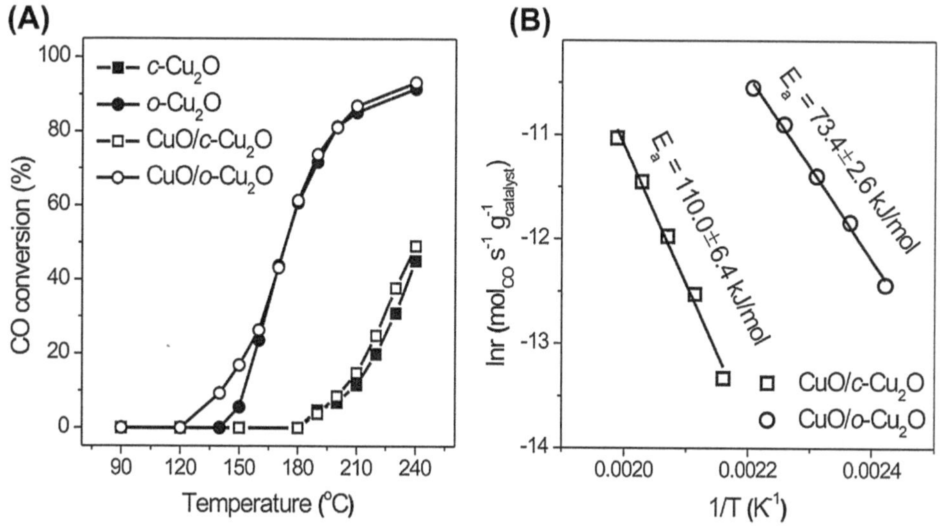

Figure 12 (A) Catalytic performance of Cu_2O cubes (*c*-Cu_2O) and octahedra (*o*-Cu_2O) and CuO thin films formed on Cu_2O cubes and octahedra (respectively denoted as CuO/*c*-Cu_2O and CuO/*o*-Cu_2O) by in situ surface restructuring in CO oxidation with excess O_2. (B) The Arrhenius plot of CO oxidation catalyzed by CuO/*c*-Cu_2O and CuO/*o*-Cu_2O. *Reproduced from reference [137]. Copyright 2011 WILEY-VCH Verlag GmbH & Co.*

(Figure 13). The CuO thin film on $Cu_2O(111)$ is terminated with three-coordinated Cu (Cu_{3c}) and three-coordinated oxygen (O_{3c}) atoms while that on $Cu_2O(100)$ is terminated only with O_{2c} atoms. On the CuO/$Cu_2O(111)$ surface (Figure 13(C)) CO and O_2 coadsorb on Cu_{3c} and O_{3c} sites form an OCOO(a) surface intermediate, the OCOO(a) surface intermediate decomposes to produce an O(a) adatom and CO_2 with an activation energy of 0.37 eV and then CO reacts with the O(a) adatom to produce CO_2 with an activation energy of 0.17 eV. Thus, the elementary step with the largest activation energy in CO oxidation catalyzed by the CuO thin film on $Cu_2O(111)$ is the decomposition of the OCOO(a) surface intermediate with an activation energy is 0.37 eV. On the CuO/$Cu_2O(100)$ surface (Figure 13(D)), CO binds with O_{2c} sites very strongly to form the CO_3(a) surface intermediate, the CO_3(a) surface intermediate decomposes to produce CO_2 and create an oxygen vacancy in CuO with an activation energy of 1.15 eV, then O_2 chemisorbs on the oxygen vacancy site in CuO to form O_2(a) that reacts with CO to produce CO_2 with an activation energy of 0.67 eV and refill the oxygen vacancy. Thus the elementary step with the largest activation energy in the CO oxidation catalyzed by the CuO thin film on $Cu_2O(100)$ is the decomposition of the CO_3(a) surface intermediate with an activation energy is 1.15 eV. These DFT calculation results agree quite well with experimental results that CO oxidation catalyzed by CuO thin film on Cu_2O octahedra (73.4 kJ/mol) proceeds with a lower apparent activation energy than that catalyzed by CuO thin film on Cu_2O cubes (110 kJ/mol). Therefore the surface restructuring process of

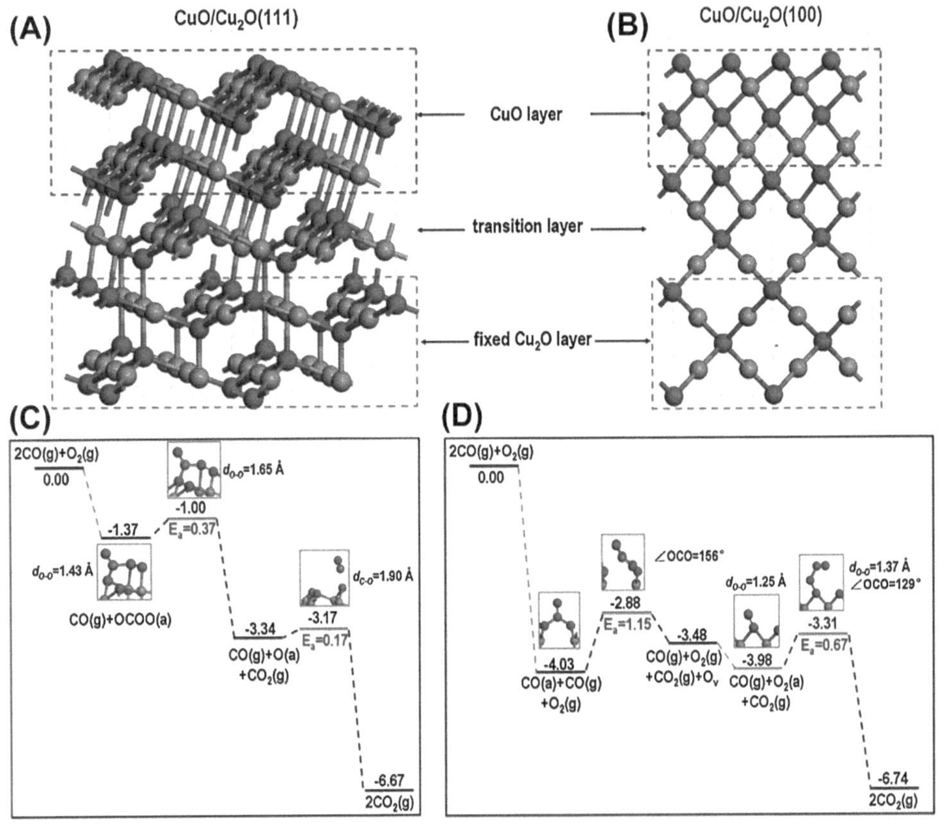

Figure 13 Optimized geometric structures of $CuO/Cu_2O(111)$ (A) and $CuO/Cu_2O(100)$ (B) surfaces, and the energy profiles of each elementary steps in CO oxidation catalyzed by $CuO/Cu_2O(111)$ (C) and $CuO/Cu_2O(100)$ (D) surfaces. The red (darker gray in print versions), pink (light gray in print versions), and gray spheres represent O, Cu, and C elements, respectively. *Reproduced from reference [137]. Copyright 2011 WILEY-VCH Verlag GmbH & Co.*

Cu_2O nanocrystals and the structure and catalytic performance of the resulting CuO thin film are controlled by the crystal plane exposed on Cu_2O nanocrystals. These findings highlight the strong connection of catalyst surface structures before and after surface restructuring. Catalysts nanoparticles commonly undergo surface restructuring to adopt the thermodynamically most stable structure under the heterogeneous catalytic reaction conditions, and the reaction temperature and atmosphere have been recognized to play decisive roles in the surface restructuring of catalyst nanoparticles. Our results reveal a novel concept of crystal plane-controlled surface restructuring of catalyst nanoparticles, not only greatly deepening the fundamental understanding of the surface restructuring phenomenon but also opening a new door for improving and designing efficient nanocatalysts.

Wang et al. studied the catalytic activity of SDS-capped 30-facet Cu_2O polyhedra exposing 24-{332} and 6-{100} crystal planes, SDS-capped 14-facet Cu_2O polyhedra exposing 8-{111} and 6-{100} crystal planes and Cu_2O cubes in CO oxidation in excess

O_2 (1% CO and 16% O_2 balanced with N_2) [106]. The surfaces of Cu_2O nanocrystals were not oxidized under the studied CO oxidation conditions. The catalytic activities of different crystal planes followed the order of {332} > {111} >> {100}, in consistence with the order of the density of coordination unsaturated Cu(I) on these three types of crystal planes. However, the effect of SDS capping ligands on the catalytic activity of Cu_2O nanocrystals was not discussed. Recently, we developed a controlled oxidation strategy to successfully remove the capping ligands such as PVP and OA on the surfaces of Cu_2O nanocrystals without the change of their morphologies and surface structures [138,139]. This enables us to study the catalytic property of Cu_2O nanocrystals without the influence of capping ligands. We studied the catalytic activity of capping-ligand-free Cu_2O cubes, octahedra, and rhombic dodecahedra in CO oxidation with stoichiometric O_2 [138]. Under such reaction conditions the Cu_2O surfaces are stable. As shown in Figure 14, the catalytic activity follows the order of octahedra >> rhombic dodecahedra > cubes, suggesting that Cu_2O {111} is most active in catalyzing CO oxidation among Cu_2O {111}, {110}, and {100} surfaces. CO-TPR results demonstrate that Cu_2O octahedra are most easily to be reduced among Cu_2O cubes, octahedra, and rhombic dodecahedra and diffuse reflectance infrared Fourier-transformed spectra show that CO chemisorption is strongest on Cu_2O octahedra at reaction temperatures. Both the reducibility and chemisorption ability toward CO of various Cu_2O nanocrystals are in consistence with their catalytic activity in CO oxidation. These observed surface chemistry and catalytic activity in CO oxidation of various Cu_2O nanocrystals can be well correlated with their exposed crystal plane and surface composition/structure. Cu_2O octahedra expose the {111} crystal plane with coordinated unsaturated Cu(I) sites and thus are most active in chemisorbing CO and catalyzing CO oxidation.

We also studied the catalytic performance of capping-ligand-free Cu_2O cubes, octahedra, and rhombic dodecahedra in propylene oxidation with molecule O_2 [139]. Cu_2O octahedra selectively exposing {111} crystal planes with coordination unsaturated Cu(I) sites are most active and meanwhile most selective toward acrolein; Cu_2O cubes selectively exposing {100} crystal planes are most selective toward CO_2; and Cu_2O rhombic dodecahedra selectively exposing {110} crystal planes are most selective toward propylene oxide. One-coordinated Cu on Cu_2O(111), three-coordinated O on Cu_2O(110), and two-coordinated O on Cu_2O(100) were identified as the active sites respectively catalyzing propylene oxidation with O_2 to acrolein, propylene oxide, and CO_2. These results reveal a crystal plane-engineering strategy of oxide catalysts not only for the innovation of selective catalysts but also for the fundamental understanding of complex heterogeneous catalytic reactions at the molecular level.

Therefore different types of crystal planes exposed on Cu_2O nanocrystals exhibit different surface compositions/structures that determine their catalytic properties in catalytic reactions. Chemisorption plays a key role in heterogeneous catalysis. The Cu(I) sites on Cu_2O surface exhibit a much stronger activity in chemisorption than the O sites, thus Cu_2O(100) surface terminated exclusively by the O sites is much less catalytically active

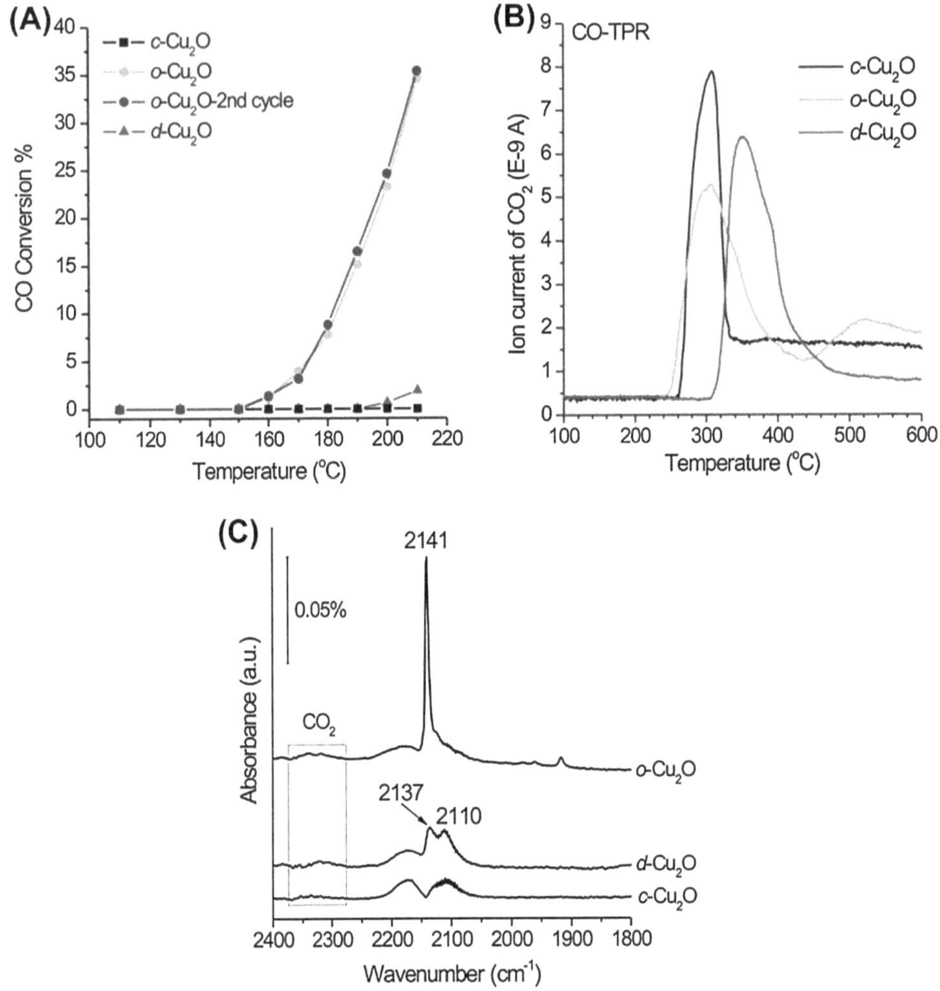

Figure 14 (A) Catalytic performance of capping-ligand-free Cu_2O cubes (*c*-Cu_2O), octahedra (*o*-Cu_2O), and rhombic dodecahedra (*d*-Cu_2O) in CO oxidation with stoichiometric O_2. (B) CO-TPR profiles of *c*-Cu_2O, *o*-Cu_2O and *d*-Cu_2O. (C) In situ DRIFTS spectra of saturated CO pulse adsorption on *c*-Cu_2O, *o*-Cu_2O, and *d*-Cu_2O at 150 °C. *Adapted from reference [138]. Copyright 2013 WILEY-VCH Verlag GmbH & Co.*

than $Cu_2O(110)$ and (111) surfaces. In the investigated photocatalytic reactions and liquid-phase organic synthesis catalytic reactions involving large reactant molecules, the adsorption of reactant molecules on the Cu_2O surface does not depend much on the coordination environment of Cu(I) sites and the adsorption capacity of the Cu_2O surface toward reactant molecules is important, thus $Cu_2O(110)$ surface exposing the highest density of Cu(I) is more catalytically active than $Cu_2O(111)$ surface, and consequently Cu_2O nanocrystals exposing the largest fraction of {110} crystal planes are most catalytically active. In the investigated gas-phase catalytic reactions operating at high temperatures, the adsorption and activation of reactant molecules on the Cu_2O surface depend on the coordination environment of Cu(I) sites and coordination unsaturated Cu(I) sites exhibit a much stronger ability than coordination saturated Cu(I) sites, thus the Cu_2O surfaces with coordination unsaturated Cu(I) sites, such as (332) and (111) surfaces, are much more catalytically active than $Cu_2O(110)$ surface only with coordination saturated Cu(I) sites, and consequently Cu_2O nanocrystals exposing the largest fraction of {332} and {111} crystal planes are most catalytically active. The crystal planes exposed on Cu_2O nanocrystals also determine their catalytic selectivity in complex heterogeneous catalytic reactions and their surface restructuring process during heterogeneous catalytic reactions. These results establish the crystal plane-surface composition/structure–catalytic property relation of Cu_2O nanocrystals at the molecular level that guides the structural design of efficient Cu_2O catalysts.

6. SUMMARY AND OUTLOOK

In summary, Cu_2O nanocrystals with well-defined morphologies exhibit strongly morphology-dependent surface chemistry and catalytic property. Thanks to their well-defined morphology, the morphology-dependent surface chemistry and catalytic property of Cu_2O nanocrystals can be well correlated to their exposed crystal planes and surface compositions/structures, particularly for uniform Cu_2O cubes, octahedra, and rhombic dodecahedra that selectively expose {100}, {111}, and {110} crystal planes, respectively. Among $Cu_2O(100)$, (111), and (110) surfaces, $Cu_2O(100)$ surface is terminated with two-coordinated O_{CUS} sites and exhibits the shortest Cu–O bond; $Cu_2O(110)$ surface is terminated with three-coordinated O_{CUS} and coordination saturated Cu_{CSA} sites; and $Cu_2O(111)$ surface is terminated with three-coordinated O_{CUS} sites, but due to its open surface structure, coordination saturated Cu_{CSA} sites and coordination unsaturated Cu_{CUS} sites in the second layer of $Cu_2O(111)$ surface are also accessible to the reactants. Therefore in chemical reactions requiring the breaking of Cu–O bond, $Cu_2O(100)$ surface is most stable, so are Cu_2O cubes; in gas-phase catalytic reactions requiring the adsorption and activation of reactant molecules, $Cu_2O(111)$ with accessible coordination unsaturated Cu(I) sites is most catalytically active, so are Cu_2O octahedra. Thus employing well-defined Cu_2O nanocrystals, the crystal plane-surface composition/structure–surface chemistry/catalytic property relation of Cu_2O catalysts at the molecular level can be successfully established. Since Cu_2O particles are enclosed by various types of crystal planes, the established surface reactivity and catalytic property of various Cu_2O crystal planes will help to understand the reactivity and catalytic property of Cu_2O particles and design the structure of efficient Cu_2O catalysts.

Above crystal plane-surface composition/structure–surface chemistry/catalytic property relation of Cu_2O catalysts learned from Cu_2O nanocrystals well exemplifies the concept of nanocrystals-based model catalysts. An advantage of nanocrystals-based model catalysts over single crystals-based model catalysts is that they can be studied under the same experimental condition as practical heterogeneous catalytic reactions and thus can bridge the "materials gap" and "pressure gap" between single crystals-based model catalysts and corresponding powder catalysts; however, several considerations must be kept in the mind when studying the structure–catalytic property relation with nanocrystals-based model catalysts. The first is the complexity of surface composition/structure of nanocrystals. Besides terrace sites, three-dimensional nanocrystals also expose a large number of edge sites and corner sites; meanwhile, nanocrystals are mostly synthesized by the wet-chemistry synthesis methods and their composition/structure are not so perfect as expected. Thus a comprehensive investigation should be adopted when studying nanocrystal-based model catalysts. The second is the surface restructuring of nanocrystals with a finite size under the catalytic reaction condition. It is thus preferred to compare the surface composition and structure of nanocrystals before and after the catalytic reaction or to perform the in situ study. The third is the capping ligand on the nanocrystal surface inherited from the wet-chemistry synthesis methods. The influences of capping ligands on the surface chemistry and catalytic properties of nanocrystals are not well understood [140]. Thus great efforts need to be devoted to preparing capping agent-free nanocrystals.

Nanocrystals with uniform composition and structure are not only novel model catalysts for the fundamental studies of structure–catalytic property relation of catalysts, but also novel candidates for highly efficient catalysts. On one hand, uniform catalyst nanocrystals exposing the largest fraction of crystal planes with the highest density of catalytic active sites are definitely the most active catalyst; on the other hand, a prerequisite of a highly selective catalyst is that the catalyst must have uniform composition and structure so as to expose a single type of required catalytic active site on the surface. Thus catalysis by nanocrystals with well-defined structure is becoming an emerging frontier of heterogeneous catalysis. With the structure–catalytic property relation learned from model catalyst studies and the advancement of controlled synthesis methods, it is promising to realize the structural design and controlled synthesis of novel efficient nanocrystal catalysts in the future.

ACKNOWLEDGMENTS

Financial supports from National Natural Science Foundation of China, Ministry of Science and Technology of China, Chinese Academy of Sciences, and Max-Planck Society via the MPG-CAS partner group program are gratefully acknowledged. Contributions from other students and coworkers (Qing Hua, Huizhi Bao, Daili Shang, Zhenhua Zhang, Zhiquan Jiang, Wenhua Zhang, and Jinlong Yang) to the research project of surface chemistry and catalysis of Cu_2O nanocrystals in my research group are also gratefully acknowledged.

REFERENCES

[1] G.A. Somorjai, Y. Li, Introduction to Surface Chemistry and Catalysis, Wiley, New York, 2008.
[2] G. Wulff, Z. Kristallogr. 34 (1901) 449–530.
[3] http://nobelprize.org/nobel_prizes/chemistry/laureates/2007/ertl-lecture.html.
[4] H.J. Freund, Chem. Eur. J. 16 (2010) 9384–9397.
[5] G. Ertl, Angew. Chem. Int. Ed. 19 (2008) 3524–3535.

[6] C.T. Campbell, Surf. Sci. Rep. 27 (1997) 1–111.
[7] V. Gorodetskii, J. Lauterbach, H.H. Rotermund, J.H. Block, G. Ertl, Nature 370 (1994) 276–279.
[8] W.X. Huang, R.S. Zhai, X.H. Bao, Langmuir 17 (2001) 3629–3634.
[9] W.X. Huang, X.H. Bao, H.H. Rotermund, G. Ertl, J. Phys. Chem. B 106 (2002) 5645–5647.
[10] G.A. Somorjai, J.Y. Park, Chem. Soc. Rev. 37 (2008) 2155–2162.
[11] F. Gao, D.W. Goodman, Annu. Rev. Phys. Chem. 63 (2012) 265–286.
[12] O. Masala, R. Seshadri, Annu. Rev. Mater. Res. 34 (2004) 41–81.
[13] C. Burda, X. Chen, R. Narayanan, M.A. El-Sayed, Chem. Rev. 105 (2005) 1025–1102.
[14] Y.N. Xia, Y.J. Xiong, B. Lim, S.E. Skrabalak, Angew. Chem. Int. Ed. 48 (2009) 60–103.
[15] G.R. Patzke, Y. Zhou, R. Kontic, F. Conrad, Angew. Chem. Int. Ed. 50 (2011) 826–859.
[16] Z. Jacek, J. Catal. 80 (1983) 263–273.
[17] J.C. Volta, J.L. Portefaix, Appl. Catal. 18 (1985) 1–32.
[18] K.B. Zhou, X. Wang, X.M. Sun, Q. Peng, Y.D. Li, J. Catal. 229 (2005) 206–212.
[19] H.X. Mai, L.D. Sun, Y.W. Zhang, R. Si, W. Feng, H.P. Zhang, H.C. Liu, C.H. Yan, J. Phys. Chem. B 109 (2005) 24380–24385.
[20] W.Q. Han, L.J. Wu, Y.M. Zhu, J. Am. Chem. Soc. 127 (2005) 12814–12815.
[21] X.W. Liu, K.B. Zhou, L. Wang, B.Y. Wang, Y.D. Li, J. Am. Chem. Soc. 131 (2009) 3140–3141.
[22] M.L. Tana, J. Zhang, H.J. Li, Y.L. Li, W.J. Shen, Catal. Today 148 (2009) 179–183.
[23] Z.L. Wu, M.J. Li, J. Howe, H.M. Meyer III, S.H. Overbury, Langmuir 26 (2010) 16595–16606.
[24] Z.L. Wu, M.J. Li, S.H. Overbury, J. Catal. 285 (2012) 61–73.
[25] J.G. Lv, Y. Shen, L.M. Peng, X.F. Guo, W.P. Ding, Chem. Commun. 46 (2010) 5909–5911.
[26] D.Y. Wang, Y.J. Kang, V. Doan-Nguyen, J. Chen, R. Küngas, N.L. Wieder, K. Bakhmutsky, R.J. Gorte, C.B. Murray, Angew. Chem. Int. Ed. 50 (2011) 4378–4381.
[27] S. Agarwal, L. Lefferts, B.L. Mojet, Chem. Cat. Chem. 5 (2013) 479–489.
[28] R. Si, J. Raitano, N. Yi, L. Zhang, S.-W. Chan, M. Flytzani-Stephanopoulos, Catal. Today 68 (2012) 68–80.
[29] R. Si, M. Flytzani-Stephanopoulos, Angew. Chem. Int. Ed. 47 (2008) 2884–2887.
[30] Y. Guan, D. Ligthart, Ö. Pirgon-Galin, J. Pieterse, R. van Santen, E. Hensen, Top. Catal. 54 (2011) 424–438.
[31] W.Q. Han, W. Wen, J.C. Hanson, X.W. Teng, N. Marinkovic, J.A. Rodriguez, J. Phys. Chem. C 113 (2009) 21949–21955.
[32] L.J. Liu, Z.J. Yao, Y. Deng, F. Gao, B. Liu, L. Dong, Chem. Cat. Chem. 3 (2011) 978–989.
[33] P. Gawade, B. Mirkelamoglu, U.S. Ozkan, J. Phys. Chem. C 4 (2010) 18173–18181.
[34] Z.L. Wu, V. Schwartz, M.J. Li, A.J. Rondinone, S.H. Overbury, J. Phys, Chem. Lett. 3 (2012) 1517–1522.
[35] S.J. Chang, M. Li, Q. Hua, L.J. Zhang, Y.S. Ma, B.J. Ye, W.X. Huang, J. Catal. 293 (2012) 195–204.
[36] Y.X. Gao, W.D. Wang, S.J. Chang, W.X. Huang, Chem. Cat. Chem. 5 (2013) 3610–3620.
[37] B.M. Choudary, R.S. Mulukutla, K.L. Klabunde, J. Am. Chem. Soc. 125 (2003) 2020–2021.
[38] K.K. Zhu, J.C. Hu, C. Kübel, R. Richards, Angew. Chem. Int. Ed. 45 (2006) 7277–7281.
[39] J.C. Hu, K.K. Zhu, L.F. Chen, C. Kübel, R. Richards, J. Phys. Chem. C 111 (2007) 12038–12044.
[40] X.W. Xie, Y. Li, Z.Q. Liu, M. Haruta, W.J. Shen, Nature 458 (2009) 746–749.
[41] L.H. Hu, K.Q. Sun, Q. Peng, B.Q. Xu, Y.D. Li, Nano Res. 3 (2010) 363–368.
[42] C.X. Xu, Y.Q. Liu, C. Zhou, L. Wang, H.R. Geng, Y. Ding, Chem. Cat. Chem. 3 (2011) 399–407.
[43] Y. Sun, P. Lv, J.Y. Yang, L. He, J.C. Nie, W.X. Liu, Y.D. Li, Chem. Commun. 47 (2011) 11279–11281.
[44] Y.H. Teng, Y. Kusano, M. Azuma, M. Haruta, Y. Shimakawa, Catal. Sci. Technol. 1 (2011) 920–922.
[45] L.H. Hu, Q. Peng, Y.D. Li, J. Am. Chem. Soc. 130 (2008) 16136–16137.
[46] X.Z. Wang, L. Ding, Z.B. Zhao, W.Y. Xu, B. Meng, J.S. Qiu, Catal. Today 175 (2011) 509–514.
[47] Y.H. Zheng, Y. Cheng, Y.S. Wang, F. Bao, L.H. Zhou, X.F. Wei, Y.Y. Zhang, Q.J. Zheng, J. Phys. Chem. B 10 (2006) 3093–3097.
[48] Q.X. Gao, X.F. Wang, J.L. Di, X.C. Wu, Y.R. Tao, Catal. Sci. Technol. 1 (2011) 574–577.
[49] X. Liu, J. Liu, Z. Chang, X. Sun, Y. Li, Catal. Commun. 12 (2011) 530–534.
[50] N. Zhao, W. Ma, Z.M. Cui, W.G. Song, C.L. Xu, M.Y. Gao, ACS Nano 3 (2009) 1775–1780.
[51] X.L. Mou, B.S. Zhang, Y. Li, L.D. Yao, X.J. Wei, D.S. Su, W.J. Shen, Angew. Chem. Int. Ed. 51 (2012) 2989–2993.
[52] J.A. van Bokhoven, Chem. Cat. Chem. 1 (2009) 363–364.

[53] Y. Li, W.J. Shen, Sci. China Chem. 55 (2012) 2485–2496.
[54] K.B. Zhou, Y.D. Li, Angew. Chem. Int. Ed. 51 (2012) 602–613.
[55] F. Zaera, Chem. Sus. Chem. 6 (2013) 1797–1820.
[56] Z.A. Qiao, Z. Wu, S. Dai, Chem. Sus. Chem. 6 (2013) 1821–1833.
[57] W.X. Huang, Top. Catal. 56 (2013) 1363–1376.
[58] C.H.B. Ng, W.Y. Fan, J. Phys. Chem. B 110 (2006) 20801–20807.
[59] P.O. Larsson, A. Andersson, Appl. Catal. B 24 (2000) 175–192.
[60] Y. Tanaka, T. Utaka, R. Kikuchi, T. Takeguchi, K. Sasaki, K. Eguchi, J. Catal. 215 (2003) 271–278.
[61] J.L.C. Fajín, M.N.D.S. Cordeiro, F. Illas, J.R.B. Gomes, J. Catal. 268 (2009) 131–141.
[62] C. Ratnasamy, J.P. Wagner, Catal. Rev. 51 (2009) 325–440.
[63] C. Baltes, S. Vukojević, F. Schüth, J. Catal. 258 (2008) 334–344.
[64] S. Zander, E.L. Kunkes, M.E. Schuster, J. Schumann, G. Weinberg, D. Teschner, N. Jacobsen, R. Schlögl, M. Behrens, Angew. Chem. Int. Ed. 52 (2013) 6536–6540.
[65] J.Q. Lu, M.F. Luo, H. Lei, X.H. Bao, C. Li, J. Catal. 211 (2002) 552–555.
[66] H. Chu, L. Yang, Q. Zhang, Y. Wang, J. Catal. 241 (2006) 225–228.
[67] B.-X. Tang, F. Wang, J.-H. Li, Y.-X. Xie, M.-B. Zhang, J. Org. Chem. 72 (2007) 6294–6297.
[68] R.A. Altman, E.D. Koval, S.L. Buchwald, J. Org. Chem. 72 (2007) 6190–6199.
[69] M. Hara, T. Kondo, M. Komoda, S. Ikeda, J.N. Kondo, K. Domen, K. Shinohara, A. Tanaka, Chem. Commun. (1998) 357–358.
[70] Z. Wang, Y. Liu, D.J. Martin, W. Wang, J. Tang, W. Huang, Phys. Chem. Chem. Phys. 15 (2013) 14956–14960.
[71] P.E. de Jongh, D. Vanmaekelbergh, J.J. Kelly, Chem. Commun. (1999) 1069–1070.
[72] A. Paracchino, V. Laporte, K. Sivula, M. Grätzel, E. Thimsen, Nat. Mater. 10 (2011) 456–461.
[73] P. McFadyen, E. Matijević, J. Colloid Interface Sci. 44 (1973) 95–106.
[74] C.-H. Kuo, M.H. Huang, Nano Today 5 (2010) 106–116.
[75] S. Sun, Z. Yang, RSC Adv. 4 (2014) 3804–3822.
[76] L. Gou, C.J. Murphy, Nano Lett. 3 (2003) 231–234.
[77] W.-C. Huang, L.-M. Lyu, Y.-C. Yang, M.H. Huang, J. Am. Chem. Soc. 134 (2012) 1261–1267.
[78] C.-H. Kuo, M.H. Huang, J. Phys. Chem. C 112 (2008) 18355–18360.
[79] X. Zhao, Z. Bao, C. Sun, D. Xue, J. Cryst. Growth 311 (2009) 711–715.
[80] H. Pang, F. Gao, Q.Y. Lu, Cryst. Eng. Comm. 12 (2010) 406–412.
[81] X. Lan, J. Zhang, H. Gao, T. Wang, Cryst. Eng. Comm. 13 (2011) 633–636.
[82] H.Y. Zhao, Y.F. Wang, J.H. Zeng, Cryst. Growth Des. 8 (2008) 3731–3734.
[83] K. Chen, D. Xue, Cryst. Eng. Comm. 14 (2012) 8068–8075.
[84] M. Leng, C. Yu, C. Wang, Cryst. Eng. Comm. 14 (2012) 8454–8461.
[85] D.-F. Zhang, H. Zhang, L. Guo, K. Zheng, X.-D. Han, Z. Zhang, J. Mater. Chem. 19 (2009) 5220–5225.
[86] I.C. Chang, P.-C. Chen, M.-C. Tsai, T.-T. Chen, M.-H. Yang, H.-T. Chiu, C.-Y. Lee, Cryst. Eng. Comm. 15 (2013) 2363–2366.
[87] H. Xu, W. Wang, W. Zhu, J. Phys. Chem. B 110 (2006) 13829–13834.
[88] Y. Xu, H. Wang, Y. Yu, L. Tian, W. Zhao, B. Zhang, J. Phys. Chem. C 115 (2011) 15288–15296.
[89] X. Zhang, Y. Xie, F. Xu, X. Liu, D. Xu, Inorg. Chem. Commun. 6 (2003) 1390–1392.
[90] X. Liang, L. Gao, S. Yang, J. Sun Adv. Mater. 21 (2009) 2068–2071.
[91] K.X. Yao, X.M. Yin, T.H. Wang, H.C. Zeng, J. Am. Chem. Soc. 132 (2010) 6131–6144.
[92] Y. Sui, W. Fu, H. Yang, Y. Zeng, Y. Zhang, Q. Zhao, Y. Li, X. Zhou, Y. Leng, M. Li, G. Zou, Cryst. Growth Des. 10 (2010) 99–108.
[93] Z. Yang, S. Sun, C. Kong, X. Song, B. Ding, J. Nanomater 2010 (2010) 1–11.
[94] Y. Zhang, B. Deng, T. Zhang, D. Gao, A.-W. Xu, J. Phys. Chem. C 114 (2010) 5073–5079.
[95] C.H. Kuo, C.H. Chen, M.H. Huang, Adv. Funct. Mater. 17 (2007) 3773–3780.
[96] M.H. Kim, B. Lim, E.P. Lee, Y. Xia, J. Mater. Chem. 18 (2008) 4069–4073.
[97] G. Prabhakaran, R. Murugan, Cryst. Eng. Comm. 14 (2012) 8338–8341.
[98] Y.-Y. Ma, Z.-Y. Jiang, Q. Kuang, S.-H. Zhang, Z.-X. Xie, R.-B. Huang, L.-S. Zheng, J. Phys. Chem. C 112 (2008) 13405–13409.
[99] H. Li, R. Liu, R. Zhao, Y. Zheng, W. Chen, Z. Xu, Cryst. Growth Des. 6 (2006) 2795–2798.

[100] Q. Chen, X. Shen, H. Gao, J. Colloid Interface Sci. 312 (2007) 272–278.
[101] J.-Y. Ho, M.H. Huang, J. Phys. Chem. C 113 (2009) 14159–14164.
[102] L. Gou, C.J. Murphy, J. Mater. Chem. 14 (2004) 735–738.
[103] D. Wang, M. Mo, D. Yu, L. Xu, F. Li, Y. Qian, Cryst. Growth Des. 3 (2003) 717–720.
[104] Z. Wang, X. Chen, J. Liu, M. Mo, L. Yang, Y. Qian, Solid State Commun. 130 (2004) 585–589.
[105] X. Wang, S. Jiao, D. Wu, Q. Li, J. Zhou, K. Jiang, D. Xu, Cryst. Eng. Comm. 15 (2013) 1849–1852.
[106] X. Wang, C. Liu, B. Zheng, Y. Jiang, L. Zhang, Z. Xie, L. Zheng, J. Mater. Chem. A 1 (2013) 282–287.
[107] M. Leng, M. Liu, Y. Zhang, Z. Wang, C. Yu, X. Yang, H. Zhang, C. Wang, J. Am. Chem. Soc. 132 (2010) 17084–17087.
[108] Y. Liang, L. Shang, T. Bian, C. Zhou, D. Zhang, H. Yu, H. Xu, Z. Shi, T. Zhang, L.-Z. Wu, C.-H. Tung, Cryst. Eng. Comm. 14 (2012) 4431–4436.
[109] S. Sun, C. Kong, S. Yang, L. Wang, X. Song, B. Ding, Z. Yang, Cryst. Eng. Comm. 13 (2011) 2217–2221.
[110] L.D. Marks, Rep. Prog. Phys. 57 (1994) 603–649.
[111] Y. Xia, Y. Xiong, B. Lim, S.E. Skrabalak, Angew. Chem. Int. Ed. 48 (2009) 60–103.
[112] G. Liu, J.C. Yu, G.Q. Lu, H.M. Cheng, Chem. Commun. 47 (2011) 6763–6783.
[113] Q. Hua, D.L. Shang, W.H. Zhang, K. Chen, S.J. Chang, Y.S. Ma, Z.Q. Jiang, J.L. Yang, W.X. Huang, Langmuir 27 (2011) 665–671.
[114] H.Z. Bao, W.H. Zhang, D.L. Shang, Q. Hua, Y.S. Ma, Z.Q. Jiang, J.L. Yang, W.X. Huang, J. Phys. Chem. C 114 (2010) 6676–6680.
[115] C.H. Lu, L.M. Qi, J.H. Yang, X.Y. Wang, D.Y. Zhang, J.L. Xie, J.M. Ma, Adv. Mater. 17 (2005) 2562–2567.
[116] C.H. Kuo, M.H. Huang, J. Am. Chem. Soc. 130 (2008) 12815–12820.
[117] J.C. Park, J. Kim, H. Kwon, H. Song, Adv. Mater. 21 (2009) 803–807.
[118] Y.M. Sui, W.Y. Fu, Y. Zeng, H.B. Yang, Y.Y. Zhang, H. Chen, Y.X. Li, M.H. Li, G.T. Zou, Angew. Chem. Int. Ed. 49 (2010) 4282–4285.
[119] M.J. Siegfried, K.S. Choi, Adv. Mater. 16 (2004) 1743–1746.
[120] M.J. Siegfried, K.S. Choi, Angew. Chem. Int. Ed. 44 (2005) 3218–3223.
[121] M.J. Siegfried, K.S. Choi, J. Am. Chem. Soc. 128 (2006) 10356–10357.
[122] Z.Y. Wang, D.Y. Luan, C.M. Li, F.B. Su, S. Madhavi, F.Y.C. Boey, X.W. Lou, J. Am. Chem. Soc. 132 (2010) 16271–16277.
[123] C.H. Kuo, Y.T. Chu, Y.F. Song, M.H. Huang, Adv. Funct. Mater. 21 (2011) 792–797.
[124] X.G. Wen, W.X. Zhang, S.H. Yang, Nano Lett. 2 (2002) 1397–1401.
[125] X.-W. Liu, Langmuir 27 (2011) 9100–9104.
[126] S. Sun, C. Kong, H. Yu, X. Song, B. Ding, Z. Yang, Cryst. Eng. Comm. 14 (2012) 40–43.
[127] Q. Hua, K. Chen, S.J. Chang, Y.S. Ma, W.X. Huang, J. Phys. Chem. C 115 (2011) 20618–20627.
[128] H.Z. Bao, Z.H. Zhang, Q. Hua, W.X. Huang, Langmuir 30 (2014) 6427–6436.
[129] Q. Hua, K. Chen, S.J. Chang, H.Z. Bao, Y.S. Ma, Z.Q. Jiang, W.X. Huang, RSC Adv. 1 (2011) 1200–1203.
[130] Z. Zheng, B. Huang, Z. Wang, M. Guo, X. Qin, X. Zhang, P. Wang, Y. Dai, J. Phys. Chem. C 113 (2009) 14448–14453.
[131] Y. Cao, J. Fan, L. Bai, F. Yuan, Y. Chen, Cryst. Growth Des. 10 (2010) 232–236.
[132] J. Shi, J. Li, X. Huang, Y. Tan, Nano Res. 4 (2011) 448–459.
[133] C.-H. Kuo, Y.-C. Yang, S. Gwo, M.H. Huang, J. Am. Chem. Soc. 133 (2011) 1052–1057.
[134] W. Zhai, F. Sun, W. Chen, L. Zhang, Z. Min, W. Li, Mater. Res. Bull. 48 (2013) 4953–4959.
[135] L. Li, C. Nan, Q. Peng, Y. Li, Chem. Eur. J. 18 (2012) 10491–10496.
[136] K. Chanda, S. Rej, M.H. Huang, Chem. Eur. J. 19 (2013) 16036–16043.
[137] H. Bao, W. Zhang, Q. Hua, Z. Jiang, J. Yang, W.X. Huang, Angew. Chem. Int. Ed. 50 (2011) 12294–12298.
[138] Q. Hua, T. Cao, H. Bao, Z. Jiang, W.X. Huang, Chem. Sus. Chem. 6 (2013) 1966–1972.
[139] Q. Hua, T. Cao, X.-K. Gu, J. Lu, Z. Jiang, X. Pan, L. Luo, W.-X. Li, W.X. Huang, Angew. Chem. Int. Ed. 53 (2014) 4856–4861.
[140] Z. Niu, Y. Li, Chem. Mater. 26 (2014) 72–83.

CHAPTER 2

Ceria Nanoshapes—Structural and Catalytic Properties

S. Agarwal[1], B.L. Mojet[1], L. Lefferts[1], A.K. Datye[2,*]
[1]Catalytic Processes and Materials, MESA+ Institute for Nanotechnology, University of Twente, AE Enschede, The Netherlands; [2]University of New Mexico, Department of Chemical & Biological Engineering and Center for Microengineered Materials, Albuquerque, NM, USA
*Corresponding author: E-mail: datye@unm.edu

1. INTRODUCTION

Metal oxide catalysts serve as catalysts and as catalyst supports for the production of renewable energy, remediation of environmental pollutants, and synthesis of chemicals [1–3]. Recent years have seen remarkable progress in the synthesis of engineered catalysts with distinct shapes and compositions tailored for achieving optimum selectivity and activity in chemical reactions [3–5]. Rare earth oxides have been widely used as structural and electronic promoters to improve activity, selectivity, and thermal stability of catalysts [2]. Cerium is the most abundant rare earth metal in the earth's crust (~66.5 ppm) and is more abundant than copper, cobalt, and lithium [6]. The main sources of cerium are the light rare earth element minerals such as bastnäsite, allanite, cerite, and monazite [7,8]. Cerium oxide (commonly known as ceria, CeO_2) is one of the most significant and applied rare earth oxides in industrial catalysis [6,9].

Cerium oxide (CeO_2) has been extensively investigated in the field of heterogeneous catalysis both as a catalyst and as a support for noble metals. This is due to its unique redox properties and high oxygen storage capacity (OSC), allowing it to quickly switch oxidation state between Ce^{4+} and Ce^{3+} in the stable fluorite structure [10,11]. Because of these redox properties, CeO_2 is used in a wide range of applications such as an ultraviolet absorber in sun blocks [12], O_2 sensors [13,14], and antioxidant in the field of biomedicine [15]. Ceria is commercially used as a catalyst support in applications such as three-way catalysts (TWC) [16] and as solid electrolytes in low-temperature solid oxide fuel cells [17,18]. Further, the oxidation of CO [19–21], NO [22], hydrocarbons [23], the low-temperature water gas shift (WGS) reaction [24–26], and steam reforming of biooil [27] have been investigated on ceria and ceria-supported catalysts.

The synthesis of CeO_2 nanoparticles of controlled morphology via hydrothermal synthesis has been adequately described in the literature, but there is uncertainty about the nature of the exposed surfaces, as well as morphological features such as pits, which have been observed primarily in CeO_2 rods. Recent advancements in aberration-corrected transmission electron microscopy (AC-TEM) make it possible to study materials at the

Catalysis by Materials with Well-Defined Structures
http://dx.doi.org/10.1016/B978-0-12-801217-8.00002-5

atomic scale (resolution ~0.8 Å). In this work, we describe the use AC-TEM and AC-HAADF (high angle annular dark field) imaging to investigate the shapes, internal defects, and the surface structures in controlled morphology CeO_2 nanoparticles (octahedra, rods, and cubes) [28]. These ceria materials were also tested for their catalytic reactivity in the WGS reaction, and the nature of the surface sites was further investigated via Fourier transform infrared (FTIR) spectroscopy.

1.1 Crystallography and Defect Structure of Ceria

Cerium oxide (CeO_2) has a cubic fluorite structure with space group F*m3m* and a unit cell parameter of 5.41 Å at room temperature [29]. The CeO_2 structure consists of a cubic close-packed array with each cerium ion (white circles) coordinated by eight oxygen ions (gray circles), vice versa, oxygen ions are surrounded by four cerium ions in a crystal unit, as illustrated in Figure 1 [30]. The importance of CeO_2 originates from its unique redox properties and high OSC, allowing it to switch between Ce^{4+} and Ce^{3+} in a stable fluorite structure. In other words, CeO_2 can undergo substantial oxygen stoichiometric changes in response to change in temperature, oxygen pressure, electric field, and presence of dopants, without undergoing a change in the fluorite crystal structure [31–38]. The transport of oxygen in the ceria lattice results in the creation of intrinsic point defects [29]. These point defects can be created by either thermal disorder or by interaction with the surrounding atmosphere. The predominant defects observed in reduced ceria are anion Frenkel pairs and anion vacancies (Figure 2) [29,39].

In the *anion Frenkel type defect*, an oxygen ion is displaced from its lattice position to an interstitial position, hence creating a vacancy at its original position and a defect at the interstitial site [39–41]. In general, the formation of these defects does not influence

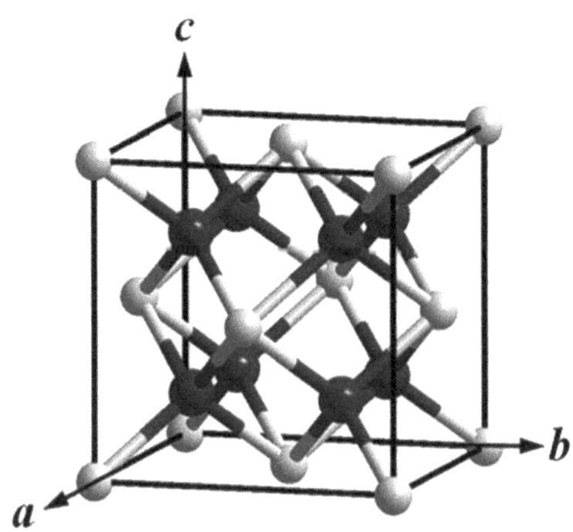

Figure 1 Fluorite structure of ceria. Dark spheres represent oxygen ions, and white spheres represent cerium ions. *Reproduced from Sun and Xue [154] with permission from the Royal Society of Chemistry.*

the overall charge and stoichiometry of the lattice. The defect type is illustrated using the Kröger and Vink defect notation as

$$O_O^X \leftrightarrow O_i'' + V_{\ddot{O}}$$

where, O_O^X and O_i'' represent oxygen ions in the lattice and interstitial positions, respectively, and $V_{\ddot{O}}$ indicates an oxygen vacancy created at the lattice site.

In the *oxygen vacancy* defect, an oxygen ion is removed from one of the lattice positions, hence creating a vacant site [42–44]. In order to maintain the charge balance in the lattice, two Ce(IV) ions reduce to Ce(III) ions. The defect formation is represented by the following equation:

$$O_O^X + 2Ce_{Ce}^X \rightarrow V_{\ddot{O}} + 2Ce_{Ce}' + \frac{1}{2}O_{2(g)}$$

Ce_{Ce}^X implies a cerium ion in its lattice position, and Ce_{Ce}' denotes the reduction of cerium from Ce(IV) to Ce(III).

In the past years, several studies based on atomistic calculations have also suggested the presence of additional defects in the ceria lattice, such as interstitial and Schottky disorder [29,41,45]. An interstitial defect is the displacement of both a cerium and two oxygen ions to form interstitial sites, for example,

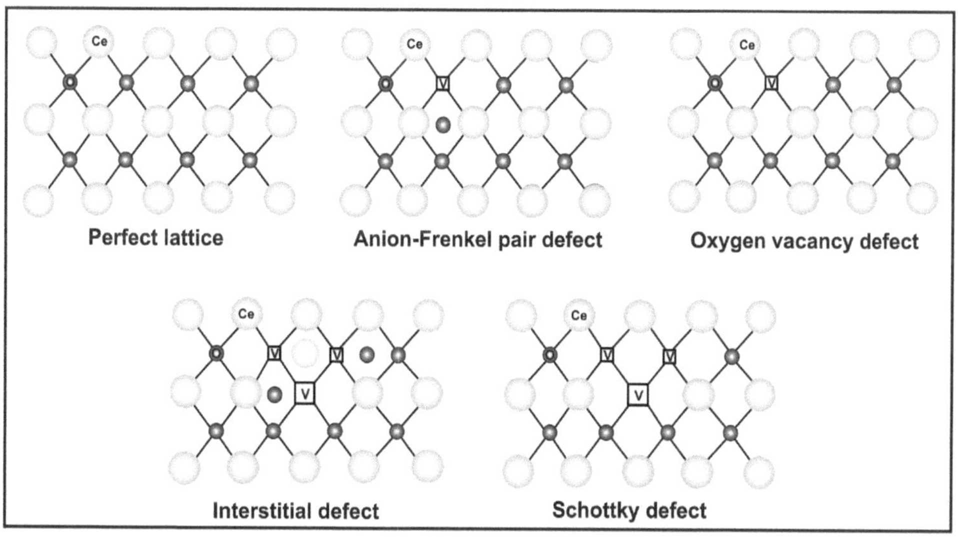

Figure 2 Possible types of point defects in the CeO_2 lattice. Key: Oxygen—dark gray circles, cerium—white circles, and vacant site—red (black in print version) square with "V."

$$2O_O^X + Ce_{Ce}^X + 2V_i^X \leftrightarrow 2O_i'' + Ce_i^{\cdots} + 2V_{\ddot{O}} + V_{Ce}''''$$

In the above equation, O_i'' and $Ce_i^{\cdots}$ indicate oxygen and cerium ions in their respective interstitial sites, V_i^x and V_{Ce}'''' represent the vacant interstitial site and vacant site created at the cerium lattice position, respectively. Finally, in the case of Schottky disorder, vacant sites are created by the removal of both the cations and anions from their lattice sites, while maintaining stoichiometry [29,45]. The following equation illustrates the defect type:

$$2O_O^X + Ce_{Ce}^X \leftrightarrow 2V_{\ddot{O}} + V_{Ce}'''' + CeO_{2(s)}$$

Due to the unique oxygen transport properties and ability to accommodate large concentration of defects, ceria is an attractive material for processes that require a constant supply of oxygen in a reductive environment.

1.2 Applications of Ceria as a Catalyst

Because of its distinctive redox properties, ceria can greatly enhance the catalytic activities for a number of important reactions when it is used as a support for transition metals [16,23,25,46]. Some of the key examples include the following:

1.2.1 Three-Way Catalysts (TWC)

To limit harmful emissions, three-way catalysts are used within petrol engines. The catalyst uses a ceramic or metallic substrate coated with metal oxides with combinations of precious metals such as platinum, palladium, and rhodium [16,47]. A three-way catalytic converter has three simultaneous functions:

1. Reduction of nitrogen oxides into elemental nitrogen.

$$2CO + 2NO \rightarrow N_2 + 2CO_2$$

2. Oxidation of carbon monoxide to carbon dioxide.

$$2CO + O_2 \rightarrow 2CO_2$$

3. Oxidation of hydrocarbons into carbon dioxide and water.

$$CxHy + (x + 0.25y)O_2 \rightarrow xCO_2 + 0.5yH_2O$$

CeO_2 is widely used as a promoter in three-way catalysts (TWCs) for treatment of toxic exhaust gases, to ensure availability of oxygen for the oxidation reactions, even during temporarily reducing conditions [16]. This application is attributed to the high OSC as well as high metal dispersion that can be easily achieved using ceria.

1.2.2 Preferential Oxidation of CO

High-purity H_2 gas is essential for fuel cell application requiring a proton exchange membrane as electrolyte (PEMFC, proton exchange membrane fuel cell) [48–51]. In PEMFC, the hydrogen is ionized to a proton, generating an electron at the anode. The proton migrates through the polymer membrane to the cathode and further reacts with O_2 to form H_2O. The CO content in the reactant gas (i.e., H_2) must be kept <100 ppm since it can poison the fuel cell electrocatalyst [51]. Industrially, hydrogen (H_2) can be produced by steam reforming, partial oxidation, and autothermal reforming of hydrocarbons [48,52]. However, during these reactions, CO_x (i.e., CO, CO_2) is also formed as a by-product. Notably, the CO content can be reduced to 0.5–1% by the reaction with steam (WGS reaction) [53]. However, the remaining concentration of CO is still too high for use in PEMFC and should be further lowered (<100 ppm) by preferentially oxidizing (PROX) CO on a suitable catalyst without excessive consumption of H_2 [51]. Based on the high activity required to remove CO while maintaining a high CO oxidation selectivity and minimum oxidation of H_2, desirable PROX catalysts generally comprise metal (Pt, Rh, Pd, Au, etc.) supported on an oxide (ceria, alumina, silica) [54–56]. So far, ceria-supported platinum (Pt), rhodium (Rh), and palladium (Pd) catalysts have been found to have remarkable activity in the low-temperature oxidation of CO [56]. This can be attributed to the CeO_2 support that can promote oxidation even in a temporarily oxygen-deficient environment due to its high OSC [50].

1.2.3 Steam Reforming of Biooil

The depletion of fossil fuel reserves has necessitated the search for an efficient and environmentally acceptable way to generate hydrogen (H_2) from sustainable resources [57,58]. One of the promising alternatives is to generate hydrogen by steam reforming of biooil, obtained via flash pyrolysis of biomass (Figure 3).

$$C_nH_mO_k + (2n-k)H_2O \rightarrow nCO_2 + \left(2n + \frac{m}{2} - k\right)H_2 \quad \Delta H^0_{298} > 0\, kJ \cdot mol^{-1}$$

Biooil is a complex mixture of oxygenates. Therefore, a wide variety of model compounds have been studied for designing and understanding appropriate catalysts for steam reforming, such as carbon monoxide, methanol, ethanol, acetic acid representing compounds in the aqueous phase, whereas phenolic compounds are typically used to mimic the organic fraction [59]. Transition metal oxides, such as CeO_2, ZrO_2, TiO_2, and mixed oxides with metal loading, seem to be good catalyst supports for steam reforming because of their tunable oxidation states, which render better stability and enhanced activity [59–62].

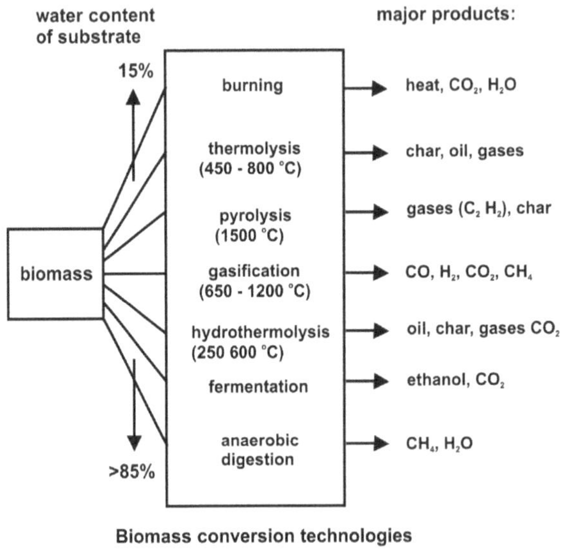

Figure 3 Scheme of the types of biomass conversion routes; note that "flash pyrolysis" is a special case of thermolysis, heating to typically 600 °C within seconds. *Reproduced from Okkerse and van Bekkum [57] with permission from the Royal Society of Chemistry.*

1.2.4 Low-Temperature WGS Reaction

WGS is a reversible exothermic process to convert CO and H_2O to generate H_2 and CO_2. The WGS reaction is usually applied downstream of the steam reforming of hydrocarbons to maximize the H_2 yield and decrease the amount of CO [63].

$$CO + H_2O \leftrightarrow CO_2 + H_2 \quad \Delta H^0_{298} = -41.1\ kJ \cdot mol^{-1}$$

The WGS reaction is exothermic; hence, the reaction equilibrium shifts to the left with increasing temperatures. Therefore, to suppress thermodynamic limitations and attain higher CO conversions, the WGS reaction is carried out in two stages. In the first stage, a high-temperature shift (HTS) is performed at 300–450 °C, followed by a low-temperature shift (LTS) reaction operated at 180–230 °C [64]. Industrial catalysts for HTS and LTS stages are Fe_2O_3 promoted with Cr_2O_3 and Cu on ZnO/Al_2O_3, respectively [10,24,65].

These traditional catalysts are unsuitable for mobile use, such as in PEMFCs. LTS catalysts are insufficiently active, whereas Cu-based LTS catalysts are pyrophoric (reactive to O_2) and require very careful engine start-up/shut-down procedures [66]. Alternative materials, such as ceria-supported metal catalysts, have received much attention as low-temperature single-step WGS catalysts [24,26,67]. It has been suggested, especially for Pt-supported catalysts, that Pt-on-ceria acts as a bifunctional catalyst in which CO is adsorbed on the metal sites and H_2O dissociation (to regenerate hydroxyl groups on ceria support) occurs on the defect sites (oxygen vacancies) on the ceria surface, increasing the catalytic activity (Figure 4) [46,68].

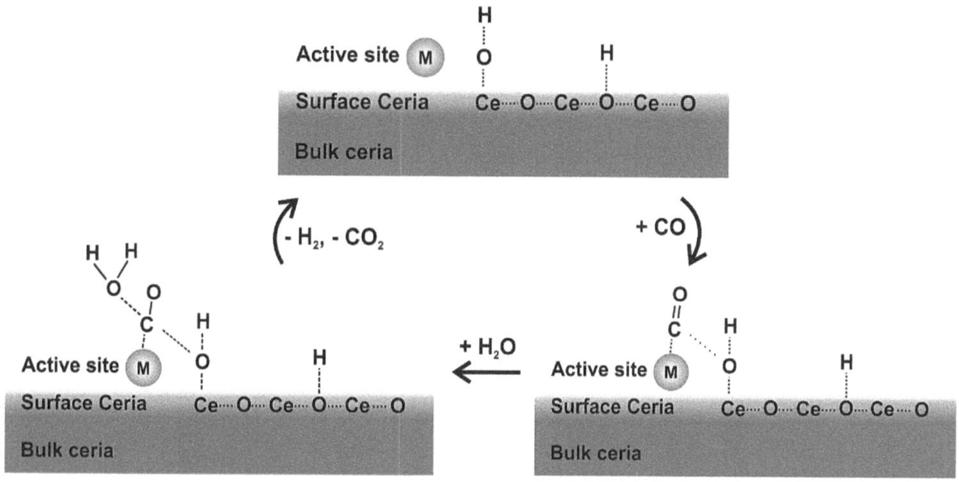

Figure 4 Low-temperature WGS reaction pathway on hydroxylated CeO_2 supported catalyst. M illustrates an active site on a metal particle.

1.3 Recent Work on the Structure and Reactivity of Ceria Nanoshapes

In recent years, many studies have reported improved activity of ceria catalysts through synthesis of ceria nanostructures having controlled morphology [5,69]. The controlled morphology generates well-defined exposed crystallographic planes, which may lead to improved catalytic activity. Ceria particles with fluorite structure (Figure 1) preferentially expose stable (111) planes [70,71]. However, by reducing the particle size of ceria to nanodimensions, contributions from the less stable (110) and (100) terminations are reported [6,9,38]. The three low index lattice planes on the surface of CeO_2 nanocrystals are shown in Figure 5 [1,72]. CeO_2 (111) has an open structure with O in the top layer followed by an accessible Ce layer, whereas, on the (110) surface, both Ce and O atoms are in the top layer, as can be seen in Figure 5 [30]. In the case of ceria (100) plane, the surface layer is terminated by O, and the Ce layer underneath is not accessible, making this surface polar and unstable [30,73]. Different planes have different numbers of nearest bonded neighbors for Ce and O on the exposed surface, for instance, in the (111) plane, Ce is bonded to 7 oxygen and oxygen to 3 cerium atoms. For the (110) and (100) planes, the Ce:O coordination number on the exposed surface is 6:3 and 6:2, respectively [1]. This leads to the relative stability of the low index ceria surfaces following the trend: (111) > (110) > (100) [72,74]. The nonpolar CeO_2 (111) surface undergoes little relaxation, whereas the other two planes can undergo significant lattice relaxations [1,72,75]. Further, the formation energies of oxygen vacancies on different surface planes of ceria vary, following the order (110) < (100) < (111) [76]. A recent study by Lin et al. provides aberration-corrected scanning transmission electron microscopy (AC-STEM) images showing the nature of the surface termination on these three low index surfaces [77].

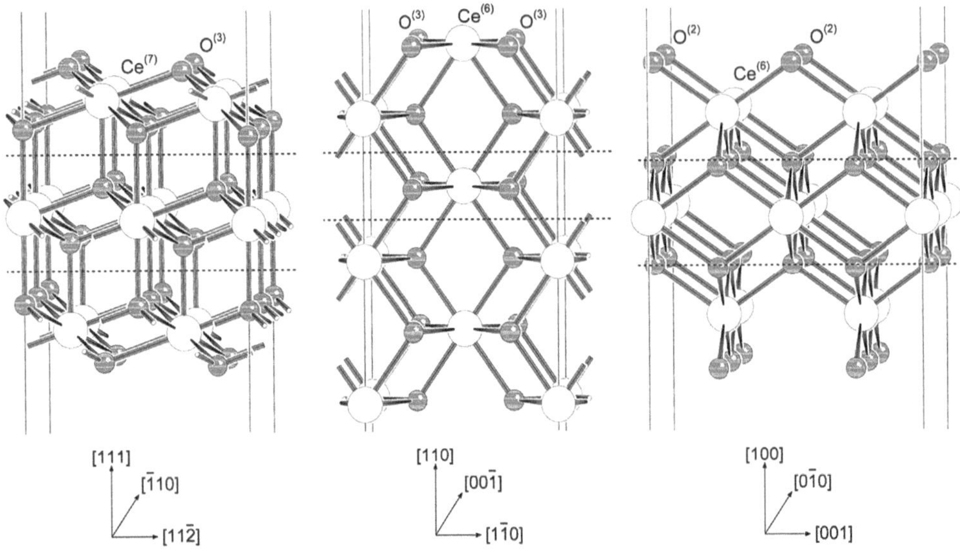

Figure 5 CeO_2 (111), (110), and (100) surfaces. Black and white spheres represent cerium and oxygen ions, respectively. *Reprinted from Ganduglia-Pirovano et al. [1]. Copyright © 2013 WILEY-VCH.*

Several fundamental investigations have been performed on epitaxial ceria thin films in ultrahigh vacuum, such as adsorption of probe molecules like CO [78], H_2O [30,79], acetone [80], acetaldehyde [81], formic acid [82], and even the WGS reaction [83]. From those studies, it is concluded that (100) surfaces are more active in adsorbing and reacting with probe molecules than the (111) surface [73]. Further, (100) planes are more susceptible to surface reconstruction and defect formation to relocate the surface charge in lattice for attaining stability [75]. While these plane-specific investigations give a glimpse into the fundamental surface-dependent behavior of CeO_2, the conditions (ultrahigh vacuum) are far from real-world catalytic conditions where the surface would be covered by surface hydroxyls and other adsorbates.

Based on the above, it can be concluded that the ceria particles with low index surface planes and easy defect formation are desirable for catalytic applications. Nanostructured ceria can be synthesized using various approaches, such as precipitation, hydrothermal, sol–gel, and surfactant-assisted synthesis (Figure 6) [6,9,38,84]. The common cerium precursors used for synthesis includes cerium(III) sulfate hydrate [$Ce_2(SO_4)_3 \cdot xH_2O$] [85], cerium(III) nitrate hexahydrate [$Ce(NO_3)_3 \cdot 6H_2O$] [20,86], ammonium cerium nitrate [$(NH_4)_2Ce(NO_3)_6$] [87,88], and cerium chloride ($CeCl_3$) [89,90]. Synthetic procedures developed in recent years have allowed the formation of ceria into desirable nanoshapes with well-defined exposed planes. For instance, ceria nanoparticles generally have octahedral shapes mainly exposing stable (111) facets [91]. Ceria cubes preferentially expose active (100) planes, while, ceria rods are thought to expose (100) and (110) surfaces, which are less stable and thus more reactive than the (111) surface

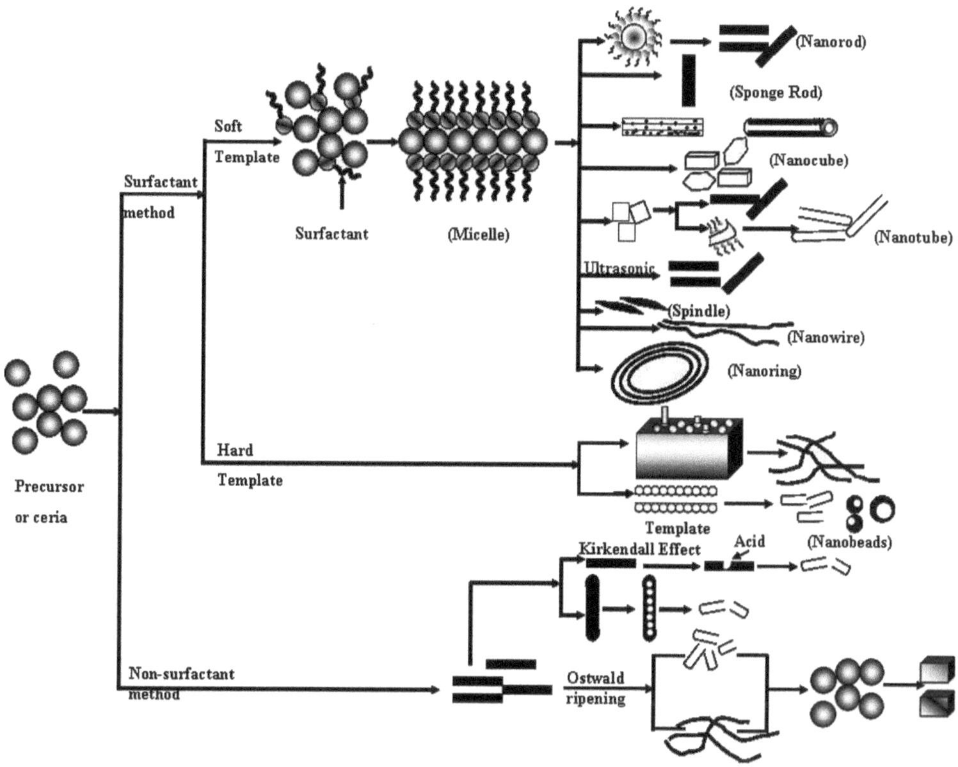

Figure 6 Synthesis pathways for the formation of CeO_2 in different shapes. *Reprinted from Lin and Choudhury [84].*

[11,20,86]. As we will discuss later, some of the previous interpretations of exposed facets on ceria nanorods need to be reconsidered in light of recent evidence. Compared with bulk ceria, these ceria nanoshapes exhibit excellent redox properties and high specific activity/selectivity due to the exposure of active surface planes, as well as having more defects and higher surface to volume ratio [6,9,76].

Activity trends have been noted in recent work with these ceria nanoshapes. For instance, ceria rods were reported to have the highest catalytic activity for the oxidation of CO (Figure 7) [11,20,21], naphthalene [92], 1,2-dichloroethane, and ethyl acetate in comparison to other nanoshapes (trend: rods > cubes > octahedra) [93]. This observation was related to the exposed planes on rods as well as surface defects such as vacancy clusters, pits, and a high degree of surface roughness. In fact, it has been suggested that the activation and transportation of active oxygen species in ceria rods is greatly enhanced due to the presence of oxygen vacancy clusters [19]. In addition, these vacancy clusters cause exposure of Ce^{3+} ions, which provide effective sites for the adsorption of CO, as illustrated in Figure 7. Hence, the existence of vacancy clusters, along with Ce^{3+} sites, positively affects the overall catalytic performance. Further, ceria cubes with (100)

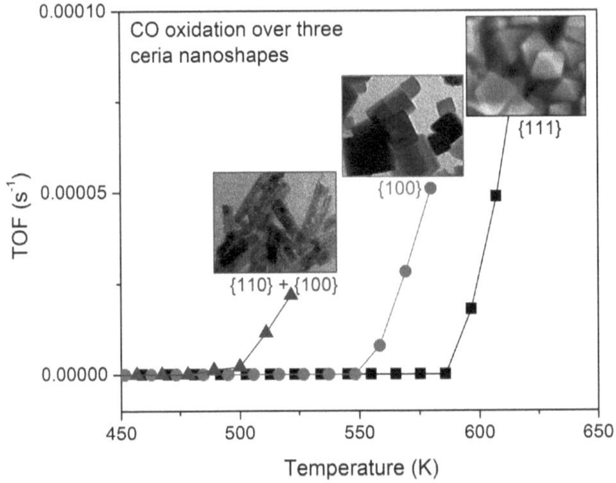

Figure 7 CO oxidation activity on ceria rods, cubes, and octahedra. Insets in plot are transmission electron microscopy (TEM) images of ceria rods, cubes and scanning electron microscopy (SEM) image of octahedra. The indicated exposed facets are those suggested by the authors, but as we show in this chapter, rods expose {111} and {100} surfaces. *Reprinted with permission from Wu et al. [21].*

exposed planes have been reported to exhibit excellent reducibility and high OSC [94]. Recently, Cassir et al. (2013) investigated hydrogen oxidation on ceria nanoshapes and reported the following reducibility trend: cube > rod > octahedron, indicating that hydrogen oxidation is dependent on the structure of the exposed ceria surfaces [70].

As mentioned in the previous section, reactions, such as WGS, CO oxidation, and steam reforming, widely use metal particles supported on cerium oxide. The performance of the catalyst is strongly influenced by the particle size, porosity, surface area, and exposed planes of the support material [9]. These properties of the support material not only influence the dispersion of metal particles but also affect the metal–support interactions. For instance, ceria rods have been found to stabilize highly dispersed gold particles of size <1 nm, while gold particles of size approximately 3 nm have been almost exclusively formed on ceria cubes [95]. Similarly, in the case of the Ag/CeO_2 catalyst, CeO_2 rods have been shown to exhibit stronger Ag–CeO_2 interaction than ceria cubes [96]. Another example includes Ni-supported on CeO_2 rods, an excellent catalytic activity and coke resistance has been reported during CO_2 reforming of methane. The enhanced activity is attributed to the increased bonding strength between the Ni metal particles with the ceria support (Figure 8), as well as an increase in the presence and transfer of lattice oxygen from the ceria support to metal particles [97]. Similar findings were also reported for Pt- and Pd-loaded ceria nanoshapes [98,99].

The literature on ceria nanoshapes is replete with different synthesis routes [9,19,20, 22,92,93,96,98,100,101] with increasing emphasis on developing structure–performance relationships of ceria nanoshapes. For example, Wu et al. (2010) reported that the

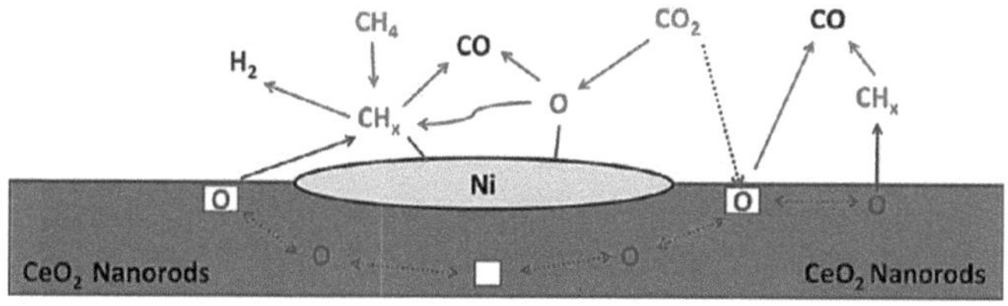

Figure 8 Catalytic mechanism for CO_2 reforming of methane over Ni-supported CeO_2 rod. *Reprinted with permission from Du et al. [97]. copyright © (2012). American Chemical Society.*

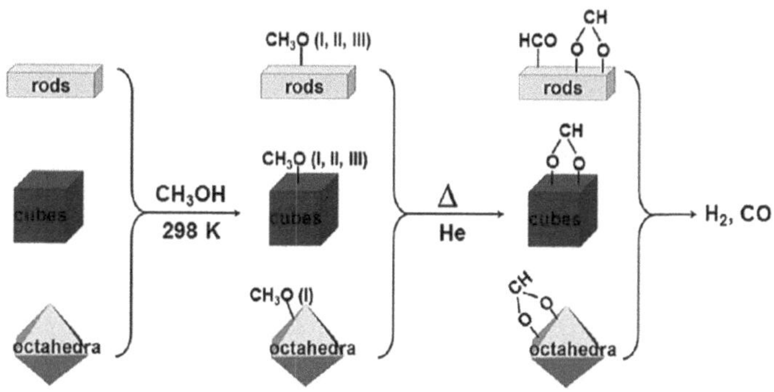

Figure 9 Schematic overview of adsorption and desorption of methanol on ceria nanoshapes. *Reprinted with permission from Wu et al. [102]. Copyright © (2012). American Chemical Society.*

adsorption of surface dioxygen species varies with the defect sites on these nanocrystals [76]. The stability and reactivity of these oxygen species are also found to vary with the exposed planes on ceria nanoforms.

The adsorption and desorption of methanol were also investigated to probe the nature of surface sites on ceria nanoshapes with well-defined exposed planes [102]. Interestingly, on the surfaces of rods (exposing (110) and (100) planes) and cubes (with (100) planes), three types of methoxy ($-CH_3O$) species, that is, on-top (I), bridging (II) and three-coordinate (III) were identified on exposure to methanol at room temperature, while exclusively on-top (I) types of methoxy groups were observed on the ceria octahedra surface (with (111) facets, Figure 9). This difference is attributed to the different Ce/O coordination ratios as well as the number of defect sites on the three nanoshapes. Furthermore, the methoxy species are found to be less stable and hence more reactive on ceria rods, resulting in formation and desorption of H_2 and CO at lower temperatures (<583 K) than on cubes and octahedra.

The suggestion that the higher catalytic activity of rods is due to the exposed (100) and (110) planes was originally made in the seminal works of Zhou [20] and Mai [86]. Indeed the synthesis approach within those papers has become the standard recipe for the preparation of rods and cubes [93,100,103–106]. Further, the conclusion from those papers that ceria rods have exposed (100) and (110) surfaces and grow along the {110} direction has been significantly relied on during the past eight years for furthering the field [19,21,76,92,96,98–100,102,104,106–108]. However, discrepancies in the literature do exist. More recently, it was suggested that ceria rods can also expose (111) surfaces as well as small amounts of (100) planes [109–111]. In addition, it was proposed that the origin of the enhanced reactivity for CO oxidation may not be connected with the surface planes, but rather with the defects seen in these rods, particularly vacancy clusters [19,22]. Another recent report proposed that the ceria rods are not single crystals but have a kind of complex core–shell structure exposing at least 50% of (110) along with the (111) plane [70]. Likewise, it has been suggested that CeO_2 cubes exhibiting (100) surfaces may actually be composed of octahedral units, implying thereby that their surfaces are composed of (111) facets [112]. In light of these uncertainties, this chapter focuses on the in-depth characterization of the ceria nanoshapes to provide an insight into the exposed planes and the surface species actually responsible for enhanced activity of nanocatalysts. We focus on the low-temperature WGS reaction.

2. SYNTHESIS AND MORPHOLOGY OF CERIA NANOSHAPES

Synthesis procedures for CeO_2 nanocrystals are well developed for generating morphologies such as cubes [86], rods [5,9,11,20,86], wires [11], tubes [6,113], octahedra [91], spindles [111], three-dimensional (3D) flower-like shapes [114,115], and spheres [9], and these materials have also been studied in catalytic reactions. The literature shows that CeO_2 morphology can play an important role in catalytic performance [5,6,9,11,21,114,115]. For instance, ceria cubes were reported to exhibit excellent reducibility and high OSC attributed to the presence of (100) planes [94]. CeO_2 rods attracted attention because they provide a high surface area. The rods are thought to expose (100) and (110) surfaces, which are less stable and thus more reactive than the (111) surface [72]. It has also been suggested that the CeO_2 rods have surface defects [19,76] such as vacancy clusters, pits, and a high degree of surface roughness. CeO_2 rods have been reported to exhibit enhanced reactivity for the oxidation of CO [11,20,21], NO [22], 1,2-dichloroethane, and ethyl acetate [93].

Previous workers used high-resolution TEM (HRTEM) to identify surface features and the morphology of the nanoparticles. Conventional HRTEM does not allow us to see the surfaces very clearly due to image delocalization and Fresnel fringes. The development of AC-TEM has pushed the resolution below 1 Å both in TEM and STEM modes [116,117]. AC-TEM was recently used to investigate the beam-induced cationic mobility on the surface of CeO_2 nanoparticles [118,119]. In our work, we used a double aberration-corrected microscope (JEOL JEM-ARM 200F) equipped with both image and probe correctors. This combination of atomic scale resolution in TEM and STEM images provides insights into the nature of the surfaces of these CeO_2 nanoshapes. Since microscopy is a local technique, we have also included

X-ray diffraction (XRD) and Brunauer–Emmett–Teller (BET) surface area measurements to provide a better average of the structure and morphology of these powders. To characterize the surface reactivity, we have used the WGS reaction as a probe reaction. In addition, FTIR spectroscopy of adsorbed CO and H_2O allows us to obtain mechanistic an insight into the surface reactions taking place on the respective CeO_2 nanoshapes. This combination of techniques allows us to present a complete picture of the nature of the surfaces in shape controlled CeO_2 nanoparticles.

CeO_2 rods and cubes were synthesized using the approach described by Mai et al. [86] while the octahedra were obtained from Sigma Aldrich as described by Agarwal [120]. All samples were calcined at 500 °C (heating rate 5 K/min) for 4 h in synthetic air (flow rate = 50 mL/min). The BET surface areas of ceria rods, octahedra, and cubes were found to be 80, 58, and 10 m^2/g, respectively. The samples do not possess microporosity, and have average pore sizes >10 nm. The pore volumes of ceria rods, octahedra, and cubes were 0.33, 0.20, and 0.05 cm^3/g, respectively. The surface area is consistent with the particle sizes observed by SEM and TEM, showing that on average the cubes are significantly larger in size than the other two samples. Figure 10 shows the XRD patterns of the three samples. The prominent CeO_2 diffraction peaks are consistent with the expected CeO_2 reflections according to ICDD card 43-1002 [121]. All three samples show some degree of asymmetry in the peak profile, as seen from a shoulder, or second peak on the left of the (311) reflection. We assign this shoulder to the existence of a bimodal particle size distribution, with the smaller particles showing a lattice constant

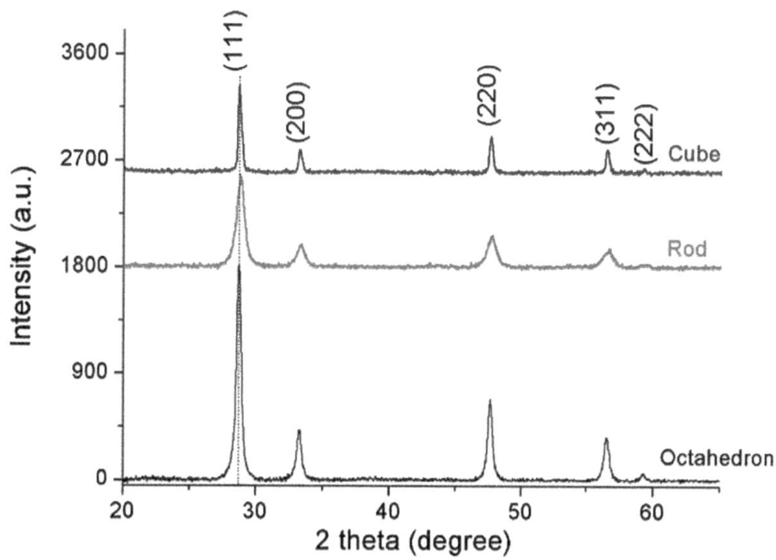

Figure 10 XRD patterns for the controlled morphology CeO_2 powders used in this study. The (111) reflection is highlighted with a vertical line and it shows that the cubes, rods, and octahedron samples have very similar lattice constants. *Reprinted from Agarwal et al. [28]. Copyright © 2013 WILEY-VCH.*

that is larger than bulk CeO_2. The lattice expansion can be caused by the presence of Ce^{3+} ions in this ionic lattice as explained by the model presented by Tsunekawa et al. [122] A bimodal particle size distribution can be clearly observed in the SEM and TEM images and hence is consistent with this interpretation. In this sample set, all the samples exhibit this bimodal particle size distribution, but we will show below that the particle size does not affect the surface termination of the CeO_2 nanoshapes. To establish the surface structure and morphology, the ceria nanoshapes were analyzed by High resolution scanning electron microscope (HR-SEM), AC-TEM, and AC-STEM, the latter in Bright Field (BF) as well as HAADF imaging modes. We also embedded the rods in epoxy and prepared cross-section samples using an ultramicrotome, which allows us to see rods end-on. We next describe the morphology of these samples, and the FTIR of adsorbed species on the surface followed by the catalytic reactivity.

In addition to XRD, Raman spectra were also taken at room temperature (ex situ) to investigate the ceria lattice vibrations of the different shapes (Figure 11) [103]. For easier comparison, the spectra have been normalized to the main peak at 465 cm^{-1}. The strong peak at 465 cm^{-1} is due to the symmetrical stretching mode of the $Ce–O_8$ crystal unit, characteristic for the fluorite lattice structure, which is in agreement with the XRD patterns [123,124]. A clear broadening of the band at 465 cm^{-1} is observed in the order octahedra < cubes < rods, which, according to literature, is due to inhomogeneous strain broadening associated with the different particle dimensions and phonon confinement [125]. Because of the different geometries of the samples, the line broadening in the Raman spectra cannot be directly related primarily to particle sizes of the materials [37].

In Figure 11, a magnified view of the smaller peaks between 200 and 700 cm^{-1} is shown. The bands below 200 cm^{-1} can be attributed to particle scattering. Two additional bands are observed at 259 and 598 cm^{-1}. The peak at 259 cm^{-1} has been assigned

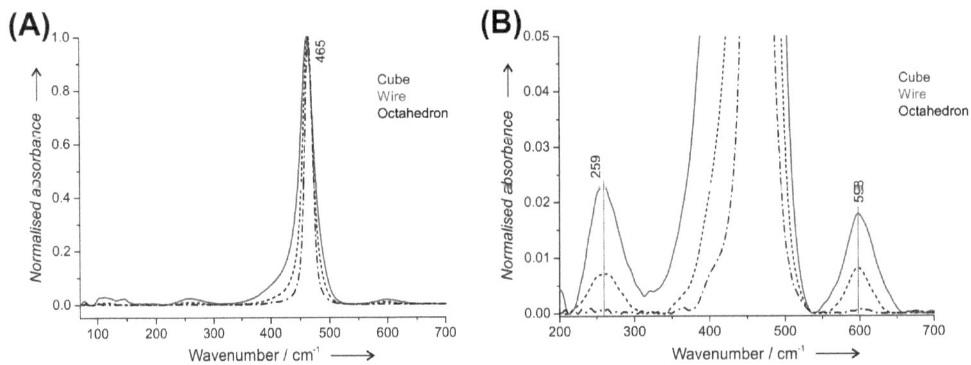

Figure 11 (A) Raman spectra (532 nm at 2 mW) of ceria cubes (blue (dark gray in print version)), rods (red (light gray in print version)), and octahedra (black) obtained at room temperature in air, normalized to the main peak at 465 cm^{-1}; (B) magnification of the wave number range 200–700 cm^{-1}. *Reprinted from Agarwal et al. [103]. Copyright © 2013 WILEY-VCH.*

previously to the displacement of oxygen atoms from ideal fluorite lattice positions [126]. The peak at 598 cm^{-1} indicates the presence of oxygen vacancies in the ceria lattice [76,127]. The presence of these peaks is a clear indication of local defects and reduced Ce^{3+} cations in the cubes and rods. It is obvious that different ceria shapes, with other surface planes exposed, have different amounts of defect sites, as was shown in the literature [76,128]. For octahedra (black spectrum in figure 11b), no oxygen vacancy or oxygen displacement bands were observed. The relative intensity of both bands increases in the order octahedra < cubes < rods, indicating the highest fraction of oxygen vacancies in ceria rods, consistent with previous studies [76,129].

2.1 Octahedra

Figure 12 shows an SEM image of ceria octahedra. Well-defined, faceted octahedra can be clearly seen. In addition, some fine-grained material is observed whose morphology cannot be determined at this magnification. As we show in the images to follow (Figures 13 and 14), the fine-grained material has a similar morphology and exposed facets as the larger particles, and the octahedral shape is the dominant one in this sample.

A high-magnification STEM image is shown in Figure 15. Since the beam diameter is about 0.8 Å in this microscope, we can easily resolve the CeO_2 crystal lattice. The BF image is analogous to an HRTEM image; hence, contrast at these high magnifications comes from phase contrast. On the other hand, the HAADF image arises from incoherent scattering at high angles and is dominated by atomic number contrast. The bright lines consequently represent rows of Ce atoms, since O atoms will be almost invisible due to their low atomic number. The complementary BF and HAADF images allow us to see clearly the morphology of the sample and the nature of the surfaces. The particles are randomly oriented; hence, they are not lined up along the low index zone axes, which is why most of them do not show crossfringes. Figure 16 shows that the CeO_2 {111} surfaces are

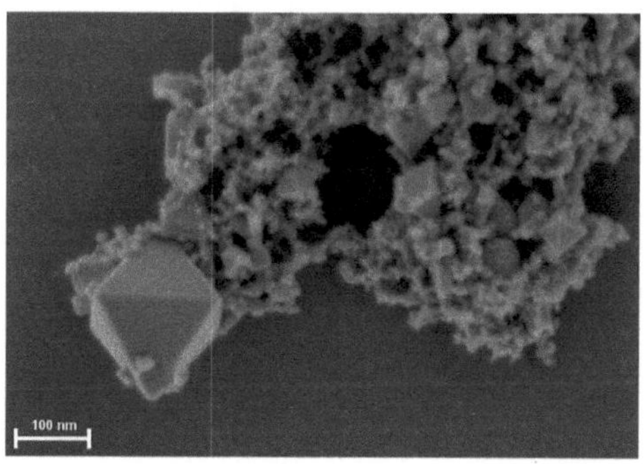

Figure 12 HRSEM image of the octahedra. Well-defined, faceted octahedra of differing sizes are present in this sample. This sample shows a distinct bimodal size distribution, with numerous particles that are much smaller than the large octahedra imaged here. *Reprinted from Agarwal et al. [28]. Copyright © 2013 WILEY-VCH.*

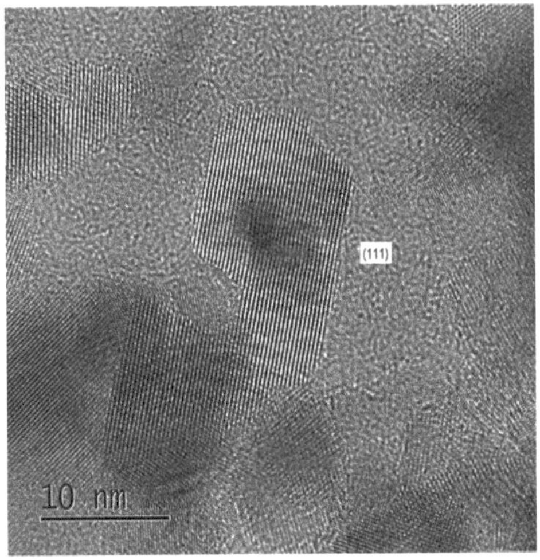

Figure 13 AC-TEM image of CeO_2 octahedra showing that the smaller particles (~10 nm) also expose prominent (111) facets. The shapes are not as well defined as the larger octahedra imaged (Figures 12 and 14).

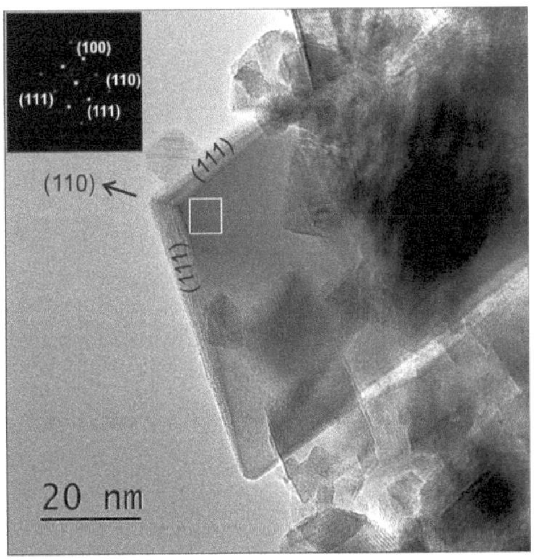

Figure 14 AC-TEM image shows that the smallest particles as well as the larger ones exhibit well-defined facets with smooth surfaces. These particles all expose prominent (111) surface facets, and the (110) facet is missing, and instead we see a sharp corner.

smooth and free from any surface steps, defects, or surface reconstruction. Since we are only dealing with CeO_2, the differences in contrast in the HAADF images can be directly related to sample thickness, and this allows for confirmation of the sample morphology. As seen in the dark field image, the prominent diamond shaped particles are thicker in the center than at the edges; hence, these images are consistent with an octahedron imaged edge-on.

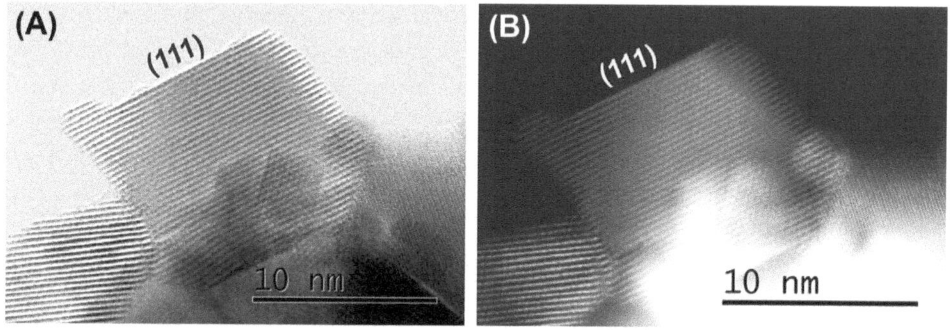

Figure 15 Higher magnification STEM HAADF images showing the nature of the (111) surface. There is no significant surface roughness evident on these. *Reprinted from Agarwal et al. [28]. Copyright © 2013 WILEY-VCH.*

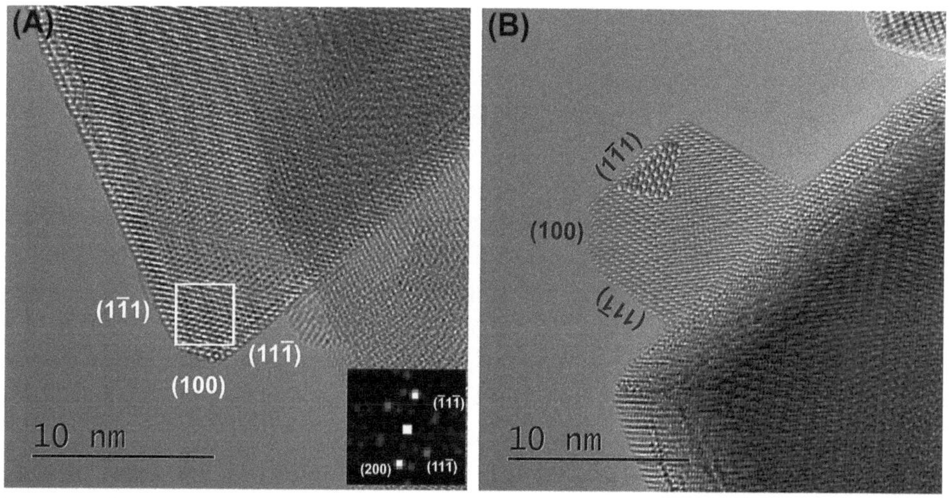

Figure 16 Aberration-corrected TEM images of ceria octahedra. These expose the (111) facets and very small (100) facets. The surface termination is abrupt, and there is no amorphous layer, or surface roughness. The inset shows the structure of the CeO_2 lattice (Ce—green (light gray in print version) and O—red (dark gray in print version) atoms) and the agreement between the white dots and the Ce atom columns. *Reprinted from Agarwal et al. [28]. Copyright © 2013 WILEY-VCH.*

A TEM image is a projection of a 3D object onto a plane. When the particle is large, some degree of surface roughness can arise because the surface along the viewing direction is not perfect over large distances. The nature of the surfaces is further confirmed by the AC-TEM images (Figure 16) where a large and small octahedron is shown. The prominent {111} facets show clearly resolved rows of Ce atom columns with small {100} facets. The inset is a model of the CeO_2 lattice showing that the AC-TEM image clearly resolves single atom columns of Ce. Approximately 20–25 images were systematically analyzed for exposed crystal planes. Care was taken to ensure that selected regions did not contain multiple overlapping particles to avoid interference when analyzing the data.

In addition to the dominant {111} planes, we also observe that the truncated ends of the octahedra have exposed {100} planes (Figure 16), as shown previously [91,130]. On the other hand, we see sharp corners where a {110} surface should have been present (Figure 14). The fact that we did not see any {110} surface facets and the fact that growth occurs along {110} direction indicates that the {110} surfaces are not stable in the ceria structure under the conditions of synthesis of the ceria octahedra.

2.2 Rods

Figure 17 shows a high-resolution SEM image of the rods. This low-magnification view shows that the sample contains a uniform distribution of rods with a width of 5–10 nm and a length of about 200 nm, that is, an aspect ratio exceeding 20. The low magnification view does not show us the cross-section of this rod, but it is clear that the primary exposed surface is along the length of the rods and the ends make a minimal contribution to the total surface area. In Figure 18, we show BF and HAADF STEM images from these rods. The insets show fast Fourier transforms (FFT) that provide information on the periodicities seen in the image. Based on the calibration of the microscope, we can index these spots as {111} planes, implying that the rods expose their {111} surfaces. The BF images also show characteristic regions of low contrast that appear to follow the crystallographic directions. The fact that these regions are dark in the HAADF image indicates less mass in those regions. These features could arise from voids within the structure, or from surface steps. As can be seen in Figure 18(A), surface steps are observed along the length of the rods. Therefore, we conclude these low contrast features represent voids as well as surface steps in the CeO_2 rods. The shapes of these voids follow the stable surfaces of the ceria structure; hence, these voids are bounded by {111} surfaces.

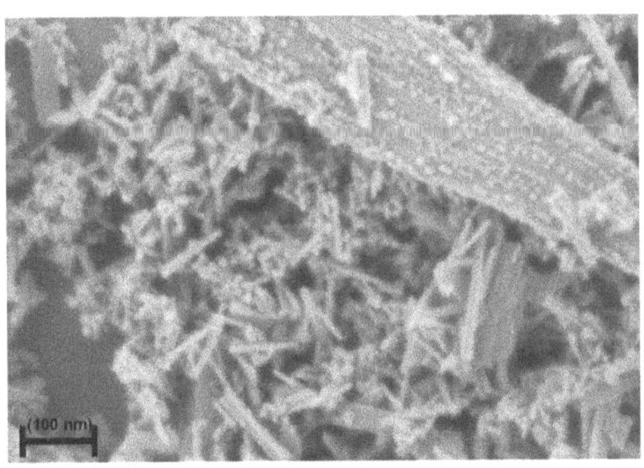

Figure 17 HRSEM image of the nanorods. The rods have aspect ratios (length/width) exceeding 20 with widths of 5–10 nm. *Reprinted from Agarwal et al. [28]. Copyright © 2013 WILEY-VCH.*

The ceria rods grow along the {110} direction. This is confirmed by the lattice fringe images seen in Figure 19(C) where we see (220) lattice fringes. We recorded numerous other images confirming the growth direction to be the {110} direction, and this is consistent with the crystallography proposed in the literature by Mai et al. [86] and Zhou et al. [20]. These authors have proposed that the rods expose (110) and (100) surfaces. However, our results show only {111} surfaces being exposed for rods prepared

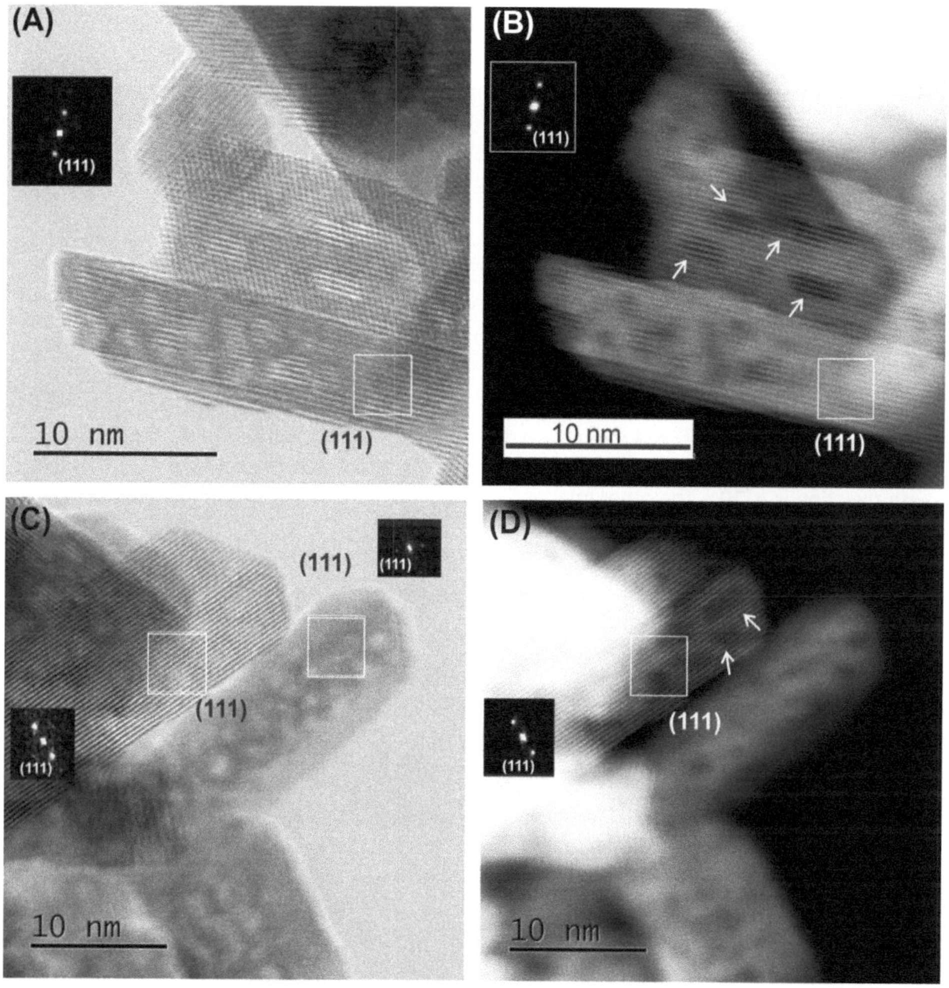

Figure 18 STEM images of ceria rods Left: BF images, Right: HAADF images. The inset in the figures is the FFT, which allows indexing of the lattice planes. The rods expose (111) surfaces and have surface steps along the length. Areas of light contrast can be seen in the BF image, and these same areas look dark in the HAADF image. The HAADF images confirm that these low contrast features are voids in the ceria rods that are bounded by (111) surfaces. The contrast in the HAADF image suggests a rectangular profile in cross-section. *Reprinted from Agarwal et al. [28]. Copyright © 2013 WILEY-VCH.*

according to Mai's recipe. In all the images we analyzed, we never saw a well-defined {110} facet in any of the rods we imaged. In order to see the top and bottom surfaces of these rods, we prepared a cross-section sample by embedding the rods in epoxy and using a microtome to prepare a section. Figure 20 shows the microtomed section where the rods are now embedded in epoxy. We reasoned that some of the rods must lie end-on so we could image their cross-sections. Based on their aspect ratio (>20), any features with significantly smaller aspect ratios must represent rods. From this image, we infer that the rods are flat, explaining why they lie on the TEM grid with the same orientation.

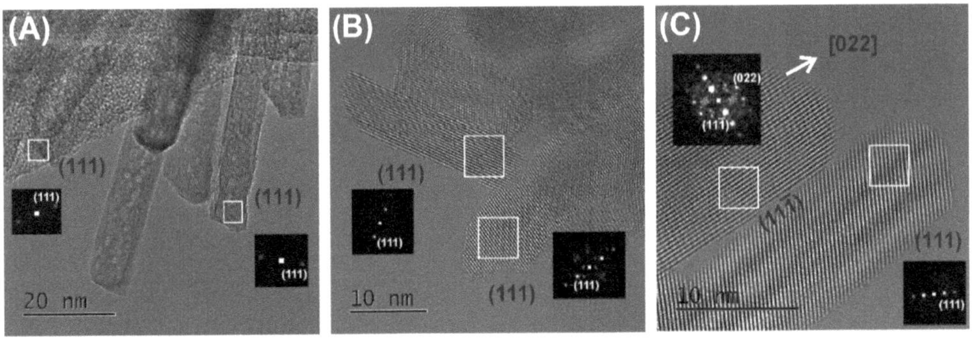

Figure 19 AC-TEM images of ceria rods at two different magnifications. Insets in the figures are FFTs of the boxed region in the images. The lattice fringes (8c) confirm that the surfaces are {111} and the growth direction is <110>. The other images likewise show only {111} surface facets. *Reprinted from Agarwal et al. [28]. Copyright © 2013 WILEY-VCH.*

Figure 20 TEM images of microtomed sections of CeO_2 rods embedded in epoxy. These sections provide an opportunity to see the rods end-on. The image on the right shows some of the rods end-on (i.e., those with a short aspect ratio), allowing us to image the rods in cross-section.

However due to the thickness of the sample along the long axis, we could not get lattice resolution in the end-on views; hence, we cannot provide direct observations of these surfaces.

In summary, we have not been able to find rods that expose a (110) surface in this sample to allow us to study the nature of this surface. We also note that while the octahedra show small {100} facets, they do not show any {110} facets; instead, we see sharp corners where the {110} facets should have been present. Further, since the rods grow along the {110} direction, the ends of the rods should also expose the {110} surfaces. But none of the rods show well-defined facets at their ends, making us believe that the {110} surface is not stable, since it represents the growth direction. To form one-dimensional structures, the surfaces that are not growing must represent stable surfaces, and be crystallographically different from the growth direction; otherwise, we should see branched rods that are not observed. Therefore, based on the AC-STEM and TEM results, we conclude that the rods also expose the stable {111} surfaces.

2.3 Cubes

Figure 21 shows a high-resolution SEM image of the cubes. Well-defined cubic crystals are seen ranging in size from 10 to 100 nm. The image also shows many smaller cubes giving rise to a bimodal particle size distribution, which may explain the asymmetric peaks seen via XRD (Figure 10). The AC-TEM images of CeO_2 cubes are shown in Figure 22. These cubes are tilted away from the cube axis, close to their {110} zone, so we see a (100) surface on the right-hand side and a (110) termination on the top. However, the actual (110) surface does not look like the structure model shown in the inset. Rather, it is corrugated and composed of (111) facets. This is perhaps the closest we can get to seeing what the (110) surface would look like, and we expect the surfaces of the rods may also look like this. On the other hand, the (100) surface shows a very good match to the structure model.

Figure 21 HRSEM image of cubes. Well-defined cubic crystals are seen ranging in size from 10 to 100 nm. *Reprinted from Agarwal et al. [28]. Copyright © 2013 WILEY-VCH.*

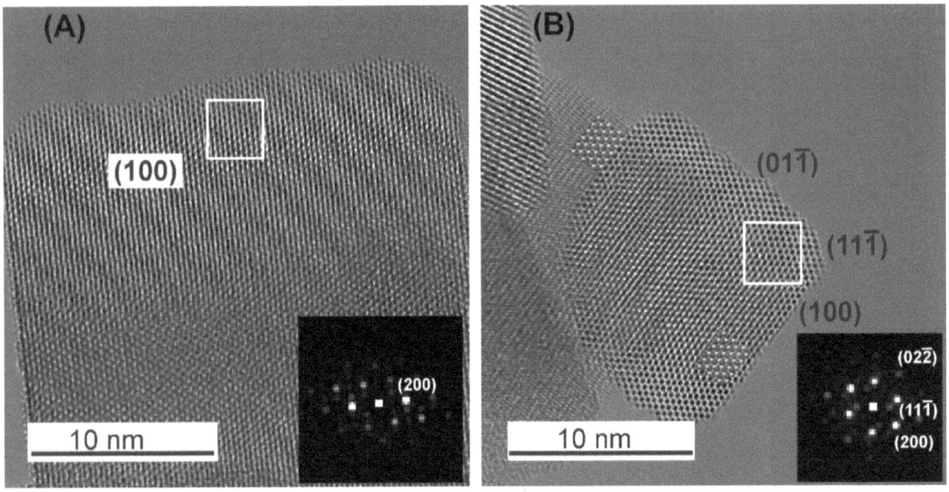

Figure 22 AC-TEM images of ceria cubes. Insets in the figures are the spatial FFT of the images. These cubes are tilted off the cube axis to a nearly [110] zone axis. In (B), we show (100) and (110) surfaces. We can see agreement with the (100) surface, but the actual (110) surface is undulating and composed of (111) facets. The lattice model (right) shows the arrangement of Ce—green (light gray in print version) and O—red (dark gray in print version) atoms.

Figure 23 shows an image of a cube oriented along the cube axis. We can see it is composed of {100} surfaces, but the corner, where we would expect a (110) surface is not well defined. We have included insets showing the atomic arrangement of the CeO_2 structure indicating that the (100) surface is not corrugated but perfectly matches what would be expected from a row of Ce atoms.

3. INFRARED SPECTROSCOPY OF ADSORBED SPECIES ON CERIA NANOSHAPES

3.1 Samples as Prepared

The ceria nanoshapes were characterized with FTIR in helium flow at 200 °C to remove physisorbed water. In Figure 24(A), the transmission FTIR spectra for the different nanoshapes taken with an empty cell as a background are shown [103]. Figure 24(A) shows that the absorbance in the background is not correlated with the amount of sample used to press the pellets. For example, the baseline of 15.5 mg of ceria rods is almost twice in intensity compared to a similar amount of cubes (14.6 mg). In general, the baseline in spectra of solid materials is determined by the light scattering capacity of the solid particles. Thus, in the present case, it can be concluded that the infrared light scattering is affected by the shape of the ceria particles. Shape specific scattering of infrared light is, for example, also known for ice crystals with different morphologies [131]. In a separate experiment (not shown), we found no correlation of peak intensities with

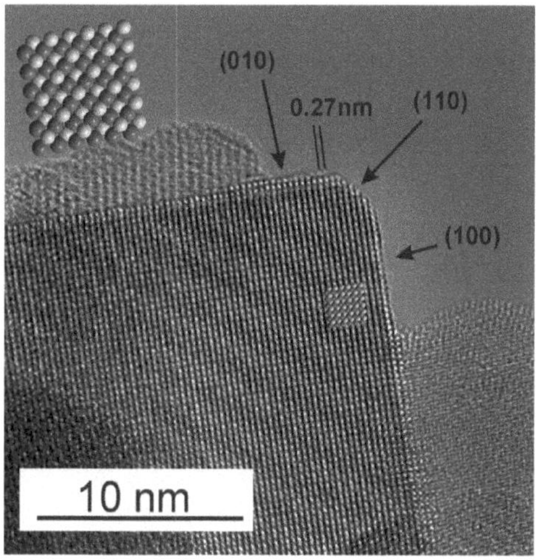

Figure 23 AC-TEM image of a cube nearly oriented along is cube axis [010]. The (100) surfaces can be clearly seen. They show single atoms steps, but no other form of surface reconstruction (i.e., into (111) facets). The (110) surface is indicated at the cube corner, but it does not appear to be a well-defined facet. The structure of CeO_2 along the [010] direction is also included as an inset, to show the agreement with the white atom contrast on the right side of the image. The higher magnification view shows the arrangement of Ce—green (light gray in print version) and O—red (dark gray in print version) atoms. *Reprinted from Agarwal et al. [28]. Copyright © 2013 WILEY-VCH.*

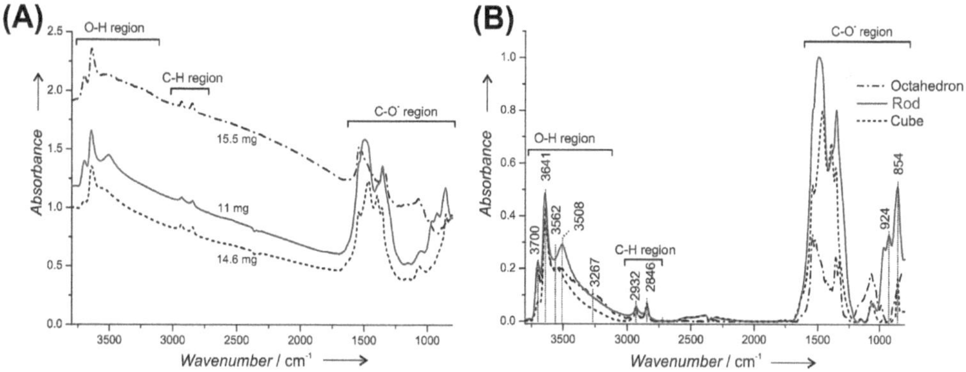

Figure 24 In situ FTIR (A) raw; (B) Baseline-corrected spectra of ceria cubes (blue (dark gray in print version)), rods (red (light gray in print version)) and octahedra (black spectrum) at 200 °C in He flow (after 30 min of flow). *Reprinted from Agarwal et al. [103]. Copyright © 2013 WILEY-VCH.*

sample weight. Thus, the absolute peak intensities cannot be normalized to the amount of sample. Nevertheless, peak ratios found in the respective spectra can be compared to each other.

The baseline-corrected spectra for the three samples are given in Figure 24(B). Comparison of the peak positions in the spectra before and after baseline correction showed that the peak maxima shifted to a lower frequency with maximum change of 2 cm^{-1}, which is within the resolution of the spectra.

Three distinct sets of peaks can be observed: O–H stretching vibrations between 3800 and 3100 cm^{-1}, C–H stretching between 3000 and 2700 cm^{-1}, $C–O^{-}_{x\ (x=2,3)}$ stretching and deformation between 1700 and 800 cm^{-1}.

In the O–H stretch range, the following bands can be distinguished for the ceria nanoshapes: isolated hydroxyls (3700 cm^{-1}), bridging hydroxyls (3641 cm^{-1}), multiple bonded hydroxyls (3550–3500 cm^{-1}), and hydrogen bridging hydroxyls (broad band 3400–3100 cm^{-1}) [132,133]. All three nanoshapes show similar O–H stretching bands although with different relative intensities. The rods and octahedra resemble each other closely in the whole hydroxyl range, but rods have a distinct peak at 3508 cm^{-1}, while octahedra have a shoulder at 3267 cm^{-1}. Cubes have a higher ratio of bridged (OH 3641 cm^{-1}) to isolated (OH 3700 cm^{-1}) hydroxyl intensity accompanied with a shoulder at 3562 cm^{-1}. This seems to be consistent with the more open structure of the (100) plane terminating cubes.

C-H stretching bands are observed at 2932 and 2846 cm^{-1} for all samples and can be assigned to formate species [134,135]. The bands have similar low intensities and peak positions for the three nanoshapes.

In addition to the C–H vibrations, high-intensity bands were observed between 1800 and 1250 cm^{-1}, which can be assigned to C–O stretch vibrations in formate and carbonate species. Clearly, these species are stable at 200 °C in He. The different types of carbonate and formate are formed as soon as CeO_2 is exposed to carbon species, such as atmospheric CO_2 [133–138]. Clearly rods and cubes have much higher integrated intensity ratios of carbonates to hydroxyl than octahedra. In addition, the spectral shapes between 1200 and 1700 cm^{-1} are different for the three samples, indicating different relative amounts of formate and carbonate species depending on ceria shape. In general, the peak positions found here are in agreement with those published previously for ceria particles [134]. However, we do not assign specific peaks individually to species at this point, because of the large overlap between them (a more thorough analysis is presented by Vayssilov et al., who give an extensive theoretical and experimental overview of CO_2 and CO adsorption on ceria surfaces [135]).

Finally, additional peaks are observed between 1250 and 850 cm^{-1}, which can be assigned to the deformation vibrations of formates and carbonates. Specifically, the band at 854 cm^{-1} is observed for rods and cubes, due to the presence of multibonded carbonates [135]. In addition, rods also show a distinct band at 924 cm^{-1}, which is absent in the other two samples. According to the literature, this peak may be assigned to oxalate types of species [136].

Clearly, the three ceria nanoshapes show different relative amounts of hydroxyls, carbonates, and formates after calcination and treatment in He at 200 °C. The differences might well arise from the different crystal planes terminating the particles, and the differences in BET surface area. Furthermore, the carbonate species are extremely difficult to remove since high temperatures are needed (>600 °C) at which the nanoparticles start to lose their specific geometries and crystal planes [21,76,133]. This demonstrates that the ceria nanoshapes already differ in their composition and vibrational spectra after synthesis and calcination; thus, before starting adsorption experiments, an observation that has been mentioned but not studied in the literature [21].

3.2 CO Adsorption and Reactivity with Water

The interaction of the surfaces of the three types of nanoshapes with CO and H_2O at 200 °C will be described in detail in this section.

3.2.1 Ceria Octahedra

After drying at 200 °C in helium (Figure 25(A), dotted black line), CO was adsorbed on the octahedra (Figure 25(A), solid red line). The black spectrum is identical to the spectrum for octahedra in Figure 24(B) and is used as a reference for comparison. Exposure to CO/He resulted in a variety of changes in the spectrum. In order to clearly see the CO-induced changes we have plotted in difference [CO–He] spectrum in Figure 25(B). The positive peaks in the difference spectrum correspond to species formed in the presence of CO, while the negative peaks correspond to species that disappeared as soon as the sample was exposed to CO. The intensity of formates and carbonates (1800–1250 cm^{-1}) increased significantly, accompanied by the appearance of the band at 854 cm^{-1} as additional evidence for carbonate formation. The C–H stretch at 2848 cm^{-1} tripled in intensity, confirming the formation of formate. The band at 2932 cm^{-1} became broader but did not change much in integrated intensity. Finally, in the O–H stretch range the peak at 3700 cm^{-1} completely disappeared. An increase of intensity was observed between 3641 and 3300 cm^{-1} resulting in a broad band with a maximum at 3586 cm^{-1}. Bands at these frequencies have been attributed to hydrogen carbonates ($CO_2(OH)^-$) [133]. Evidently, the O–H species giving rise to the 3641 cm^{-1} band are not reactive with CO.

The formation of carbonates upon CO exposure is in accordance with recently published experiments at room temperature [21]. However, the frequencies reported for these species are at significantly higher wave number (between 1616 and 1700 cm^{-1}) than the ones we observe in the current study. The formation of formates from the interaction of CO with hydroxyls is in agreement with other studies on ceria particles [21,135,138,139]. Formate band positions are similar as in the literature, although bands vary in position by 30 cm^{-1}, suggesting that the samples are similar but not identical [21]. Furthermore, the disappearance of the 3700 cm^{-1} band has been assigned to reduction of the Ce^{4+} to Ce^{3+} when removing this specific OH species [133,135].

After exposure to CO, the cell was flushed with helium to check the stability of the adsorbed species; obviously, gas phase CO (2143 cm^{-1}) disappeared. Approximately 10%

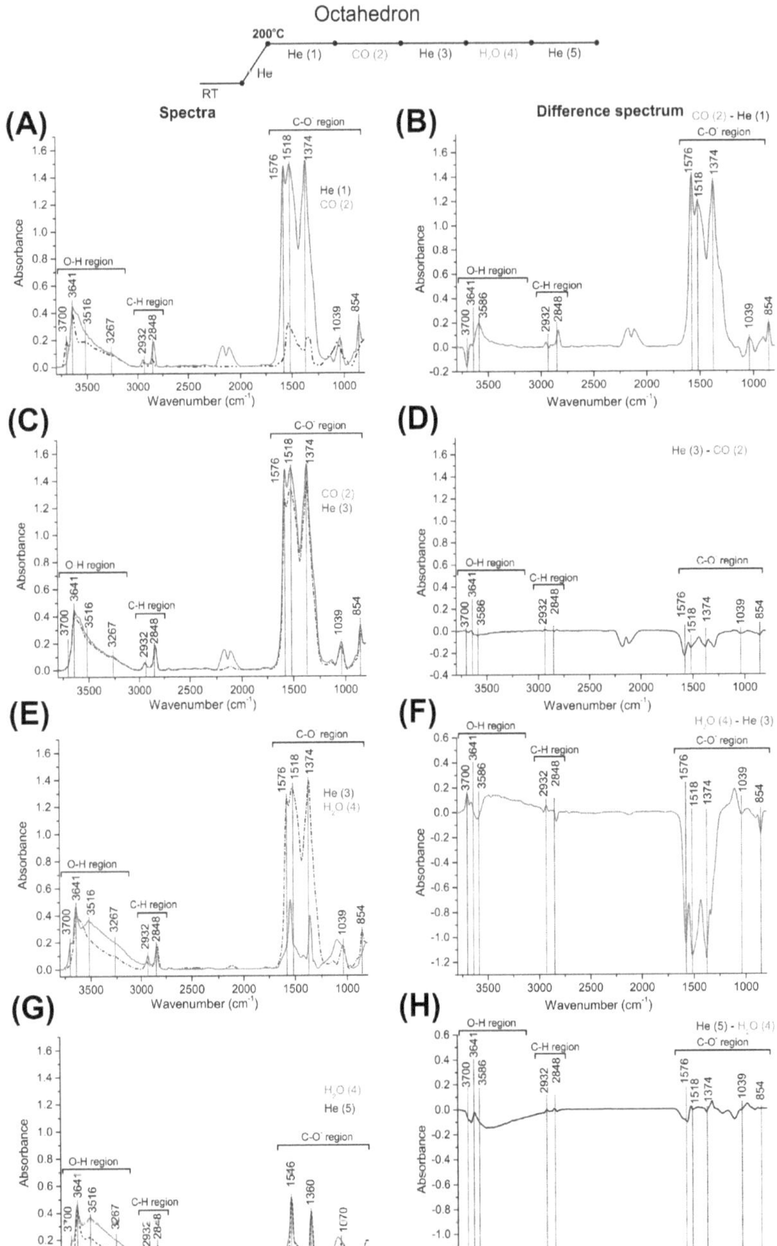

Figure 25 Baseline-corrected in situ FTIR spectra at 200 °C of ceria octahedra (A) in He flow (black dash-dot spectrum), followed by 33 vol% CO/He (red (dark gray in print version) solid line); (B) He-subtracted CO spectrum (red (dark gray in print version) solid line); (C) in 33 vol% CO/He (red (dark gray in print version) solid line) followed by He(3) Flow (green (black in print version) dash-dot spectrum); (D) He(3) subtracted CO spectrum (green (black in print version) solid line); (E) Comparison between the sample in He(3) (green (black in print version) dash-dot line) with subsequent exposure to H_2O(4) (orange (light gray in print version) solid line) and (F) Corresponding subtraction of He(3) and H_2O(4), (G) H_2O(4) (solid orange (light gray in print version) line) and after flushing with He(5) (blue (black in print version) dotted line), (H) corresponding subtraction of He(5) and H_2O(4). *Adapted from Agarwal [120].*

of the formed surface species disappeared upon helium flow (dash-dotted line in Figure 25(C)). Subsequently, the sample was exposed to H_2O/He at 200 °C, and the resulting spectrum is given in Figure 25(E), solid line. In Figure 25(F), the accompanying [H_2O–He] difference spectrum is shown. Upon water addition, most of the species formed during CO treatment disappeared, as can clearly be seen between 1800 and 1200 cm^{-1}. The polycarbonate band at 854 cm^{-1} disappears, suggesting complete removal of carbonates from the ceria octahedra. The small band at 2932 cm^{-1} increased, while the formate C–H stretch at 2848 cm^{-1} slightly decreased, due to water. The integrated intensities of the two bands hardly changed. This indicates that the bands are not likely to be originating from a combination of carbonate and formate vibrations as has been suggested in the literature [133,135,140], since almost all carbonates were removed by water. Consequently, we conclude that these bands belong to the stable formate species on the surface, in good agreement with the literature [139]. In addition, the O-H stretch band at 3700 cm^{-1} starts to reappear upon exposure of water, while the band at 3586 cm^{-1} decreases, both pointing at reoxidation of the ceria by water [133]. Finally, the increase in intensity in the 3500–3000 cm^{-1} window is due to H-bonded water. Flushing with He (Figure 25(G) and (H)) induces only minor effects, including removal of H-bonded water.

3.2.2 Ceria Rods

In Figure 26(A), the spectra for ceria rods after CO adsorption and the sample as prepared in He at 200 °C can be seen. In the difference spectrum (Figure 26(B)), the changes induced upon adsorption of CO can be seen more clearly. Unlike the observations on the octahedra, two negative –OH peaks can be observed for ceria rods at 3700 and 3641 cm^{-1} in Figure 26(B), caused by the interaction of CO with the active –OH species. As a result of this interaction, three more distinct –OH bands are now visible that indicate the presence of other types of –OH groups after CO adsorption. Furthermore, no hydrogen carbonates (3583 cm^{-1}) were found for ceria rods. As a result of the interaction between CO and ceria hydroxyls, formates and carbonates were formed on ceria rods as can be concluded from the C–H stretch bands (2932 and 2846 cm^{-1}) and the C–O stretch bands from formates and carbonates between 1700 and 1200 cm^{-1}.

Flushing with He after CO adsorption causes only minor effects (Figure 26(C) and (D)). Obviously, CO in gas phase disappears, whereas the formate and carbonate peaks decrease only slightly.

Upon exposure to H_2O, these adsorbates decompose and –OH groups are partially restored on the catalyst (Figure 26(E) and (F)). Further, the band at 857 cm^{-1} decreased in intensity on switching from CO to H_2O, indicating the decomposition of carbonates on the surface. Again, flushing with He (Figure 26(G) and (H)) results in only minor changes.

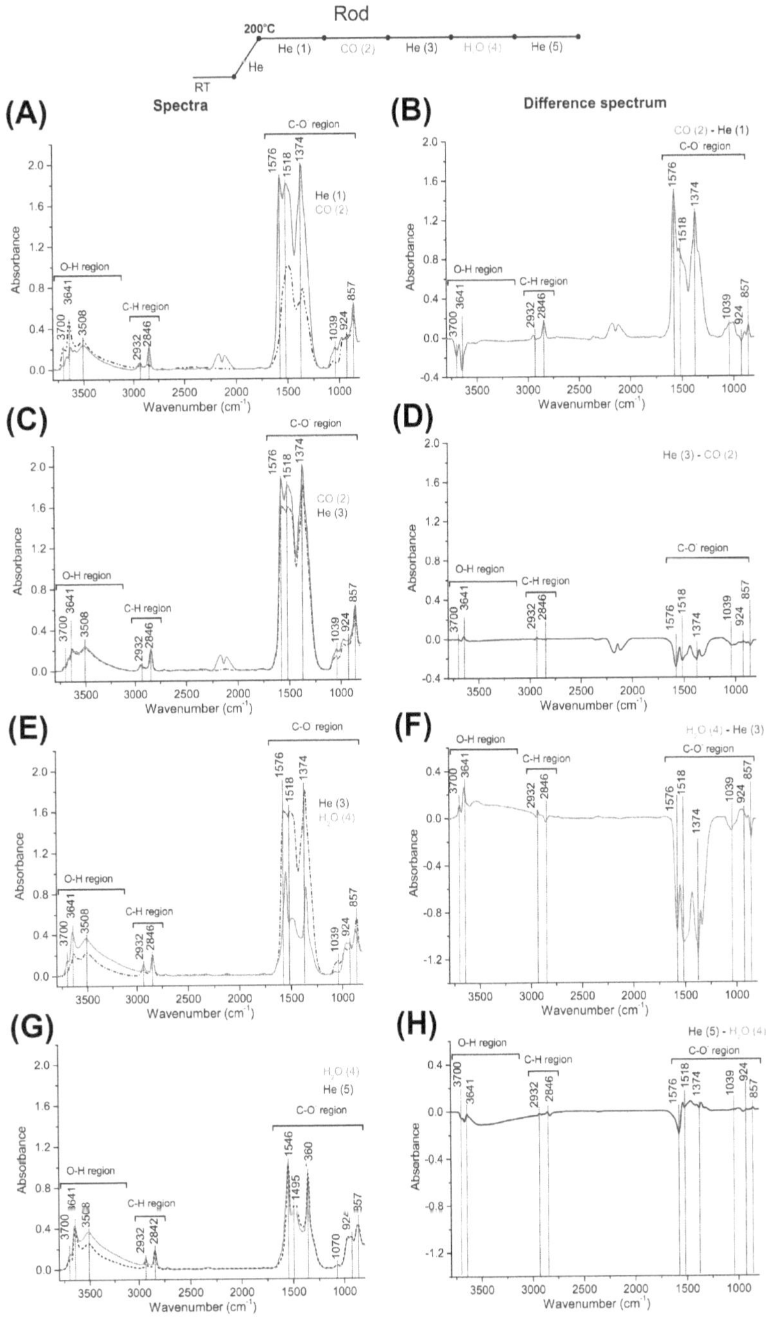

Figure 26 Baseline-corrected in situ FTIR spectra at 200°C of ceria rods (A) in He flow (black dash-dot-dotted spectrum), followed by 33 vol% CO/He (red (dark gray in print version) solid line); (B) He-subtracted CO spectrum (red (dark gray in print version) solid line); (C) in 33 vol% CO/He (red (dark gray in print version) solid line) followed by He(3) flow (green (black in print version) dash-dot spectrum); (D) He(3) subtracted CO spectrum (green (black in print version) solid line); (E) Comparison between the sample in He(3) (green (black in print version) dash-dotted line) with subsequent exposure to H_2O(4) (orange (light gray in print version) solid line); (F) Corresponding subtraction of He(3) and H_2O(4), (G) H_2O(4) (solid orange (light gray in print version) line) and after flushing with He(5) (blue (black in print version) dashed line); (H) Corresponding subtraction of He(5) and H_2O(4). *Adapted from Agarwal [120].*

3.2.3 Ceria Cubes

The FTIR spectra obtained for the fresh ceria cubes (black dash-dot-dotted spectrum) and after CO adsorption (red solid line) are shown in Figure 27(A), and the corresponding difference spectrum in Figure 27(B). A sharp decrease of –OH intensity was observed, especially for the band at 3641 cm^{-1}. Similar to the rod sample, the three specific hydroxyl peaks decrease in intensity and split into five distinct O–H vibrations after CO exposure, see Figure 27(A). At the same time, carbonates and formates are formed, similar but with different relative intensities as compared to rods and octahedra, as will be discussed in the next section. The multicoordinated carbonate peak appears in cubes at 867 cm^{-1}, which is approximately 10 cm^{-1} higher than for rods and octahedra. At the same time, the band at 854 cm^{-1} slightly decreased, suggesting the formation of a slightly altered multicoordinated carbonate species, possibly caused by mutual interaction of these species. No bicarbonate was observed at around 3600 cm^{-1}, similar to rods but different as compared to octahedra. Subsequent flushing with He (Figure 27(C) and (D)) gives minor changes; a small fraction of the carbonate and formate species decompose.

On subsequent treatment with H_2O, the peaks related to adsorbed formates and carbonates largely decreased (1700–1200 cm^{-1}) (Figure 27(E) and (F)) pointing toward their decomposition, and reforming most of the active –OH groups. The integrated intensity for the peaks at 2932 and 2842 cm^{-1} remained constant, whereas a clear decrease was observed for the 867 cm^{-1} peak.

Flushing with He (Figure 27(G) and (H)) results not only in the removal of a small fraction of formate and carbonate species, but also an increase in intensity is observed at 1395 cm^{-1} and to a lesser extent at 1474 and 1337 cm^{-1}.

3.3 Discussion

XRD (Figure 10) and Raman spectroscopy (Figure 11) confirm the general observation that ceria nanoshapes exhibit a distorted fluorite structure, with an increasing number of oxygen vacancies from octahedra to cubes to rods, in agreement with findings reported in the literature [76]. The presence of vacancies in rods and cubes is supported by the FTIR spectra in helium (Figure 24) showing that the integrated intensity ratio of carbonates to OH groups in rods and cubes is significantly higher compared to that in octahedra. The formation of carbonates on ceria is due to the interaction with CO_2 from the ambient, resulting in the reduction of ceria to Ce^{3+} and hence the creation of oxygen vacancies [141–143].

The experimental results obtained at 200 °C are described in the previous section. We repeated these observations at 350 °C because these samples become catalytically active for WGS at 350 °C (see the next section). For a more detailed description, we refer to a recent publication [144]. A comparison of the results obtained at these temperatures is presented in Figure 28, in the form of the difference spectra that were introduced in Figures 25–27 for all three shapes, and we now show these difference plots at these two temperatures.

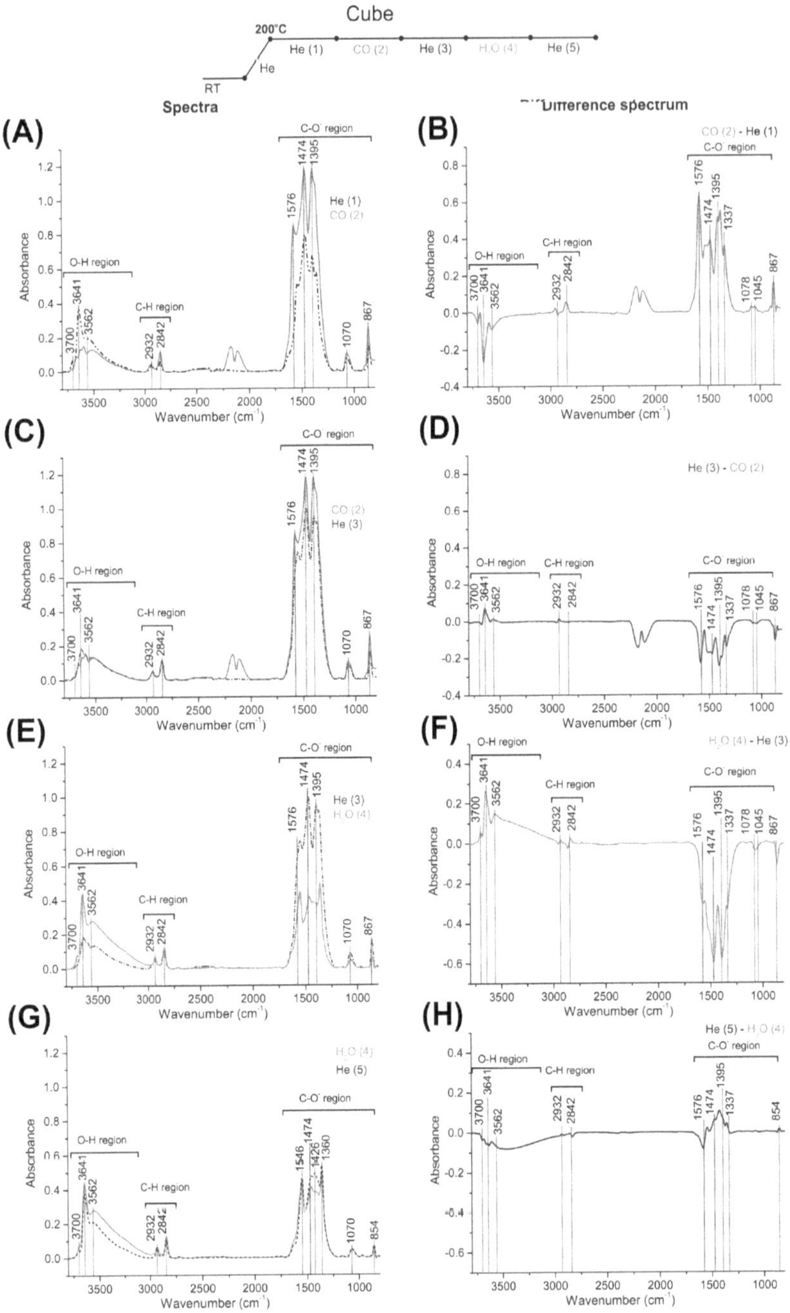

Figure 27 Baseline-corrected in situ FTIR spectra at 200 °C of ceria (A) in He flow (black dash-dot-dotted spectrum), followed by 33 vol% CO/He (red (dark gray in print version) solid line); (B) He-subtracted CO spectrum (red (dark gray in print version) solid line); (C) in 33 vol% CO/He (red (dark gray in print version) solid line) followed by He(3) flow (green (black in print version) dash-dot spectrum); (D) He(3) subtracted CO spectrum (green (black in print version) solid line); (E) Comparison between the sample in He(3) (green (black in print version) dash-dotted line) with subsequent exposure to H_2O(4) (orange (light gray in print version) solid line); (F) Corresponding subtraction of He(3) and H_2O(4), (G) H_2O(4) (solid orange (light gray in print version) line) and after flushing with He(5) (blue (black in print version) dashed line), (H) Corresponding subtraction of He(5) and H_2O(4). *Adapted from Agarwal [120].*

An intriguing observation is that although the nanoshapes have initially similar, but not identical, -OH vibrations (Figure 24), their reactivity toward CO at 200 °C is remarkably different. Figure 28(A) and (B) compares the difference spectra [CO–He] after CO adsorption for the three nanoshapes at 200 °C and 350 °C, respectively. At 200 °C, all samples show a clear decrease in the band for isolated hydroxyls at 3700 cm^{-1}, indicating partial reduction of ceria to Ce^{3+} by CO [133], via reaction of an isolated OH group. For ceria octahedra, only a minority of the other –OH groups interact with CO, and a clear band of hydrogen carbonates is observed at 3586 cm^{-1}. For rods and cubes, it is clear that the band at 3641 cm^{-1}, previously assigned as bridged –OH [132,133], significantly decreased upon CO adsorption showing a higher reactivity of the –OH species at 3641 cm^{-1}, as compared to octahedra. The reactivities of these species are evidently determined by the specific ceria nanoshape. Furthermore, both rods and cubes do not show formation of hydrogen carbonates, in contrast to octahedra (3586 cm^{-1}). Finally, the –OH band at 3562 cm^{-1} interacted with CO on cubes only, which was not observed on octahedra and rods. The observations are quite different at 350 °C (Figure 28(B)); not only octahedra but also rods seem to form bicarbonates, whereas the extent of consumption of OH is significantly smaller as compared to 200 °C, which is in line with the observation that formates and carbonates are being formed to a lesser degree.

Figure 28(A) shows also difference spectra [CO–He] for the three nanoshapes in the 1800–800 cm^{-1} range at 200 °C. It is immediately apparent that, relative to the initial amount of –OH present (Figure 24), different amounts of formates and carbonates are formed, in the order cubes << rods < octahedra, please note absolute intensities cannot be directly compared as discussed in Section 3.1. On octahedra an intense broad band is observed at 1518 cm^{-1}, which is related to hydrogen–carbonate formation in agreement with the band at 3586 cm^{-1}. For cubes and rods, many different bands and shoulders are observed between 1576 and 1337 cm^{-1}, with different relative intensities. At 350 °C (Figure 28(B)) a similar observation is made that the formation of carbonates and formates is much smaller on cubes than on rods and octahedra; note the difference in the scale of the Y-axis in Figure 28(A) and (B). This is mainly caused by the difference in the surface area of the samples, similar to the observations at 200 °C. Figure 28(C) and (D) shows that thermal decomposition of the species formed in CO is much more significant at 350 °C than at 200 °C, obviously. At 350 °C, a significant fraction of the formates seem to decompose on CO removal, also partly recovering the OH groups involved.

It is recognized that nanoshapes have different exposed crystal planes on their surfaces, but the nature of exposed surfaces on these shapes has been debated [11,86,145–148]. It was originally proposed that rods expose (110) and (100) crystal planes, whereas octahedra mainly expose (111) and cube (100) surfaces. In Section 2, we have described our observations that lead to the conclusion, contrary to the initial assignments, that rods primarily expose (111) surfaces. These exposed crystal planes differ in their cerium to oxygen ratio in the topmost layer, for instance the (111) plane has a Ce:O ratio of 7:3, as

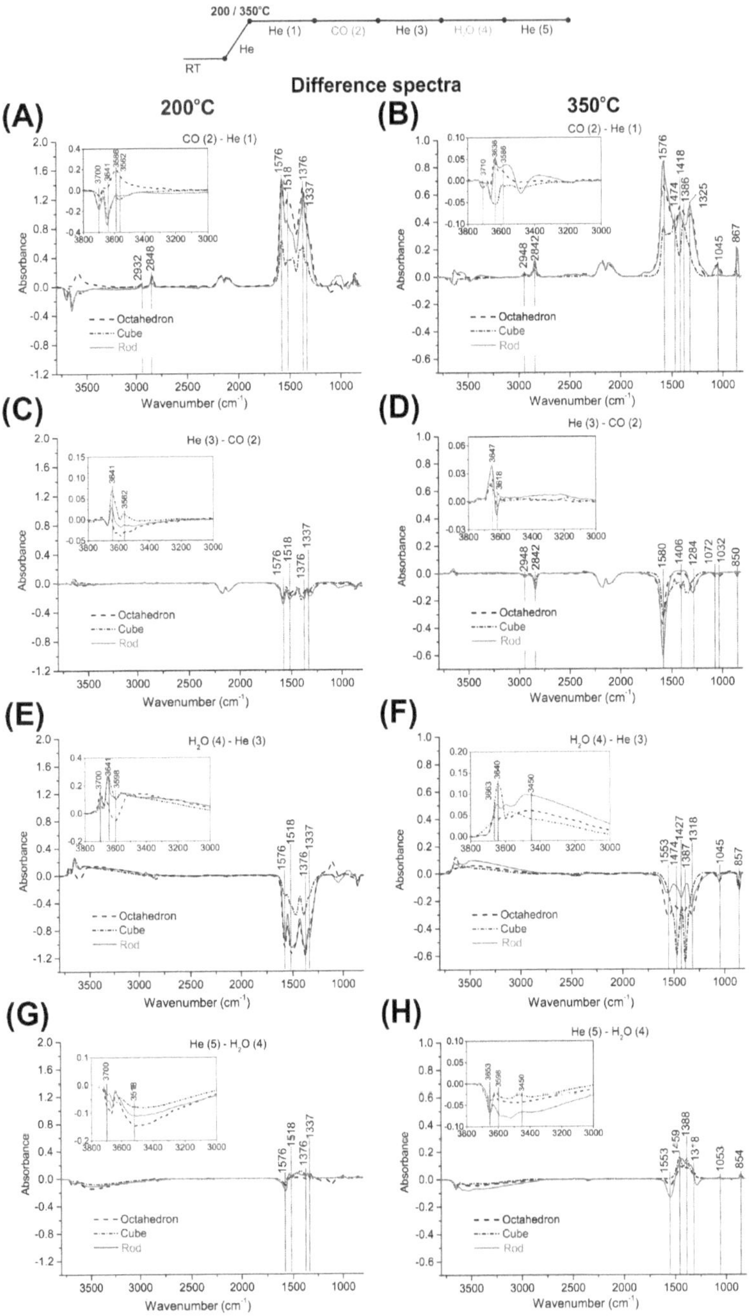

Figure 28 In situ FTIR subtracted spectrum of three ceria nanoshapes at 200 °C (left plots) and 350 °C (right plots) obtained by subtracting (A) and (B) He (1) from CO (2); (C) and (D) CO (2) from He (3); (E) and (F) He (3) from H_2O (4); (G) and (H) H_2O (4) from He (5). Insets in all the plots are expanded subtracted spectra of hydroxyl regions (3800–3000 cm^{-1}). *Adapted from Agarwal [120].*

compared to 6:3 and 6:2 for the (110) and (100) planes, respectively [1]. We recognize that these surfaces can reconstruct in the presence of adsorbates. In addition, the presence of transition metals on the ceria surface can dramatically influence the reactivity. Undoped single crystal ceria thin films are inactive toward CO oxidation at low temperatures and show poor reducibility [149,150]. The presence of defects in ceria nanoshapes has been shown to improve oxygen migration from the bulk to the surface of ceria [39,76,151], thus affecting the reducibility.

The surface hydroxyl groups observed on ceria in this study show similar vibrational stretch frequencies, especially for the band at 3641 cm^{-1}. Based on the same stretch frequency, one would expect a similar reactivity toward CO, independent of the nanoshape, but this is not true. Clearly, there must be another explanation for the distinctly different behavior of –OH (3641 cm^{-1}) on octahedra compared to rods and cubes at 200 °C. The main difference between ceria octahedra on the one hand and cubes and rods on the other hand is the absence of defects in octahedra. We propose that the hydroxyl reactivity on ceria nanoshapes is determined by the presence of these vacancies via a concerted reaction mechanism, which involves the hydroxyl group and a vacancy. Thus, it is not only the surface orientation that determines the reactivity of ceria nanoshapes but also the existence of defects and enhanced oxygen transport from the bulk, which plays a significant role in the overall behavior of ceria nanoparticles.

A recent study reported the formation energy of carbonates from CO on different surface planes of ceria following the order (100) (−74 kcal mol^{-1}) > (110) (−45 kcal mol^{-1}) >> (111) (−6 kcal mol^{-1}) [129]. Consequently, carbonate formation is not favored on the (111) planes of the octahedra [21,128,135,152,153], explaining why only a small amount of carbonates are found on octahedra (Figure 24).

The energy for carbonate formation on specific crystal planes is related to the atomic arrangement of the exposed crystal planes and the reducibility of the ceria nanoshape [153]. Since two oxygen atoms are required from the surface, the distance between oxygen atoms, and their mobility is a key parameter. On (110) and (100) surfaces, oxygen displacement and accompanying reduction of ceria as a result of carbonate formation is favorable, which is in agreement with the observations in Figure 28(A) and (B): Octahedra and rods form bicarbonate at 350 °C in contrast to cubes, at 200 °C this is only observed on octahedra. The hydrogen–carbonate formation on octahedra indicates that on (111) planes –OH groups with a relatively high reactivity are involved instead of, or in addition to, lattice oxygen.

The reactivity of the formate and (bi)carbonate species with H_2O is shown in Figure 28(E) and (F). The bands at 3586 and 1518 cm^{-1} on octahedra, assigned to hydrogen carbonate, disappeared upon exposure to H_2O at 200 °C. For ceria cubes, almost all hydroxyl species recovered, while for rods and octahedra, the recovery is not complete [103]. The missing hydroxyls are converted into formate species that are stable at 200 °C in the presence of water. Cubes show almost complete regeneration of the –OH groups, and only a minor amount of remaining formate species (the positive and negative peaks in Figure 28(A) and (E) are similar in size). In contrast, octahedra and rods contain much more formate that does not react with water at 200 °C, as the negative C–H peaks (3000–2800 cm^{-1}) in Figure 28(E) are significantly smaller than the positive intensities in Figure 28(A).

At 350 °C, the results seem quite different as can be seen in Figure 28(B) and (F). Formates and carbonates on cubes seem even more reactive with water as compared to octahedra and rods. It is likely that in the case of cubes, carbonate present on the as-prepared sample, exposed to ambient, contributes. The observed reactivity with water follows a similar order as reported for CO oxidation of these materials [21,44,86]. In those studies, the higher oxidation rates were attributed to the increased OSC of ceria nanoshapes due to altered lattice oxygen mobility and reactivity.

As can be seen in the difference spectrum at 350 °C (Figure 28(H)), negative peaks at 1553 cm^{-1} and positive peaks at 1459, 1388, 1318 cm^{-1} are observed. These peaks are clearly associated with carbonates and not with formate species, since no CH peaks were observed in the 3000−2800 cm^{-1} region. Based on the peak positions, we assign negative and positive peaks to bidentate and polydentate carbonate species, respectively [140]. From this observation, it seems that rearrangements of carbonate species on ceria octahedra and rods occurred, which resulted in the formation of stable polycarbonates at the expense of bidentate carbonates. This agrees well with the theoretical prediction of Vayssilov et al., who have reported the transition of bidentate carbonates in the vicinity of oxygen vacancies to stable polycarbonates due to the enhanced structural flexibility next to the oxygen vacancy defects [135]. In addition, a change in the dipole moment of the carbonate C–O bond upon desorption of water may also affect the peak position and intensity of carbonates. For cubes (blue spectrum), an increase in intensity of polydentate carbonate peaks (1459, 1388 cm^{-1}) was observed with a subtle decrease in bidentate species (1553 cm^{-1}). Note that any influence of He treatment after exposure to water has only a minor effect at 200 °C, like removal of hydrogen-bonded water and slight rearrangement of peaks in carbonate region (Figure 28(G)). At 350 °C (Figure 28(H)), the effect of the rearrangement of carbonate species is more pronounced than at 200 °C.

The reducibility and oxygen mobility might as well affect the reactivity toward water of the ceria nanoshapes. The results presented show that ceria nanoshapes have a distinct influence on the activation of water and the reaction of adsorbed carbonates and formates with water. This study suggests a large impact of ceria particle shape on reactions involving water, such as the WGS reaction. In summary, rods and octahedra are similar and differ from cubes with respect to

- formation of bicarbonates at 350 °C on rods and octahedra, not on cubes.
- carbonates and formates more reactive with water on cubes, than on rods and octahedra. Specifically, the extent of regeneration of –OH species and decomposition of carbonates in the presence of water is higher for cubes in comparison to that of the other two nanoshapes.

These observations are in line with the HRTEM result in Section 2 that rods and octahedra expose similar low index planes, quite different from cubes, exposing (100) facets. As discussed above, some properties observed do not follow this trend, which is assigned to the fact that rods contain much more defect and rough surfaces, as was also observed with HRSEM.

4. WGS REACTIVITY MEASUREMENTS

To investigate catalytic behavior, samples were tested in the WGS reaction. WGS catalytic measurements were performed for hydrogen pretreated ceria nanoshapes (Figure 29). The activity of the samples was expressed in moles of CO per meter square of CeO_2 to be able to correlate surface structure with activity. The activity for all samples increases with temperature. Surprisingly, the specific CO conversion rates for octahedra and rods are identical, while those for cubes are much more active at temperatures >200 °C. These activity trends are consistent with the observations from AC-TEM on the exposed surfaces of the three ceria shapes. The similar specific catalytic activities for rods and octahedra can be explained by the dominance of exposed {111} surfaces. The higher catalytic activity, generally expressed as CO conversion per gram of catalyst, for rods reported in the literature [11,20,21] thus may be caused by the high surface area rather than the type of exposed crystal planes. On the other hand, ceria cubes expose the highly active {100} surfaces, which clearly show enhanced specific WGS activity compared to the stable {111} surfaces.

The AC-TEM results in this study show that rods and octahedra expose {111} surface planes that result in similar catalytic WGS activity per meter square. In addition, the type of hydroxyl species present and the surface interaction with CO on octahedra and rods are similar, which is consistent with the identical WGS activity per

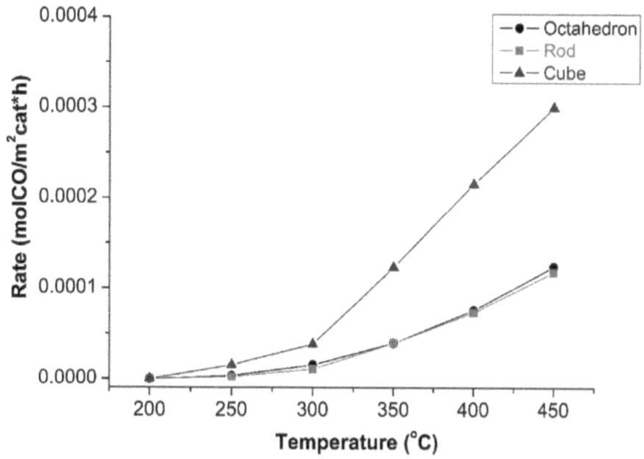

Figure 29 Rate of CO conversion per meter square during the WGS reaction for rods (red (light gray in print version) squares), octahedra (black circles), and cubes (blue (dark gray in print version) triangles) between 200 and 450 °C after H_2 pretreatment at 500 °C.

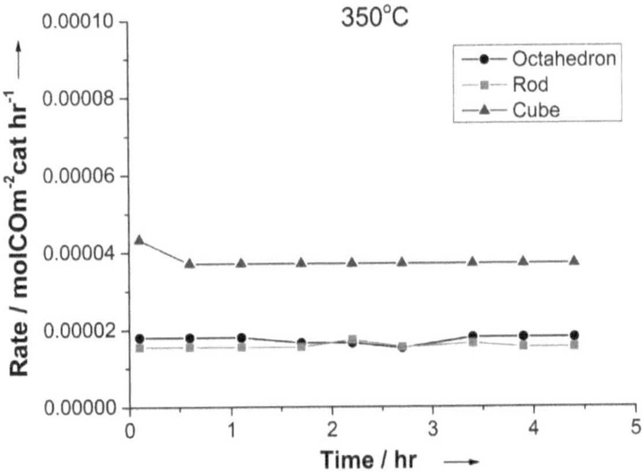

Figure 30 Rate of CO conversion per meter square as a function of time during WGS reaction for rods (red (light gray in print version) squares), octahedra (black circles), and cubes (blue (dark gray in print version) triangles) between 200 and 450 °C after H_2 pretreatment at 500 °C. *Reprinted from Agarwal et al. [103]. Copyright Wiley-VCH (2013).*

meter square (Figure 29). Cubes consist of more reactive {100} surface planes, which have different –OH bands and interaction with CO, resulting in a higher WGS reactivity per meter square. We studied the reactivity over longer term and found very little deactivation (Figure 30) confirming that the surfaces of ceria are stable at these reaction conditions.

5. SUMMARY

This chapter presents a review of the synthesis methods, characterization, and reactivity of CeO_2 nanoshapes. By using aberration-corrected electron microscopy, we were able to show the surface termination of rods, cubes, and octahedra. We found that both rods and octahedra expose stable (111), surfaces while cubes are composed primarily of (100) facets. The rods contain internal voids and surface steps as revealed clearly in HAADF images. The exposed surfaces are consistent with the observed reactivity patterns and, when normalized to surface area, the WGS reactivities of ceria octahedra and rods were very similar, but the cubes were more reactive. Likewise, in situ FTIR showed that ceria rods and octahedra exhibit similar spectra for –OH groups and carbonates and formates formed upon exposure to CO, while for ceria cubes, clear differences were observed. These results provide definitive information on the nature of the exposed surfaces in these CeO_2 nanostructures and their influence on reactivity for the WGS reaction. We note that the exposed surface facets do not fully explain the reactivity patterns, since internal defects and oxygen vacancy formation play a significant role. When transition metals are deposited on these ceria nanoshapes, it is found that the rods are generally the most active. Clearly, this is an aspect that deserves further study.

ACKNOWLEDGMENTS

The research was supported by DOE grant DE-FG02-05ER15712 and NSF grant OISE 0730277, Partnership for International Research and Education. S.A. acknowledges the ADEM Innovation Lab for financial support (Project Number 30961318). The authors also gratefully acknowledge M.A. Smithers for HRSEM and TEM imaging and Karin Altena-Schildkamp for BET measurements performed at University of Twente. We would like to thank Ruben Lubkemann and Bert Geerdink for technical assistance. Electron microscopy was performed at the Center for High Resolution Transmission Electron Microscopy at the Nelson Mandela Metropolitan University, and at Sasol Technology in Sasolburg, South Africa. The National Research Foundation, Sasol, and the Department of Science and Technology in South Africa are acknowledged for financial support. Additional SEM work was performed at the University of New Mexico.

REFERENCES

[1] M.V. Ganduglia-Pirovano, A. Hofmann, J. Sauer, Surf. Sci. Rep. 62 (6) (2007) 219–270.
[2] G.A.M. Hussein, J. Anal. Appl. Pyrolysis 37 (2) (1996) 111–149.
[3] H. Idriss, M.A. Barteau, Active sites on oxides: from single crystals to catalysts, in: H.K. Bruce, C. Gates (Eds.), Advances in Catalysis, Academic Press, 2000, pp. 261–331.
[4] F. Zaera, ChemSusChem 6 (10) (2013) 1797–1820.
[5] K. Zhou, Y. Li, Angew. Chem. Int. Edition 51 (3) (2012) 602–613.
[6] C. Sun, H. Li, L. Chen, Energy Environ. Sci. 5 (9) (2012) 8475–8505.
[7] Cerium, http://reehandbook.com/cerium.html# downloaded at 18:13 on 15/01/2014, 2013.
[8] K. Schermanz, Mining, production, application and safety issues of cerium-based materials, in: Catalysis by Ceria and Related Materials, World scientific, 2002, pp. 1–13.
[9] D. Zhang, X. Du, L. Shi, R. Gao, Dalton Trans. 41 (48) (2012) 14455–14475.
[10] K.G. Azzam, I.V. Babich, K. Seshan, L. Lefferts, Appl. Catal. B 80 (1–2) (2008) 129–140.
[11] Tana, M. Zhang, J. Li, H. Li, Y. Li, W. Shen, Catal. Today 148 (1) (2009) 179–183.
[12] J.F.D. Lima, R.F. Martins, C.R. Neri, O.A. Serra, Appl. Surf. Sci. 255 (2009) 9006–9009.
[13] N. Izu, W. Shin, N. Murayama, S. Kanzaki, Sensors and Actuators, B 87 (2002) 95–98.
[14] H.J. Beie, A. Gnorich, Sensors and Actuators, B 4 (1991) 393–399.
[15] A.S. Karakoti, N.A. Monteiro-Riviere, R. Aggarwal, J.P. Davis, R.J. Narayan, W.T. Self, J. McGinnis, S. Seal, JOM 60 (3) (2008) 33–37.
[16] A.F. Diwell, R.R. Rajaram, H.A. Shaw, T.J. Truex, The role of ceria in three-way catalysts, in: A. Crucq (Ed.), Studies in Surface Science and Catalysis, Elsevier, 1991, pp. 139–152.
[17] C. Xia, M. Liu, Solid State Ionics 152–153 (2002) 423–430.
[18] B.C.H. Steele, J. Power Sources 49 (1994) 1–14.
[19] X. Liu, K. Zhou, L. Wang, B. Wang, Y. Li, J. Am. Chem. Soc. 131 (9) (2009) 3140–3141.
[20] K. Zhou, X. Wang, X. Sun, Q. Peng, Y. Li, J. Catal. 229 (1) (2005) 206–212.
[21] Z. Wu, M. Li, S.H. Overbury, J. Catal. 285 (1) (2012) 61–73.
[22] L. Liu, Y. Cao, W. Sun, Z. Yao, B. Liu, F. Gao, L. Dong, Catal. Today 175 ((1) (2011) 48–54.
[23] A. Trovarelli, C. de Leitenburg, M. Boaro, G. Dolcetti, Catal. Today 50 (2) (1999) 353–367.
[24] R.J. Gorte, S. Zhao, Catal. Today 104 (1) (2005) 18–24.
[25] R.J. Gorte, AIChE J. 56 (5) (2010) 1126–1135.
[26] G. Jacobs, P. Patterson, L. Williams, D. Sparks, B. Davis, Catal. Lett. 96 (1–2) (2004) 97–105.
[27] S. Czernik, R. Evans, R. French, Catal. Today 129 (3'Äì4) (2007) 265–268.
[28] S. Agarwal, L. Lefferts, B.L. Mojet, D.A.J.M. Ligthart, E.J.M. Hensen, D.R.G. Mitchell, W.J. Erasmus, B.G. Anderson, E.J. Olivier, J.H. Neethling, A.K. Datye, ChemSusChem 6 (10) (2013) 1898–1906.
[29] A. Trovarelli, Structural properties and nonstoichiometric behaviour of CeO2, in: Catalysis by Ceria and Related Materials, World Scientific, 2002, pp. 15–50.
[30] D. Mullins, P. Albrecht, F. Calaza, Top. Catal. 56 (15–17) (2013) 1345–1362.
[31] F. Esch, S. Fabris, L. Zhou, T. Montini, C. Africh, P. Fornasiero, G. Comelli, R. Rosei, Science 309 (5735) (2005) 752–755.

[32] P. Gao, Z. Kang, W. Fu, W. Wang, X. Bai, E. Wang, J. Am. Chem. Soc. 132 (12) (2010) 4197–4201.
[33] G.S. Herman, Surf. Sci. 437 (1–2) (1999) 207–214.
[34] A. Nakajima, A. Yoshihara, M. Ishigame, Phys. Rev. B 50 (18) (1994) 13297–13307.
[35] J. Paier, C. Penschke, J. Sauer, Chem. Rev. (Washington, DC, U. S.) 113 (6) (2013) 3949–3985.
[36] B.W. Sheldon, V.B. Shenoy, Phys. Rev. Lett. 106 (21) (2011) 216104.
[37] T. Taniguchi, T. Watanabe, N. Sugiyama, A.K. Subramani, H. Wagata, N. Matsushita, M. Yoshimura, The J. Phys. Chem. C 113 (46) (2009) 19789–19793.
[38] Q. Yuan, H.-H. Duan, L.-L. Li, L.-D. Sun, Y.-W. Zhang, C.-H. Yan, J. Colloid Interface Sci. 335 (2) (2009) 151–167.
[39] E. Mamontov, T. Egami, J. Phys. Chem. Solids 61 (8) (2000) 1345–1356.
[40] W.C. Chueh, S.M. Haile, Philos. Trans. A Math. Phys. Eng. Sci. 368 (1923) (2010) 3269–3294.
[41] T.X.T. Sayle, M. Molinari, S. Das, U.M. Bhatta, G. Möbus, S.C. Parker, S. Seal, D.C. Sayle, Nanoscale 5 (13) (2013) 6063–6073.
[42] B. Choudhury, A. Choudhury, Mater. Chem. Phys. 131 (3) (2012) 666–671.
[43] P.L. Land, J. Phys. Chem. Solids 34 (11) (1973) 1839–1845.
[44] T.X.T. Sayle, S.C. Parker, C.R.A. Catlow, Surf. Sci. 316 (3) (1994).
[45] A. Walsh, S.M. Woodley, C.R.A. Catlow, A.A. Sokol, Solid State Ionics 184 (1) (2011) 52–56.
[46] K.G. Azzam, I.V. Babich, K. Seshan, L. Lefferts, J. Catal. 251 (1) (2007) 153–162.
[47] J. Kašpar, P. Fornasiero, N. Hickey, Catal. Today 77 (4) (2003) 419–449.
[48] T.V. Choudhary, D.W. Goodman, Catal. Today 77 (1–2) (2002) 65–78.
[49] L.F. Brown, Int. J. Hydrogen Energy 26 (4) (2001) 381–397.
[50] A. Faur Ghenciu, Curr. Opin. Solid State Mater. Sci. 6 (5) (2002) 389–399.
[51] O. Pozdnyakova, D. Teschner, A. Wootsch, J. Kröhnert, B. Steinhauer, H. Sauer, L. Toth, F.C. Jentoft, A. Knop-Gericke, Z. Paál, R. Schlögl, J. Catal. 237 (1) (2006) 1–16.
[52] D.L. Trimm, Z.I. Önsan, Catal. Rev. 43 (1–2) (2001) 31–84.
[53] J.N. Armor, Appl. Catal. A: Gen. 176 (2) (1999) 159–176.
[54] N. Bion, F. Epron, M. Moreno, F. Mariño, D. Duprez, Top. Catal. 51 (1–4) (2008) 76–88.
[55] S.Y. Chin, O.S. Alexeev, M.D. Amiridis, J. Catal. 243 (2) (2006) 329–339.
[56] A. Wootsch, C. Descorme, D. Duprez, J. Catal. 225 (2) (2004) 259–266.
[57] C. Okkerse, H. van Bekkum, Green Chem. 1 (2) (1999) 107–114.
[58] R. Trane, S. Dahl, M.S. Skjøth-Rasmussen, A.D. Jensen, Int. J. Hydrogen Energy 37 (8) (2012) 6447–6472.
[59] B. Matas Güell, I. Babich, K.P. Nichols, J.G.E. Gardeniers, L. Lefferts, K. Seshan, Appl. Catal. B 90 (1–2) (2009) 38–44.
[60] C. Rioche, S. Kulkarni, F.C. Meunier, J.P. Breen, R. Burch, Appl. Catal. B 61 (1–2) (2005) 130–139.
[61] R.W. Snell, B.H. Shanks, ACS Catal. 3 (4) (2013) 783–789.
[62] S. Wan, T. Pham, S. Zhang, L. Lobban, D. Resasco, R. Mallinson, AIChE J. 59 (7) (2013) 2275–2285.
[63] D.S. Newsome, Catal. Rev. 21 (2) (1980) 275–318.
[64] W. Ruettinger, O. Ilinich, R.J. Farrauto, J. Power Sources 118 (1–2) (2003) 61–65.
[65] R. Burch, Phys. Chem. Chem. Phys. 8 (47) (2006) 5483–5500.
[66] K.G. Azzam. A Further Step toward H2 in Automobile : Development of an Efficient Bi-functional Catalyst for Single Stage Water Gas Shift. Ph.D., 2008. University of Twente.
[67] G. Jacobs, E. Chenu, P.M. Patterson, L. Williams, D. Sparks, G. Thomas, B.H. Davis, Appl. Catal. A 258 (2) (2004) 203–214.
[68] K.G. Azzam, I.V. Babich, K. Seshan, L. Lefferts, J. Catal. 251 (1) (2007) 163–171.
[69] A. Tschope, W. Liu, M. Flytzanistephanopoulos, J.Y. Ying, J. Catal. 157 (1) (1995) 42–50.
[70] T. Désaunay, G. Bonura, V. Chiodo, S. Freni, J.P. Couzinié, J. Bourgon, A. Ringuedé, F. Labat, C. Adamo, M. Cassir, J. Catal. 297 (0) (2013) 193–201.
[71] R. Wang, S.I. Mutinda, Chem. Phys. Lett. 517 (4–6) (2011) 186–189.
[72] Z. Yang, T.K. Woo, M. Baudin, K. Hermansson, J. Chem. Phys. 120 (16) (2004) 7741–7749.
[73] P.M. Albrecht, D.R. Mullins, Langmuir 29 (14) (2013) 4559–4567.
[74] M. Nolan, S. Grigoleit, D.C. Sayle, S.C. Parker, G.W. Watson, Surf. Sci. 576 (1–3) (2005) 217–229.
[75] H. Nörenberg, J.H. Harding, Surf. Sci. 477 (1) (2001) 17–24.

[76] Z. Wu, M. Li, J. Howe, H.M. Meyer, S.H. Overbury, Langmuir 26 (21) (2010) 16595–16606.
[77] Y. Lin, Z. Wu, J. Wen, K.R. Poeppelmeier, L.D. Marks, Nano Lett. 14 (1) (2013) 191–196.
[78] D.R. Mullins, S.H. Overbury, J. Catal. 188 (2) (1999) 340–345.
[79] M. Molinari, S.C. Parker, D.C. Sayle, M.S. Islam, J. Phys. Chem. C 116 (12) (2012) 7073–7082.
[80] S.D. Senanayake, W.O. Gordon, S.H. Overbury, D.R. Mullins, J. Phys. Chem. C 113 (15) (2009) 6208–6214.
[81] D.R. Mullins, P.M. Albrecht, J. Phys. Chem. C 117 (28) (2013) 14692–14700.
[82] S.D. Senanayake, D.R. Mullins, J. Phys. Chem. C 112 (26) (2008) 9744–9752.
[83] J.A. Rodriguez, S. Ma, P. Liu, J. Hrbek, J. Evans, M. Perez, Science 318 (5857) (2007) 1757–1760.
[84] K.-S. Lin, S. Chowdhury, Int. J. Mol. Sci. 11 (9) (2010) 3226–3251.
[85] G. Wang, L. Wang, X. Fei, Y. Zhou, R.F. Sabirianov, W.N. Mei, C.L. Cheung, Catal. Sci. Technol. 3 (10) (2013) 2602–2609.
[86] H.-X. Mai, L.-D. Sun, Y.-W. Zhang, R. Si, W. Feng, H.-P. Zhang, H.-C. Liu, C.-H. Yan, The J. Phys. Chem. B 109 (51) (2005) 24380–24385.
[87] S.K. Pahari, N. Sutradhar, A. Sinhamahapatra, P. Pal, A.B. Panda, New J. Chem. 35 (7) (2011) 1460–1465.
[88] J.J. Schneider, M. Naumann, C. Schäfer, A. Brandner, H.J. Hofmann, P. Claus, Beilstein J. Nanotechnol. 2 (2011) 776–784.
[89] N. Bugayeva, Mater. Res. Soc. Symp. Proc. 876E (2005) 195–200.
[90] W. Wang, J.Y. Howe, Y. Li, X. Qiu, D.C. Joy, M.P. Paranthaman, M.J. Doktycz, B. Gu, J. Mater. Chem. 20 (36) (2010) 7776–7781.
[91] Z.L. Wang, X. Feng, J. Phys. Chem. B 107 (49) (2003) 13563–13566.
[92] L. Torrente-Murciano, A. Gilbank, B. Puertolas, T. Garcia, B. Solsona, D. Chadwick, Appl. Catal., B 132–133 (0) (2013) 116–122.
[93] Q. Dai, H. Huang, Y. Zhu, W. Deng, S. Bai, X. Wang, G. Lu, Appl. Catal., B 117-118 (0) (2012) 360–368.
[94] Y. Zhiqiang, Z. Kebin, L. Xiangwen, T. Qun, L. Deyi, Y. Sen, Nanotechnology 18 (18) (2007) 185606.
[95] M.B. Boucher, S. Goergen, N. Yi, M. Flytzani-Stephanopoulos, Phys. Chem. Chem. Phys. 13 (7) (2011) 2517–2527.
[96] S. Chang, M. Li, Q. Hua, L. Zhang, Y. Ma, B. Ye, W. Huang, J. Catal. 293 (0) (2012) 195–204.
[97] X. Du, D. Zhang, L. Shi, R. Gao, J. Zhang, J. Phys. Chem. C 116 (18) (2012) 10009–10016.
[98] W.-Q. Han, W. Wen, J.C. Hanson, X. Teng, N. Marinkovic, J. Rodriguez, J. Phys. Chem. C 113 (52) (2009) 21949–21955.
[99] N. Singhania, E.A. Anumol, N. Ravishankar, G. Madras, Dalton Trans. 42 (2013) 15343–15354.
[100] R. Si, M. Flytzani-Stephanopoulos, Angew. Chem. Int. Edition 47 (15) (2008) 2884–2887.
[101] R.W. Snell, S.H. Hakim, J.A. Dumesic, B.H. Shanks, Appl. Catal. A 464–465 (0) (2013) 288–295.
[102] Z. Wu, M. Li, D.R. Mullins, S.H. Overbury, ACS Catal. 2 (11) (2012) 2224–2234.
[103] S. Agarwal, L. Lefferts, B.L. Mojet, ChemCatChem 5 (2) (2013) 479–489.
[104] A. Ciftci, D.A.J.M. Ligthart, P. Pastorino, E.J.M. Hensen, Appl. Catal. B 130–131 (0) (2013) 325–335.
[105] Y. Guan, D.A.J.M. Ligthart, Ö. Pirgon-Galin, J.Z. Pieterse, R. Santen, E.M. Hensen, Top. Catal. 54 (5–7) (2011) 424–438.
[106] G. Yi, H. Yang, B. Li, H. Lin, K.-i. Tanaka, Y. Yuan, Catal. Today 157 (1–4) (2010) 83–88.
[107] N. Yi, R. Si, H. Saltsburg, M. Flytzani-Stephanopoulos, Energy Environ. Sci. 3 (6) (2010) 831–837.
[108] J. Han, H.J. Kim, S. Yoon, H. Lee, J. Mol. Catal. A: Chem. 335 (1'Äì2) (2011) 82–88.
[109] I.I. Soykal, B. Bayram, H. Sohn, P. Gawade, J.T. Miller, U.S. Ozkan, Appl. Catal. A 449 (0) (2012) 47–58.
[110] N. Ta, J. Liu, S. Chenna, P.A. Crozier, Y. Li, A. Chen, W. Shen, J. Am. Chem. Soc. 134 (51) (2012) 20585–20588.
[111] S. Wang, L. Zhao, W. Wang, Y. Zhao, G. Zhang, X. Ma, J. Gong, Nanoscale 5 (12) (2013) 5582–5588.
[112] S. Asahina, S. Takami, T. Otsuka, T. Adschiri, O. Terasaki, ChemCatChem 3 (6) (2011) 1038–1044.
[113] W.-Q. Han, L. Wu, Y. Zhu, J. Am. Chem. Soc. 127 (37) (2005) 12814–12815.
[114] H. Li, G. Lu, Q. Dai, Y. Wang, Y. Guo, Y. Guo, ACS Appl. Mater. Interfaces 2 (3) (2010) 838–846.
[115] L.-S. Zhong, J.-S. Hu, A.-M. Cao, Q. Liu, W.-G. Song, L.-J. Wan, Chem. Mater. 19 (7) (2007) 1648–1655.

[116] S.J. Pennycook, Ultramicroscopy 123 (2012) 28–37.
[117] O.L. Krivanek, M.F. Chisholm, M.F. Murfitt, N. Dellby, Ultramicroscopy 123 (2012) 90–98.
[118] U.M. Bhatta, I.M. Ross, T.X.T. Sayle, D.C. Sayle, S.C. Parker, D. Reid, S. Seal, A. Kumar, G. Möbus, ACS Nano 6 (1) (2012) 421–430.
[119] G. Möbus, Z. Saghi, D.C. Sayle, U.M. Bhatta, A. Stringfellow, T.X.T. Sayle, Adv. Funct. Mater. 21 (11) (2011) 1971–1976.
[120] S. Agarwal. Surface Chemistry of Tailored Ceria Nanoparticles: Interaction with CO and H2O. Ph.D., 2014. Univ. of Twente, Enschede. p. 169.
[121] JCPDS, International Centre for Diffraction Data, 2003.
[122] S. Tsunekawa, K. Ishikawa, Z.Q. Li, Y. Kawazoe, A. Kasuya, Phys. Rev. Lett. 85 (16) (2000) 3440–3443.
[123] V.G. Keramidas, W.B. White, J. Chem. Phys. 59 (3) (1973) 1561–1562.
[124] I. Kosacki, T. Suzuki, H.U. Anderson, P. Colomban, Solid State Ionics 149 (1–2) (2002) 99–105.
[125] J.E. Spanier, R.D. Robinson, F. Zhang, S.-W. Chan, I.P. Herman, Phys. Rev. B 64 (24) (2001) 245407.
[126] B.M. Reddy, G.K. Reddy, L. Katta, J. Mol. Catal. A: Chem. 319 (1–2) (2010) 52–57.
[127] J.R. McBride, K.C. Hass, B.D. Poindexter, W.H. Weber, J. Appl. Phys. 76 (4) (1994) 2435–2441.
[128] T.X.T. Sayle, S.C. Parker, C.R.A. Catlow, J. Chem. Soc. Chem. Commun. 0 (14) (1992) 977–978.
[129] M. Nolan, G.W. Watson, J. Phys. Chem. B 110 (33) (2006) 16600–16606.
[130] J.P.Y. Tan, H.R. Tan, C. Boothroyd, Y.L. Foo, C.B. He, M. Lin, J. Phys. Chem. C 115 (9) (2011) 3544–3551.
[131] P. Yang, H. Wei, H.-L. Huang, B.A. Baum, Y.X. Hu, G.W. Kattawar, M.I. Mishchenko, Q. Fu, Appl. Opt. 44 (26) (2005) 5512–5523.
[132] A. Badri, C. Binet, J.-C. Lavalley, J. Chem. Soc. Faraday Trans. 92 (23) (1996) 4669–4673.
[133] C. Binet, M. Daturi, J.-C. Lavalley, Catal. Today 50 (1999) 207–225.
[134] C. Binet, A. Jadi, J.-C. Lavalley, J. de chim. physique 89 (1992) 1779–1797.
[135] G.N. Vayssilov, M. Mihaylov, P.S. Petkov, K.I. Hadjiivanov, K.M. Neyman, J. Phys. Chem. C 115 (47) (2011) 23435–23454.
[136] C. Binet, A. Badri, M. Boutonnet-Kizling, J.-C. Lavalley, J. Chem. Soc. Faraday Trans. 90 (7) (1994) 1023–1028.
[137] C. Li, Y. Sakata, T. Arai, K. Domen, K. Maruya, T. Onishi, J. Chem. Soc. Faraday Trans. 1 85 (4) (1989) 929–943.
[138] C. Li, Y. Sakata, T. Arai, K. Domen, K. Maruya, T. Onishi, J. Chem. Soc. Faraday Trans. 1 85 (6) (1989) 1451–1461.
[139] T. Shido, Y. Iwasawa, J. Catal. 136 (2) (1992) 493–503.
[140] C. Li, K. Domen, K.-i. Maruya, T. Onishi, J. Catal. 125 (2) (1990) 445–455.
[141] L.G. Appel, J.G. Eon, M. Schmal, Catal. Lett. 56 (1998) 199–202.
[142] T. Staudt, Y. Lykhach, N. Tsud, T. Skala, K.C. Prince, V. Matlin, J. Libuda, J. Phys. Chem. C 115 (2011) 8716–8724.
[143] T. Staudt, Y. Lykhach, N. Tsud, T. Skála, K.C. Prince, V. Matolín, J. Libuda, J. Catal. 275 (2010) 181–185.
[144] S. Agarwal, L. Lefferts, X. Zhu, E.J.M. Hensen, and B.L. Mojet, J. Phys. Chem. C. (2014 Under review).
[145] T.X.T. Sayle, B.J. Inkson, A. Karakoti, A. Kumar, M. Molinari, G. Mobus, S.C. Parker, S. Seal, D.C. Sayle, Nanoscale 3 (4) (2011) 1823–1837.
[146] C. Sun, H. Li, Z. Wang, L. Chen, X. Huang, Chem. Lett. 33 (6) (2004) 662–663.
[147] A. Vantomme, Z.-Y. Yuan, G. Du, B.-L. Su, Langmuir 21 (3) (2004) 1132–1135.
[148] D. Zhang, H. Fu, L. Shi, C. Pan, Q. Li, Y. Chu, W. Yu, Inorg. Chem. 46 (7) (2007) 2446–2451.
[149] H. Cordatos, T. Bunluesin, J. Stubenrauch, J.M. Vohs, R.J. Gorte, J. Phys. Chem. 100 (2) (1996) 785–789.
[150] J. Stubenrauch, J.M. Vohs, J. Catal. 159 (1996) 50–57.
[151] Y.M. Chiang, E.B. Lavik, I. Kosacki, H.L. Tuller, J.Y. Ying, Appl. Phys. Lett. 69 (2) (1996) 185–187.
[152] M. Huang, S. Fabris, J. Phys. Chem. C 112 (23) (2008) 8643–8648.
[153] M. Nolan, J.E. Fearon, G.W. Watson, Solid State Ionics 177 (35–36) (2006) 3069–3074.
[154] C. Sun, D. Xue, Phys. Chem. Chem. Phys. 15 (34) (2013) 14414–14419.

CHAPTER 3

The Characterization and Structure-Dependent Catalysis of Ceria with Well-Defined Facets

Amanda K.P. Mann[1,2,*], Zili Wu[2,3], Steven H. Overbury[2,3,†]
[1]Analytical Sciences, Merck Research Laboratories, Rahway, NJ, USA; [2]Chemical Science Division, Oak Ridge National Laboratory, Oak Ridge, TN, USA; [3]Center for Nanophase Materials Sciences, Oak Ridge National Laboratory, Oak Ridge, TN, USA
†Corresponding author: E-mail: overburysh@ornl.gov

1. INTRODUCTION

Cerium oxide (CeO_{2-x}, $0 \leq x \leq 0.5$) is a material that exhibits rich redox and acid–base chemistries and has thus been used in high-impact academic and industrial research programs alike for many applications, most notably as a component of catalytic converters. CeO_2 is also used as a heterogeneous catalyst or catalyst support to promote a number of important reactions [1–3]. It has a cubic fluorite crystal structure, and Ce is the most abundant of the lanthanide elements. CeO_2 has high oxygen mobility and has an excellent oxygen storage capacity (OSC) as its crystal structure can accommodate a high number of oxygen vacancies [4,5]. It exhibits both acid and base sites on its surface in close proximity, allowing it to promote a diverse array of reactions [6–8]. The combination and interplay of these characteristics make cerium oxide an ideal catalyst or support material for many reactions, including oxidation of CO [9–11], organics [12–15], and ammonia [16], reduction and hydrogenation of CO_2 [17,18], NO [19,20], nitriles [21], and organics [22–26], steam reforming of ethanol [27,28], dehydration reactions [29,30], ketonization [31–34], and coupling reactions [35–38]. As such, it finds use as a component in many applications, including three-way catalysts [39–41], oxygen sensors [42,43], oxygen permeation membrane systems [44,45], and fuel cells [46,47]. The structure-dependent properties of CeO_2, which have remained poorly understood to this point due to the use of CeO_2 particles with ill-defined structures, can be elucidated via the use of single-crystalline particles with defined surfaces.

The synthesis and designing of catalysts with high efficiencies, selectivities, and long lifetimes are the cornerstone of many scientific pursuits. Heterogeneous catalysis is a surface-based process, and therefore, understanding the mechanisms of adsorption and reaction of reactants is important, yet challenging, as the surfaces of solids, such as CeO_2, can be quite diverse, containing many types of active sites that can change over the course of the reaction. As a result, achieving high selectivity and understanding the

*Merck address is author's current address.

Catalysis by Materials with Well-Defined Structures
http://dx.doi.org/10.1016/B978-0-12-801217-8.00003-7

reaction mechanisms are often challenges. Producing particles with controlled shapes that expose different surface atom arrangements, or facets, in combination with decreasing the size of the particles for catalysis to the nanoscale has been shown to be an effective tool to enhance the properties of catalysts by increasing selectivity, longevity, and/or activity of heterogeneous catalysts [48–54]. However, in order to reveal the size- and shape-dependent properties for the adsorption and catalytic reaction of various reactants on CeO_2, fundamental studies with high-quality samples are required. This understanding can then guide the design of advanced catalytic materials.

One strategy to understand the effect of the catalytic surface-dependent properties of CeO_2 is to study single-crystalline films. The use of these films has many advantages, including the fact that they utilize oriented single crystals with a small number of defects and minimal contributions from corner and edge sites. The reactions are performed under controlled ultrahigh vacuum (UHV) conditions that allow for analysis at low coverages of reactants with a number of instruments that are difficult or impossible to utilize in ambient pressures, such as scanning tunneling microscopy (STM) or X-ray photoelectron spectroscopy (XPS). Single-crystalline CeO_2 (100) and (111) films have been utilized to study many CeO_2-based catalytic reactions, including the adsorption and reaction of methanol [55], formic acid [56], acetone [57], ethylene glycol [58], and acetaldehyde [59,60], among others [61–63]. Nevertheless, single-crystalline films have low surface areas, and UHV conditions are substantially different from relevant, practical reactor conditions.

Nanoparticles can achieve high surface areas and be used in conditions similar to those used in the industry; however, it is still desirable to achieve single-crystal materials with oriented surfaces in order to understand the fundamental transformations of reactants in catalysis. In order to combine the advantages of single-crystalline films and nanoparticles, it becomes necessary to produce single-crystalline, shape-controlled nanoparticles. Several synthetic strategies have been developed to generate CeO_2 nanoparticles with different shapes, including colloidal [64] and hydrothermal syntheses [65–69], and have been previously reviewed [70–72]. The characterization and utilization of these shape-controlled CeO_2 nanoshapes, including cubes, rods, and octahedra for heterogeneous catalysis are the focus of this review.

2. CHARACTERIZATION OF SIZE- AND STRUCTURE-DEPENDENT PROPERTIES OF CeO_2 NANOPARTICLES

2.1 Size-Dependent Properties

Before the shape-dependent properties of CeO_2 nanoparticles can be examined, it is first important to survey the effects of limiting the size of materials to the nanoscale (i.e., at least one dimension is between 1 and 100 nm). Limiting the size of a particle results in an increase in surface area, and therefore increases the percentage of atoms on the surface

[73]. A higher surface area per mass results in the enhancement of surface-dependent properties, such as shape and size, relative to the bulk-dependent properties. In addition, limiting the particle size to the nanoscale can include the introduction of quantum effects [74–78].

Several groups have examined the effects of size and surface area on the reducibility of CeO_2 particles. Laachir et al. modified the surface area of CeO_2 particles by changing the calcination conditions and quantified how the degree of ceria reduction changed as a function of the surface area [79]. They found that the reduction of CeO_2 with various surface areas began at similar temperatures, but the CeO_2 particles with a higher surface area underwent a greater degree of reduction at low temperatures compared to the particles with a lower surface area. In addition, they noted that the reoxidation of ceria is a facile process, and did not display surface area-dependent properties. The authors supported their conclusions using in situ magnetic susceptibility measurements, Fourier transform infrared spectroscopy (FTIR), XPS, and temperature-programmed reduction (TPR) [79]. Zhou and Huebner investigated the effect of ceria particle size on the lattice relaxation of CeO_2 particles via the formation of oxygen vacancies using a combination of powder X-ray diffraction (XRD) and modeling. They reported that the oxygen vacancies increased by two orders of magnitude when particle size was reduced from 60 to 4 nm [80]. In other words, it is expected that the reducibility of CeO_2 is directly related to its particle size, that is, the nanosized particles are easier to reduce than larger CeO_2 particles.

Chemically, CeO_2 reducibility is directly related the to the acid–base properties of the surface. Reduction of the Lewis acid Ce^{4+} sites results in the formation of less acidic Ce^{3+} sites. Therefore, these differences in reducibility and acidity of the CeO_2 nanoparticles should have significant implications for redox and acid/base-mediated catalytic reactions. Soykal et al. studied the reducibility of CeO_2 as it is related to particle size during steam reforming of ethanol and found that more reduction occurred in the smaller particles, promoting higher conversion at 450 °C [81]. They also found that reduction of the surface via ethanol adsorption resulted in a decrease in acidic sites on the surface. Igarashi and coworkers examined how particle size changed the selectivity for the reaction of butanediols on CeO_2 particles. For example, they found that smaller CeO_2 particles catalyzed the C–C bond scission of 1,3-butanediol to form methanol and ethanol, whereas larger CeO_2 particles selected for dehydration to form 3-buten-2-ol and tran-2-buten-1-ol [82]. However, recent studies warn that differences observed in the reducibility of CeO_2 can be related to the damage incurred by the use of X-ray radiation from instrumentation used to probe the oxidation state of cerium such as XPS and X-ray absorption near edge structure (XANES) and conclusions relating to the particle size and reducibility should be made with caution [69]. In addition, many of these studies highlight the differences between CeO_2 particles with two different sizes or surface areas. A more complete picture of the effects of size and surface area on CeO_2 properties and catalysis could be obtained via a study comparing particles with more sizes, that is, at least three sizes to allow for a more clear elucidation of trends. In addition to changes in reducibility and catalytic reactivity, CeO_2 nanostructures have been shown to have other size-dependent properties, including lattice relaxation [80,83], electronic conductivity [84,85], Raman shifts [86], phase transition boundaries [87], and absorption edge energies [88].

2.2 Structural Properties, Coordination Numbers, and Surface Stability

Three low index CeO_2 facets have been widely studied in various applications and calculations for CeO_2, including (111), (100), and (110), and nanoshapes expressing these facets will be the focus of this chapter. Structural models for each are shown in Figure 1 [89]. The surface stabilities of the different facets have been modeled by density functional theory (DFT) calculations and found to have the following stability of (111) > (110) > (100) [90]. The thermodynamically stable CeO_2 (111) surface has seven coordinated Ce cations and is expected to be oxygen terminated which is supported by noncontact atomic force microscopy [91] and ion scattering spectroscopy measurements [92]. The surface O anions on the CeO_2 (111) surface would be three coordinate. CeO_2 (110) surfaces are expected to have six-coordinate Ce cations and three-coordinate oxygen anions. Few studies have been performed to study the CeO_2 (110) surface; however, it has been predicted by calculations that it will reduce the surface energy by undergoing vertical relaxations [93]. A CeO_2 (100) surface is slightly more complicated, as an ideal, bulk-terminated surface could have either oxygen atoms or Ce atoms in the top surface, leaving a polar, and therefore, unstable surface [94]. Nörenberg and coworkers studied the CeO_2 (100) surfaces using scanning tunneling microscope (STM) and low-energy electron diffraction, and found that the surface was oxygen terminated, and underwent reconstruction of the oxygen layer to reduce the surface energy [95]. The surface has been found through calculations that it can undergo reconstruction to overcome charge instability by removing half of the oxygen atoms (Figure 1(D)) [96]. The reconstructed CeO_2 (100) surface shown in Figure 1(D) consists of six-coordinate Ce and two-coordinate oxygen.

For purposes of learning structure–activity relationships, it would be desirable for the nanoshapes to be monofaceted. For example, ideal cubes would expose six {100} facets, while octahedra would expose only {111} facets. However, shape-controlled nanoparticles often have defects or truncations that deviate from this model. Therefore, it is important to be able to determine the extent of faceting on their surfaces, edges, and corners. High-resolution electron microscopy (HREM) and tomography are powerful tools to give an insight into the atomic arrangement or morphological structure of CeO_2 nanoshapes. Florea et al. performed tomographic analysis of CeO_2 cubes, octahedra, and wires to give an insight into the nature of the surface terminations and faceting (Figure 2) [97]. For example, they found that CeO_2 cubes predominately expose six {100} facets, with {110} truncations at the edges and {111} truncations at the corners, while octahedra expose eight {111} facets with a very small amount of {110}

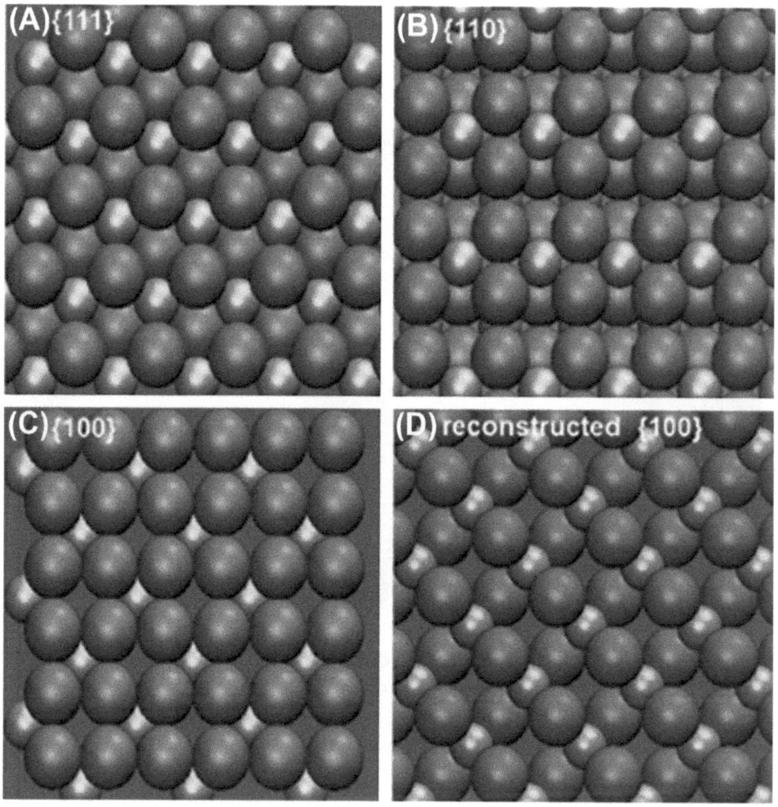

Figure 1 Top-down view of the structural models for (A) fully oxidized CeO_2 (111), (B) fully oxidized CeO_2 (110), (C) fully oxidized ideal CeO_2 (100), and (D) fully oxidized reconstructed CeO_2 (100). Red (dark gray in print versions) and light blue (light gray in print versions) balls represent oxygen and cerium atoms, respectively. *From Ref. [89]. Reprinted with permission from (Z. Wu, M. Li, D.R. Mullins, S.H. Overbury, Probing the surface sites of CeO_2 nanocrystals with well-defined surface planes via methanol adsorption and desorption, ACS Catal. 2 (2012) 2224–2234). Copyright (2012) American Chemical Society.*

truncations at the corners. Previously, CeO_2 rods were proposed to expose {110} and {100} facets [10,67,98]. However, tomographic studies found that the rods are porous and the surface is rough, resulting in a highly defective surface that can have pentagonal or hexagonal geometries depending on the synthesis conditions and diameter of the rods [97]. The porous nature of the rods and a large number of defects, such as oxygen vacancies, likely contribute to and complicate the assignment of rod faceting [99]. Liu et al. found dependence upon the precursor utilized in the hydrothermal synthesis of rods; the use of $CeCl_3$ promoted {100} and {111} planes, whereas the use of $Ce(NO_3)_3$ as a precursor resulted in {100} and {110} planes being predominantly exposed [9]. Importantly, these synthesis-dependent morphologies may result in discrepancies in the literature and confusion when assigning structure-dependent properties.

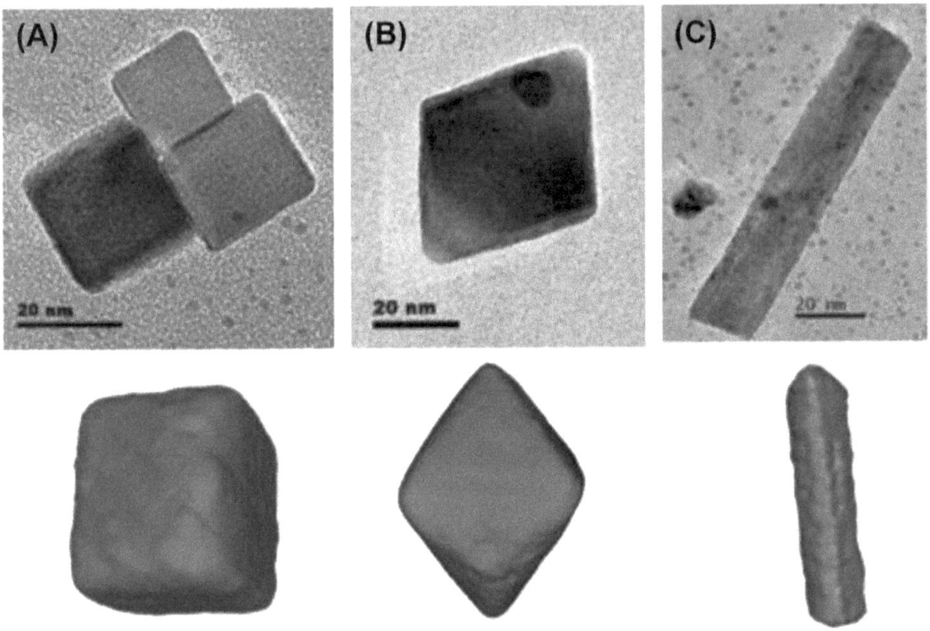

Figure 2 Representative transmission electron microscopy (TEM) images and the corresponding 3D models, illustrating the three typical morphologies of CeO_2 nanoparticles prepared by the solvothermal method: (A) cubes; (B) octahedrons; and (C) rods. *From Ref. [97]. Reprinted with permission from (I. Florea, C. Feral-Martin, J. Majimel, D. Ihiawakrim, C. Hirlimann, O. Ersen, Three-dimensional tomographic analyses of CeO_2 nanoparticles, Cryst. Growth Des. 13 (2013) 1110–1121). Copyright (2013) American Chemical Society.*

Elegant studies using HREM have been used to elucidate the atomic surface structure of CeO_2 {100} cubes whose analysis is complicated by the need of the polar (100) surface to undergo reconstruction to reduce the surface energy. It has been previously suggested that the reorganization of the surface occurs by the removal of half of the surface oxygen atoms. In an aberration-corrected high-resolution transmission electron microscopy study (HRTEM), Bhatta et al. observed extensive Ce cation hopping on laterally extended {100} surfaces of cuboid nanoparticles and observed cationic reconstruction [100]. They propose reconstruction into linear CeO chains, triggered by mobile oxygen and vacancies, and suggest that easily extractible O that can lead to high oxidative catalytic activity. Lin et al. were able to directly image both the oxygen and cerium surface atoms of CeO_2 cubes using aberration-corrected HREM, giving an additional insight into the surface termination [101]. Importantly, they found that the cubes dominantly exposed CeO_2 {100} surfaces with a range of Ce, O, and reduced CeO_{2-x} terminations (Figure 3), indicating that the model proposing the reconstruction of the surface by the removal of exactly half of the surface oxygen atoms is too simple. Instead, surface energy is likely minimized by expressing a combination of surface terminations that

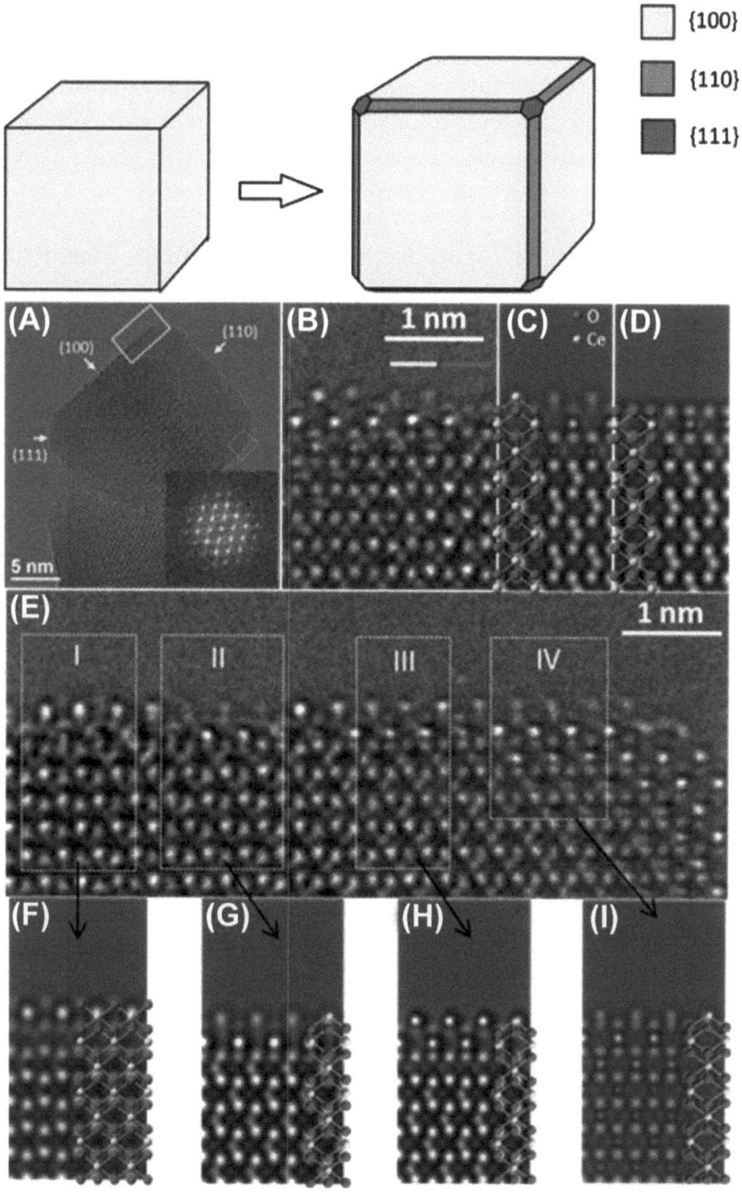

Figure 3 Top: Exposed facets of CeO_2 cubes. The left one is a simplified shape, and the right one is more realistic. Besides the predominantly exposed {100} facets, {110} and {111} facets exist as minor facets. Bottom: The {100} surfaces of CeO_2 cubes. (A) An HREM image on a typical CeO_2 nanocube at the [110] zone axis with the fast Fourier transform (FFT) of the region of interest (the area highlighted with white box) shown in the inset. (B) A magnified HREM image of the (100) surface of (A), highlighted with the blue (gray in print versions) box. (C)–(D) Simulated HREM images of (B) using Ce and O terminations, respectively. The atomic models are directly overlaid on the atom positions of the HREM images. (E) A magnified HREM image of the (100) surface of (A), highlighted with the white box. (F) A simulated image of region I in (E) using the ($\sqrt{2}\times\sqrt{2}$)R45° reconstructed CeO model (overlaid). (G)–(I) Simulated HREM images of region II, III, and IV in (E) using a Ce-terminated model. *From Ref. [101]. Reprinted with permission from (Y. Lin, Z. Wu, J. Wen, K.R. Poeppelmeier, L.D. Marks, Imaging the atomic surface structures of CeO_2 nanoparticles, Nano Lett. 14 (2013) 191–196). Copyright (2013) American Chemical Society.*

reduce the surface dipole. In addition, they found that CeO_2 cube surfaces exposed small amounts of {110} and {111} facets at the edges and corners, respectively. This study highlights that the surface structures of shape-controlled nanoparticles are quite complex. Results that give atomic-level information of the surface structure are critical to the understanding of catalytic reaction mechanisms and will give a deeper level of understanding to structure–function relationships.

2.3 Structure-Dependent Vacancies, Defects, and Acid–Base Properties

It is important to evaluate the type, number, and density of defects present on various CeO_2 nanoshapes, as defects can often modify the catalytic activity by promoting different selectivity or activities [102–104]. For example, Lawrence et al. increased the density of defects present on CeO_2 rods using a low-pressure thermal activation process and found that a high density of defects increased the activity for CO oxidation over CeO_2 rods that had not been treated [105]. Wu et al. performed in situ Raman spectroscopy to monitor the adsorption of O_2 on reduced CeO_2 nanoshapes and probed the amount and densities of defects on CeO_2 cubes, rods, and octahedra [106]. The visible Raman spectra of CeO_2 rods, which have the smallest diameters of the nanoshapes, showed a broadened F_{2g} peak at $462\,cm^{-1}$ compared to the cubes and octahedra (Figure 4(A)). The broadening can be attributed to size-dependent effects, which include phonon confinement and inhomogeneous strain broadening [86]. UV Raman, however, is more sensitive to defects in CeO_2 due to resonance enhancement of the D band (Figure 4(B)). The ratio of the D band intensity at $592\,cm^{-1}$ to the F_{2g} band intensity (I_D/I_{F2g}) relates to the relative amounts of intrinsic defect sites for the various CeO_2 nanoshapes, specifically, the higher the ratio, the higher the number of defects in the sample. Using this technique, rods were found to have the highest number of defects, and the octahedra the least (Figure 4(B)). The D band at $592\,cm^{-1}$ can reasonably be assigned to displaced interstitial oxygen (Frenkel-type) defects rather than to reduced Ce^{3+} sites. This assignment is based on the high thermal stability of the defects, previous observation of such defects by neutron diffraction [107], and the observation that O_2 treatments did not remove the defects (i.e., the oxygen vacancy is charge balanced by interstitial oxygen).

In addition, UV-resonance Raman and in situ Raman of O_2 adsorption was performed on the reduced nanoshapes. Upon reduction, a new shoulder band in the UV-Raman spectrum at $560\,cm^{-1}$ is observed and can be assigned to oxygen vacancies associated with Ce^{3+} sites. In contrast to the peak at $592\,cm^{-1}$, the peak disappears upon O_2 treatment. The adsorption of O_2 can result in the generation of superoxide, O_2^-, or peroxide sites, O_2^{2-}, and the formation of these species can be monitored by visible Raman spectroscopy. Peroxide formation indicates the presence of two-electron defect sites and the higher the Raman shift, the higher the degree of clustering. The most two-electron defect sites were observed on the rods, followed by cubes, then octahedra. Superoxide, correlating to one-electron defect sites, and peroxide bands were observed

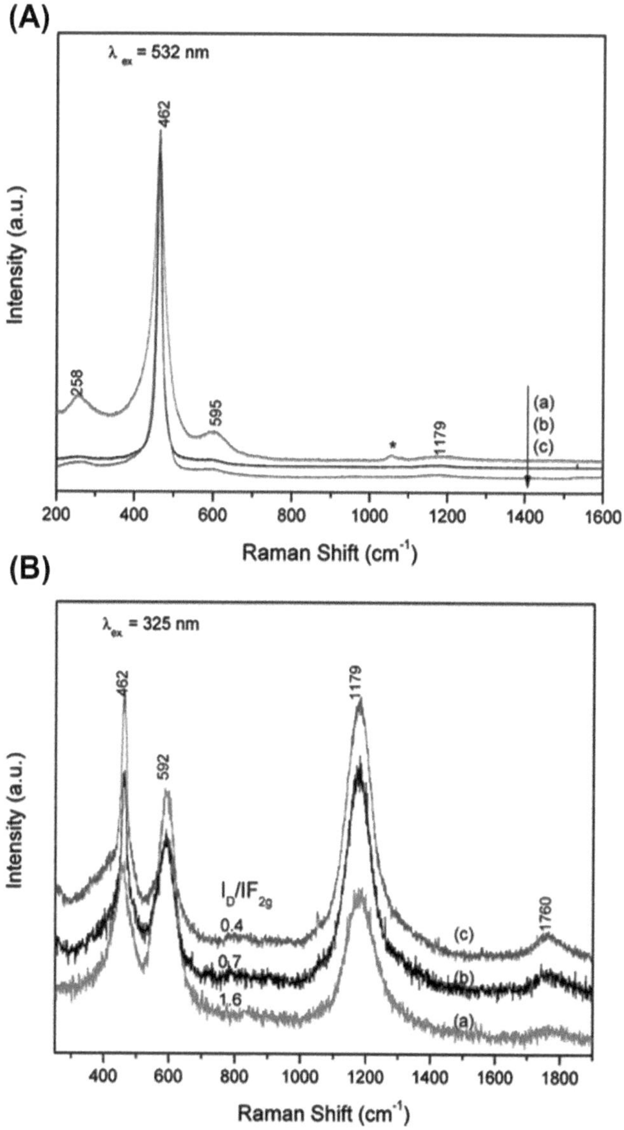

Figure 4 (A) Visible ($\lambda_{ex}=532$ nm) and (B) UV ($\lambda_{ex}=325$ nm) Raman spectra of ceria (A) rods, (B) cubes, and (C) octahedra. Asterisks (*) represent Raman bands due to trace nitrate/phosphate species in ceria. The relative intensity ratio of I_D/I_{F2g} inserted in the plots in (B) follows the sequence rods > cubes > octahedra, indicating that the rods have the most intrinsic defect sites while octahedra have the least. *From Ref. [106]. Reprinted with permission from (Z. Wu, M. Li, J. Howe, H.M. Meyer, S.H. Overbury, Probing defect sites on CeO_2 nanocrystals with well-defined surface planes by Raman spectroscopy and O_2 adsorption, Langmuir 26 (2010) 16595–16606). Copyright (2010) American Chemical Society.*

on the CeO_2 rods, indicating that the rods had both isolated and clustered vacancies upon reduction. In contrast to the rods, no superoxide species associated with one-electron defect sites were observed on CeO_2 cubes or octahedra. The Raman shifts for the peroxide species that formed on the CeO_2 cubes were predominately indicative of isolated two-electron defects. CeO_2 octahedra have a small amount of both clustered and isolated two-electron defect sites. It is expected that reduction should only generate isolated defects sites on CeO_2 {111}, so the clustered defect sites might form at the edges and corners.

As CeO_2 undergoes reduction, it is advantageous to understand the ease of reduction and how and where defects form on the various surfaces. Nolan et al. used DFT to calculate the electronic structure of vacancies for the various surfaces, and found that the electrons are expected to be localized at Ce ions near the vacancy rather than be distributed across all the Ce ions for all the low index facets. They also calculated that the CeO_2 (110) surface would have the lowest vacancy formation, followed by CeO_2 (100) and then CeO_2 (111) [108]. Studies to examine defect formation on CeO_2 (111) thin films have been performed by Nörenberg and coworkers using STM [109]. Interestingly, the authors found that defects form in triangular clusters, rather than randomly oriented clusters. Further reduction of the surface results in clusters that form in a line. Esch and coworkers used high-resolution STM to study the structure and location of CeO_2 (111) defects and determined that electrons were localized on Ce ions at oxygen vacancies [110]. They also found that the formation of subsurface oxygen vacancies was critical for the formation of linear rows or clusters of surface vacancies. Nörenberg and coworkers also studied the CeO_2 (100) surface and found new surface structures with increased reduction [95].

When determining the structure and other characteristics of various surfaces, the adsorption of probe molecules can give an insight into the nature of the surface. In particular, methanol is an advantageous probe molecule to identify the character of surface sites and defects of CeO_2 particles. Methanol adsorbs and dissociates to form methoxy (and a proton) on all ceria surfaces, but adsorption sites differ depending on the coordination numbers of surface cerium cations and location of defects. Such shape-dependent methanol adsorption was found on CeO_2 rods, cubes, and octahedra using in situ IR and Raman spectroscopy [89]. Specifically, methanol can undergo dissociation to form four different binding motifs to Ce cations, first as an on-top methoxy (Type I), as a bridging methoxy, with (Type II) or without (Type II′) an adjacent oxygen vacancy, and finally as a three-coordinate methoxy species (Type III) as shown in Figure 5. Based upon the structures in Figure 1, CeO_2 {111} facets are primarily expected to dissociate methanol to form on-top (Type I) methoxy, whereas on-top methoxy are not expected to form on CeO_2 {100} faces since steric hindrance from surface oxygen should inhibit its formation [89]. Instead, reconstructed CeO_2 {100} is anticipated primarily to form bridged methoxy groups. Finally, on CeO_2 {110}, methanol is expected to disassociate into on-top methoxy groups. In Figure 5, the observed distributions of the different bonding sites estimated from FTIR are tabulated.

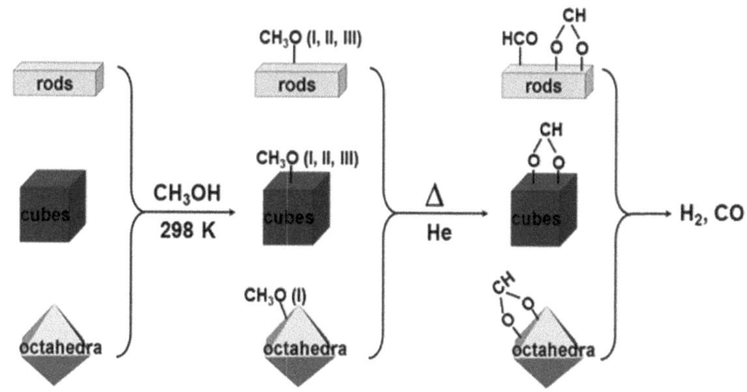

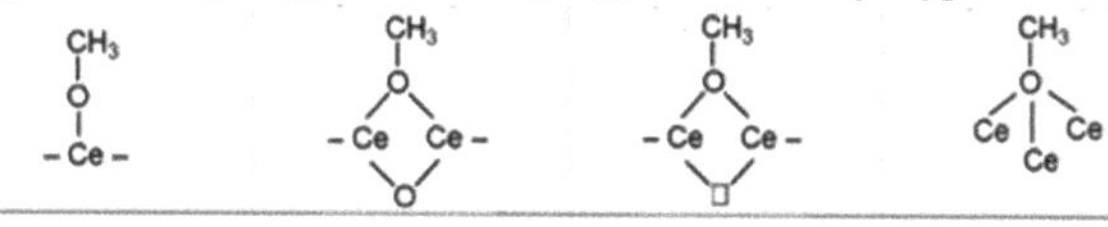

	Type I methoxy	Type II methoxy	Type II' methoxy	Type III methoxy
Rods {110}	75%	3%	12%	10%
Cubes {100}	43%	5%	40%	12%
Octahedra {111}	100%	None	None	None

Figure 5 Illustration of the moieties observed on the various CeO_2 nanoshapes under different conditions and the percentage of the different methoxy types forming from methanol adsorption on the three CeO_2 nanoshapes at room temperature. *From Ref. [89]. Adapted with permission from (Z. Wu, M. Li, D.R. Mullins, S.H. Overbury, Probing the surface sites of CeO_2 nanocrystals with well-defined surface planes via methanol adsorption and desorption, ACS Catal. 2 (2012) 2224–2234). Copyright (2012) American Chemical Society.*

On-top bonding was observed for all shapes, while bridging and three-coordinate methoxy were observed only on rods and cubes. Although 40% of the methoxy occupy bridging sites on {100}, there are substantial amounts of on-top and three-coordinate methoxy. These are consistent with the mobile and disordered nature of the oxygen anions, discussed above for the {100} surfaces, which may even move in response to methoxy adsorption. CeO_2 rods primarily form on-top methoxy (75%) and also had moderate amounts of bridging Type II′ (12%) and three-coordinate (10%) methoxy groups. The presence of Type II′ methoxy groups on CeO_2 rods and cubes is consistent with that in studies indicating that they possess a higher number of oxygen vacancies compared to CeO_2 octahedra [106]. Three-coordinate sites are not expected on ideal surfaces, but it is notable that CeO_2 rods and cubes exhibit similar small, percentages (10–12%). These may result from truncations exposing different facets, steps, or oxygen vacancies. The absence of Type III on octahedra suggests a low number of surface oxygen vacancies on {111} surfaces.

The adsorption and reaction of methanol on the various nanoshapes were probed by temperature-programmed desorption (TPD). All the nanoshapes produced dehydrogenation products, H_2, and CO, as well as smaller amounts of methanol, H_2O, and formaldehyde. In situ IR spectroscopy indicated that formate was the primary intermediate to H_2 and CO formation, though the CeO_2 rods also produced formyl. In addition, it was found that a large amount of carbon-containing species were retained on the surface as formate and carbonate species, especially for the rods. It is important to note that methanol adsorption and TPD results differed in comparison to those of studies performed on CeO_2 (100) and (111) films [55]. Instead, the TPD outcomes on CeO_2 nanoshapes mirrored the results obtained on defective and/or reduced CeO_{2-x} (100) and (111) films. This effect can be attributed to a higher number of defects intrinsically present in CeO_2 nanoshapes compared to that in oriented films, and the ability of products and intermediates to readsorb in reactor environments, highlighting the importance of complementary UHV and industry-relevant catalytic reactor condition studies.

In addition to detecting coordination environments and defect sites, methanol adsorption and dissociation give an insight into the acid and base properties of the cerium oxide surface. In particular, a basic oxygen site is needed to abstract the proton from methanol, while an acidic site is needed to stabilize the methoxy group [55]. Cerium oxide has been shown to possess an amphoteric surface, with strongly basic surface O^{2-} groups and weakly acidic cerium ions in close proximity, and thus, methanol dissociation is facile [6]. However, when considering shape-controlled CeO_2 nanoshapes, surface coordination numbers differ according to the facet exposed. The coordination numbers are intrinsically linked to the acidity and basicity of the surface Ce and O atoms. For example, a lower coordination number for a surface oxygen atom would be expected to result in a higher basicity, and thus, {100} facets would be predicted to be more basic than {111} or {110}. Similarly, a lower coordination number for surface cerium atoms would be expected to correlate to a higher acidity, and consequently, {100} and {110} facets would be expected to be more acidic than {111} facets. Therefore, nanoshapes are expected to exhibit structure-dependent acid and base properties. The redox properties of the materials are also expected to affect the acid–base properties, as the reduction of cerium oxide results in a decrease in the Lewis acidity of the Ce ions. Indeed the interplay between acid–base chemistry and redox properties of ceria shapes illustrates the complex nature of structure-dependent properties and highlights the importance of fundamental studies. In addition, the ability to form oxygen vacancies on the surface is related to the surface structure and energies required to stabilize those vacancies. Methanol adsorption caused a significant degree of reduction on CeO_2 rods and cubes as observed via in situ Raman spectroscopy [89]. Reduction of CeO_2 surface is expected to occur via the disproportionation condensation of two hydroxyl groups formed upon dissociation of methanol, resulting in the formation of H_2O and an oxygen vacancy. The degree of reduction was most significant on CeO_2 rods, followed by cubes, and none on the octahedra, consistent with oxygen vacancy formation energies, which will be discussed next.

2.4 Shape-Dependent Redox Properties

As mentioned previously, the redox properties of CeO_2 can vary with particle size, and similarly, the redox properties of CeO_2 are expected to vary with shape. Several studies have been performed to examine how shape affects the redox properties of the CeO_2, examining properties such as the oxygen mobility and OSC. Using isotopically labeled O_2, it was determined that the order of oxygen exchangeability is rods > cubes >> octahedra (Figure 6) [111]. The OSC can be tested by reducing a CeO_2 sample, and then measuring the O_2 uptake in a series of pulses. Mai et al. found that CeO_2 cubes had the highest OSC, followed by rods [67]. The octahedra displayed a much lower OSC than the cubes or rods did. Importantly, the authors also found that the experimental OSC for the rods and cubes was

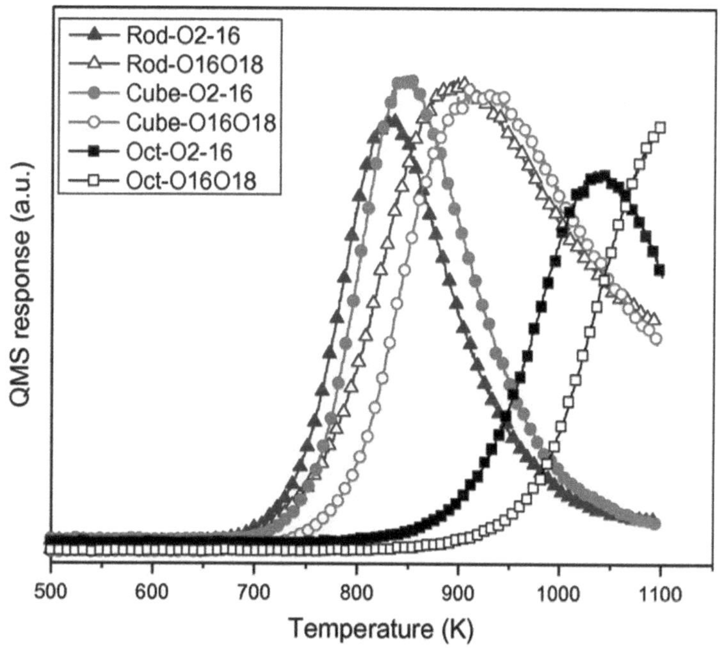

Figure 6 QMS profiles of O_2 isotopomers, $^{16}O_2$ and $^{16}O^{18}O$, during oxygen isotopic exchange over CeO_2 rods, cubes, and octahedra. *From Ref. [111]. Reprinted from (Z. Wu, M. Li, S.H. Overbury, On the structure dependence of CO oxidation over CeO_2 nanocrystals with well-defined surface planes, J. Catal. 285 (2012) 61–73). Copyright (2012), with permission from Elsevier.*

significantly higher than the expected OSC for the various shapes if only the surface was involved in the OSC. This result suggests that the bulk structure also contributes to the OSC activity. Also, it is interesting to note that they found the OSC did not mirror the oxygen vacancy formation energies, which have been calculated to follow the order of {110} < {100} < {111} [108]. Wang et al. showed that CeO_2 two-dimensional (2D) nanoplates with (100) surfaces had a higher OSC activity than nanoparticles and nanospheres and attributed the increased OSC over the CeO_2 spheres with the 2D nature of the plates [112]. These structure-dependent redox properties have significant effects on the catalytic reactivity and selectivity of CeO_2 nanoshapes, and selected examples will be highlighted in the next section.

3. HETEROGENEOUS CATALYSIS REACTIONS ON NANOSHAPED CeO_2

3.1 CO Oxidation

CO oxidation is a particularly relevant reaction for CeO_2 as it proceeds via a Mars–van Krevelen mechanism where the rate determining step is the removal of lattice O from CeO_2 to oxidize CO [113]. Therefore, the reactivity for CO oxidation is directly related to the capability of the CeO_2 to form oxygen vacancies. In one of the first studies to examine the effect of shape on CO oxidation with CeO_2, Zhou et al. compared the reactivity of CeO_2 rods and nanoparticles without defined shapes [98]. They found that CeO_2 rods exhibited higher CO oxidation capabilities than did nanoparticles without shape-control. Pan et al. found that shape, crystallinity, and surface area affected the reactivity of CeO_2 nanoshapes for CO oxidation, which had the following reactivity: nanotubes > rods with a long aging time > nanowires > rods with a short aging time > nanoparticles > cubes at 300 °C [114]. In particular, a combination of high surface area, a large number of active surface sites, and high crystallinity contributed to a higher CO conversion. Tana et al. examined the morphology-dependent properties of CeO_2 nanowires and rods and CeO_2 nanoparticles that were proposed to expose {111} facets for CO oxidation as well, and found that the order of conversion was wires > rods > nanoparticles and related the observed conversions to the exposed facets [10].

We have provided an additional insight into these studies by comparing CO oxidation on CeO_2 cubes, rods, and octahedra, and combining these results with Raman and in situ IR spectroscopy studies to determine the reaction mechanism [111]. The CO oxidation light-off temperature follows the order of the oxygen vacancy formation energies, that is, rods < cubes < octahedra, as expected if the activity follows the order of reducibility (Figure 7). It was also noted that the room temperature adsorption of CO had structure-dependent transformations. First, the amount of CO adsorbed at Ce^{4+} coordinatively unsaturated sites (*cus*) was the greatest on the rods, followed by cubes, then octahedra. In addition, carbonate species are observed to form on all three nanoshapes; however, the type of carbonate species varied. Specifically, the cubes and

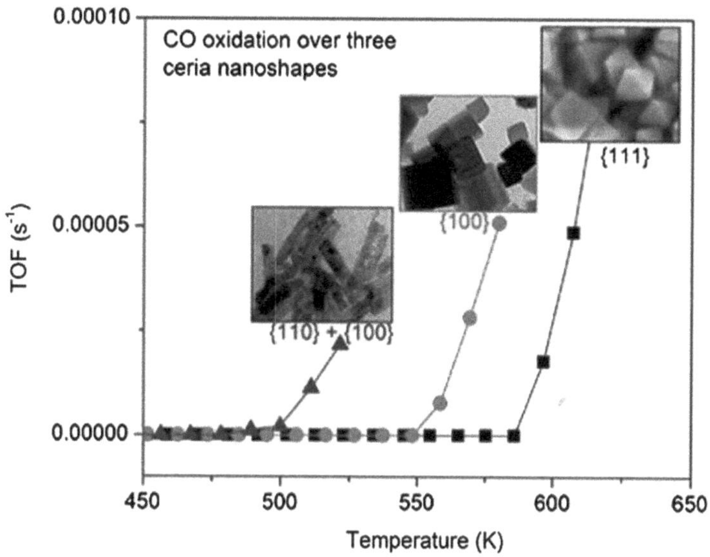

Figure 7 Comparison of CO turnover frequency (TOF) (converted CO per surface O per s) over three ceria shapes. *From Ref. [111]. Reprinted from (Z. Wu, M. Li, S.H. Overbury, On the structure dependence of CO oxidation over CeO_2 nanocrystals with well-defined surface planes, J. Catal. 285 (2012) 61–73). Copyright (2012), with permission from Elsevier.*

rods formed bicarbonate, bidentate, and bridged carbonate species, whereas the octahedra formed bidentate and unidentate carbonate species.

CeO_2 cubes and rods showed lower temperature CO_2 evolution at 570K, which is associated with the reaction of CO with lattice oxygen. Temperature-programmed isotopic exchange reactions were performed by cofeeding $^{18}O_2$ and $C^{16}O$, and it was verified via in situ IR and quadrupole mass spectrometry (QMS) that CO reacts with lattice oxygen to produce CO_2, while the cofed O_2 reoxidizes the surface. However, it was also found that CO_2 production also had several other competing mechanistic pathways, including water-gas shift (WGS) reactions of CO with surface OH groups and the formation of carbonaceous material and its subsequent oxidation. Notably, it was found that H_2 is produced concurrently with CO_2 and expected to be formed via a WGS reaction of CO with surface hydroxyl groups accounting for 50% of the CO_2 produced at lower temperatures. A higher temperature CO_2 peak is observed at 740K, and is produced by a combination of the WGS reaction and CO oxidation reaction with lattice oxygen occurring simultaneously.

CO temperature-programmed reduction (CO-TPR) with in situ diffuse reflectance infrared spectroscopy (DRIFTS) was performed to probe the surface species formed during the reaction (Figure 8). As the CeO_2 rods, cubes, and octahedra were heated, formate species were observed, formed via the reaction of surface OH and CO, supporting the fact that H_2 and CO_2 form through a WGS reaction mechanism. The rods also exhibited bands associated with formyl species, which were expected to be

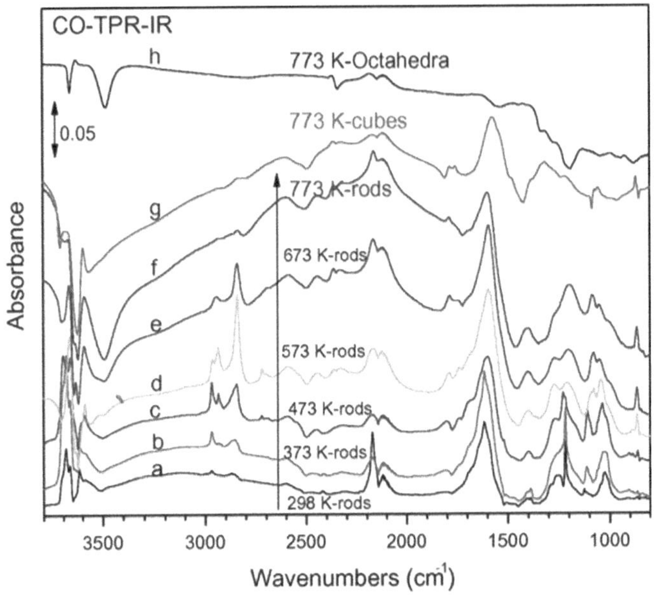

Figure 8 IR spectra collected during CO-TPR of ceria rods at different temperatures (A)–(F) and cubes (G) and octahedra (H) at 773 K. Spectra are referenced to the background spectra collected at the indicated temperature during cooling down from the pretreatment process of the ceria sample. *From Ref. [111]. Reprinted from (Z. Wu, M. Li, S.H. Overbury, On the structure dependence of CO oxidation over CeO_2 nanocrystals with well-defined surface planes, J. Catal. 285 (2012) 61–73). Copyright (2012), with permission from Elsevier.*

intermediates to the formation of formate species. At higher temperatures, carbonate species were observed on the rods and cubes, but not the octahedra. In situ Raman spectroscopy was used to compliment the FTIR data. Raman spectroscopy indicated that a carbonaceous material was formed on the surface of the nanoshapes after reaction, which was present in a higher amount on the rods than on the cubes and octahedra. This carbonaceous species likely forms via CO disproportionation, where two CO molecules react to form CO_2 and C, and it was found that the material could be removed with O_2 treatment. This reaction to form this carbonaceous material is expected to require Ce^{3+} sites, suggesting that these sites are more abundant on the surface of the rods.

3.2 Selective Ethanol Oxidation

The study of the reactions of C_2 and higher molecules containing H significantly increases the number of potential products as well as the complexity of reaction mechanisms. However, the catalytic transformations of these molecules are of great importance and relevance to industrial processes. The catalytic reaction of ethanol presents an interesting model system to study as ethanol can undergo dehydration or dehydrogenation to form ethylene or acetaldehyde, respectively, oxidation to yield acetic acid, coupling to generate ethyl acetate, or C–C bond scission to produce methane, CO, or CO_2. Li et al. studied the mechanism of structure-related reactivity and selectivity for the oxidation of ethanol on CeO_2 using nanoshapes in combination with TPD, temperature-programmed surface reaction (TPSR), steady-state reactions, and DRIFTS [115]. In situ DRIFTS indicates that room temperature adsorption results in the formation of ethoxide via deprotonation of ethanol on the surface of the nanoshapes, observed strongly on the rods and cubes, but weakly on the octahedra. Ethoxide groups bind as monodentate and bidentate species on both CeO_2 cubes and rods. In addition, acetate peaks were observed on the cubes and rods. Upon heating, the amount of acetate increased and the ethoxide peaks decreased. However, the octahedra show ethoxide loss at a lower temperature than the rods and cubes. While acetate is formed on the surface of the octahedra upon heating, the bands are weak, likely because the majority of ethoxide has already desorbed.

The TPD results showed that all the nanoshapes evolved ethanol, ethylene, acetaldehyde, acetic acid, CO, CO_2, methane, and ethyl acetate and exhibited a similar selectivity for acetaldehyde (28–30%) as shown in Figure 9, which is expected to form via scission of the α-CH bond of ethoxide. However, the nanoshapes displayed different selectivities for several other products, including ethanol, ethylene, and methane in TPD. For example, ethylene, which is proposed to be generated via an E1 unimolecular CO cleavage and a base catalyzed β-CH scission, was formed as 36% of the product yield on the CeO_2 rods, compared to 28.5% on the cubes and just 17% on the octahedra. The selectivity to form acetaldehyde versus ethylene is likely related to the O coordination, availability of H, and vacancies, and indicates that this reaction mechanism displays surface structure dependence.

In addition, using steady-state reactions in which ethanol was cofed with O_2, they were able to determine the variation in product selectivity at several temperatures. For example, it was found that at 350 °C, octahedra had the highest selectivity to produce acetaldehyde at >80% selectivity compared to approximately 60–70% for cubes and rods. At this temperature, the product obtained at the next highest selectivity was CO_2, with the most being generated on cubes, followed by rods, then octahedra, but almost no ethylene was observed in the product stream. At temperatures >350 °C on the rods and >400 °C on the cubes and octahedra, acetaldehyde formation decreases and the generation of other products including ethylene, H_2, CO, CH_4, ethyl acetate, and CO_2 increases. In this temperature range, the energy barriers to perform C–C bond scission, mostly likely of surface acetate, are overcome. The formation of H_2 and CH_4 indicates that the cofed O_2 stream is completely consumed. Ethylene and ethyl acetate, which could form via an esterification reaction of ethanol with acetic acid, are also observed in this temperature region [115]. Importantly, this study highlights the strong structure dependence of ethanol adsorption, desorption, and reaction on CeO_2 nanoshapes related to their acidity and basicity, vacancies, coordination numbers, and surface H availability.

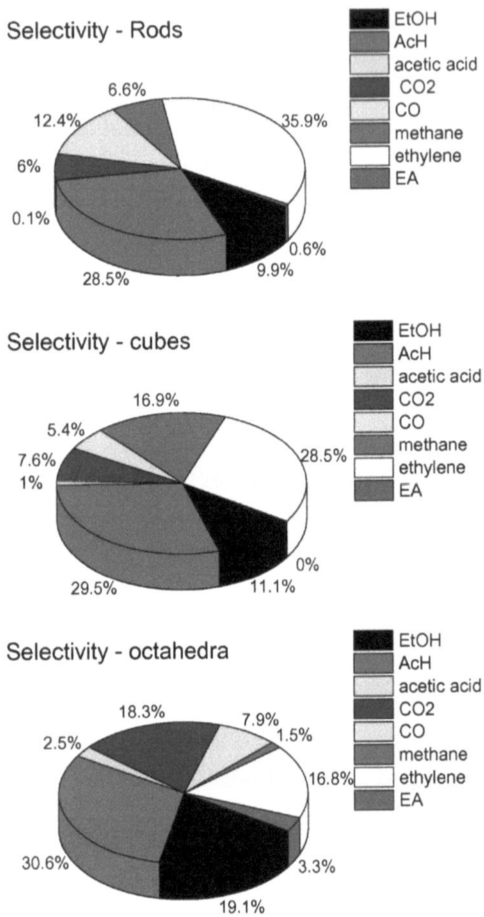

Figure 9 Selectivities resulting from adsorbing ethanol on ceria octahedra, cubes, and rods from TPD product profiles (corrected for carbon numbers, QMS mass overlap, and QMS response). *From Ref. [115]. Reprinted from (M. Li, Z. Wu, S.H. Overbury, Surface structure dependence of selective oxidation of ethanol on faceted CeO_2 nanocrystals, J. Catal. 306 (2013) 164–176). Copyright (2013), with permission from Elsevier.*

Zhou et al. also probed the oxidation of ethanol on CeO_2 rods and cubes, but with the goal of eliminating volatile organic compounds (VOCs) [116]. Similar to Li et al., they found that acetaldehyde is the major oxidation product when coflowing ethanol and air at lower temperatures, with smaller amounts of ethyl acetate and acetic acid being formed. However, at higher temperatures, CO_2 becomes the dominant product, and the conversion of ethanol significantly increases. They determined that the CeO_2 rods had a higher activity for ethanol oxidation, with ethanol conversion >99% by 170 °C, and a higher selectivity to produce CO_2. In contrast, the CeO_2 cubes had a lower CO_2 selectivity and required a higher reaction temperature, with 95% of the ethanol converted at

325 °C. The increased reactivity for the rods is attributed to the ease of reduction of the Ce^{4+} active sites on the rods to form Ce^{3+} while oxidizing ethanol. Cofeeding air then allows for oxygen to dissociate on the surface and reform the Ce^{4+} active site [116].

3.3 Coupling Reactions of Acetaldehyde

The adsorption and reaction of acetaldehyde on various CeO_2 nanoshapes are of interest as acetaldehyde has been shown to undergo coupling reactions to form products such as crotonaldehyde, crotyl alcohol, butadiene, butene, benzene, and acetone [35,36]. In addition, acetaldehyde can be reduced to form ethanol, oxidized to generate acetic acid, or undergo C–C bond scission to form CH_4, CO, or CO_2. Previously, Mullins et al. showed that the adsorption and reaction of acetaldehyde on CeO_{2-x} (111) and (100) surfaces had significant differences depending on the nature of the surface exposed [60,117]. For example, on CeO_2 (111), acetaldehyde primarily binds as η^1-acetaldehyde and either desorbs as acetaldehyde or reacts to form ethylene, acetylene, and H_2 [117]. In contrast, acetaldehyde binds as η^2-acetaldehyde on CeO_2 (100) and produces CO, CO_2, H_2O, acetylene, and a small amount of the coupling product, crotonaldehyde [60]. However, UHV experiments would not be expected to be selective for coupling products if the surface coverage is low, and therefore, the examination of the adsorption and reaction of acetaldehyde on CeO_2 nanoshapes could provide structure-dependent information and an insight into the selectivity and reactivity of acetaldehyde on CeO_2.

Mann et al. studied the reaction of acetaldehyde on CeO_2 octahedra, wires, and cubes, using TPD, temperature-programmed surface reactions (TPSR), and in situ DRIFTS [118]. DRIFTS showed that the room temperature adsorption of acetaldehyde produced η^1-acetaldehyde on the octahedra, coupling products on the wires, and enolate and acetate species and coupling products on the cubes. In TPD, the major desorption products for all shapes consisted of acetaldehyde, CO, and CO_2. In addition, crotonaldehyde was observed as a minor product from TPD on all nanoshapes. Crotonaldehyde is expected to be formed on CeO_2 via an aldol condensation mechanism, in which adsorbed acetaldehyde forms an enolate moiety, and then couples with a second acetaldehyde molecule to produce one equivalent of crotonaldehyde and water.

Interestingly, ethanol was formed as a major product on the wires and as a minor product on the cubes, and only as a trace product on the octahedra. Ethanol is expected to form via a Cannizzaro disproportionation reaction in which two acetaldehyde molecules disproportionate to form an ethoxide and an acetate group via a nucleophilic attack by surface oxygen, followed by a hydride shift [48,119]. The acetate groups can undergo ketonization to form acetone and CO_2, or may decompose to CH_4, CO, and CO_2. Importantly, basic surface oxygen atoms are needed to perform this reaction. The increased production of ethanol and acetate groups on cubes compared to that on octahedra implies a more basic character in {100} than the {111}, consistent with the lower coordination numbers of surface oxygen on {100} versus {111}. Therefore, the basicity of the surface is a significant factor in the structure-dependent selectivity. Further, ethanol production can result in the generation of other coupling products, as it can reduce crotonaldehyde to produce crotyl alcohol, 1,3-butadiene, and 2-butene. Importantly, these products are observed on cubes and wires in TPSR, but noticeably absent on the octahedra. Therefore, a more basic surface results in the formation of ethanol, crotyl alcohol, 1,3-butadiene, and 2-butene in addition to crotonaldehyde.

3.4 Catalytic Oxidation of Ethyl Acetate and 1,2-Dichloroethane for VOC Degradation

Dai and coworkers explored the oxidation of ethyl acetate and 1,2-dichloroethane on CeO_2 rods, cubes, and octahedra in the presence of O_2 [120]. For both reactants, the authors found that CeO_2 rods showed the highest conversion at the lowest temperatures for equal weights of catalysts, followed by cubes and octahedra. However, the CeO_2 rods have a much higher surface area than cubes and octahedra (56, 21, and $10\,m^2\,g^{-1}$, respectively), and this result may suggest that CeO_2 cubes are actually more active on a per surface area basis. The oxidation of ethyl acetate by CeO_2 rods produced CO_2 and H_2O, with small amounts of ethanol as a by-product. Interestingly, at certain conditions, the CeO_2 octahedra produced CO as a main product, rather than CO_2, consistent with incomplete oxidation of ethyl acetate. This result could be due to the lower OSC and oxygen mobility of the CeO_2 {111} facet. The authors proposed that the mechanism for 1,2-dichloroethane consists of the adsorption of the molecule at Ce^{3+} or Ce^{4+} surface sites, where it undergoes dissociation, splitting the C–Cl bond. They propose that ethyl acetate adsorbs only near Ce^{3+} surface sites, which can dissociate the molecule. Next, both the reactants can undergo oxidation via a Mars–van Krevelen mechanism, in which surface lattice oxide attacks the reactant. Finally, the cofed oxygen can adsorb and reoxidize the surface. This example again highlights the importance of the faceting and defects in providing high selectivity and conversion.

3.5 CeO_2 Nanoshapes for the Total Oxidation of Naphthalene

Turrente-Murciano and coworkers showed the shape dependence of the catalytic total oxidation of naphthalene to CO_2 by comparing CeO_2 rods with cubes [121]. The selectivity to form CO_2 does not significantly vary between the CeO_2 nanoshapes; however, the activity normalized to surface area follows the order of rods > cubes > nanoparticles. The authors noted that the reactivity correlated with the reported reactivity for CO oxidation for the various shapes, and related the reactivity to the concentration of oxygen vacancies and Ce^{3+}/Ce^{4+} ratios for the various facets, and supported these conclusions with XPS and Raman spectroscopy [121].

3.6 Selective Oxidation of Toluene

Lv et al. utilized CeO_2 cubes synthesized using oleic acid as a structure-directing agent and CeO_2 nanoparticles coated with oleic acid postsynthesis to selectively oxidize toluene to benzaldehyde in water using an autoclave, and monitored the process by GC [122]. The authors found that the use of CeO_2 cubes resulted in the production of benzaldehyde; however, CeO_2 nanoparticles with {111} and {100} facets resulted in the formation of benzyl alcohol, benzaldehyde, and benzoic acid. In fact, the selectivity to produce benzoic acid by the nanoparticles increased as the reaction time was increased. Importantly, the authors found that the oleic acid played a critical role in the oxidation of toluene. If the reaction temperature was increased, thereby removing the oleic acid ligands, the reactivity was lowered, and similarly, after reusing the catalysts, the amount of ligand was decreased, and the catalyst was deactivated. However, reintroduction of the ligand via the addition of sodium oleate restored the catalytic activity. The authors propose that the oleic acid plays a dual role. First, it provides a hydrophobic medium for the toluene to access the surface of CeO_2, and thereby controlling the diffusion and orienting the methyl group of the toluene toward the surface. Second, it can mediate the active oxygen species to help facilitate the oxidation of toluene. The authors propose that the facets exposed by the different CeO_2 nanoshapes dictate which active oxygen species are available. For the cubes, only one type of active oxygen species is available, resulting in the selectivity to generate benzaldehyde. The nanoparticles without shape control provide multiple types of active oxygen, and thus produce benzoic acid, benzaldehyde, and benzyl alcohol. This example provides an insight into how not only shape control but also surfactants and surface modifiers can mediate and modify the catalytic selectivity of shape-controlled CeO_2 nanoshapes.

3.7 WGS Reactions

The WGS reaction is an important transformation for the generation of H_2 as an energy source from CO and H_2O. Han et al. studied the WGS reaction using synchrotron-based XRD and XANES on a heterogeneous CeO_2 sample consisting of 1D rods and nanotubes (~80%) with exposed {110} and {100} facets, and nanoparticles (~20%) [123]. Previous studies using 3D nanoparticles or bulk CeO_2 samples showed insignificant activity for the reaction, yet Han et al. showed that the 1D particles were effective for catalyzing the reaction, especially at high temperatures. This result was attributed to the 1D nature and high surface area of the particles that significantly increased the reducibility of the CeO_2 in a reducing environment, thereby increasing the catalytic activity of the particles for the WGS reaction. In addition, the exposed facets likely contribute to enhanced reducibility, as CeO_2 {110} and {100} facets exhibit lower oxygen vacancy formation energies than do CeO_2 {111} facets [108]. The reducibility in this case was measured by examining the lattice parameter, which increases as Ce^{4+} is reduced to Ce^{3+} (Figure 10). The authors showed that Pd loading further increased the reducibility of the catalyst. Indeed, CeO_2 has been shown to be a better

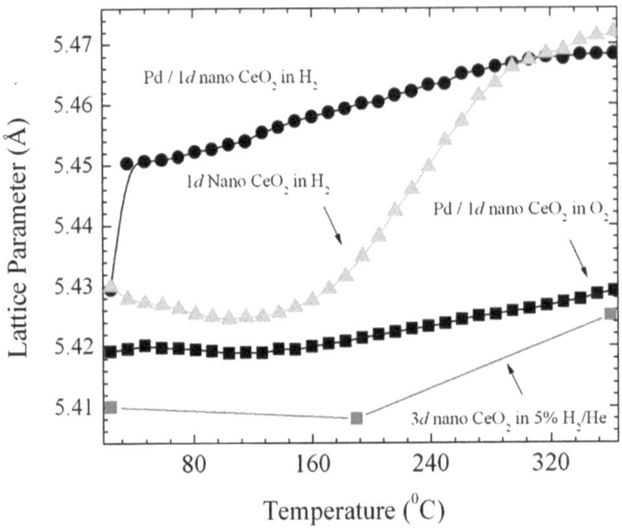

Figure 10 Ceria lattice parameter changes during the temperature ramping of 1D-ceria, Pd-loaded 1D-ceria, and 3D-nanoceria. *From Ref. [123]. Reprinted with permission from (W.-Q. Han, W. Wen, J.C. Hanson, X. Teng, N. Marinkovic, J.A. Rodriguez, One-dimensional ceria as catalyst for the low-temperature water–gas shift reaction, J. Phys. Chem. C 113 (2009) 21949–21955). Copyright (2009) American Chemical Society.*

support than a catalyst in the WGS reaction. Of particular interest is a study performed by Si et al. on CeO_2 nanoshapes loaded with Au nanoparticles [124]. The authors found that a high lattice strain of the various CeO_2 crystal planes associated with the different shapes correlated to the Au-loaded nanoshapes having a higher activity for the WGS reaction, with rods having the highest activity, followed by polyhedra, then cubes.

4. FUTURE DIRECTIONS

As has been shown, CeO_2 nanoshapes with controlled facets can result in enhanced reactivities and improved selectivities for various catalytic reactions. These modifications in reactivities can be attributed to shape-dependent factors such as differences in the acid and base properties, surface energies and geometries, vacancy formation energies, and redox properties, among others. The elucidation of shape-dependent properties on the catalytic activity is important for the design of optimized catalysts and fundamental understanding of reaction mechanisms. This goal requires detailed studies that elucidate the structure-dependent characteristics, such as those highlighted in Section 2. The combination of these studies with reaction surveys, such as those highlighted in Section 3, paint a complex picture that highlights how structure-dependent properties are often related. It can be difficult to completely deconvolute their effects on catalytic selectivity and reactivity. However, these studies also emphasize that the ability to control the shape and structure of nanoparticles is a powerful method for promoting efficient and desirable reactions and for probing surface structure relationships.

Despite the advances made in understanding how shape affects the reactivity of CeO_2 catalysts, studies utilizing bare CeO_2 samples can still suffer from issues such as deactivation of the catalyst, low conversions, and low selectivities. One way to address these issues is to design advanced materials for catalysis by exploiting the benefits of the various shape-controlled CeO_2 nanoshapes, while enhancing or utilizing those properties with additives or dopants to improve selectivity. Several examples have already been published that demonstrate the effect of the shape-control of the support. For example, the effects of shaped-controlled CeO_2 rods and octahedra as supports for vanadia catalysts have been demonstrated in the oxidative dehydrogenation of i-butane [125]. Importantly, it was found that rods lowered the activation energy of the oxidation reaction, correlated to the oxygen vacancy formation energies of the nanoshapes. In ethanol steam reforming using cobalt deposited on CeO_2 cubes and rods, it was found that Co/CeO_2 cubes had a higher activity, and mirrored the intrinsic activity for steam reforming of bare CeO_2 nanoshapes (cubes > rods) [126]. Cu–Pd supported on CeO_2 nanoplates with a new hexagonal crystal structure displayed high selectivity for the catalytic hydrogenation of substituted nitroaromatics without sacrificing conversion under mild conditions [127]. Several other reactions using shape-controlled CeO_2 nanoshapes have been probed, including CuO/CeO_2 for NO reduction by CO [128], Au on CeO_2 nanoshapes for WGS reactions [124], and CO oxidation [129], among others. These examples highlight how the beneficial properties of CeO_2 can be exploited for enhanced catalytic reactivity, and how the shape of CeO_2 plays a fundamental role, even as a support. These modified CeO_2 samples give a new level of control to modify the acidity, basicity, and reducibility of the catalysts, and provide another approach to investigate reaction pathways and the role of acid–base site cooperativity.

5. CONCLUSIONS

The shape-dependent properties, such as acidity, basicity, reducibility, and the location and number of defects, have been examined for CeO_2 shapes with {111}, {100}, and {110} facets. Octahedra have fewer defects than the cubes and rods, and the distribution and type (one electron versus two electron) varies for the different facets. The oxygen vacancy formation energies follows the order of rods < cubes < octahedra. Shape-controlled CeO_2 has been tested for many catalytic reactions, including the oxidation of ethanol, CO, toluene, and naphthalene, and the WGS reaction. The various shapes provided different reactivities and selectivities for these reactions. The ability to perform fundamental studies that probe shape-dependent properties such as the acid and base sites on the surface, the reducibility, surface energies, and defect concentrations and densities provide guidance on why these reactions exhibit the observed reactivities and selectivities for various reactions. Further fundamental studies utilizing shape-controlled CeO_2 particles for characterization, reaction studies, or as supports for catalysts present a promising future for the design of advanced catalysts.

ACKNOWLEDGMENTS

This research is supported by the U.S. Department of Energy, Office of Science, Basic Energy Sciences, Chemical Sciences, Geosciences, and Biosciences Division. Part of the work including synthesis, XRD, TEM, and scanning electron microscopy was conducted at the Center for Nanophase Materials Sciences, which is a DOE Office of Science User Facility.

REFERENCES

[1] L. Vivier, D. Duprez, ChemSusChem 3 (2010) 654–678.
[2] J. Beckers, G. Rothenberg, Green Chem. 12 (2010) 939–948.
[3] Z.-A. Qiao, Z. Wu, S. Dai, ChemSusChem 6 (2013) 1821–1833.
[4] E. Aneggi, M. Boaro, C.D. Leitenburg, G. Dolcetti, A. Trovarelli, J. Alloys Compd. 408–412 (2006) 1096–1102.
[5] M.V. Ganduglia-Pirovano, A. Hofmann, J. Sauer, Surf. Sci. Rep. 62 (2007) 219–270.
[6] C. Binet, M. Daturi, J.-C. Lavalley, Catal. Today 50 (1999) 207–225.
[7] D. Martin, D. Duprez, J. Mol. Catal. A: Chem. 118 (1997) 113–128.
[8] Y. Li, D. He, Q. Zhu, X. Zhang, B. Xu, J. Catal. 221 (2004) 584–593.
[9] X. Liu, K. Zhou, L. Wang, B. Wang, Y. Li, J. Amer. Chem. Soc. 131 (2009) 3140–3141.
[10] Tana, M. Zhang, J. Li, H. Li, Y. Li, W. Shen, Catal. Today 148 (2009) 179–183.
[11] G.S. Zafiris, R.J. Gorte, J. Catal. 143 (1993) 86–91.
[12] A. Yee, S.J. Morrison, H. Idriss, J. Catal. 186 (1999) 279–295.
[13] D. Enache, D. Knight, G. Hutchings, Catal. Lett. 103 (2005) 43–52.
[14] W. Kuang, Y. Fan, Y. Chen, Catal. Today 68 (2001) 75–82.
[15] M. Palacio, P.I. Villabrille, G.P. Romanelli, P.G. Vázquez, C.V. Cáceres, Appl. Catal. A 359 (2009) 62–68.
[16] S.M. Lee, H.H. Lee, S.C. Hong, Appl. Catal. A 470 (2014) 189–198.
[17] Y. Wang, F. Wang, Y. Chen, D. Zhang, B. Li, S. Kang, X. Li, L. Cui, Appl. Catal. B 147 (2014) 602–609.
[18] A. Trovarelli, C. Deleitenburg, G. Dolcetti, J.L. Lorca, J. Catal. 151 (1995) 111–124.
[19] G. Qi, R.T. Yang, R. Chang, Appl. Catal. B 51 (2004) 93–106.
[20] P. Bera, K.R. Priolkar, P.R. Sarode, M.S. Hegde, S. Emura, R. Kumashiro, N.P. Lalla, Chem. Mater. 14 (2002) 3591–3601.
[21] J.B. Branco, D. Ballivet-Tkatchenko, A.P. de Matos, J. Mol. Catal. A: Chem. 307 (2009) 37–42.
[22] Y. Azizi, C. Petit, V. Pitchon, J. Catal. 256 (2008) 338–344.
[23] A. Sepúlveda-Escribano, F. Coloma, F. Rodríguez-Reinoso, J. Catal. 178 (1998) 649–657.
[24] T. Marzialetti, J.L.G. Fierro, P. Reyes, Catal. Today 107–108 (2005) 235–243.
[25] M. Chong, D-g. Cheng, L. Liu, F. Chen, X. Zhan, Catal. Lett. 114 (2007) 198–201.
[26] A. Saadi, Z. Rassoul, M.M. Bettahar, J. Mol. Catal. A: Chem. 164 (2000) 205–216.
[27] J. Kugai, V. Subramani, C. Song, M.H. Engelhard, Y.-H. Chin, J. Catal. 238 (2006) 430–440.
[28] D. Srinivas, C.V.V. Satyanarayana, H.S. Potdar, P. Ratnasamy, Appl. Catal. A 246 (2003) 323–334.
[29] S. Sato, R. Takahashi, T. Sodesawa, N. Honda, J. Mol. Catal. A: Chem. 221 (2004) 177–183.
[30] V. Solinas, E. Rombi, I. Ferino, M.G. Cutrufello, G. Colón, J.A. Navío, J. Mol, Catal. A: Chem. 204–205 (2003) 629–635.
[31] M.A. Hasan, M.I. Zaki, L. Pasupulety, Appl. Catal. A 243 (2003) 81–92.
[32] O. Nagashima, S. Sato, R. Takahashi, T. Sodesawa, J. Mol. Catal. A: Chem. 227 (2005) 231–239.
[33] M. Gliński, J. Kijeński, A. Jakubowski, Appl. Catal. A 128 (1995) 209–217.

[34] O. Nagashima, S. Sato, R. Takahashi, T. Sodesawa, T. Akashi, Appl. Catal. A 312 (2006) 175–180.
[35] H. Idriss, C. Diagne, J.P. Hindermann, A. Kiennemann, M.A. Barteau, J. Catal. 155 (1995) 219–237.
[36] J. Raskó, J. Kiss, Appl. Catal. A 287 (2005) 252–260.
[37] G. Postole, B. Chowdhury, B. Karmakar, K. Pinki, J. Banerji, A. Auroux, J. Catal. 269 (2010) 110–121.
[38] S. Carrettin, J. Guzman, A. Corma, Angew. Chem. Int. Ed. 44 (2005) 2242–2245.
[39] S. Bernal, G. Blanco, J.J. Calvino, J.A.P. Omil, J.M. Pintado, J. Alloys Compd. 408–412 (2006) 496–502.
[40] J. Kašpar, P. Fornasiero, N. Hickey, Catal. Today 77 (2003) 419–449.
[41] R. Monte, J. Kašpar, Top. Catal. 28 (2004) 47–57.
[42] H.J. Beie, A. Gnörich, Sens. Actuators B 4 (1991) 393–399.
[43] N. Izu, W. Shin, I. Matsubara, N. Murayama, Sens. Actuators B 94 (2003) 222–227.
[44] X. Yin, L. Hong, Z.-L. Liu, J. Membr. Sci. 268 (2006) 2–12.
[45] X. Qi, Y.S. Lin, C.T. Holt, S.L. Swartz, J. Mater. Sci. 38 (2003) 1073–1079.
[46] M.A. Scibioh, S.-K. Kim, E.A. Cho, T.-H. Lim, S.-A. Hong, H.Y. Ha, Appl. Catal. B 84 (2008) 773–782.
[47] L. Adijanto, A. Sampath, A.S. Yu, M. Cargnello, P. Fornasiero, R.J. Gorte, J.M. Vohs, ACS Catal. 3 (2013) 1801–1809.
[48] M.A. Barteau, Chem. Rev. 96 (1996) 1413–1430.
[49] K.M. Bratlie, H. Lee, K. Komvopoulos, P. Yang, G.A. Somorjai, Nano Lett. 7 (2007) 3097–3101.
[50] K. An, G.A. Somorjai, ChemCatChem 4 (2012) 1512–1524.
[51] M. Valden, X. Lai, D.W. Goodman, Science 281 (1998) 1647–1650.
[52] J.M. Vohs, Chem. Rev. 113 (2012) 4136–4163.
[53] M. Frank, M. Baumer, Phys. Chem. Chem. Phys. 2 (2000) 3723–3737.
[54] R. Schlögl, S.B. Abd Hamid, Angew. Chem. Int. Ed. 43 (2004) 1628–1637.
[55] P.M. Albrecht, D.R. Mullins, Langmuir 29 (2013) 4559–4567.
[56] S.D. Senanayake, D.R. Mullins, J. Phys. Chem. C 112 (2008) 9744–9752.
[57] S.D. Senanayake, W.O. Gordon, S.H. Overbury, D.R. Mullins, J. Phys. Chem. C 113 (2009) 6208–6214.
[58] T.L. Chen, D.R. Mullins, J. Phys. Chem. C 115 (2011) 13725–13733.
[59] F.C. Calaza, Y. Xu, D.R. Mullins, S.H. Overbury, JACS 134 (2012) 18034–18045.
[60] D.R. Mullins, P.M. Albrecht, J. Phys. Chem. C 117 (2013) 14692–14700.
[61] F. Calaza, T.-L. Chen, D. Mullins, S. Overbury, Top. Catal. 54 (2011) 56–69.
[62] D.R. Mullins, S.D. Senanayake, T.L. Chen, J. Phys. Chem. C 114 (2010) 17112–17119.
[63] J. Stubenrauch, E. Brosha, J.M. Vohs, Catal. Today 28 (1996) 431–441.
[64] R. Yang, L. Guo, J. Mater. Sci. 40 (2005) 1305–1307.
[65] T. Taniguchi, K.-i. Katsumata, S. Omata, K. Okada, N. Matsushita, Cryst. Growth Des. 11 (2011) 3754–3760.
[66] L. Yan, R. Yu, J. Chen, X. Xing, Cryst. Growth Des. 8 (2008) 1474–1477.
[67] H.-X. Mai, L.-D. Sun, Y.-W. Zhang, R. Si, W. Feng, H.-P. Zhang, H.-C. Liu, C.-H. Yan, J. Phys. Chem. B 109 (2005) 24380–24385.
[68] Z.L. Wang, X. Feng, J. Phys. Chem. B 107 (2003) 13563–13566.
[69] M.N. Revoy, R.W.J. Scott, A.P. Grosvenor, J. Phys. Chem. C 117 (2013) 10095–10105.
[70] Q. Yuan, H.-H. Duan, L.-L. Li, L.-D. Sun, Y.-W. Zhang, C.-H. Yan, J. Colloid Interface Sci. 335 (2009) 151–167.
[71] D. Zhang, X. Du, L. Shi, R. Gao, Dalton Trans. 41 (2012) 14455–14475.
[72] K.-S. Lin, S. Chowdhury, Int. J. Mol. Sci. 11 (2010) 3226–3251.
[73] A.I. Frenkel, C.W. Hills, R.G. Nuzzo, J. Phys. Chem. B 105 (2001) 12689–12703.
[74] M.G. Bawendi, M.L. Steigerwald, L.E. Brus, Annu. Rev. Phys. Chem. 41 (1990) 477–496.
[75] H. Weller, Adv. Mater. 5 (1993) 88–95.
[76] V. Maheshwari, J. Kane, R.F. Saraf, Adv. Mater. 20 (2008) 284–287.
[77] G. Markovich, C.P. Collier, S.E. Henrichs, F. Remacle, R.D. Levine, J.R. Heath, Acc. Chem. Res. 32 (1999) 415–423.
[78] T. Hyeon, Chem. Commun. (2003) 927–934.
[79] A. Laachir, V. Perrichon, A. Badri, J. Lamotte, E. Catherine, J.C. Lavalley, J. El Fallah, L. Hilaire, F. Le Normand, E. Quemere, G.N. Sauvion, O. Touret, J. Chem. Soc. Faraday Trans. 87 (1991) 1601–1609.

[80] X.-D. Zhou, W. Huebner, Appl. Phys. Lett. 79 (2001) 3512–3514.
[81] I.I. Soykal, H. Sohn, D. Singh, J.T. Miller, U.S. Ozkan, ACS Catal. 4 (2014) 585–592.
[82] A. Igarashi, N. Ichikawa, S. Sato, R. Takahashi, T. Sodesawa, Appl. Catal. A 300 (2006) 50–57.
[83] X.-D. Zhou, W. Huebner, H.U. Anderson, Appl. Phys. Lett. 80 (2002) 3814–3816.
[84] Y.M. Chiang, E.B. Lavik, I. Kosacki, H.L. Tuller, J.Y. Ying, Appl. Phys. Lett. 69 (1996) 185–187.
[85] E.B. Lavik, I. Kosacki, H.L. Tuller, Y.M. Chiang, J.Y. Ying, J. Electroceram 1 (1997) 7–14.
[86] J.E. Spanier, R.D. Robinson, F. Zhang, S.-W. Chan, I.P. Herman, Phys. Rev. B: Condens. Matter 64 (2001) 245407.
[87] Z. Wang, S.K. Saxena, V. Pischedda, H.P. Liermann, C.S. Zha, Phys. Rev. B: Condens. Matter 64 (2001) 012102.
[88] S. Tsunekawa, T. Fukuda, A. Kasuya, J. Appl. Phys. 87 (2000) 1318–1321.
[89] Z. Wu, M. Li, D.R. Mullins, S.H. Overbury, ACS Catal. 2 (2012) 2224–2234.
[90] M. Nolan, S. Grigoleit, D.C. Sayle, S.C. Parker, G.W. Watson, Surf. Sci. 576 (2005) 217–229.
[91] Y. Namai, K-i. Fukui, Y. Iwasawa, J. Phys. Chem. B 107 (2003) 11666–11673.
[92] D.R. Mullins, P.V. Radulovic, S.H. Overbury, Surf. Sci. 429 (1999) 186–198.
[93] T.X.T. Sayle, S.C. Parker, C.R.A. Catlow, Surf. Sci. 316 (1994) 329–336.
[94] J. Conesa, Surf. Sci. 339 (1995) 337–352.
[95] H. Nörenberg, J.H. Harding, Surf. Sci. 477 (2001) 17–24.
[96] M. Baudin, M. Wójcik, K. Hermansson, Surf. Sci. 468 (2000) 51–61.
[97] I. Florea, C. Feral-Martin, J. Majimel, D. Ihiawakrim, C. Hirlimann, O. Ersen, Cryst. Growth Des. 13 (2013) 1110–1121.
[98] K. Zhou, X. Wang, X. Sun, Q. Peng, Y. Li, J. Catal. 229 (2005) 206–212.
[99] Y. Guan, D.A.J. M. Ligthart, Ö. Pirgon-Galin, J.Z. Pieterse, R. Santen, E.M. Hensen, Top. Catal. 54 (2011) 424–438.
[100] U.M. Bhatta, I.M. Ross, T.X.T. Sayle, D.C. Sayle, S.C. Parker, D. Reid, S. Seal, A. Kumar, G. Möbus, ACS Nano 6 (2011) 421–430.
[101] Y. Lin, Z. Wu, J. Wen, K.R. Poeppelmeier, L.D. Marks, Nano Lett. 14 (2013) 191–196.
[102] J. Maier, Angew. Chem. Int. Ed. 32 (1993) 313–335.
[103] J.S. Spendelow, Q. Xu, J.D. Goodpaster, P.J.A. Kenis, A. Wieckowski, J. Electrochem. Soc. 154 (2007) F238–F242.
[104] G.A. Somorjai, Surf. Sci. 299–300 (1994) 849–866.
[105] N.J. Lawrence, J.R. Brewer, L. Wang, T.-S. Wu, J. Wells-Kingsbury, M.M. Ihrig, G. Wang, Y.-L. Soo, W.-N. Mei, C.L. Cheung, Nano Lett. 11 (2011) 2666–2671.
[106] Z. Wu, M. Li, J. Howe, H.M. Meyer, S.H. Overbury, Langmuir 26 (2010) 16595–16606.
[107] E. Mamontov, T. Egami, J. Phys. Chem. Solids 61 (2000) 1345–1356.
[108] M. Nolan, S.C. Parker, G.W. Watson, Surf. Sci. 595 (2005) 223–232.
[109] H. Nörenberg, G.A.D. Briggs, Surf. Sci. 424 (1999) L352–L355.
[110] F. Esch, S. Fabris, L. Zhou, T. Montini, C. Africh, P. Fornasiero, G. Comelli, R. Rosei, Science 309 (2005) 752–755.
[111] Z. Wu, M. Li, S.H. Overbury, J. Catal. 285 (2012) 61–73.
[112] D. Wang, Y. Kang, V. Doan-Nguyen, J. Chen, R. Küngas, N.L. Wieder, K. Bakhmutsky, R.J. Gorte, C.B. Murray, Angew. Chem. Int. Ed. 50 (2011) 4378–4381.
[113] M. Huang, S. Fabris, J. Phys. Chem. C 112 (2008) 8643–8648.
[114] C. Pan, D. Zhang, L. Shi, J. Fang, Eur. J. Inorg. Chem. 2008 (2008) 2429–2436.
[115] M. Li, Z. Wu, S.H. Overbury, J. Catal. 306 (2013) 164–176.
[116] G. Zhou, B. Gui, H. Xie, F. Yang, Y. Chen, S. Chen, X. Zheng, J. Ind. Eng. Chem. 20 (2014) 160–165.
[117] T.L. Chen, D.R. Mullins, J. Phys. Chem. C 115 (2011) 3385–3392.
[118] A.K.P. Mann, Z. Wu, F.C. Calaza, S.H. Overbury, ACS Catal. 4 (2014) 2437–2448.
[119] J. Zhou, D.R. Mullins, Surf. Sci. 600 (2006) 1540–1546.
[120] Q. Dai, H. Huang, Y. Zhu, W. Deng, S. Bai, X. Wang, G. Lu, Appl. Catal. B 117–118 (2012) 360–368.
[121] L. Torrente-Murciano, A. Gilbank, B. Puertolas, T. Garcia, B. Solsona, D. Chadwick, Appl. Catal. B 132–133 (2013) 116–122.
[122] J. Lv, Y. Shen, L. Peng, X. Guo, W. Ding, Chem. Commun. 46 (2010) 5909–5911.

[123] W.-Q. Han, W. Wen, J.C. Hanson, X. Teng, N. Marinkovic, J.A. Rodriguez, J. Phys. Chem. C 113 (2009) 21949–21955.
[124] R. Si, M. Flytzani-Stephanopoulos, Angew. Chem. Int. Ed. 47 (2008) 2884–2887.
[125] Z. Wu, V. Schwartz, M. Li, A.J. Rondinone, S.H. Overbury, J. Phys. Chem. Lett. 3 (2012) 1517–1522.
[126] I.I. Soykal, B. Bayram, H. Sohn, P. Gawade, J.T. Miller, U.S. Ozkan, Appl. Catal. A 449 (2012) 47–58.
[127] Y. Zhang, F. Hou, Y. Tan, Chem. Commun. 48 (2012) 2391–2393.
[128] L. Liu, Z. Yao, Y. Deng, F. Gao, B. Liu, L. Dong, ChemCatChem 3 (2011) 978–989.
[129] G. Yi, Z. Xu, G. Guo, K.-i. Tanaka, Y. Yuan, Chem. Phys. Lett. 479 (2009) 128–132.

CHAPTER 4

Gold Clusters and Nanoparticles Stabilized by Nanoshaped Ceria in Catalysis

Yejun Guan[1,*], Weiyu Song[2], Emiel J.M. Hensen[2]
[1]Shanghai Key Laboratory of Green Chemistry and Chemical Processes, Department of Chemistry, East China Normal University, Shanghai, China; [2]Laboratory of Inorganic Materials Chemistry, Schuit Institute of Catalysis, Department of Chemical Engineering and Chemistry, Eindhoven University of Technology, Eindhoven, The Netherlands
*Corresponding author: E-mail: yjguan@chem.ecnu.edu.cn

1. INTRODUCTION

This chapter will focus on the effect of ceria morphology on the structure and catalytic activity of ceria-supported gold catalysts in CO oxidation and water–gas shift (WGS) reactions. The discovery by Haruta and co-workers that very small gold particles supported on transition metal oxides such as titania are active for CO oxidation even below room temperature has led to an impressive research effort [1]. A comprehensive book on gold catalysis is available [2], which seems already somewhat outdated due to rapid developments in the field with new insights about working mechanism of gold catalysts, novel approaches to synthesize more active systems and characterize supported gold at the atomic scale, and a rapidly expanding scope of catalytic reactions catalyzed by gold. Significant scientific debate is still unfolding on the role of gold, its active state, and the importance of the support. Although it is generally assumed that nanosizing is pivotal to explain gold's unique catalytic performance, it is becoming increasingly clear that the interface between gold and the support is of key importance to many catalytic reactions, either by stabilizing the active gold species or by participating in the reaction directly.

One of the major aspects of metal–support interactions is the surface structure of the support. Ceria is here a clear case in point, not only because well-defined ceria nanoshapes can be prepared but also because it constitutes a key component in several industrial catalysts [3]. Ceria's widespread applicability mainly originates from its outstanding oxygen storage capacity. This is causally related to the ease in forming and healing oxygen vacancies at the surface of ceria, which involves the bond-breaking and -forming process between surface O and Ce atom. This property is associated to particle size of the ceria support but, more importantly, also to the surface termination structure of the ceria support as discussed in Chapters 5, 6, and 8. Gold supported on ceria shows promising catalytic performance in low-temperature CO oxidation [4], sometimes as high as Au/Fe_2O_3 and Au/TiO_2 catalysts. Gold on nanosized ceria prepared by deposition–precipitation and

Catalysis by Materials with Well-Defined Structures
http://dx.doi.org/10.1016/B978-0-12-801217-8.00004-9

dried at 373K is ~100 times more active than conventional Au/CeO_2 catalysts [4], pointing to the synergistic effect of gold and ceria surface redox property. Besides CO oxidation, Au/CeO_2 is one of the best catalysts for low-temperature WGS reaction, as first reported by Fu et al. [5]. Of determining importance appears to be the size of the metal nanoparticles or clusters, the ability of the ceria to provide oxygen atoms during catalytic reactions, and the influence of the oxidation state of the gold atoms at the interface perimeter. To shed deep insight into the support effect, researchers investigated model systems, i.e., by deposition of gold on ceria nanoshapes with well-defined surface structure prepared by a hydrothermal synthesis technique [6]. In this way, ceria with different morphologies can be prepared, i.e., cubes, rods, and spheres, in such quantities that systematic catalysis studies can be undertaken. These materials have unique structures with different surface planes presented, namely {111}, {110}, and {100}, which show distinctly different redox properties. Defects, which come either as cerium or oxygen vacancies, are essential in stabilizing isolated Au atoms, clusters, or nanoparticles. The electronic properties of Au species are determined by the reduction degree of the ceria surface, which is associated with the exact termination of the surface plane and the involved reaction atmosphere. Many techniques including HRTEM, XPS, FTIR, and EXAFS have been used to elucidate the state of gold, either in the conventional ex situ manner or under in situ conditions. Although strong morphology-dependent reactivity in CO oxidation and the WGS reaction has been recognized, accurate structure–performance relations for gold on ceria with different morphologies have not been formulated yet, among others due to the complexity to unequivocally determine the state of active gold. Therefore, theoretical studies based on first principle methods provide a manner to validate certain hypotheses made by experimentalists [7]. In this chapter, we have summarized the experimental and theoretical progress achieved in this field in the past 15 years. We first reviewed the experimental methods to synthesize ceria nanoparticles with different morphology. We then introduce various characterization techniques to study the support and its interaction with gold and discuss experimental observations focusing mainly on CO oxidation and the WGS reaction. In each section, we also discussed the corresponding theoretical efforts dedicated to understanding these observations.

2. GOLD SPECIES ON CERIA NANOSTRUCTURES

2.1 Interaction between Gold and Ceria Surface

Despite intensive research on ceria-supported gold catalysts, the surprising catalytic activity of Au/CeO_2 remains a strongly debated topic. Some important aspects that have been widely discussed involve the electronic state of Au species (positively charged/cationic, neutral, or negatively charged/anionic), the nuclearity of Au (isolated, clustered atoms, or nanoparticles), the structure of ceria (polycrystalline vs. preferentially exposing certain facets), and the Au/ceria interface [2]. All these aspects result in the current view that ceria not only acts as the support for gold but is more often than not involved in the catalytic reactions. As a support, nanostructured ceria contains defect sites, mainly oxygen vacancies. On these sites, the gold species may be strongly attached and show different electronic properties from gold nanoparticles [8–10]. Much effort including experimental and theoretical methods has been made to clarify the interaction between gold and the ceria surface [11–16]. Figure 1 shows a model structure of a Au cluster on a ceria surface, supported by EXAFS results [16]. For a fresh catalyst, the Au–O distance has been found in the range of 1.96–2.21 Å by several groups. Such differences may be the result of the location of Au on different terminating facets of ceria. For instance, DFT calculations predict: (1) $R_{Au\text{–}O}=2.05$ Å for Au on top of CeO_2 {111}; (2) $R_{Au\text{–}O}=2.33$ Å for Au in a CeO_2 {111} cerium vacancy; (3) $R_{Au\text{–}O}=2.11$ Å for Au in a CeO_2 {110} cerium vacancy; and (4) after a first Au atom adsorbed in a CeO_2 {111} oxygen vacancy, the next adsorbed Au lies at $R_{Au\text{–}O}=2.12$ Å from the nearest oxygen. It is in principle also useful to elucidate Au–Ce and Au–Au distances to understand the structure of these catalysts. In general, this is not straightforward due to the averaging nature of the EXAFS techniques, although in isolated cases it has been possible to see the Au–Ce backscatterer for isolated gold cations on the ceria surface [14]. It has been argued that the cationic Au layer strongly attached to the ceria {111} or {110} planes acts a glue of the metallic Au top layers and the underlying ceria, which is not observed for {100} planes. Guzman et al. [11] showed that the presence of cationic Au species is also related to the nanosized nature of the support (3.4 nm), since no such Au_n–O–Ce bond was observed on conventional ceria with very large particle size (>17 nm).

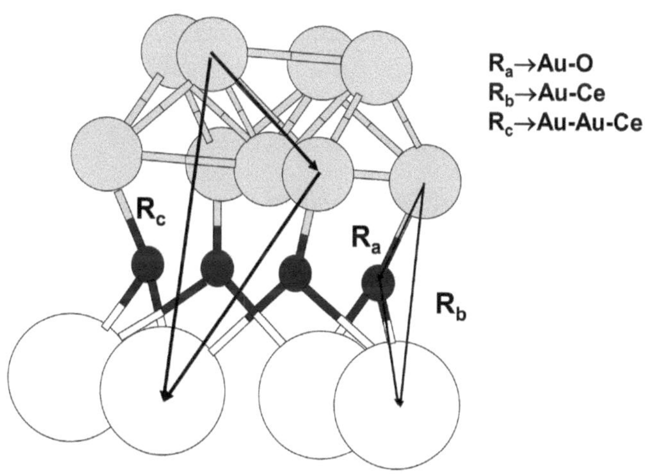

Figure 1 Model of the gold–ceria interface: the distances Au–O, Au–Ce, and Au–Ce relative to the three-body Au–Au–Ce term are shown. The drawing does not take into account surface relaxation. Gold atoms in yellow (light gray in print version), oxygen atoms in red (dark gray in print version), and cerium atoms in white. *Adapted with permission from Ref. [16]. Copyright (2012) American Chemical Society.*

Baron et al. [17] compared the interaction of gold with CeO_2 {111} thin films and CeO_x nanoparticles. They found that Au nanoparticles mainly formed at the step edges of ceria presumably through the decoration of point defects upon deposition. The formation of cationic $Au^{\delta+}$ on CeO_2 {111} films was kinetically limited and was attributed to the interaction of the gold adatoms with defects (oxygen vacancies) at low temperatures (~100 K), which easily transformed into metallic at 300 K. In contrast, "cationic" Au species were formed in abundance on nano-CeO_x and exhibited enhanced thermal stability. Majimel and co-workers [18] investigated the behavior of supported gold nanostructures with different sizes on the {111} surface of CeO_2 nanoparticles under the electron beam, where the high intensity of the electron beam induced formation of oxygen vacancies on the ceria surface. On a reduced ceria {111} surface, small-sized gold nanoparticles (<5 nm) shrunk layer by layer and migrated to the ceria surface, in which case a strong Au_n–O–Ce interaction would be expected.

These surface oxygen vacancies have also been studied by Raman spectroscopy in combination with HRTEM by Vindigni et al. [19], who proposed three families of oxygen vacancies. The first family locates at the external surfaces and readily reacts with O_2 with the formation of O_2^{2-} species. The second family locates in the highly nonstoichiometric surface layer formed by a Au-catalyzed reduction in H_2 at unusually low temperature. The third type of oxygen vacancy is located inside the ceria.

Introducing a second component (La_2O_3 or Gd_2O_3) into ceria may increase the density of oxygen vacancies, therefore leading to stronger interaction between gold and the support as for instance observed for Au/La_2O_3(Gd_2O_3)-CeO_2 catalysts prepared via deposition–precipitation method [5,20]. Upon cyanide leaching, most metallic gold (up to 90% of the total gold content) can be removed with very fine clusters or atomically dispersed Au remaining. These cationic Au were proposed to be formed in the heating process via a lattice exchange of Ce^{4+} by atomic Au^{3+}. The amount of Au^{3+} might be correlated to the amount of surface oxygen defect sites of ceria.

2.2 Morphology-Dependent Synthesis of Ceria Materials

In the past decades, remarkable progress has been achieved in the preparation of ceria nanostructures with different morphologies [21–23]. The general synthesis strategy is oriented growth directed by crystallographic structure of cerium-based materials with or without the aid of an appropriate surface modifier. One of the common methods is hydrothermal/solvothermal treatment [24], which involves the use of an autoclave under controlled temperature and autogeneous pressure of a solution. Typically, such solutions used in hydrothermal preparation of cerias contain aqueous NaOH, urea, or NH_3. Single crystalline and uniform ceria nanopolyhedra, nanorods, and nanocubes were prepared at gram scale by this method by variation of the temperature and the NaOH concentrations. As shown in Figure 2, ceria rods have typically widths of 6–10 nm with a broad length distribution between 30 and 200 nm [15]. The ceria nanocubes exhibit a broader

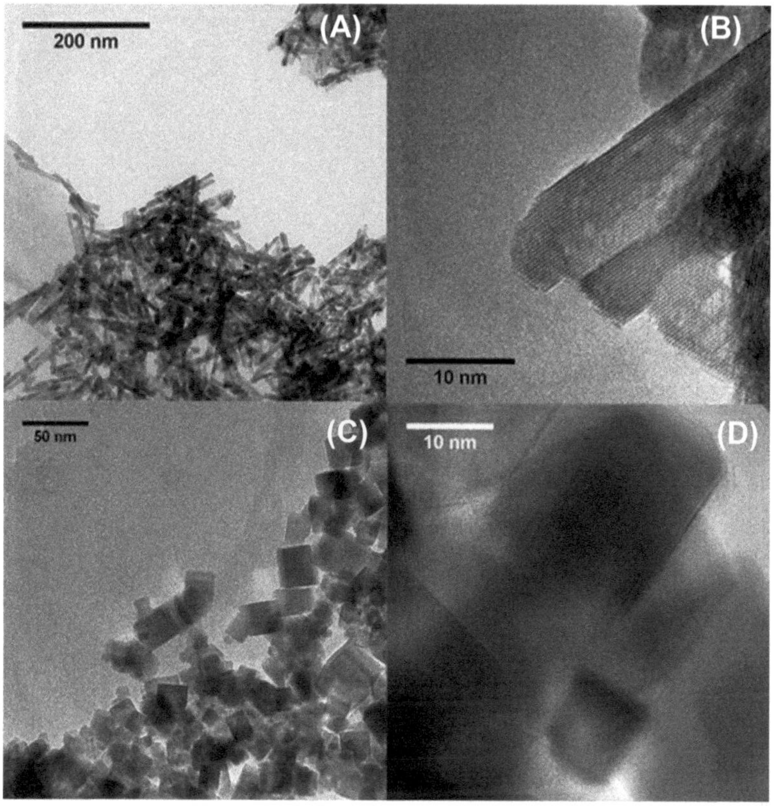

Figure 2 Transmission electron micrographs of (A, B) CeO_2(rod) and (C, D) CeO_2(cube). *Reprinted from Ref. [15].*

size distribution from 20 to 60 nm. The exposed surfaces for CeO_2 nanorods have been assigned to either {110} planes [24–26] or {111} planes [27,28] depending on the synthesis conditions. The *d*-spacing of the nanorod shape of 0.31 nm suggested that the dominant facet is {111} for the samples we have synthesized, although the peer review process forced the statement that the exposed surface is the {110} plane. A recent high-resolution TEM study (Figure 3) in collaboration with Datye [29] confirms that these rods are mainly enclosed by {111} surfaces.

The surface of CeO_2(rod) exhibits a defective structure, which is thought to be due to the presence of oxygen vacancies in the surface. The {100} plane with a *d*-spacing of 0.27 nm dominates the surface of CeO_2(cube). The crystallite sizes determined by application of Scherrer's equation are 8 and 35 nm for CeO_2(rod) and CeO_2(cube), respectively. The reason for the lower crystallite size of the CeO_2(rod) is the lower synthesis temperature than that employed for CeO_2(cube). This difference is consistent with the BET surface areas of $24\,m^2\,g^{-1}$ for CeO_2(cube) and $110\,m^2\,g^{-1}$ for CeO_2(rod) [15]. Mai

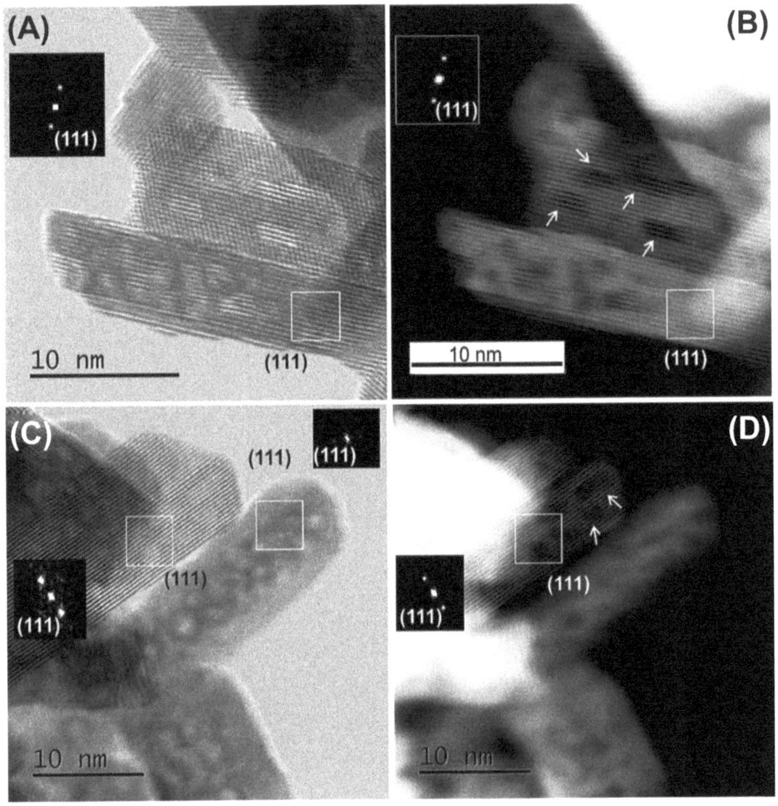

Figure 3 STEM images of CeO_2 rods. (A, C) BF images, (B, D) HAADF images. The inset in the figures is the FFT that allows indexing of the lattice planes. The rods expose (111) surfaces and have surface steps along the length. Areas of light contrast can be seen in the BF image and these same areas look dark in the HAADF image. The HAADF images confirm that these low contrast features are voids in the CeO_2 rods that are bounded by (111) surfaces. The contrast variation in the HAADF image suggests a rectangular profile in the cross section. *Reprinted from Ref. [29] with permission of John Wiley & Sons, Inc. (2013).*

et al. [24] considered that three key parameters are responsible for the shape evolution: the formation of hexagonal $Ce(OH)_3$ intermediate species, $Ce(OH)_3$ transformation into CeO_2 at elevated temperature, and the base concentration. The whole hydrothermal process is assumed to be ruled by the well-known dissolution/recrystallization mechanism. Anisotropic $Ce(OH)_3$ nuclei form immediately when the Ce^{3+} ions are mixed with NaOH solution. When hydrothermally treated in low OH^- concentrations ($C_{NaOH} < 1\,mol\,L^{-1}$) at a temperature of 373 K, the dissolution/recrystallization rate is slow, resulting in high driving force for the anisotropic growth of the $Ce(OH)_3$ nuclei. Under this condition, CeO_2 nanopolyhedra with enclosing {111} and {100} facets were formed. When the OH^- concentration was high enough ($C_{NaOH} > 1\,mol\,L^{-1}$), the dissolution/recrystallization rate was considerably promoted so as to drive the $Ce(OH)_3$ nuclei to grow anisotropically. Accordingly, pure CeO_2 nanorods were obtained at high base concentrations ($C_{NaOH} > 6\,mol\,L^{-1}$). When the hydrothermal temperature was increased to 453 K, $Ce(OH)_3$ became unstable and was readily oxidized into CeO_2. As a result, CeO_2 nanocubes with enclosing {100} facets were produced at high base concentration. Further studies suggested that the anions in the solution (like Cl^-, NO_3^-, SO_4^{2-}, and PO_4^{3-}) have a great influence on the morphology and properties of CeO_2 nanostructures. A common finding is also that the nanorods derived from a $Ce(OH)_3$ precursor appear porous or at least defective in nature in contrast to the dense ceria nanocubes. This difference may be tentatively attributed to the elimination of water from the hydroxide nanorod precursor, which does not occur when calcining the oxide nanocube precursor.

2.3 Gold Species Stabilized on Rod and Cube Ceria

When supported on ceria with different planes exposed, gold nanoparticles tend to exhibit distinctly different electronic properties. Si and Flytzani-Stephanopoulos [6] investigated the state of Au nanoparticles supported on nanoscale ceria with different morphologies (rod, cube, and polyhedron) by XPS. High-resolution XPS spectra show a distinct binding energy shift of the Au 4f peaks between the rod/polyhdedron and the cube samples. Cationic Au species (Au^+ and Au^{3+}) dominated on the rod or polyhedron samples, while metallic gold (Au^0) was predominant on the ceria nanocubes. The cationic Au species strongly bind to the surface and were resistant to the cyanide leaching. In contrast, weakly bonded Au^0 particles were fully removed from the nanocube surface. The weak bonding strength between Au^0 and nanocube can be explained by (1) its much larger crystalline size (25–50 nm) and (2) dominance of the {100} facets, which is much less defective than the other surfaces. The preparation of stable CeO_2 nanocubes with sizes in the range of 5–15 nm yet remains a great challenge.

To shed more light on the leaching ability of Au nanoparticles on CeO_2 catalysts, we have also prepared Au/CeO_2 catalysts with different morphologies and compared their performance as catalysts before and after cyanide leaching in different reactions [15]. Figure 4 displays representative transmission electron micrographs of gold nanoparticles supported by CeO_2(rod) and CeO_2(cube) at a metal loading of 2 wt%. The size of the gold particles is in the range of 2–8 nm. The average particle size is 4.6 ± 1.5 and 4.7 ± 1.5 nm for Au/CeO_2(rod) and Au/CeO_2(cube), respectively. The values are close to the particle size of 4.0 nm determined for a 2 wt% Au supported on polyhedral ceria [14]. Most gold nanoparticles have a pyramidal or polyhedron shape, containing metallic atoms in the top layers and oxidic atoms in the interface [18,30,31]. The top layers appeared to be enclosed by {111} and {100} facets [31]. It should be noted that the amount of oxidic Au atoms in the interface is likely related to the oxygen vacancies on the ceria surface according to hydrogen consumption data measured during TPR [6,15].

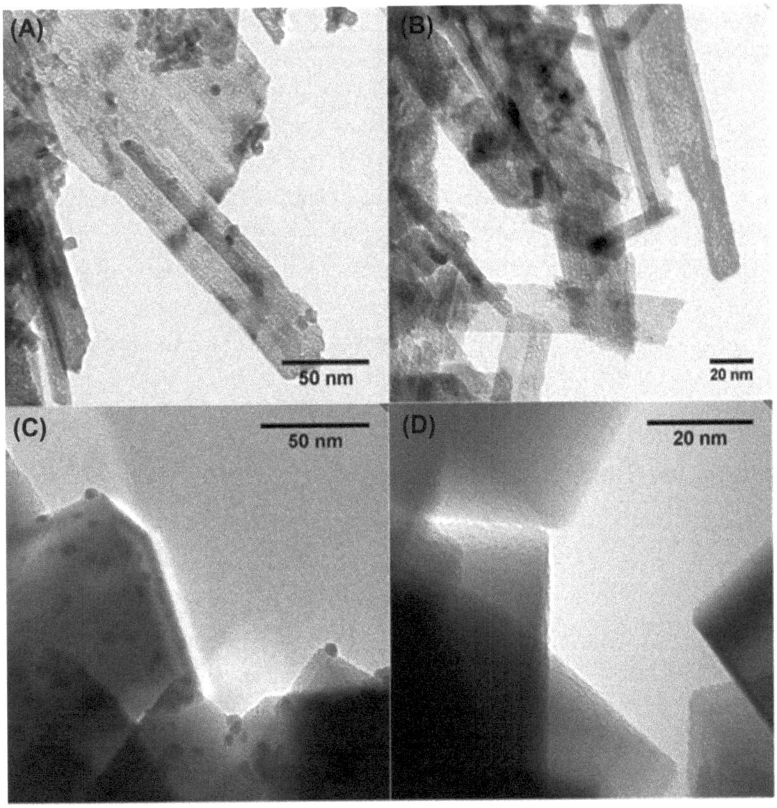

Figure 4 Transmission electron micrographs of (A) Au/CeO_2(rod), (B) Au/CeO_2(rod)-CN, (C) Au/CeO_2(cube), and (D) Au/CeO_2(cube)-CN. *Reprinted from Ref. [15].*

3. IN SITU CHARACTERIZATION OF GOLD SPECIES SUPPORTED ON CERIA ROD AND CUBE

The electronic properties of gold supported on ceria vary with the morphology of the support. FTIR using CO as a probe molecule is a powerful technique which can reveal the surface electronic property of Au as well as the interaction between Au and ceria surface. The most informative observations provided by IR spectroscopy are the frequency of the ν_{CO} bands and the stability of the species that they present [32]. Generally, the higher the ν_{CO}, the higher the oxidation state of the gold site [33–35]. Bands in the 2210–2130 cm^{-1} region are typically assigned to CO adsorbed to cationic Au, namely Au^{3+} and Au^{+}. Bands close to 2100 cm^{-1} are attributed to neutral Au nanoparticles, with a blue shift (to around 2110 cm^{-1}) taking place on positively charged nanoparticles and a red shift (to around 2030 cm^{-1}) on negatively charged ones, which critically depends on the support and catalyst pretreatment. Figure 5 compares the FTIR spectra after CO adsorption at 80 K for various ceria-supported gold catalysts [15]. The spectrum in the

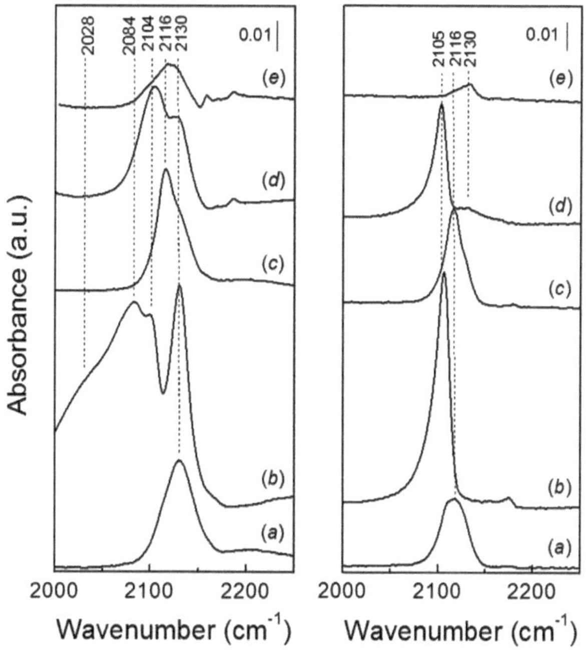

Figure 5 FTIR spectra of adsorbed CO at 80 K of (left) Au/CeO_2(rod) and (right) Au/CeO_2(cube): (A) fresh calcined catalyst, (B) reduction at 393 K, (C) reoxidation at 423 K, (D) reduction at 773 K, and (E) reoxidation at 313 K. *(Reprinted from [15].)*

carbonyl region for the fresh Au/CeO_2(rod) catalysts exhibits a broad band centered around 2130 cm^{-1} with shoulders at 2116 and 2207 cm^{-1}. The band at 2130 cm^{-1} is most likely assigned to CO adsorption to Ce^{3+} centers [36,37]. The band at 2116 cm^{-1} is due to adsorption of CO to metallic gold. CO stretching band in the range of 2106–2118 cm^{-1} relate to CO adsorption to corner sites. The band at 2207 cm^{-1} is likely due to CO adsorbed to Au^{3+}. In contrast, the corresponding IR spectrum of the calcined Au/CeO_2(cube) catalyst only showed a strong band at 2117 cm^{-1}, indicative of the presence of neutral gold nanoparticles. The bands around 2130 and 2207 cm^{-1} present in the spectra of Au/CeO_2(rod) are not observed, which suggests that no cationic gold species are present in this sample. The absence of the band at 2130 cm^{-1} also shows that the surface of ceria nanocubes does not contain Ce^{3+} species (or oxygen vacancies). The substantial difference in the IR spectra between Au/CeO_2(cube) and Au/CeO_2(rod) suggests the presence of very finely dispersed gold clusters in strong interaction with the rod structure.

Upon exposure of the reduced catalysts to O_2 at 313 K, the band at 2116 cm^{-1} appeared again at the expense of the one around 2105 cm^{-1} in Au/CeO_2(rod) and Au/CeO_2(cube). This effect can be understood by the presence of surface oxygen adatoms. Another observation is the disappearance of the bands below 2100 cm^{-1}. This could be due to the oxidation of the very fine clusters of gold on the ceria surface in Au/CeO_2(rod). To investigate the stability of the gold nanoparticles against sintering at high temperature, the sample was further reduced at 773 K. In the case of Au/CeO_2(rod) the bands at 2084 and 2103 cm^{-1} were recovered, but with a much lower intensity. This suggests that the gold clusters and particles have substantially sintered. Expectedly, the same shift of the band at 2117 cm^{-1} toward lower wavelength was observed after high-temperature reduction of Au/CeO_2(cube). Also in this case the intensity is much lower, suggesting that the nanoparticles have grown in size.

Figure 6 contains the FTIR spectra for the leached Au/CeO_2 catalysts. The spectrum of Au/CeO_2(rod)-CN after reduction at 393 K contains only a weak feature around 2130 cm^{-1}. After oxidation this band disappears. Reduction at 523 K results in the appearance of a dominant band around 2130 cm^{-1} due to surface Ce^{3+} states with a shoulder in the region 2105–2115 cm^{-1} due to reduced gold clusters. A weak feature around 2082 cm^{-1} is also observed. Reoxidation at 313 K results in the disappearance of all bands, indicating that the small gold particles were oxidized. Reduction at 773 K results in bands at 2103 and 2130 cm^{-1}. Clearly, these particles are still very small, because reoxidation at 313 K erodes all bands. The corresponding spectra of Au/CeO_2(cube)-CN show no bands for the fresh catalyst and after reduction at 393 K. This agrees with the absence of

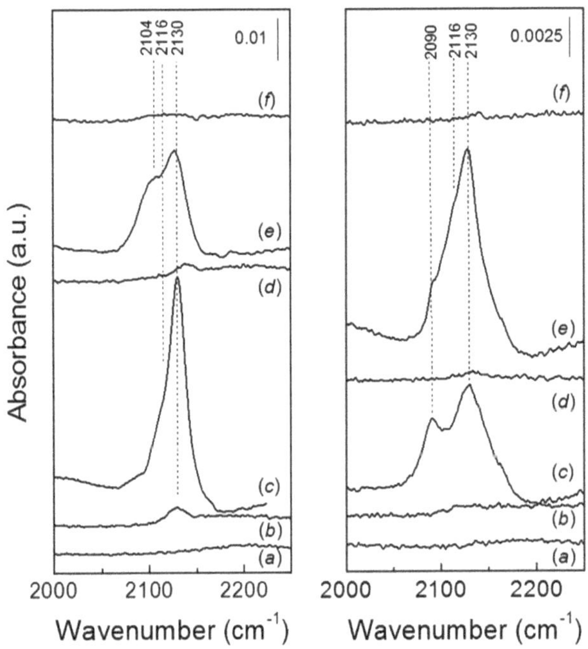

Figure 6 FTIR spectra of adsorbed CO at 80 K of (left) Au/CeO_2(rod)-CN and (right) Au/CeO_2(cube)-CN: (A) fresh calcined catalyst, (B) reduction at 393 K, (C) reduction at 423 K, (D) reoxidation at 423 K, (E) reduction at 773 K, and (F) reoxidation at 313 K. *Reprinted from Ref. [15].*

a TPR feature below 513 K for this catalyst. After reduction at 523 K two prominent bands are observed with a much lower intensity than the other catalysts. The one at 2128 cm^{-1} is due to Ce^{3+} and the one at 2090 cm^{-1} should be due to very finely dispersed gold clusters. Similar to Au/CeO_2(rod)-CN, these bands erode upon reoxidation, even after high-temperature reduction. The observation of a CO IR band at 2090 cm^{-1} at this very low Au loading (<0.01 wt%) points to a very high gold dispersion.

X-ray absorption spectroscopy is a useful technique to obtain detailed information about the oxidation state and structure of small supported gold particles that are impossible to detect using transmission electron microscopy [38,39], for instance, the isolated gold cations remained in close interaction with ceria surface on a Au/CeO_2 catalyst after NaCN leaching process. Au L_{III} near-edge spectra at room temperature, drying and reduction at various temperatures for Au/CeO_2(rod), Au/CeO_2(rod)-CN, and Au/CeO_2(cube) are collected in Figure 7. The dried Au/CeO_2(rod) catalysts contain a whiteline at ~4 eV above the edge, which is typical for Au^{3+} compounds. This feature is more intense for the leached catalyst, indicating that a larger proportion of the gold atoms is in the cationic form. The whiteline is nearly absent for Au/CeO_2(cube). Quantification of the amount of cationic gold was carried out by fitting the near-edge spectra by linear combinations of the XANES spectra of a gold foil and that of Au/CeO_2(sphere)-CN [14]. The latter catalyst exclusively contained cationic gold as was evident from the EXAFS data analysis [14]. The fraction of cationic gold in these materials as a function

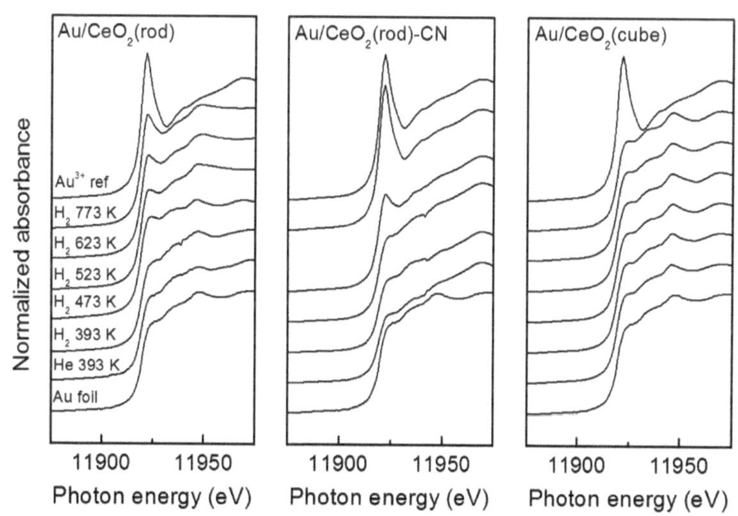

Figure 7 Near-edge spectra at the Au L_{III} edge of (left) Au/CeO_2(rod), (middle) Au/CeO_2(rod)-CN, and (right) Au/CeO_2(cube). Top and bottom spectra are those of fresh Au/CeO_2(rod)-CN and a Au foil, respectively. The remaining spectra were obtained (from bottom to top) from heating in He at 393 K, reduction at 393 K, reduction at 473 K, reduction at 523 K, reduction at 623 K, and reduction at 773 K. *Reprinted from Ref. [15].*

of the pretreatment conditions was calculated. Au/CeO_2(rod)-CN only contains cationic gold after preparation. Reduction at 393 K in H_2 leads to a substantial decrease of the fraction of cationic gold. After reduction at 473 K only 9% cationic gold is left. To completely reduce the gold a temperature of 773 K is required. The fraction of cationic gold in Au/CeO_2(rod) is initially much lower and then decreases similarly to Au/CeO_2(rod)-CN to very low levels at 773 K. The fresh Au/CeO_2(cube) catalyst contains nearly 90% reduced gold atoms. This amount is nearly independent of the reduction temperature. Similar structures and changes have been observed for Au/CeO_2(sphere).

Table 1 shows the fit parameters of the Fourier transformed k^3-weighted EXAFS functions for Au/CeO_2(rod), Au/CeO_2(rod)-CN, and Au/CeO_2(cube). The FT functions of the EXAFS functions and the fits for the Au/CeO_2(rod) catalysts are shown in Figure 8. The FT EXAFS spectrum of dried Au/CeO_2(rod) contains a Au–Au shell at a coordination distance (R) of 2.84 Å with a coordination number (CN) of 7.3. It also contains a small contribution of an oxygen backscatterer. These data point to the presence of metallic, yet not fully reduced gold nanoparticles with a particle size of about 3–4 nm. Reduction at 523 K results in the disappearance of the Au–O shell and a small increase of the Au–Au shell. Further reduction at 773 K induces a slightly higher CN and a longer bond distance. The spectrum of dried Au/CeO_2(rod)-CN is completely different and can only be fitted by the inclusion of a Ce backscatterer next to a Au–O shell. The first coordination shell contains ~3 oxygen atoms at $R = 1.99$ Å. The features at larger distance likely suggest a Au–Ce shell structure at $R = 3.27$ Å. The structural parameters point to the presence of gold substituting for a lattice cerium cation in the CeO_2\{111\} surface [10]. Reduction at 523 K results in a complete change of the structure around the gold atoms. The fit parameters show a Au–Au contribution at $R = 2.74$ Å with CN = 4.0 and a small contribution of oxygen at $R = 1.95$ Å. The Ce shell has disappeared. In agreement with the XANES results, this implies that the gold cations were reduced and have transformed into very small metallic clusters. Upon reduction of the catalyst at 773 K, CN and R for the Au–Au shell increase to 7.2 and 2.78 Å, respectively. These changes point to sintering of the small clusters to larger particles. The decreasing contribution of oxygen neighbors may be attributed to the decreasing interaction with the support or more complete reduction. The EXAFS fit parameters of Au/CeO_2(cube) reduced at 523 and 773 K evidence the presence of well-reduced gold nanoparticles of around 4 nm.

Upon cyanide leaching of these materials, the gold loading has decreased substantially [15]. The leached Au/CeO_2(rod) catalyst contains 0.38 wt% Au. The gold loading in Au/CeO_2(cube)-CN is lower than the detection limit of 0.01 wt%. The electron micrographs of both leached catalysts do not show gold nanoparticles anymore, indicating that leaching is effective in removing the large nanoparticles in the parent materials. In order to obtain more insight into nature of the interaction of gold with the ceria surfaces, the leaching procedure was carried out for a set of ceria nanorod- and nanocube-supported catalysts with varying gold loading. The amount of gold retained by CeO_2(rod) upon leaching does not strongly depend on the gold loading of the parent

Table 1 Fit parameters of k^3-weighted EXAFS spectra at the Au L_{III} edge of Au/CeO_2 catalysts after different pretreatments

Sample	Treatment	EXAFS fit parameters				
		Shell	*R* (Å)	CN	$\Delta\sigma^2$ (Å^2)	E_0 (eV)
Au/CeO_2(rod)	He, 393 K	O	1.96	0.7	0.005	−6.2
		Au	2.82	7.3	0.012	
	H_2, 523 K	Au	2.82	7.6	0.010	−7.5
	H_2, 773 K	Au	2.84	8.0	0.011	−7.1
Au/ CeO_2(rod)-CN	He, 393 K	O	1.99	3.4	0.007	−8.9
		Ce	3.27	3.8	0.012	
	H_2, 523 K	O	1.95	0.2	0.001	−1.6
		Au	2.74	4.0	0.013	
	H_2, 773 K	O	2.06	0.3	0.006	−5.0
		Au	2.78	7.2	0.015	
Au/ CeO_2(cube)	H_2, 523 K	Au	2.86	8.7	0.009	−8.6
	H_2, 773 K	Au	2.86	8.5	0.010	−7.9

$\Delta k = 2.5–10.4\,Å^{-1}$; estimated error in $R \pm 0.02\,Å$, $N \pm 20\%$, $\Delta\sigma^2 \pm 10\%$; normalized residual of fits is between 30% and 36%.
Reprinted from Ref. [15].

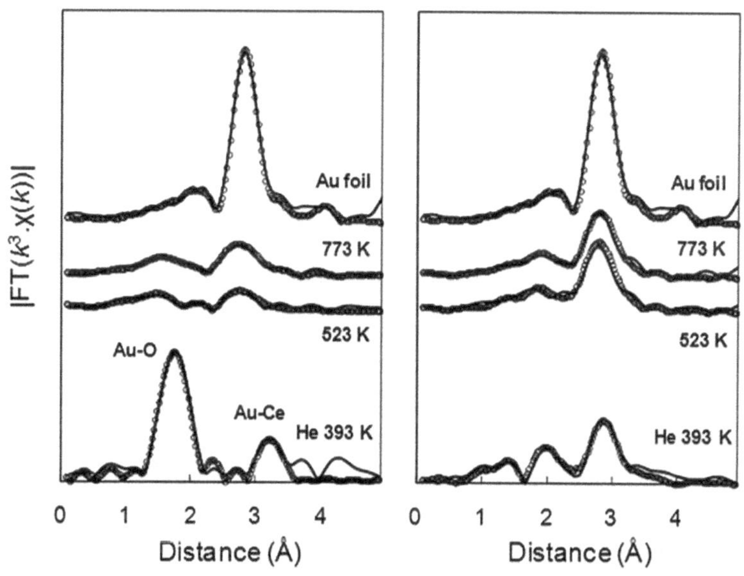

Figure 8 Experimental (solid line) and fitted (dotted points) FT EXAFS functions of (left) Au/CeO_2(rod)-CN and (right) Au/CeO_2(rod) after drying at 393 K and reduction at increasing temperatures. The corresponding spectrum of a Au foil is included. *Reprinted from Ref. [15].*

material. For a set of Au/CeO_2(cube) catalyst gold was completely removed by cyanide leaching. The strong metal–support interaction brought about by calcination of the catalyst precursor is responsible to stabilize the Au species resistant to leaching.

4. MORPHOLOGY-DEPENDENT CATALYTIC ACTIVITY OF Au/CeO_2

4.1 CO Oxidation

In general, the particle size of ceria is of great importance to the CO oxidation activity of Au nanoparticles. Carrettin et al. [40] prepared a ceria material with average particle size of 3.3 nm and surface area as high as 180 $m^2 g^{-1}$. Gold nanoparticles on this material showed increased activity for CO oxidation by two orders of magnitude compared with those supported on conventional ceria nanoparticles of larger size (e.g., d = 15.9 nm and S = 70 $m^2 g^{-1}$). This result was explained by the synergistic effect of nano-Au and nano-ceria by enhancing the formation of highly reactive superoxide and peroxide oxygen species at one-electron defect sites on the support. In contrast, there is no consistent conclusion on the effect of ceria morphology on the catalytic activity of Au/CeO_2 catalysts for CO oxidation. Huang et al. [41] firstly reported the catalytic activity of gold supported on ceria nanorod with {111} planes exposed. The ceria nanorods were prepared by hydrothermal treatment of large agglomerated particles in 5 or 10 M KOH solution at 453 K for different periods (4–45 h). Longer duration of the hydrothermal treatment improves the purity and enhances the aspect ratio of ceria nanorods. Gold was loaded on the nanorods by impregnation followed by reduction with $NaBH_4$ solution. This nanorod-supported Au catalysts showed lower activity than conventionally prepared Au/CeO_2 nanoparticles. In another report [42], a higher CO oxidation activity was noted for Au/CeO_2(rod) than for Au/CeO_2(particle). The temperature for obtaining 50% CO conversion on Au/ceria-rod (268 K) was substantially lower than that of Au/ceria-particle (308 K). At 278 K, the rate of CO oxidation over Au/ceria-rod was 4.02 $mol_{CO}\,g_{Au}^{-1}\,h^{-1}$, i.e., about 26 times higher than that for Au/CeO_2(particle) (0.15 $mol_{CO}\,g_{Au}^{-1}\,g^{-1}$). This activity difference was explained by the special single crystalline CeO_2 with 1D-like morphology predominantly exposing {100} and {110} planes, which was far more reactive than polycrystalline CeO_2 powders in CO oxidation [25].

We compared the activities of CO oxidation on Au nanoparticles supported on rod and cube shapes of ceria [15]. The nanostructured ceria supports were not active in CO oxidation at temperatures below 373 K. The turnover frequencies (TOF) followed the trend Au/CeO_2(rod) (216 h^{-1}) > Au/CeO_2(rod)-CN (83 h^{-1}) > Au/CeO_2(cube) (3 h^{-1}) at 313 K. The leached counterpart of Au/CeO_2(cube) did not show any activity. To calculate the TOFs, the dispersion for the nonleached catalysts was determined by TEM analysis, while that of Au/CeO_2(rod)-CN was assumed to be one. The Au/CeO_2(rod) contained also very dispersed gold cations or clusters that could not be imaged by EM analysis. As these cations were stable under the oxidizing reaction conditions, the TOF for Au/CeO_2(rod) should be a lower bound. Nonetheless, the activity of Au/CeO_2(rod) was substantially higher than that of Au/CeO_2(rod)-CN. Moreover, Au/CeO_2(rod) was much more active than Au/CeO_2(cube), which contained fully reduced gold nanoparticles of the same dispersion. Yi et al. [42] also found that Au/CeO_2-rod presented higher activity than Au/CeO_2-cube in (preferential) oxidation of CO. Apparently, Au/CeO_2(cube) has the lowest CO oxidation activity in catalysts with various morphologies. The activity difference between Au/CeO_2(rod) and Au/CeO_2(particle) needs further study due to the complexity of surface defects presented by different morphologies. It is worth noting that the exact pretreatment procedure has pronounced effect on CO oxidation activity [2,15].

4.2 Computational Studies on CO Oxidation by Au/CeO_2

Despite a substantial number of experimental studies on CO oxidation by ceria-supported gold catalysts, there remains much that is unclear about the nature of active site and specifically on the nuclearity of gold, be it atomically dispersed or in the form of clusters or nanoparticles [43]. The issue of the reaction mechanism is also not resolved and it is expected to depend strongly on the nature of the active site. Various experimental studies have provided evidence for the direct participation of oxygen of the ceria support, yet this has not been fully understood from theoretical modeling studies [44–50]. A third aspect, which is addressed in recent modeling efforts, is the influence of the ceria support morphology on the surface reactivity [15,27–29]. In this section, we will review main theoretical studies dedicated to these aspects.

Au_1/CeO_2 {111}: Carmellone and Fabris [44] explored CO oxidation for two structural models, one involving a Au ion adsorbed on the CeO_2 {111} surface and the other one involving replacement of a Ce ion in the same surface by Au (Figure 9). For a model consisting of a single Au atom on CeO_2 {111}, the oxidation reaction is controlled by spillover from the Au^+ adatom to the oxide support, forming a metastable Au-COO* configuration in which the CO molecule bridges between the Au^+ and the closest lattice O* atom. The calculated activation barrier for this process is 0.86 eV.

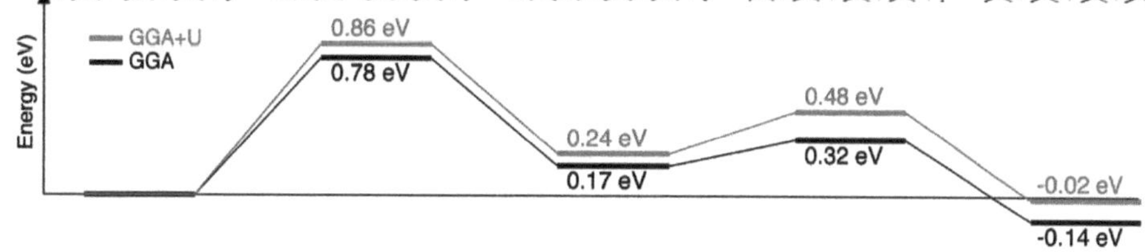

Figure 9 CO oxidation on Au_1/CeO_2 (111). *Reprinted with permission from Ref. [44]. Copyright (2009) American Chemical Society.*

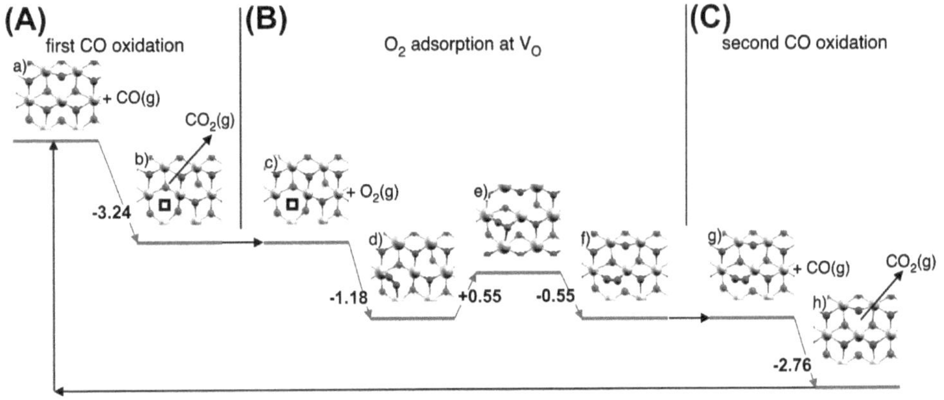

Figure 10 CO oxidation on $Au_xCe_{1-x}O_2$ model. *Reprinted with permission from Ref. [44]. Copyright (2009) American Chemical Society.*

Oxidation and CO_2 desorption take advantage of the oxygen-buffering capacity of ceria and lead to the formation of a surface O vacancy with an activation energy of only 0.24 eV. However, during this reaction, the gold atom diffuses into the vacancy and becomes negatively charged, preventing further adsorption of molecular CO or O_2. In essence, this mechanism predicts that the catalyst would be deactivated. For the doped model ($Au_xCe_{1-x}O_2$ solid solutions in which Au substitutes for Ce lattice sites, Figure 10), higher activity is predicted compared with the single Au^+ adatom supported on the stoichiometric oxide surface. Such substitutional systems can sustain a catalytic cycle in which the oxidation state of the dispersed Au ions remains positive. The oxidation steps result in the formation of surface O vacancies and are assisted by reversible Au^{3+}/Au^+ and Ce^{4+}/Ce^{3+} oxidation/reduction steps. Molecular oxygen can heal such vacancies, forming activated O adspecies and leading to a local rearrangement of the AuO_4 unit with relatively low barrier. Molecular CO will react with such O adspecies spontaneously, forming CO_2 and recovering the initial stoichiometry of the $AuCe_{1-x}O_2$ system.

Au_{13}/CeO_2 {111}: Compared with the single Au atom on the stoichiometric ceria surface that deactivates due to Au migration, Carmellone and Fabris [44] proposed that the cohesive energy in a Au cluster might inhibit deactivation. This possibility was explored in detail by Kim et al. [45] These authors identified a similar reaction pathway: CO adsorbed at an interfacial Au site, followed by reaction with a surface oxygen and desorption of CO_2. A problem of this mechanism is that the desorption energy of CO_2 lies between 122 and 209 kJ mol^{-1}. This relates to the high Ce–O bond strength in the CeO_2 {111} surface. This contrasts the Au_1/CeO_2{111} case [44], in which desorption is much more facile. The difference originates from the compensating effect of migration of the Au atom into the oxygen vacancy during CO_2 desorption, which stabilizes the final state. However, this migration step may lead to deactivation of the single Au atom.

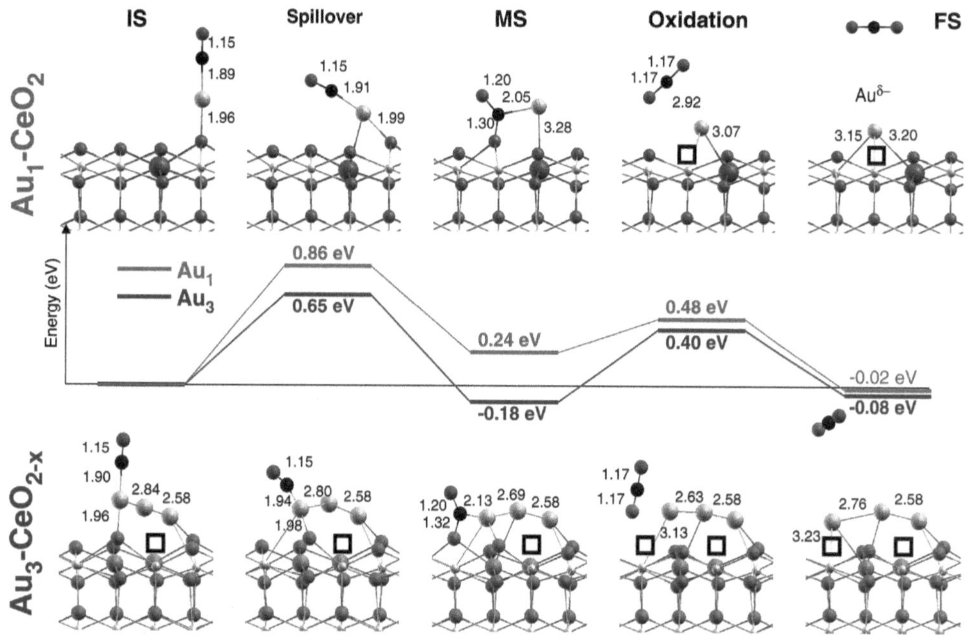

Figure 11 Calculated reaction pathway and corresponding energetics for oxidation of the first CO molecule by a supported Au_3 cluster on CeO_2 (111). *Reprinted with permission from Ref. [46]. Copyright (2013) American Chemical Society.*

Au_3/CeO_2 {111} for CO oxidation: Recent result from Fabris' group discusses another reaction path for CO oxidation on a Au_3/CeO_{2-x} {111} model [46]. The Au_3 cluster is adsorbed on one vacancy site of CeO_2 {111} surface (Figure 11). The authors stressed the so-named fluxionality of this small Au_3 cluster. On the one hand, having a three-atom cluster hinders migration of gold into the oxygen vacancy site that would cause deactivation as for Au_1/CeO_2 {111}, yet on the other hand the desorption energy for CO_2 is only 56 kJ mol^{-1}, which compares favorably to values for Au_{13}/CeO_2 {111}. After CO_2 desorption, the Au_3 cluster interacts with the vacancy site slightly, facilitating CO_2 desorption but leaving the possibility for O_2 adsorption. The authors pointed out the role of dynamics at the Au–oxide interface in explaining reactivity of ceria-supported gold catalysts.

Henkelman and co-workers found that the desorption energy of CO_2 on a stepped-CeO_2 {111} (the reader is referred to Ref. [48] for structures) containing a Au_{12} cluster is relatively low at 40 kJ mol^{-1} [48]. The reason can be understood from the changes in the Ce coordination number. The desorption of CO_2 involves the cleavage of O–Ce bond. Lower Ce coordination number in the stepped surface lowers the CO_2 desorption energy. This work is an interesting case showing how defects of the ceria support influence the CO oxidation activity. It has been found that the CeO_2 termination by {111}, {110}, and {100} planes occurring in nanorod, nanocube, and polyhedral shapes has a strong effect on the CO oxidation activity and this will be the focus of the remainder of this section.

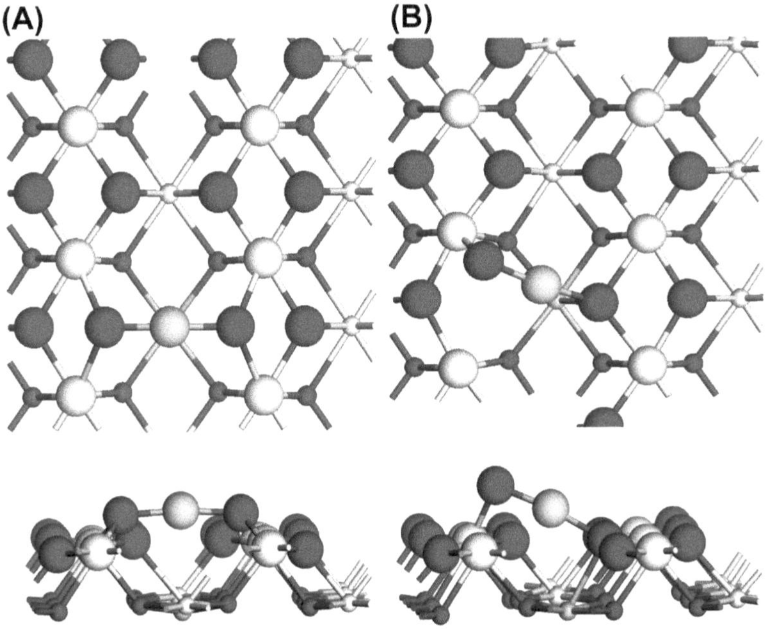

Figure 12 Structures of Au adsorbed on the CeO_2 (110) surface. *Reprinted with permission from Ref. [49]. Copyright (2013) American Chemical Society.*

Single Au/CeO_2 {110} [49]: Previous theoretical works have concluded that the most stable adsorption site for a single Au atom is the bridge site between two surface O atoms on CeO_2 {110} surface. Yet, a more stable structure was identified, which involves migration of one of the surface O atoms to a bridging position between Ce and Au (Figure 12). The migrated O atom is reactive toward CO oxidation. Although CO adsorption to the Au atom is not possible, the reaction can proceed via an Eley–Rideal mechanism (Figure 13) between gas-phase CO and the migrated O atom of the ceria support (activation barrier: 45 kJ mol^{-1}; reaction energy: −28 kJ mol^{-1}). This result provides support for the speculation made in several experimental studies that the reactive oxygen species is located at the interface between gold and nanostructured CeO_2. After CO_2 desorption, O_2 can adsorb at the interface between Au and the resulting vacancy site. It takes 88 kJ mol^{-1} to dissociate adsorbed O_2. One of the O atoms fills the earlier created vacancy, while the other one is located on the bridge site between Au and Ce. This O adatom can be easily reduced by CO (reaction barrier is 28 kJ mol^{-1}), completing the reaction cycle. An important question to be answered is why this mechanism does not work for single Au supported on CeO_2 {111} surface [44]. The key step here is the adsorption of O_2 on Au/CeO_{2-x} {110}, which essentially prevents deactivation that occurs in Au/CeO_{2-x} {111}. The electronic structure of the Au atom adsorbed close to the vacancy sites of CeO_2 {110} and CeO_2 {111} can be used to understand this difference (Figure 14). On the CeO_2 {110} surface, the Au atom slightly migrates toward the vacancy site. The oxidation state of this Au atom is negative (−0.65 *e*) as determined

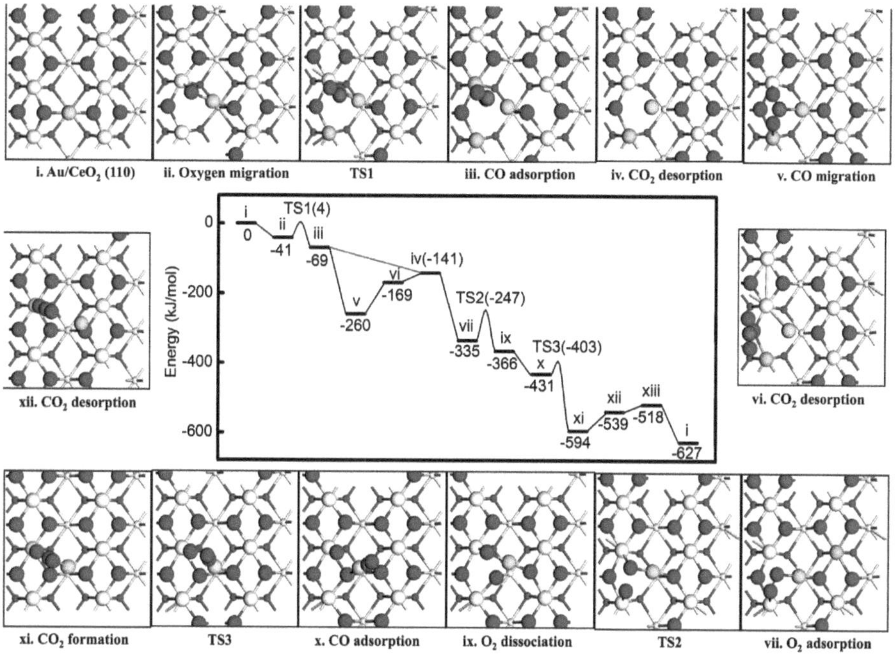

Figure 13 Reaction energy diagram and structures of intermediate and transition states for CO oxidation on Au/CeO_2 (110). *Reprinted with permission from Ref. [49]. Copyright (2013) American Chemical Society.*

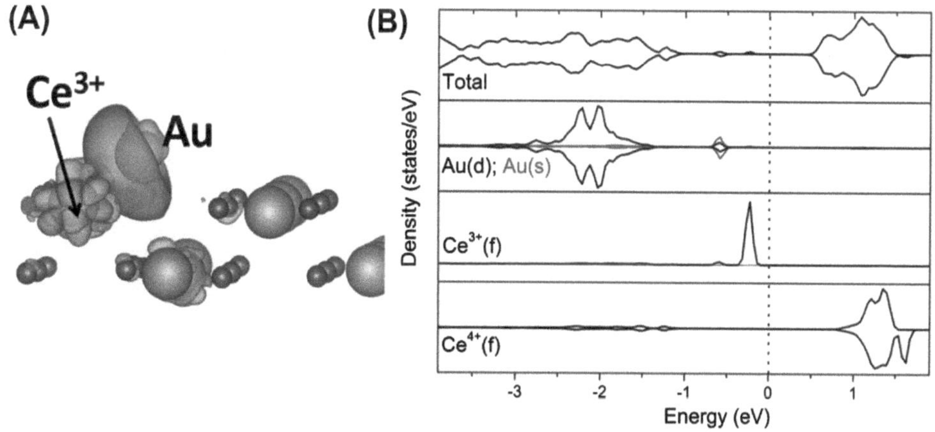

Figure 14 (A) Electron density difference upon adsorption of an Au atom on CeO_{2-x} (110) (yellow (dark gray in print version) and blue (light gray in print version) colors represent increasing and decreasing electron densities, respectively); (B) total DOS and atom-resolved projected DOS analysis. *Reprinted with permission from Ref. [49]. Copyright (2013) American Chemical Society.*

by Bader charge analysis. The binding of the Au atom on the O vacancy site results in a strong rearrangement of the charges at the Au–CeO_2 interface. Detailed analysis for this state shows that charge transfer occurs from the reduced ceria surface to the Au atom.

The result is a partially reduced ceria surface with one Ce^{3+} ion. The transferred electron is localized in the Au 6s orbital. This results in a negatively charged Au atom and formation of a directional Au–Ce covalent bond. The electronic state of Au of Au/CeO_{2-x} {111} is similar to that of Au/CeO_{2-x} {110}: the Bader charge of Au is −0.59 *e*, close to the value reported in a previous study (−0.60 *e*). Indeed, CO can also not adsorb on Au in Au/CeO_{2-x} {111}. This is in sharp contrast with the strong adsorption of O_2 (194 kJ mol^{-1}) at the interface site in Au/CeO_{2-x} {110}. The elongation of the O–O bond from 1.21 Å in gas phase to 1.46 Å points to its strong activation. The two O atoms are coordinated to Ce (d_{O-Ce} = 2.24 Å) and Au (d_{O-Au} = 2.01 Å). Based on these findings and in the light of those obtained for Au/CeO_2 {111}, it may be concluded that the geometry of the ceria surface has a strong influence on its ability to be involved in O_2 adsorption at the interface between Au and Ce (Figure 15). The CeO_{2-x} {110} surface is more open than the CeO_{2-x} {111} surface. In the former surface, the Ce and O atoms lie in the same plane, whereas in the latter, the Ce atoms are in the second layer. As a

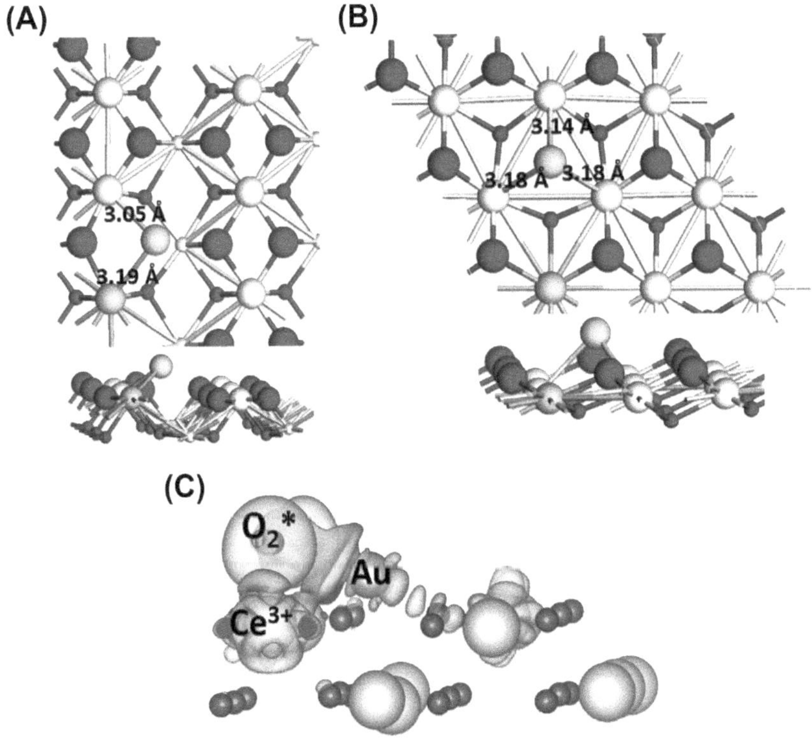

Figure 15 Structure of Au adsorbed on ceria surfaces containing one O vacancy for (A) CeO_{2-x} (110) and (B) CeO_{2-x} (111) (only the top two layers of the ceria model are shown; see caption of Figure 1 for color scheme); (C) electron density difference after and before adsorption of O_2 on Au/CeO_{2-x} (110). *Reprinted with permission from Ref. [49]. Copyright (2013) American Chemical Society.*

result, O_2 can adsorb at the interface between Au and surface vacancy site by coordination to a Ce surface atom in CeO_{2-x} {110}. Electron density difference analysis shows that there is strong accumulation of electrons between O_2* to Au and Ce^{3+}, indicative of formation of a strong bond. This bonding is not possible on Au/CeO_{2-x} {111}, essentially because of the repulsive force between O_2 and the surface O in the top layer. The result is that O_2 adsorption is inhibited on Au/CeO_{2-x} {111}.

Au nanocluster supported on CeO_2 {110} [50]: In contrast to a single Au atom supported on CeO_2 {110}, CO can adsorb on a CeO_2 {110} supported Au nanocluster at different sites with adsorption energies ranging from −5 to −65 kJ mol^{-1} (Figure 16). For the interfacial adsorption model, CO adsorbs on top of a Au atom with an adsorption energy of −26 kJ mol^{-1}. The reaction with a neighboring O atom of the ceria support, which results in adsorbed CO_2, proceeds with negligible barrier. This reaction is strongly exothermic by −146 kJ mol^{-1}. The removal of one surface O atom from ceria results in reduction of one of the surface Ce^{4+} ions to Ce^{3+}. Product CO_2 is only weakly adsorbed to the interface between the Au nanorod and the ceria support by −3 kJ mol^{-1}. Since desorption of CO_2 results in an oxygen vacancy in the ceria support, one more Ce^{4+} ion is reduced to Ce^{3+}. Molecular oxygen from the gas phase adsorbs on the resulting vacancy site with an energy of −135 kJ mol^{-1}. The bond distance in the adsorbed O_2

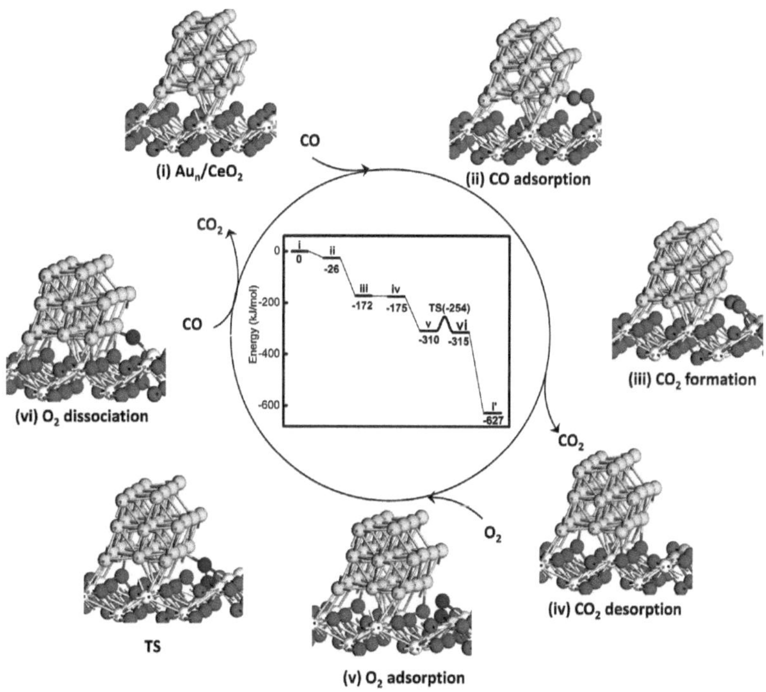

Figure 16 CO oxidation on Au$_n$/CeO$_2$ (110). *Reproduced from Ref. [50] with permission from The Royal Society of Chemistry.*

molecule is slightly elongated to 1.45 Å (gas-phase O–O bond distance is 1.23 Å). No spin magnetic moment is observed for adsorbed O_2, pointing to the formation of O_2^{2-}. This means that electrons are transferred from the ceria support to the antibonding orbital of adsorbed O_2. In accordance with this, one Ce^{3+} is oxidized to Ce^{4+}. It takes 56 kJ mol^{-1} to dissociate the adsorbed O_2. The generated atomic O is then easily reduced by CO. For CeO_2 {111} surface supported Au nanocluster as explored by Kim et al., the desorption energy of CO_2 was found to be very large, amounting to 122 kJ mol^{-1} for a clean ceria support up to 209 kJ mol^{-1} for a defective ceria support. Thus, on CeO_2 {111} the Mars-van Krevelen mechanism will not occur for a Au nanocluster. In contrast, on CeO_2 {110} supported Au nanocluster, CO adsorbed to the Au nanocluster can easily react with a lattice O atom of the underlying support. The product CO_2 is only very weakly bound to the surface.

We will now discuss the dependence of CO oxidation of the surface termination of the ceria support to which the Au nanoparticle is attached. The Mars–van Krevelen mechanism successfully explains direct participation of lattice O in the CO oxidation as observed in several experimental works. For the Au_n/CeO_2 {110}, it is found that CO_2 desorption proceeds without barrier and with a slightly exothermic desorption energy (−3 kJ mol^{-1}). This is due to the specific adsorption mode of CO_2 which is unfavorable. These data are to be compared to those for Au/CeO_2 {111} models [44,45]. For a single Au atom on CeO_2 {111}, CO adsorbs to the Au atom and then reacts with a lattice O atom of the ceria surface [44]. The product CO_2 desorbs into gas phase with an exothermic desorption energy of 23 kJ mol^{-1}. However, after desorption of CO_2, the Au atom migrates into the oxygen vacancy site created upon CO_2 formation and desorption and thus becomes negatively charged. Then, the gold atom cannot adsorb CO or O_2 anymore. For CeO_2 {111} surface-supported Au clusters [45], the desorption energy of CO_2 was found to be very large, amounting to 122 kJ mol^{-1} for a clean ceria support up to 209 kJ mol^{-1} for a defective ceria support. In contrast, CO_2 desorption is exothermic by 72 kJ mol^{-1} for a single Au atom on CeO_2 {110} [49]. The main difference with the CeO_2 {111} surface is that the Au atom will not migrate into the oxygen vacancy and, instead, O_2 will be strongly adsorbed the top layer is composed of both Ce and O ions. The attractive interaction between surface Ce and O_2 molecule makes the adsorption possible. This inhibits catalyst deactivation. From these discussion, we understand that CO adsorbed to a Au nanocluster on CeO_2 {110} can easily react with a lattice O atom of the support. The important corollary of these findings is that the mechanisms explored for single Au and Au clusters on CeO_2 {111} cannot explain the involvement of ceria support O atoms via a Mars-van Krevelen mechanism. For the more open CeO_2 {110} surface, both single Au atoms and Au nanoparticles can catalyze CO oxidation using O atoms of the subjacent ceria support.

Both isolated Au and a Au nanocluster showed much higher CO oxidation activity when adsorbed on CeO_2 {110} compared to the cases where they are adsorbed on CeO_2 {111}. On Au_1/CeO_2 {110}, no CO adsorption was found. So CO can only react with the support oxygen via an Eley–Rideal mechanism. The calculated reaction barrier is 45 kJ mol^{-1}. On Au_n/CeO_2 {110}, CO can absorb on the interface between Au nanocluster and ceria support. The adsorbed CO then reacts with surface oxygen via Langmuir–Hinshelwood mechanism with negligible barrier. Thus, we speculate that the active site for low-temperature CO oxidation reaction is a metallic Au nanocluster, instead of atomically dispersed single-site Au.

4.3 WGS Reaction

The WGS reaction is industrially important for the upgrading of hydrocarbon-based fuel reformate streams. As this reaction is limited by the WGS equilibrium at typical reaction temperatures, it has become custom to carry out the WGS reaction at the industrial scale in two stages: one at higher temperature with typically iron and the other one at lower temperature with copper as the main catalytic metal. However, for generation of CO-free H_2 as for instance required for PEM fuel cell applications the activity of these commercial catalysts is too low. The very high catalytic activity of supported gold catalysts was among others reported by Fu et al. [5,20]. In these works, the role of atomically dispersed Au species was stressed based on cyanide leaching experiments that remove metallic gold nanoparticles. Such isolated cationic gold atoms have also been implicated in explaining WGS catalytic data for TiO_2- and Fe_2O_3-supported gold catalysts [51–53]. Yang et al. [51] employed UV treatment to stabilize isolated gold atoms on TiO_2 and contended that these catalyze the WGS reaction at low temperature. For Au/Fe_2O_3, Allard et al. [52,53] showed that atomic gold species are strongly bound to the support, even after redox heat treatments and exposure to the WGS reaction conditions. These elegant studies have led to extensive studies on the nature of the active gold species under WGS reaction conditions (single-site Au atoms vs. clustered Au atoms; ionic vs. metallic) in the past decade.

Recent studies suggested that the high activity of ceria supported-Au catalysts is more likely ascribed to the presence of metallic clusters or nanoparticles. Kim and Thompson [54] performed a similar leaching procedure as Fu et al. and found that the reaction rate before leaching was significantly higher than after leaching. On this basis, they argued that nanocrystalline gold plays an important role in the WGS reaction. A similar conclusion was drawn by Behm's group [55]. This debate has attracted much attention and advanced in situ characterization techniques have been employed to elucidate the nature of gold in the working catalyst, particularly by use of in situ time-resolved X-ray absorption spectroscopy [15,56,57]. Data on $Au/Ce(Gd)O_2$ and $Au/CeO_x/TiO_2$ catalysts strongly point to transformation of cationic to metallic gold under WGS conditions.

The morphology effect of ceria on the WGS activity of Au/CeO_2 catalyst was first reported by Si and Flytzani-Stephanopoulos [6]. In this study, a strong shape/crystal plane effect of CeO_2 on the gold-ceria activity for the WGS reaction has been identified. Rod-like ceria enclosed by presumed {110} and {100} planes is most active for gold stabilization/activation. We carried out a systematical study on this issue and the results are given in the following. Figure 17 shows the CO conversion during the WGS reaction as a function of temperature for Au/CeO_2(rod) and Au/CeO_2(rod)-CN. The CO conversion of Au/CeO_2(cube) was below 2% and did not depend on temperature. The catalysts were reduced at 453 K prior to reaction. At 473 K the CO conversion was 2.5% for Au/CeO_2(rod). With temperature the CO conversion increased to about 18% at 673 K. Following the subsequent decrease of the reaction temperature, the activities became slightly lower than during the increasing temperature mode. The initial CO conversion of Au/CeO_2(rod)-CN at 473 K was about two times higher than that of Au/CeO_2(rod). The CO conversion increased to about 15% at a temperature of 553 K. At this temperature, the conversion difference with Au/CeO_2(rod) was a factor of two. A further increase of the temperature led to severe deactivation and only after reaction above 633 K a small activity increase was noted. During the decreasing temperature branch, the activity of the catalyst was substantially lower than during the increasing temperature branch. To compare the intrinsic activities of these two catalysts, we assume that the dispersion of Au/CeO_2(rod) is that of the fresh catalysts (4.6 nm, $D=28\%$). Reduction at 523 K of Au/CeO_2(rod)-CN results in a coordination number of 4. This implies that (nearly) all gold atoms can participate in the catalysis ($D=100\%$). A rough approximation shows that Au/CeO_2(rod)-CN is about four times more active per

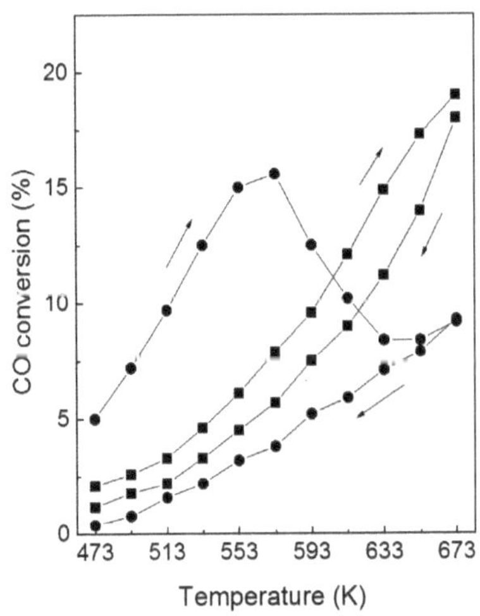

Figure 17 CO conversion during WGS reaction as a function of temperature of a mixture of 40 vol% N_2, 20 vol% H_2O, 20 vol% H_2, 10 vol% CO, and 10 vol% CO_2 of (circles) Au/CeO_2(rod) and (squares) Au/CeO_2(rod)-CN (arrows indicate temperature gradient). *Reprinted from Ref. [15].*

surface gold atom than Au/CeO_2(rod) at a temperature of 533 K. This difference is likely due to the dependence of the oxygen surface coverage as a function of the particle size. Small particles can become more easily oxidized, as is also evident from the above-described CO IR results. The importance of ceria in participating the WGS reaction has also been noted by adding 20 ppm H_2S to the reactor feed. This had a detrimental effect on the catalytic activity and no conversion of CO was observed anymore. This activity loss is expected because S is one of the well-known species which strongly adsorbs to Ce vacancies. It has been shown that the surface defects and O vacancies of ceria surface play a vital role in H_2O adsorption and activation [58–62].

Further experiments concentrated on the role of cationic gold in the WGS reaction. In principle, AuO_x in the Au/CeO_2 catalysts can be fully reduced upon exposure to H_2 at 473–773 K. Therefore, cationic Au species do not exist under WGS reaction conditions [56]. In situ X-ray absorption spectroscopy recorded at 373 and 473 K during the WGS reaction also shows similar results on Au/CeO_2(rod). Whereas calcined Au/CeO_2(rod)-CN only contained cationic gold (Figure 18), exposure of this catalyst to an atmosphere of CO/H_2O for 15 min at 373 K resulted in complete reduction of the gold phase. The reduction is more extensive than upon treatment in H_2 at 393 K. Treatment in 20% O_2/He led to reoxidation of only 20% of the gold atoms. This shows that the initially isolated gold cations have most likely agglomerated into small Au clusters, which can only be partially reoxidized [13]. Under similar conditions Au/CeO_2(rod) contained only metallic gold. EXAFS spectra recorded for Au/CeO_2(rod) after prolonged WGS at

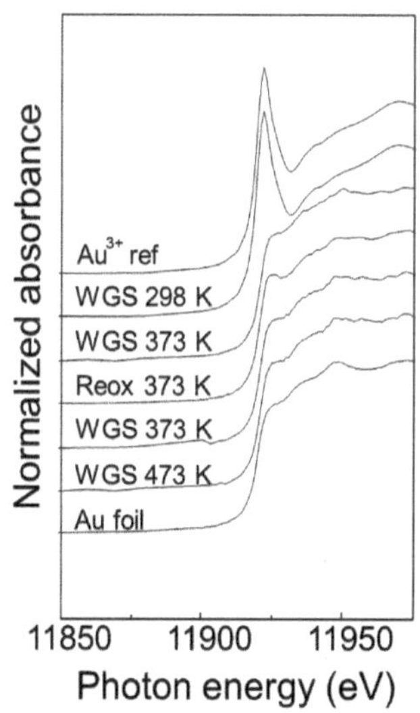

Figure 18 Near-edge spectra at the Au L_{III} edge of Au/CeO_2(rod)-CN during the WGS reaction and upon reoxidation. Spectra of fresh Au/CeO_2(rod)-CN and a Au foil are shown for reference. *Reprinted from Ref. [15].*

473 K showed a constant Au–Au coordination number of 7.4, which agrees well with the results reported in Table 1. A similar analysis for Au/CeO_2(rod)-CN gave a Au–Au coordination number of 4.4 under WGS conditions. These results show that the gold phases in these materials under WGS conditions are very similar to the gold phases after reduction in H_2. Based on the CO conversion measured during the in situ EXAFS experiments, we again find a higher TOF of 2.2 mol_{CO}/mol_{Au}.min for Au/CeO_2(rod)-CN than the value of 1.3 mol_{CO}/mol_{Au}.min for Au/CeO_2(rod).

The morphology effect of ceria on steam reforming of methanol, in which the WGS reaction also plays a role, has also been observed by Flytzani-Stephanopoulos and co-workers [63]. By comparing the activities of small amounts (~1 at%) of gold deposited on CeO_2 with rod and cube shapes, and the former catalyst showed at least an order of magnitude higher activity than the latter one. The authors proposed that strongly bounded Au_n–O–Ce species are the active sites, which are largely observed on {110} surface of CeO_2 rod but present only in negligible concentrations on the {100} surfaces of CeO_2 cube. Kinctics study suggested that the apparent activation energies did not depend on the shape or surface of ceria [64]. Therefore, this reaction is probably structure insensitive over the Au/CeO_2 catalysts.

4.4 Computational Studies on WGS by Au/CeO_2

From literature, three mechanisms may be distinguished for the WGS reaction: redox, formate, and carboxyl mechanisms [65]. The ceria support itself is well known to exhibit the ability to store and release oxygen under oxidizing and reducing conditions, respectively. Based on this, Flytzani-Stephanopouls and co-workers prefer to describe reactivity of Au/CeO_2 in terms of the redox mechanism with CO being adsorbed to gold and oxidized by an O atom of ceria followed by reoxidation of the support by water to give hydrogen [66]. The formate mechanism was first proposed by Shido and Iwasawa and involves the reaction of CO with surface OH groups to form the HCOO group intermediate [67,68]. Based on spectroscopic investigations, several groups have provided further support for this mechanism [55,69–72]. However, Meunier et al. [73] investigated the WGS activity of $Au/Ce(La)O_2$ by operando diffuse reflectance IR spectroscopy and found that the rate of formate decomposition was 60-fold smaller than the rate of CO_2 formation, pointing to the minor role of surface formate in the reaction mechanism. Accordingly, it may be expected that formate is a spectator species. As an alternative, the carboxyl mechanism was proposed, although no clear experimental evidence for involvement of the COOH intermediate in the WGS reaction has been presented yet.

Theoretically, several structural models have been used to determine the mechanism of the WGS reaction for gold on ceria. Based on a Au_4/CeO_{2-x} {111} model, Liu et al. [74] proposed the dissociation of H_2O to adsorbed H and OH, CO reaction with OH to give adsorbed COOH, COOH dissociation to give CO_2 and adsorbed H, and, finally, recombination of the two H atoms on Au to form H_2. Rodriguez et al. explored the same mechanism on a larger Au_{29} cluster [75]. Water dissociation was found to be the rate-limiting step and the authors suggested that the reduced ceria support facilities H_2O dissociation. Following this line, the authors performed extensive work on oxide-supported metals to show the importance of the metal–support interface in the WGS reaction [76]. It is worthwhile to mention that for Cu {111}, Gokhale et al. explored the same mechanism, except for the COOH dissociation step: instead of direct dissociation of H to the Cu surface, COOH dissociates H to a OH surface intermediate [77]. A similar dissociation step was studied by Liu et al. for Au {111} [78]. Based on a Au_n/CeO_2 {111} model, Chen et al. compared redox and formate mechanisms [79]. They found that it is energetically unfavorable to break the O–H bond of the ceria-adsorbed OH group, which is a key step in both mechanisms. Their following work proposed a feasible alternative carboxyl mechanism [80].

WGS reaction on Au_1/CeO_2 {110} [81]: As discussed above, a single Au atom prefers to adsorb on the bridge site between two surface O atoms on CeO_2 {110}. One of these surface O atoms prefers to migrate to the bridge site between Ce and Au, generating a highly reactive Au–O_s species, which may be easily reduced by CO. This will generate a surface oxygen vacancy neighboring to Au. Such oxygen vacancy between the Au adatom and ceria surface plays an important role in the water activation step [76,80]. Water adsorbs at the interface between Au and the defective ceria surface with an adsorption energy of 61 kJ mol^{-1} (Figure 19) [81]. Water dissociation proceeds with a low barrier of 45 kJ mol^{-1} to produce an H atom on the Au atom and an OH group attached to the Ce^{3+} site from ceria support. This reaction is exothermic by 20 kJ mol^{-1}.

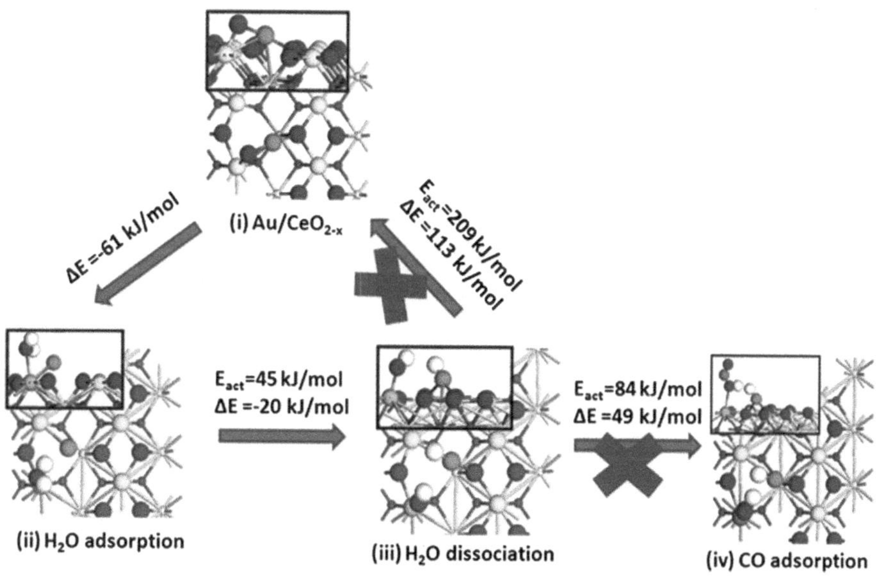

Figure 19 Reaction diagram of WGS on Au_1/CeO_2 (110). *Reprinted with permission from Ref. [81]. Copyright (2014) American Chemical Society.*

The redox mechanism for H_2 formation involves the reaction of ceria-bound OH with Au-bound H to give H_2. The barrier for this step is 209 kJ mol^{-1} with a positive reaction energy of 113 kJ mol^{-1}. On Au_n/CeO_2 {111} [79], the reaction energy for the H_2 formation step was reported to be between 108 and 134 kJ mol^{-1}. The reason for the high barrier of this step is the strong O–H bond energy. The same argument was put forward by Hu and co-workers [79] to explain the unfavorable energetics for CO inserting into the OH group to form HCOO (formate mechanism). This reaction also involves OH bond cleavage. The high barrier for H_2 formation suggests that the redox mechanism is only relevant at high temperatures, supporting speculations based on experiment [73]. To sum up, the strong O–H bound makes the direct redox mechanism difficult at low temperatures. The same applies to the formate mechanism as it also involves O–H bond cleavage [79]. As an alternative, the carboxyl mechanism in which the ceria-adsorbed OH group reacts with CO was examined for the Au_1/CeO_2 {110} model. However, no CO adsorption state could be identified for the mononuclear Au site. Alternatively, the Eley–Rideal mechanism in which CO from the gas phase directly reacts with the surface OH group to form COOH was examined. It turns out the barrier for this step is 84 kJ mol^{-1} with the final state less stable by 49 kJ mol^{-1} than the initial one. If we take into consideration the effect of entropy of CO in gas phase, the free energy barrier for this process would be much larger than this value. For example, taking into account the entropy of CO at room temperature (196.67 kJ mol^{-1}), we predict an overall barrier for this process of 143 kJ mol^{-1}. On this basis, we exclude the possibility for such reaction at low temperatures. To conclude, for the Au_1/CeO_2 {110} model the redox mechanism will only be operative at relatively high temperatures because of the strong O–H bond of the hydroxyl bound to ceria. This precludes recombinative H_2 formation with the H atom adsorbed to the single Au atom. The same applies to the formate mechanism. As the single Au atom cannot adsorb CO due to its negative charge, we predict that the carboxyl mechanism is also not operative in the low-temperature WGS reaction for this single Au atom model.

WGS reaction on Au_n/CeO_2 {110} [81]: For the gold cluster model on the CeO_2 {110} surface, CO adsorbs at the interface between the gold cluster and the support surface (Figure 20). This leads to spontaneous reaction of CO with a ceria surface O atom and desorption as CO_2, generating one oxygen vacancy at the interface. Water adsorbs molecularly at the interface between the oxygen vacancy and the Au cluster with its two H atoms pointing to a surface O atom and the Au cluster. The adsorption energy is 38 kJ mol^{-1}. It takes only 5 kJ mol^{-1} to dissociate one of the OH bonds, forming two hydroxyl groups on the ceria support. The final state is 14 kJ mol^{-1} more stable than the adsorbed state of water. Direct formation of H_2 from this state costs a prohibitive energy of 220 kJ mol^{-1}. This result is consistent with the difficulty in cleaving O–H bonds found for the monoatomic Au/CeO_2 {110} model. An alternative path involves migration of the H of OH to the Au cluster followed by reaction with OH on the ceria

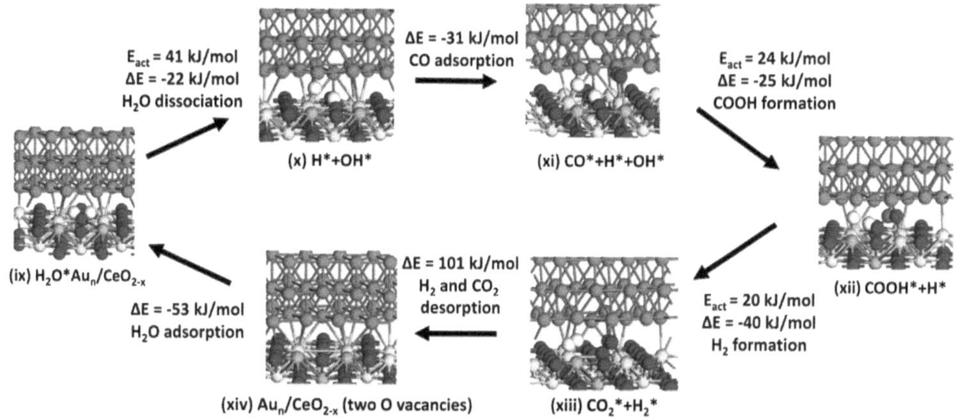

Figure 20 Reaction scheme for the WGS reaction on the Au_n/CeO_2 (110) model. *Reprinted with permission from Ref. [81]. Copyright (2014) American Chemical Society.*

support. The barrier for H migration is 112 kJ mol^{-1} and it is endothermic by 102 kJ mol^{-1}. The migrated H atom is located at the bridge site of two Au atoms. It then reacts with the other OH group of the ceria support to form H_2 with a reaction barrier of 144 kJ mol^{-1} and a reaction energy of 118 kJ mol^{-1}. These barriers for H_2 formation are relatively high, so that this pathway would also require relatively high reaction temperatures. An important corollary of these calculations is that the redox mechanism will only be relevant at higher reaction temperatures and it does not depend on the size of the gold phase, i.e., single-site Au, small Au_3, Au_{10}, or large Au clusters would tend to give rather similar barriers for the most difficult step, namely H_2 formation [79].

Instead of direct OH bond cleavage, the carboxyl mechanism was considered for the Au cluster model. In this pathway, CO first adsorbs on the Au cluster neighboring to the OH group with an adsorption energy of −38 kJ mol^{-1}. The reaction proceeds by COOH formation between adsorbed CO and OH fragments. The barrier for this process is 31 kJ mol^{-1} with a favorable exothermic reaction energy of −25 kJ mol^{-1}. COO–H then dissociates the H to the OH group locating at ceria support, forming a COOH–OH complex. The barrier for this path is only 13 kJ mol^{-1}, significantly lower than the first one. The final state is more stable by 3 kJ mol^{-1} than the initial one. It takes 23 kJ mol^{-1} to desorb the CO_2 molecule into the gas phase, leaving water on the two surface oxygen vacancies at the interface between the Au cluster and ceria. Water will then dissociate one of the H atoms to the Au cluster, leaving an OH group behind on the ceria support. The barrier for water dissociation is 41 kJ mol^{-1} and the reaction energy is −22 kJ mol^{-1}. The reaction continues by adsorption of another CO at the Au cluster site (E_{ads} = 31 kJ mol^{-1}). CO will then easily react with this remaining OH group, forming COOH with activation and reaction energies of 24 and −25 kJ mol^{-1}, respectively. The H atom of COOH points to the H dissociated from H_2O in the previous step (state xii).

These two H atoms form H_2, leaving CO_2 attached to the ceria support. The barrier for this last step is 20 kJ mol^{-1} and the final state is 40 kJ mol^{-1} more stable than the initial state. It takes 101 kJ mol^{-1} to desorb H_2 and CO_2 into gas phase. After desorption of H_2 and CO_2, two oxygen vacancies are present on the ceria support. Water then adsorbs at the interface site between the Au cluster and the defective ceria support, closing the catalytic cycle.

A significant discussion has arisen in the literature on the nuclearity of the gold phase (single Au atom vs. clustered Au atoms) in Au/CeO_2 WGS catalysts. Two key elementary steps may be distinguished in the WGS reaction. The first one is the formation of oxygen vacancies on the ceria surface, which are essential for the activation of water. Oxygen vacancies are generated by reduction with gas-phase CO. For the Au_1/CeO_2 {110} model, this poses a problem as the negatively charged Au adatom cannot adsorb CO. An alternative way of removing the surface oxygen is by an Eley–Rideal mechanism (E_{act} = 45 kJ mol^{-1}; $E_{reaction}$ = −28 kJ mol^{-1}), whose reaction rate is expected to be low, however, due to its low reaction probability. For a Au_1/CeO_2 {111} model [49], CO adsorption may occur as long as the gold is positively charged. This study reported that upon CO_2 formation and desorption, the Au ion will migrate to the resulting oxygen vacancy site to block site for water activation. Moreover, the Au atom will become negatively charged at this location, preventing further CO adsorption. This initiating step will proceed much more easily on the Au_n/CeO_2 {110} model, because CO can adsorb on the gold phase and will react with negligible barrier with a ceria surface oxygen atom to form CO_2. Thus, for the oxygen vacancy formation step, gold clusters are expected to be more active than a single gold atom. The second key step in the WGS reaction is the reaction of CO with OH to form COOH. On the Au_1/CeO_2 {110} model, water dissociation leads to one H atom on gold and one in the form of OH on ceria. Similar to the initial Au_1/CeO_2 {110} model, no CO adsorption was found on Au after water dissociation. The activation barrier via the Eley–Rideal mechanism is 84 kJ mol^{-1} and reaction energy is 49 kJ mol^{-1}, implying low reaction rate at relatively low temperatures. CO can adsorb on the Au cluster in Au_n/CeO_2 {110} after water dissociation, so that the reaction of adsorbed CO with the OH group of ceria support will proceed. By analyzing the two key steps in WGS, we predict that a Au cluster should be much more active than a single-site Au atom. In essence, this difference relates to the charge state of the isolated Au atom. It precludes CO adsorption, which is a prerequisite for both oxygen vacancy formation and COOH formation.

The current proposed mechanisms for WGS can be categorized into two groups: one involving O–H bound cleavage (redox mechanism and formate mechanism) and the other one involving COOH as an intermediate (carboxyl). Alternatively, OH reacts with CO to form COOH. In the present contribution, we considered the redox mechanism as representative for the first type of mechanisms. All the available theoretical studies predict high barrier for the OH* + H* step to form H_2 (the key step in the redox mechanism): 130 kJ mol^{-1} for Au_3/CeO_2 {111} [79], 106 kJ mol^{-1} for Au_{10}/CeO_2 {111} [79], and, in the present work, 209 kJ mol^{-1} for Au_1/CeO_2 {110} and 144 kJ mol^{-1} for Au_n/CeO_2 {110}. On the contrary, computed barriers for the COOH formation step are typically much lower: 10 kJ mol^{-1} for Au_4/CeO_2 {111} [74], 62 kJ mol^{-1} for Au_3/CeO_2 {111} [79], and 31 kJ mol^{-1} for Au_n/CeO_2 {110}. Based on these data, we conclude that the dominant reaction path of WGS under low-temperature conditions is the carboxyl mechanism.

An essential part of this work is the comparison of two key elementary steps for the WGS reaction: oxygen vacancy formation and COOH formation. By comparing the kinetics of these two steps for different ceria surface termination, we may contribute to understanding experimentally observed reactivity trends as a function of ceria morphology. For a Au_n/CeO_2 {111} model, a previous study has shown that the Mars-van Krevelen mechanism in which adsorbed CO reacts with ceria surface O is inhibited by high desorption energy (E_{des} = 122–209 kJ mol^{-1}) because of the strong Ce–O bond strength in CeO_2 {111} surface [45]. With the weaker Ce–O bond in Au_n/CeO_2 {110} and due to the specific CO_2 adsorption mode, CO_2 desorption only costs −3 kJ mol^{-1}, generating one oxygen vacancy [45]. It has also been observed that the CO oxidation rate strongly depends on the morphology of the ceria support. Thus, we expect that the ceria surface termination will have significant influence on the oxygen vacancy formation, and thus, on the water activation step. The COOH formation step involves reaction of adsorbed CO with ceria surface OH. On Au_n/CeO_2 {110}, the activation barrier for this step is 31 kJ mol^{-1} with a reaction energy of −25 kJ mol^{-1}. On Au_n/CeO_2 {111} [80], the activation barrier is 63 kJ mol^{-1} with a reaction energy 27 kJ mol^{-1}. The substantial difference in the kinetics is consistent with the experimental observation that gold supported on nanorod ceria, which mainly exposes {110} and {100} surfaces, is much more active than gold on polyhedra of ceria with mainly {111} and {100} surfaces. The results indicate that both the oxygen vacancy formation energy and the OH reactivity contribute to differences in the WGS activity. Facile oxygen vacancy leads to facile water activation; higher activity of the OH group leads to easier formation of COOH.

5. CONCLUSIONS

Gold supported on ceria has been investigated as a potential catalyst in great detail for a number of technologically important reactions such as CO oxidation and the WGS reaction. Important aspects of such catalysts are the strong metal–support interactions, the importance of the gold–ceria interface in catalytic reactions, and the strong dependence of the nature of gold and its reactivity on the surface termination of ceria. Using experimental and theoretical modeling studies, much progress has been achieved in understanding the structure–activity relations of gold species and ceria nanoparticles.

Well-defined ceria nanostructures with various morphologies present different ratios of surface planes such as {111}, {110}, and {100} with different intrinsic reactivity, leading also to different constitutions of the supported gold phase. The exact nature of reactivity of a particular surface plane affects the coordination and electronic properties of Au, among others by the presence of defects in the ceria support. In this regard, the {111} and {110} facets are known to result in relatively strong Au_n–O–Ce bonds, leading to high dispersion and high catalytic activity of gold in CO oxidation, the WGS reaction, and selective oxidation. The exact mechanism of these reactions depends on the nuclearity of the gold phase as well as the reactivity of the support. The effect of defects at steps and edges of faces in ceria particles has been less widely investigated and deserves further attention. The exact termination of the various ceria surface planes under reactive conditions, possibly containing different densities of hydroxyl groups of varying reactivity, is another topic which should be studied in more detail.

REFERENCES

[1] M. Haruta, T. Kobayashi, H. Sano, N. Yamada, Chem. Lett. 2 (1987) 405–408.
[2] G.C. Bond, C. Louis, D.T. Thompson, Catalysis by Gold, Catalytic Science Series, vol. 6, Imperial College Press, London, 2006.
[3] A. Trovarelli, P. Fornasiero, Catalysis by Ceria and Related Materials, second ed., Imperial College Press, London, 2013.
[4] J. Guzman, S. Carrettin, A. Corma, J. Am. Chem. Soc. 127 (2005) 3286–3287.
[5] Q. Fu, H. Saltsburg, M. Flytzani-Stephanopoulos, Science 301 (2003) 935–938.
[6] R. Si, M. Flytzani-Stephanopoulos, Angew. Chem. Int. Ed. 47 (2008) 2884–2887.
[7] A. Asthagiri, M.J. Janik, Computational Catalysis, Royal Society of Chemistry, Cambridge, 2014.
[8] C.J. Zhang, A. Michaelides, D.A. King, S.J. Jenkins, J. Phys. Chem. C 113 (2009) 6411–6417.
[9] C.J. Weststrate, R. Westerstrom, E. Lundgren, A. Mikkelsen, J.N. Andersen, A. Resta, J. Phys. Chem. C 113 (2009) 724–728.
[10] N.J. Castellani, M.M. Branda, K.M. Neyman, F. Illas, J. Phys. Chem. C 113 (2009) 4948–4954.
[11] J. Guzman, S. Carrettin, J.C. Fierro Gonzalez, Y. Hao, B.C. Gates, A. Corma, Angew. Chem. Int. Ed. 44 (2005) 4778–4781.
[12] D. Tibiletti, A. Amieiro-Fonseca, R. Burch, Y. Chen, J.M. Fisher, A. Goguet, C. Hardacre, P. Hu, D. Thompsett, J. Phys. Chem. B 109 (2005) 22553–22559.
[13] W.L. Deng, A.I. Frenkel, R. Si, M. Flytzani-Stephanopoulos, J. Phys. Chem. C 112 (2008) 12834–12840.
[14] Y.J. Guan, E.J.M. Hensen, Phys. Chem. Chem. Phys. 11 (2009) 9578–9582.
[15] Y.J. Guan, M. Ligthart, G.D. Elzing, J.A.Z. Pieterse, R.A. van Santen, E.J.M. Hensen, Top. Catal. 54 (2011) 424–438.
[16] A. Longo, L.F. Liotta, G. Pantaleo, F. Giannici, A.M. Venezia, A. Martorana, J. Phys. Chem. C 116 (2012) 2960–2966.
[17] M. Baron, O. Bondarchuk, D. Stacchiola, S. Shaikhutdinov, H.J. Freund, J. Phys. Chem. C 113 (2009) 6042–6049.
[18] J. Majimel, M. Lamirand-Majimel, I. Moog, C. Feral-Martin, M. Treguer-Delapierre, J. Phys. Chem. C 113 (2009) 9275–9283.
[19] F. Vindigni, M. Manzoli, A. Damin, T. Tabakova, A. Zecchina, Chem. Eur. J. 17 (2011) 4356–4361.
[20] Q. Fu, A. Weber, M. Flytzani-Stephanopoulos, Catal. Lett. 77 (2001) 87–95.
[21] C.W. Sun, H. Li, L.Q. Chen, Energy Environ. Sci. 5 (2012) 8475–8505.
[22] D.S. Zhang, X.J. Du, L.Y. Shi, R.H. Gao, Dalton Trans. 41 (2012) 14455–14475.
[23] N. Ta, J.Y. Liu, W.J. Shen, Chin. J. Catal. 34 (2013) 838–850.

[24] H.X. Mai, L.D. Sun, Y.W. Zhang, R. Si, W. Feng, H.P. Zhang, H.C. Liu, C.H. Yan, J. Phys. Chem. B 109 (2005) 24380–24385.
[25] K.B. Zhou, X. Wang, X.M. Sun, Q. Peng, Y.D. Li, J. Catal. 229 (2005) 206–212.
[26] X.W. Liu, K.B. Zhou, L. Wang, B.Y. Wang, Y.D. Li, J. Am. Chem. Soc. 131 (2009) 3140–3141.
[27] P.X. Huang, F. Wu, B.L. Zhu, X.P. Gao, H.Y. Zhu, T.Y. Yan, W.P. Huang, S.H. Wu, D.Y. Song, J. Phys. Chem. B 109 (2005) 19169–19174.
[28] N. Ta, J. Liu, S. Chen, P.A. Crozier, Y. Li, A. Chen, W. Shen, J. Am. Chem. Soc. 134 (2012) 20585–20588.
[29] S. Agarwal, L. Lefferts, B.L. Mojet, M. Ligthart, E.J.M. Hensen, D.R.G. Mitchell, W.J. Erasmus, B.G. Anderson, E.J. Olivier, J.H. Neethling, A.K. Datye, ChemSusChem 6 (2013) 1898–1906.
[30] T. Akita, K. Tanaka, M. Kohyama, J. Mater. Sci. 43 (2008) 3917–3922.
[31] T. Uchiyama, H. Yoshida, Y. Kuwauchi, S. Ichikawa, S. Shimada, M. Haruta, S. Takeda, Angew. Chem. Int. Ed. 50 (2011) 10157–10160.
[32] M. Mihaylov, H. Knozinger, K. Hadjiivanov, B.C. Gates, Chem. Ing. Tech. 79 (2007) 795–806.
[33] F. Menegazzo, M. Manzoli, A. Chiorino, F. Boccuzzi, T. Tabakova, M. Signoretto, F. Pinna, N. Pernicone, J. Catal. 237 (2006) 431–434.
[34] A. Chiorino, M. Manzoli, F. Menegazzo, M. Signoretto, F. Vindigni, F. Pinna, F. Boccuzzi, J. Catal. 262 (2009) 169–176.
[35] F. Vindigni, M. Manzoli, A. Chiorino, F. Boccuzzi, Gold Bull. 42 (2009) 106–112.
[36] A. Laachir, V. Perrichon, A. Badri, J. Lamotte, E. Catherine, J.C. Lavalley, L. Hilarie, F. Lenormand, E. Quemere, G.N. Sauvion, O. Touret, J. Chem. Soc. Faraday Trans. 87 (1991) 1601–1609.
[37] H. Daly, J. Ni, D. Thompsett, F.C. Meunier, J. Catal. 254 (2008) 238–243.
[38] V. Aguilar-Guerrero, B.C. Gates, Chem. Commun. (2007) 3210–3212.
[39] J.T. Miller, A.J. Kropf, Y. Zha, J.R. Regalbuto, L. Delannoy, C. Louis, E. Bus, J.A. van Bokhoven, J. Catal. 240 (2006) 222–234.
[40] S. Carrettin, P. Concepcion, A. Corma, J.M.L. Nieto, V.F. Puntes, Angew. Chem. Int. Ed. 43 (2004) 2538–2540.
[41] X.S. Huang, H. Sun, L.C. Wang, Y.M. Liu, K.N. Fan, Y. Cao, Appl. Catal. B: Environ. 90 (2009) 224–232.
[42] G. Yi, Z. Xu, G. Guo, K. Tanaka, Y. Yuan, Chem. Phys. Lett. 479 (2009) 128–132.
[43] M. Flytzani-Stephanopoulos, B.C. Gates, Annu. Rev. Chem. Biomol. Eng 3 (2012) 545–574.
[44] M.F. Carmellone, S. Fabris, J. Am. Chem. Soc. 131 (2009) 10473–10483.
[45] H.Y. Kim, H.M. Lee, G. Henkelman, J. Am. Chem. Soc. 134 (2012) 1560–1570.
[46] P. Ghosh, M.F. Camellone, S. Fabris, J. Phys. Chem. Lett. 4 (2013) 2256–2263.
[47] H.Y. Kim, G. Henkelman, J. Phys. Chem. Lett. 3 (2012) 2194–2199.
[48] H.Y. Kim, G. Henkelman, J. Phys. Chem. Lett. 4 (2013) 216–221.
[49] W.Y. Song, E.J.M. Hensen, J. Phys. Chem. C 117 (2013) 7721–7726.
[50] W.Y. Song, E.J.M. Hensen, Catal. Sci. Tech. 3 (2013) 3020–3029.
[51] M. Yang, L.F. Allard, M. Flytzani-Stephanopoulos, J. Am. Chem. Soc. 135 (2013) 3768–3771.
[52] L.F. Allard, M. Flytzani-Stephanopoulos, S.H. Overbury, Microsc. Microanal. 16 (2010) 375–385.
[53] L.F. Allard, A. Borisovich, W. Deng, R. Si, M. Flytzani-Stephanopoulos, S.H. Overbury, J. Electron Microsc. 58 (2009) 199–212.
[54] C.H. Kim, L.T. Thompson, J. Catal. 244 (2006) 248–250.
[55] R. Leppelt, B. Schumacher, V. Plzak, M. Kinne, R.J. Behm, J. Catal. 244 (2006) 137–152.
[56] X. Wang, J.A. Rodriguez, J.C. Hanson, M. Perez, J. Evans, J. Chem. Phys. 123 (2005) 221101–221105.
[57] R. Si, J. Tao, J. Evans, J.B. Park, L. Barrio, J.C. Hanson, Y.M. Zhu, J. Hrbek, J.A. Rodriguez, J. Phys. Chem. C 16 (2012) 23547–23555.
[58] L. Kundakovic, D. Mullins, S.H. Overbury, Surf. Sci. 457 (2000) 51–62.
[59] M.A. Henderson, C.L. Perkins, M.H. Engelhard, S. Thevuthasan, C.H.F. Peden, Surf. Sci. 526 (2003) 1–18.
[60] M. Fronzi, S. Piccinin, B. Delley, E. Traversa, C. Stampfl, Phys. Chem. Chem. Phys. 11 (2009) 9188–9199.
[61] D.R. Mullins, P.M. Albrecht, T.L. Chen, F.C. Calaza, M.D. Biegalski, H.M. Christen, S.H. Overbury, J. Phys. Chem. C 116 (2012) 19419–19428.
[62] B.H. Chen, Y.S. Ma, L.B. Ding, L.S. Xu, Z.F. Wu, Q. Yuan, W.X. Huang, J. Phys. Chem. C 117 (2013) 5800–5810.

[63] N. Yi, R. Si, H. Saltsburg, M. Flytzani-Stephanopoulos, Appl. Catal. B: Environ. 95 (2010) 87–92.
[64] N. Yi, R. Si, H. Saltsburg, M. Flytzani-Stephanopoulos, Energy Environ. Sci. 3 (2010) 831–837.
[65] G. Bond, Gold Bull. 42 (2009) 337–342.
[66] M. Flytzani-Stephanopoulos, Acc. Chem. Res. 47 (2014) 783–792.
[67] T. Shido, Y. Iwasawa, J. Catal. 136 (1992) 493–503.
[68] T. Shido, Y. Iwasawa, J. Catal. 141 (1993) 71–81.
[69] G. Jacobs, L. Williams, U. Graham, G.A. Thomas, D.E. Sparks, B.H. Davis, Appl. Catal. A: Gen. 252 (2003) 107–118.
[70] G. Jacobs, P.M. Patterson, D.E. Sparks, B.H. Davis, Appl. Catal. A: Gen. 269 (2004) 63–73.
[71] G. Jacobs, S. Khalid, P.M. Patterson, D.E. Sparks, B.H. Davis, Appl. Catal. A: Gen. 268 (2004) 255–266.
[72] G. Jacobs, E. Chenu, P.M. Patterson, L. Williams, D. Sparks, G. Thomas, B.H. Davis, Appl. Catal. A: Gen. 258 (2004) 203–214.
[73] F.C. Meunier, D. Reid, A. Goguest, S. Shekhtman, C. Hardacre, R. Burch, W. Deng, M. Flytzani-Stephanopoulos, J. Catal. 247 (2007) 277–287.
[74] Z.P. Liu, S.J. Jenkins, D.A. King, Phys. Rev. Lett. 94 (2005) 196102.
[75] J.A. Rodriguez, P. Liu, H. Hrbek, J. Evans, M. Perez, Angew. Chem. Int. Ed. 46 (2007) 1329–1332.
[76] J.A. Rodriguez, J.C. Hanson, D. Stacchiola, S.D. Senanayake, Phys. Chem. Chem. Phys. 15 (2013) 12004–12025.
[77] A.A. Gokhale, J.A. Dumestic, M. Mavrikakis, J. Am. Chem. Soc. 130 (2008) 1402–1414.
[78] X.M. Liu, Z.M. Ni, P. Yao, Q. Xu, J.H. Mao, Q.Q. Wang, Acta Phys. Chim. Sin. 26 (2010) 1599–1606.
[79] Y. Chen, J. Cheng, P. Hu, H.F. Wang, Surf. Sci. 602 (2008) 2828–2834.
[80] Y. Chen, H.F. Wang, R. Burch, C. Hardacre, P. Hu, Faraday Discuss. 152 (2011) 121–133.
[81] W.Y. Song, E.J.M. Hensen, ACS Catal. 4 (2014) 1885–1892.

CHAPTER 5

Gold/Ceria: The Making of a Robust Catalyst for Fuel Processing and Hydrogen Production

Nan Yi*, Maria Flytzani-Stephanopoulos
Department of Chemical and Biological Engineering, Tufts University, Medford, MA, USA
*Corresponding author: E-mail: yimethanol@gmail.com

1. INTRODUCTION

The last decade has witnessed an explosion of research on gold-based nanocatalysts for a variety of chemical processes, including hydrogen production, environmental pollutant abatement, and selective oxidation and hydrogenation reactions to name but a few of the intended applications. For a review of gold catalysis, see references [1–5]. For some of these reactions, the catalysis is exclusively done by the gold, either in the form of few atom clusters in solution [6] or single-atom sites, ligated to supports with –O ligands (e.g., Au–O_x on ceria [7] for the water–gas shift (WGS) reaction or Au on NaY [8,9] and MgO [10] for the CO oxidation reaction at room temperature). For a practical catalyst, the active sites should be suitably anchored on supports to retain stability in long-term use, and thus achieve maximal utilization efficiency. In this context, the choice of the support is crucial; i.e., some supports will be better dispersants of gold than others, and it is important to understand this property for proper catalyst design. A large number of supports have been examined for each and every reaction found to be catalyzed by gold. Consensus is being built slowly, especially for reactions like the CO oxidation [9], due to the extreme sensitivity of gold to the reaction conditions, and has been challenging even for one and the same support, e.g., for Au/TiO_2. Diametrically opposite findings have been reported and this has caused confusion in the literature, remaining to this date, 25 years after Haruta first reported the Au/TiO_2 system as an extremely active catalyst for the CO oxidation reaction [11]. To add complexity, gold, if hydrolyzed in highly alkaline solutions, is as active for the CO oxidation in unsupported form [12], as when supported on TiO_2 under similar conditions [13].

The situation is considerably simpler for the WGS reaction, as recently pointed out in a review of the relevant literature by Flytzani-Stephanopoulos [5]. A number of supports may be used to disperse gold atomically, and stabilize it in its active form as Au–O_x–S species, where S denotes the metal ion of the oxide support, e.g., Ce, Fe, Ti, serving as the ligand to Au through –O bonding [9]. These atomically dispersed gold species with –O and –OH

Catalysis by Materials with Well-Defined Structures
http://dx.doi.org/10.1016/B978-0-12-801217-8.00005-0

groups vicinal to gold are active for the WGS reaction as well as for other reactions, such as the oxidative coupling of methanol and the steam reforming of methanol, through different mechanisms than the WGS reaction. Clearly, this recent appreciation may allow us to prepare gold and other metals in stable atomic dispersions on a variety of supports, maximizing the atomic efficiency of the catalyst. The key effort will be in the stabilization of these structures over a broader temperature range than what has been accomplished to date.

In this chapter, we review the gold on ceria supports for a few reactions of interest to energy processes, especially hydrogen production. We also show the atomically dispersed gold on ceria to be active for "green" chemicals production, e.g., through the oxidative coupling of methanol to methylformate. Ceria is selected because it is the dispersant support of choice for metals that operate at least in part as oxidized centers, a list that includes the platinum group metals (PGM) in the three-way catalyst (TWC) and in the fluid catalytic cracking (FCC) catalyst [14]. Other than the PGM, Cu, Ni, and Au metals, to mention but a few, atomically dispersed on ceria have been examined for potential applications to fuel processing, SO_2 reduction [15,16], and N_2O decomposition [17,18]. The unique properties of ceria as a catalyst and catalyst support have been reviewed by several researchers in the field, such as Trovarelli [19], Gorte [20], and Dai et al. [21]. Here, we briefly point out the important properties of ceria as a support for the catalytic reactions noted earlier, and then especially focus on the gold/ceria system. The latter is a powerful oxidation catalyst, but also a catalyst system that allows the gold to display its own unique properties, by stabilizing it at the right active state for the selected reactions. When ceria itself is a catalyst for the same reactions, it is interesting to understand the effect of the metal and determine whether a bifunctional (cooperative) mechanism is truly present.

2. CERIA: STRUCTURE, SYNTHESIS METHODS, CHARACTERIZATION, AND CATALYSIS

As mentioned earlier, there are a number of excellent reviews of ceria which should be consulted for a detailed review of ceria and nanoceria properties [19–23]. For catalysis applications, cerium(IV) oxide (or ceria), especially ceria doped with zirconia, lanthana, and gadolinia, is a robust material that can be taken to temperatures as high as 800–1000°C without total loss of its oxygen storage capacity (OSC) or its surface area. These are useful properties for application to the automotive exhaust catalyst, the high-temperature water gas shift (WGS) reaction [24], as well as to electrocatalysis, e.g., in solid oxide fuel cells [25]. In recent years, interest in ceria-supported metals for processes occurring at lower temperatures has been intense, due in most part to the realization that nanoscale ceria has a plethora of useful defect sites that may be harnessed for catalysis. At the same time, the progress made in synthesizing ceria nanoshapes in a controlled way makes it possible to design such materials to catalyze structure-sensitive reactions on ceria, and to tune the reaction selectivity of gold or other metals atomically anchored on ceria, with retention of high activity.

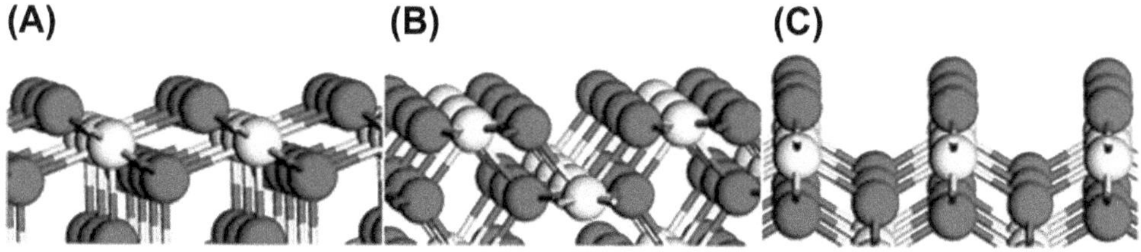

Figure 1 Bulk truncated structures (side view) of CeO_2 (A) (111), (B) (110), and (C) (100) surface. *Reproduced from Ref. [26]. Copyright 2013, the Royal Society of Chemistry.*

Structure. The electron configuration of cerium (Ce) is $4f^25d^06s^2$, which determines the oxidation state of cerium as +3 and +4. The typical corresponding oxide composition is in the range of Ce_2O_3–CeO_2. Pure ceria, or cerium dioxide, CeO_2, has the fluorite crystal structure. It has the face-centered cubic unit cell with the space group F_m3_m. In its typical structure, each cerium cation is coordinated by eight equivalent nearest-neighbor oxygen anions at the corner of a cube, each anion being tetrahedrally coordinated by four cations [26], shown in Figure 1.

Upon partial reduction of ceria, oxygen vacancies are formed. In this reaction, the oxidation state of Ce changes reversibly from Ce(IV) to Ce(III), following the reaction:

$$O_o^x + 2Ce_{Ce}^x \rightarrow V_{\ddot{O}} + 2Ce_{Ce}^{'} + 0.5\ O_2$$

according to the Kroger–Vink notation. This process allows ceria to store or release oxygen, depending on the oxygen partial pressure. The OSC is central to oxidation catalysis applications of ceria [27].

Theoretical molecular simulation is a useful tool to help us understand the structure and relative stabilities of the CeO_2 surfaces [28–30]. The three low-index surfaces, namely {111}, {110}, and {100}, exhibit different properties. The surface structure of each plane is strongly perturbed upon reduction, and the extraction of an oxygen vacancy is associated with a reduction of two neighboring Ce(IV) species to Ce(III) [26]. Specifically, for the {111} surface, all exposed Ce ions are seven-coordinated and all surface O ions are bonded to three Ce ions. It is the most compact structure and its surface atoms have the lowest coordination deficiency. The {110} surface presents six-coordinated Ce and three-coordinated O. The {100} surface has six-coordinated Ce and two-coordinated O. Surface energies calculated using molecular dynamics suggest the order as $\{111\} < \{110\} < \{100\}$, with the {111} being the most stable plane [31].

Synthesis Methods. The preparation of ceria has been extensively explored over the past two decades. Among the different synthesis approaches most prevalent are precipitation, hydrothermal synthesis, and solid-state methods. The overall goal is to obtain the desired structure in a controlled manner, typically amounting to decreasing the less reactive crystal planes and increasing the more reactive ones. In general, liquid-phase synthesis is widely used to prepare ceria with controlled morphology [32,33]. Synthesis parameters, including precursors, temperature, and postprecipitation treatments, play an important role in controlling the shape of ceria nanostructures. Of course, industrially, rather simple protocols are used to produce high surface area polycrystalline ceria at low cost.

Precursors: Ce^{3+} and Ce^{4+} compounds are two choices for cerium precursors. Usually, Ce^{3+} precursors, for example, $Ce(NO_3)_3$, are used to grow one-dimensional nanostructures, like ceria nanorods [34]. However, Ce^{4+} precursors, like $(NH_4)_2Ce(NO_3)_6$, are applied to prepare spherical nanoparticles [26,35]. Even with the same cerium cation, the choice of anions has been reported to affect the shape of ceria. For example, $CeCl_3{\cdot}7H_2O$ leads to the formation of ceria nanowires of 40 nm in diameter and 3–10 μm in length. Instead, the use of Ce(III) nitrate or phosphate gives ceria nanorods, but of different aspect ratios, depending on the choice made [36].

The choice of a base for precipitation, for example, NaOH vs. NH_4OH, is not important, if the ceria crystals do not grow hydrothermally (in an autoclave) [26,34]. Ceria nanoparticles are then formed without shape selectivity. For hydrothermal synthesis, the choice of base becomes important. The combination of $Ce(NO_3)_3$ and ammonia gives spherical ceria nanoparticles in the size of 12 nm. Replacing ammonia with a stronger base, NaOH for example, produces ceria nanocubes. The treatment temperature also affects the final ceria nanoshapes. Increasing the treatment temperature from 100 °C to 180 °C, the final shape of ceria will change from nanorods to nanocubes [34].

Thermal aging, although detrimental to the surface area, has been demonstrated to affect the crystal morphology by promoting irregular truncation of edges and corners and development of more reactive surface combinations in all crystal shapes [37]. The hydrothermal process was suggested to be governed by a dissolution–recrystallization mechanism, shown in Figure 2. $Ce(OH)_3$ nuclei formed when the Ce^{3+} was added into the NaOH solution. Under lower OH^- concentration ($C_{NaOH} < 1$ mol/L) and lower temperature (~100 °C), the dissolution–recrystallization process is slow and leads to the formation of CeO_2 nanopolyhedra with enclosing {111} and {100} facets. When the OH^- concentration was increased, the formation rate of dissolution–recrystallization was greatly promoted. Thus $Ce(OH)_3$ nuclei grew anisotropically and ceria nanorods with enclosing {110} and {100} formed. Ceria nanocubes were produced under higher OH^- concentrations and much higher temperatures [34].

Characterization. X-ray diffraction (XRD), Raman spectroscopy, and high resolution transmission electron microscopy (HRTEM) provide direct evidence of the different concentration of crystalline planes due to the alteration of synthesis parameters. At the same time, the formation of oxygen vacancies due to the reversible reduction/oxidation reaction of Ce^{4+} and Ce^{3+}, can be characterized with the aid of probe molecules, such as isotopic-labeled oxygen, CO, and methanol. Thus, Fourier transform infrared spectroscopy (FTIR) and temperature-programmed reactions coupled with mass spectrometry are useful given their sensitivity to the shifts of the vibration frequencies and desorption temperatures of adsorbates and products.

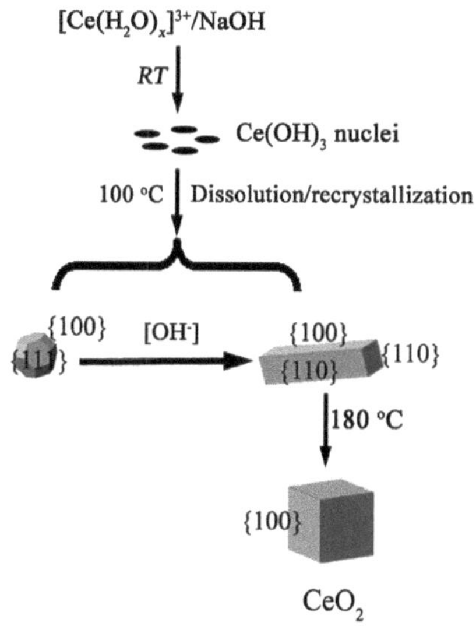

Figure 2 Schematic diagram for the shape-selective synthesis of CeO_2 nanopolyhedra, nanorods, and nanocubes. *Reproduced from Ref. [34]. Copyright 2005, the American Chemical Society.*

Oxygen Storage Capacity (OSC). Theoretical calculations of the OSC values of {100}, {110}, and {111} surfaces of CeO_2 are, respectively, 5.7, 4.0, and 6.6 μmol O/m^2 [38]. For hydrothermally prepared ceria, the measured value of OSC for ceria nanorods is 554 μmol O/g, higher than that of ceria nanopolyhedra (318 μmol O/g) and ceria nanocubes (353 μmol O/g). Compared with bulk ceria, the ceria nanoshapes also have higher OSC due to the significant increase of their surface areas. To exclude the surface area effect on the OSC, the measured OSC values are scaled by the surface area. The surface areas are 62.8 m^2/g for nanopolyhedron, 60.8 m^2/g for nanorod, and 33.2 m^2/g for nanocube, respectively. In units of μmol O/m^2, the OSC follows the order nanocubes (~) nanorods > nanopolyhedra. This can be explained by the fact that ceria nanopolyhedra are enclosed by eight {111} and six {100} planes, the nanorods are enclosed by four {110} and two {100} planes, while the nanocubes are enclosed by six {100} planes. Controlled preparations of nanocrystalline ceria have also shown that below ~6 nm the OSC of ceria increases dramatically [39]. This is accompanied by an increase of the concentration of Ce^{3+} ions, as measured by X-ray photoelectron spectroscopy (XPS) and X-ray absorption near-edge structure (XANES) [39,40]. The presence of Ce^{3+} is important to understand the catalytic performance of ceria.

Ceria has the fluorite structure and features a strong Raman peak at ~465/cm corresponding to its F_{2g} symmetry. The nanorods present a much broader 462/cm peak than nanocubes and nanooctahedra [41]. This size dependence is consistent with the XRD and transmission electron microscopy (TEM) findings that nanorods usually have a smaller size compared with that of nanocubes and nanooctahedra [34]. Wu et al. [41] have combined Raman spectroscopy and O_2 adsorption to probe the defect sites on ceria. They identified two different types of oxygen defect sites on ceria nanoshapes, namely superoxide on one-electron defect sites and peroxide on two-electron defect sites, which is related to a different clustering degree depending on the type of ceria nanoshapes. Those two distinct oxygen sites were assigned to the Raman peaks at 825 and 1131/cm. These oxygen species were further confirmed by FTIR examination of isotopic exchange between gaseous O_2 and lattice oxygen of ceria [42]. Periodic density functional theory calculations also confirmed the dioxygen species adsorbed on CeO_2 and predicted that the reduced surfaces are energetically more favorable than the unreduced surfaces for oxygen adsorption [43].

Dosing the surface of ceria nanoshapes with probe molecules, such as carbon monoxide [44] and methanol [45,46], would provide insight into the catalytic properties, especially when those molecules are the intended reactants. CO-TPR combined with isotopic oxygen exchange experiments, has revealed that the lattice oxygen reactivity and mobility depend strongly on the surface planes exposed [44]. CO_2 desorption profiles obtained from CO-TPR over all three ceria nanoshapes is shown in Figure 3(A). It exhibits two reduction temperature regimes, namely low temperature (<850 K) and high temperature (>850 K). The more intense and broader low-temperature CO_2 peaks at 570 and 740 K were observed over ceria nanorods, which suggested the presence of large amount of reactive oxygen on the surface of ceria nanorods. A low-temperature CO_2 peak (570 K) with extension to higher temperature occurred over ceria nanocubes, while a rather broad high temperature peak at 740 K was observed over ceria nanooctahedra. Figure 3(B) shows that the temperature of H_2 evolution roughly coincides with the temperature of CO_2 production in the higher temperature range (650–850 K). The formation of hydrogen could be explained by the reactions between surface OH groups and CO via the WGS reaction. CO-FTIR further suggested that carbonate species are similar in nature on all the ceria nanoshapes, but they are bound much stronger on ceria nanorods than on ceria nanocubes and ceria nanooctahedra [44]. The analysis of lattice strain regarding ceria nanoshapes also revealed that the nanorods have the highest lattice strain, followed by the polyhedra, while the nanocubes are the least strained [47].

CH_3OH-TPD. Methanol is another probe molecule, which can be used to distinguish the reactivity of each crystal face of ceria [45,46]. When methanol was dosed onto ceria single crystal surfaces, namely, CeO_2 {111} and CeO_2 {100}, methanol temperature-programmed desorption (TPD) identified the high-structure sensitivity of the reaction [45]. On CeO_2 {111} surfaces, adsorbed methanol dissociated on oxygen vacancy to produce surface methoxide, which was further converted into formaldehyde and water. A different reactivity was found on the CeO_2 {100} surface, due to the abundance of exposed Ce^{3+} cations, which drives the methoxide to completely dehydrogenate to CO and H_2. Pretreatment also affects the product distribution over ceria surfaces [46]. Formaldehyde and water desorption followed methanol adsorption on the fully oxidized CeO_2 {111} surface, while CO and H_2 desorption took place when the CeO_2 {111} surface was pre-reduced.

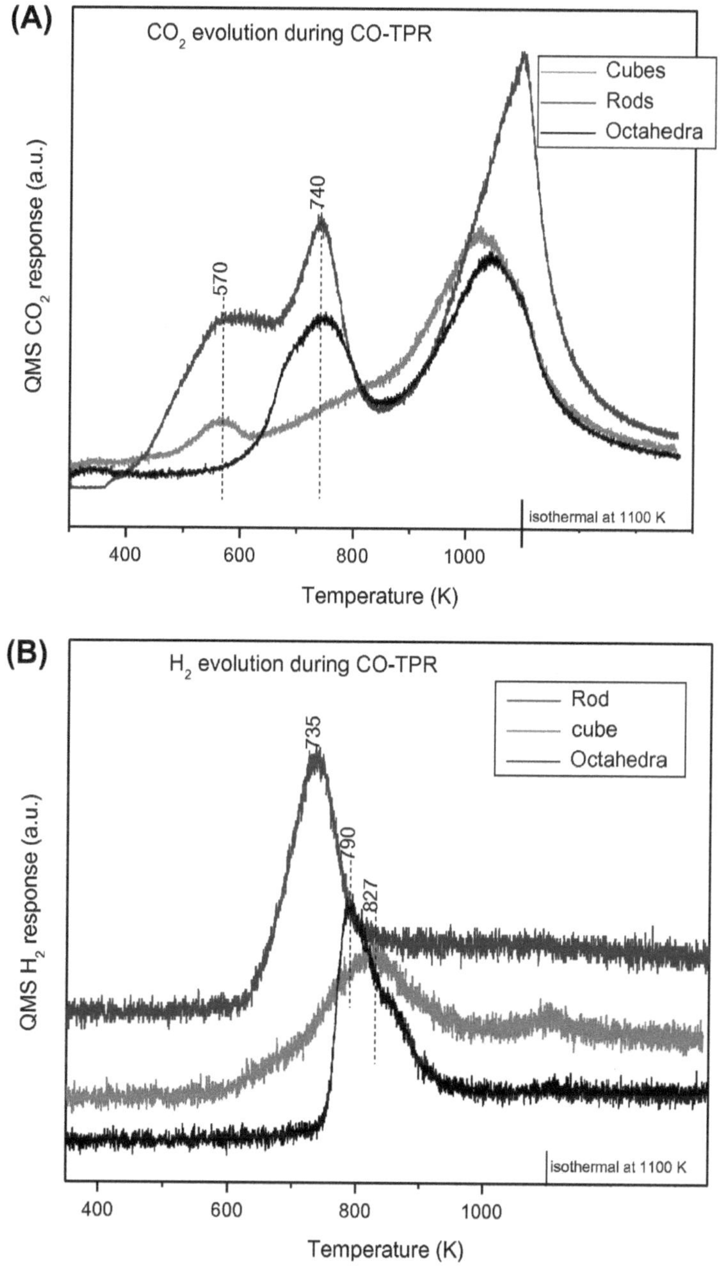

Figure 3 CO_2 (A) and H_2 (B) evolution during CO-TPR over three ceria nanoshapes. *Reproduced from Ref. [44]. Copyright 2012, Elsevier.*

The combination of methanol FTIR and TPD has been applied to investigate the interaction between methanol and ceria nanoshapes, including rods, cubes, and octahedral [48]. Three different types of methoxy species, namely, on-top, bridging, and three-coordinated, were identified. Hydrogen and carbon monoxide are the main products on those three ceria nanoshapes. The desorption temperature and relative ratio of hydrogen and carbon monoxide depend on the type of ceria nanoshape and the number of surface oxygen defect sites. Spectroscopy studies suggested hydrogen and carbon monoxide are formed by dehydrogenation of the methoxy species via formate species as one of the reaction intermediates. Some formate and possible carbonate species are retained on the surface of ceria nanoshapes after high-temperature desorption. This can explain the higher-than-stoichiometric H/CO ratio obtained from analysis of methanol-TPD results.

Oxidation Catalysis. Ceria was originally selected as an additive to the TWC to improve the performance of the PMG over alumina [27], by better dispersing the platinum metals and suppressing their sintering upon aging [49]. Zirconia-doped ceria was found to increase the OSC of ceria and is the additive of choice in the TWC to this date. While the high OSC may not be fully retained after the harsh aging treatment required for the TWC, the application of ceria has been extended to other reactions that take place at lower temperatures [50]. Fundamental studies also benefit from the availability of such materials, especially for structure-sensitive reactions, as the methanol reactions mentioned earlier. The increased surface oxygen part of the OSC can also facilitate the anchoring of metal atoms on ceria and improve their catalytic activity in the supported metal/ceria systems. An example of nanoscale ceria as a better catalyst over microcrystalline ceria is soot combustion, where the activity was further correlated to the type of ceria planes exposed [51]. In another example, CO oxidation [52] on ceria nanorods with well-defined {110} and {100} planes showed higher activity than the "classical" ceria nanoparticles with {111} planes exposed. In general, the activity followed the order nanorods > nanocubes > nanooctahedra. A later study by the same group [53] has further refined the reactivity of nanorods for the CO oxidation reaction. Ceria nanorods with {111} and {100} planes exposed were more active than ceria nanorods with {110} and 100} planes although their surface areas were similar. This was attributed to different types and distributions of oxygen vacancies on the nanorods.

3. METAL–CERIA INTERACTION

While this chapter focuses on the Au/CeO_2 as a powerful catalyst, the interaction of gold and ceria resembles that of the PGM, and other metals on ceria, hence a short survey of this interaction from key reports in the literature is in order.

Unique properties of ceria and doped ceria make it an excellent metal support for a wide number of catalytic applications; first and foremost is the availability of surface oxygen species which ceria can supply to the metal site. The PGM on ceria are activated at low temperatures for CO and hydrocarbon oxidation [14], and the WGS reactions [7,20], as a result of this oxygen availability [14,50]. Other metals, such as Cu [54–57] and Ni [16,56], have been found to be similarly activated by the surface oxygen of ceria for a variety of redox reactions, including methane combustion, and the SO_2 reduction by CO [58]. Metals, such as Zr [34] and Ni [59], can also introduce strain into the ceria lattice, inducing the formation of defects. Ceria can incorporate the dopant elements in different coordination modes, in turn perturbing the dopant chemical properties. Yi et al. [59] reported the microstrain of NiO lattice due to the formation of $Ce_xNi_{1-x}O_y$ solid solutions that played a significant role in methane combustion reaction.

From studies of inverse structures, i.e., ceria nanoparticles deposited on metal films [60,61], the metal–ceria interface has unambiguously been identified as the locus of the active sites for the CO oxidation and the WGS reaction. However, the type of the interfacial interaction has remained vague and largely unexplored to this day. What are the active sites at the interface? Are they M–O_x–Ce species? Undercoordinated metal atoms? What is the role of ceria in this interaction? Zhou et al. [61] reported a large amount (~38%) of Ce^{3+} at the interface of CeO_2 islands deposited on gold microcrystalline films, but could not probe the state of gold due to the presence of the majority species, i.e., metallic gold.

In a recent paper, Cargnello et al. [62] reported another way to prove that Pt, Pd, and Ni metals on ceria are active through the perimeter (or just the corner) atoms of the respective metal nanoparticles adhering to the ceria surfaces. They show a size-independent CO oxidation reaction rate and correlate the latter to the fraction of available perimeter (or corner) atoms for each nanoparticle size. The latter were prefabricated using colloidal methods to certain sizes before depositing them on ceria. Their findings are in agreement with a previous report by Zhou et al. [61] who used alternating gold and ceria nanolayers of different thickness but the same number of interfaces to form "nanotowers" on Si wafers using vapor deposition in an E-beam apparatus. The CO oxidation reaction rate was correlated with the length of the interfaces; it was independent of the metal nanolayer size. Figure 4 shows these findings. However, neither of these two studies could identify the active sites at the metal–ceria interface, which presumably are the metal atoms ligated through –O bonds to the support.

In their well-cited 2003 paper, Fu et al. [7] showed that the interaction of ceria and platinum or gold is important and atomically dispersed Pt(Au)–O_x species on the ceria surface catalyze the WGS reaction. These are residual atoms from the removal (by cyanide leaching) of the metal nanoparticles, initially present on ceria, and from additional isolated metal atoms (cations) on the surface strongly bound to the ceria, and resisting leaching. Thus, efficient design of these types of catalysts should avoid particle formation and focus only on atomic dispersions of the metals on ceria [9].

Studies of metal adatoms on ceria single crystal surfaces can improve our understanding of the interaction between metals and ceria. The metal center, as MO_x, is stabilized through oxygen binding to the various facets of ceria [63]. This picture as shown in Figure 5 has emerged as a more precise depiction of the metal–ceria support interaction, as discussed in a recent review [9].

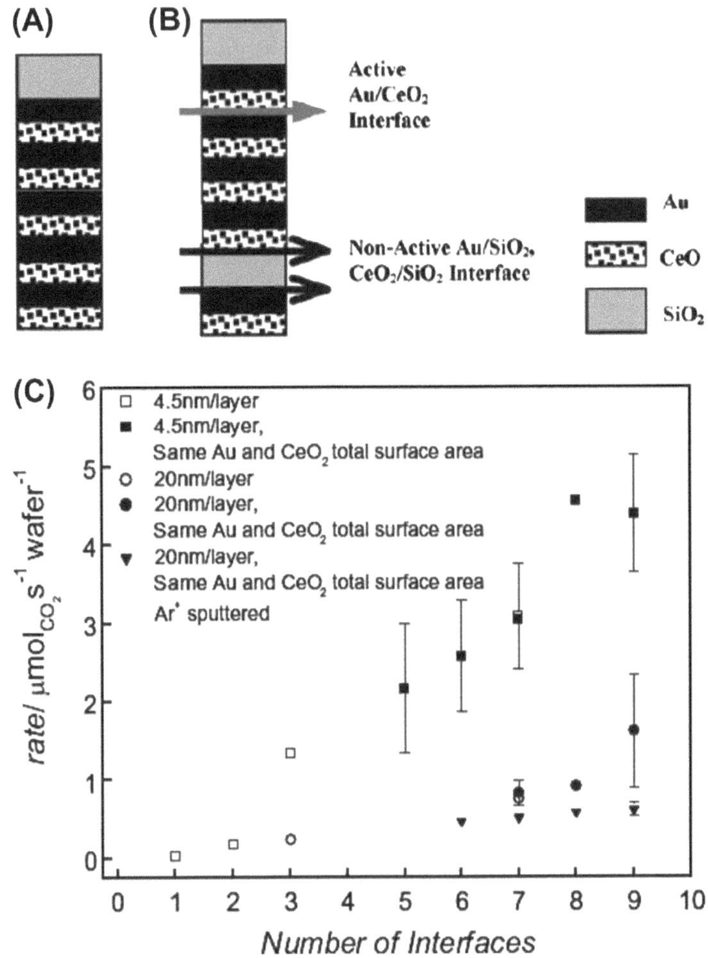

Figure 4 (Top) Sketch of a single nanotower Au/CeO_2 with the same Au and CeO_2 surface areas and different numbers of active interfaces: (A) 9 and (B) 8. (C) CO oxidation rate as a function of number of Au–CeO_2 interface in a nanotower. *Reproduced from Ref. [61]. Copyright 2009, Wiley.*

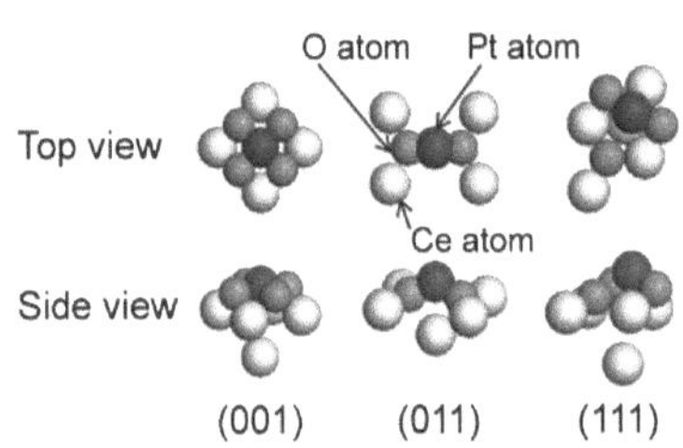

Figure 5 Predicted Pt anchoring site structure at (001), (011), and (111) surfaces of ceria. *Reproduced from reference [59]. Copyright 2010, Elsevier.*

Constrained ab initio thermodynamic simulation was applied to study the nature of small Pt_n (n = 1–10) on the stoichiometric and partially reduced CeO_2 {111} surfaces [64]. This agrees well with the experimental evidence that the presence of Pt clusters promoted the reducibility of the ceria surface. Under the WGS reaction conditions, oxygen vacancies and vacancy clusters are thermodynamically stable, while oxygen adatoms are not stable on the Pt cluster. Noninterfacial Pt sites are prone to be covered by CO molecules. But the presence of these co-adsorbed CO species does not change the redox behavior of the ceria surface significantly.

The comparison of the Group IB metals (Cu, Ag, and Au) on CeO_2 {111} has indicated that the energetically favorable configuration for Cu and Ag adsorption is at the threefold O-top(O_d) site, whereas for Au adsorption is the O-bridge (O_u–O_u) site. Cu atom exhibits the largest adsorption energy, followed by Ag and then Au. Adsorption is accompanied by a charge transfer process, with the adsorbed metal atom oxidized and one of the neighboring Ce^{4+} ions reduced [65].

4. GOLD/CERIA CATALYSTS: THEORETICAL STUDIES, PREPARATION METHODS, CHARACTERIZATION, AND CATALYSIS

The Au/CeO_2 system is the newest of all the ceria-supported metals, as gold saw its comeback to catalysis only in the late 1980s after the pioneering work of Giuseppe Parravano and his students on gold catalysts in the 1960s [66]. An exponential growth of published reports on gold and the plethora of the reactions it catalyzes has taken place over the last 25 years. Au/CeO_2 is a powerful oxidation catalyst, vying the activity of the PGM/CeO_2. The CO oxidation reaction was first studied on Au/CeO_2 by Hoflund and co-workers [67]. This system was soon followed by a detailed structural and kinetic study by Liu and Flytzani-Stephanopoulos [54]. The latter reference also established the Au/CeO_2 as an active catalyst for methane oxidation. Our understanding of the gold/ceria activity at that time was incomplete. We could account only in part for the Au–O_x–Ce interaction because we had no knowledge of the presence of sub-nm and atomic gold on the ceria, and we were not aware of the importance of these species as the active sites for these reactions. In 2001, Fu et al. [35] first reported that the Au/CeO_2 system was also an excellent low-temperature WGS catalyst. Some interesting correlations were obtained in that work. Nanoscale ceria was superior to bulk ceria, and the gold loading correlated with activity for reasons that became evident and reported a few years later [7,68]. Indeed, the WGS reaction takes place only at the interfacial Au–O_x–Ce sites, the number of which increases with gold loading when there is a lot of surface oxygen on the ceria, a property of the La-doped and nanocrystalline ceria used in that work. Removing the gold nanoparticles by NaCN solutions at high pH left nothing but the strongly interacting Au–O_x–Ce atomic species on the surface. But the mass or surface area-specific reaction rates were invariant to the massive (90%) loss of gold. Clearly, the metallic gold particle surfaces did not provide any active sites for this reaction. Similar results were found for Pt/ceria [7,69].

Since ceria can anchor and stabilize oxidized Pt (see Figure 4) or Au species on its surface, other reactions that operate on oxidized metals will be catalyzed by this system. For example, the oxidative self-coupling of methanol to methylformate takes place in $CH_3OH + O_2$ gas mixtures over oxidized gold [70] and has been shown to occur over the oxidized gold atoms on ceria and zinc oxide [71]. It is therefore important for practical catalyst designs to review the preparation methods that would lead to ceria (or other oxide) surfaces with a maximum amount of atomic Au–O_x sites. This is undertaken in the section following the theoretical studies of the Au/CeO_2 system.

Theoretical studies. Considerable progress has been made in theoretical investigations of gold-based catalysts [72]. The first theoretical computation exploring the adsorption of gold on ceria in configurations active for the WGS reaction was reported by Liu et al. in 2005 [73]. Single gold adatoms were considered on both the stoichiometric and the partially reduced CeO_2 {111} surfaces. A single gold atom anchored atop an oxygen atom was a preferred location, as shown in Figure 6. The same authors also explored up to four gold adatoms on CeO_2 {111}. The first gold adatom was assumed to occupy an oxygen vacancy site, and the others progressively occupied the adjacent sites lying nearly atop oxygen atoms. They concluded that once the number of adsorbed gold atoms exceeded the number of available oxygen vacancies, clusters would grow preferentially around the gold prefilled site, rather than undergoing nucleation on the patches of the stoichiometric surface.

More recently, the calculations were redone: on stoichiometric CeO_2 {111} surface, the most stable adsorption site of Au bridges two oxygen atoms based on GGA + U results [74]. Castellani et al. [75] have investigated the binding of gold adatoms at step sites of ceria. Their model incorporated both convex and concave step morphology.

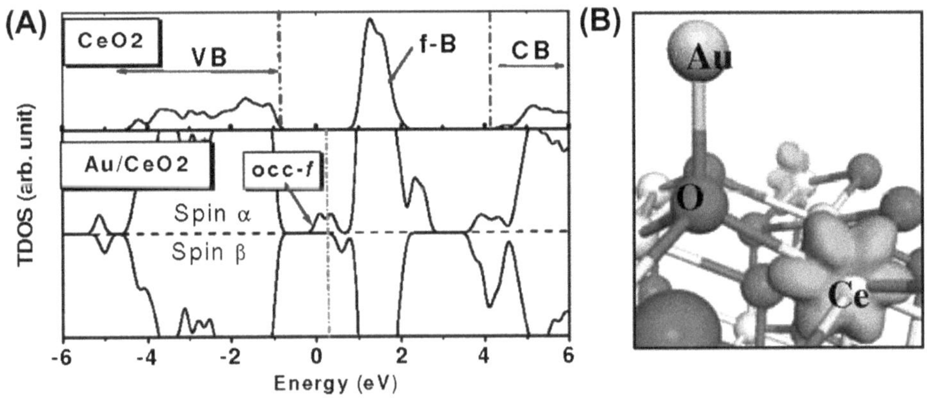

Figure 6 (A) Total density of states of CeO_2 (111) and of Au/CeO_2(111). (B) Three-dimensional contour plot of the spin density of the Au/CeO_2(111). *Reproduced from Ref. [73]. Copyright 2005, the American Physical Society.*

Calculation results identified a trivial degree of charge transfer between surface and adsorbate and suggested gold adsorption in most of the step-related sites is unperturbed relative to the planar surface.

Since the reaction gas and temperature control the reducibility of the catalyst surface, under certain reducing conditions, it is to be expected that the single (oxidized) gold atom will be destabilized and diffuse to form clusters. Thus, it is of interest to also explore the behavior of gold clusters on ceria surfaces.

By applying the GGA + U methodology, Zhang et al. [76] have studied gold clusters of up to 11 atoms on the CeO_2 {111} surface. The gold atom was anchored at a single oxygen vacancy, and the stability of the structure against unexpected rearrangement was tested after adding each new gold atom. The three-dimensional clusters were suggested to be energetically favored in general. Also the result implied that gold clusters may exhibit *fcc*/*hcp* polymorphism. These structures may be relevant for some reactions, but this must be evaluated for the various potential mechanisms.

Preparation Methods. The properties of gold catalysts, such as the gold particle size, the oxidation state, and the interaction between gold and supports, strongly depend on the preparation methods [77]. A variety of preparation methods are available, including impregnation (IMP), co-precipitation (CP), and deposition–precipitation (DP). The primary goal of those synthesis approaches is to maximize the interaction between the gold precursor and the supports.

Aurochloric acid ($HAuCl_4$) is widely used as the gold source because of its relatively low cost compared with other gold compounds. However, residual chloride ions have been found to promote mobility and agglomeration of gold species during heat treatments [78]. The extent of hydroxylation of gold changes with the pH value of solution, as shown in Figure 7 [79]. Special considerations are required if ammonia is used as a base to prepare the gold catalysts, because the chemistry of fulminating gold is a safety issue [80]. For example, the $[Au(NH_3)_4]^{3+}$ complex can only be obtained by very slow addition of concentrated ammonia to a solution of $HAuCl_4$ while keeping the value of pH lower than 5.

IMP is the simplest method to prepare gold catalysts. However, additional steps are required to remove the residual Cl^-, thus it is not a practical method to obtain well-dispersed gold particles. CP is a better method to get catalysts with smaller particles. But it is difficult to control the rates of precipitation of the gold and support metal hydroxides [81]. A good alternative has been found in the urea gelation co-precipitation (UGC) method which has often been used to prepare active gold–ceria catalysts. The UGC method produces a smaller ceria particle size (~4 nm), an effect of gold content, some of which remains in the bulk after calcination. The amount of gold on the surface during reaction depends on the gas composition and heat treatment used [7,35,68].

The DP method is widely used for preparing gold supported on oxide catalysts, aimed at controlling the gold particles to smaller sizes. Gold hydroxide is supposed to precipitate on the oxide supports [82]. The preparation variables, such as pH,

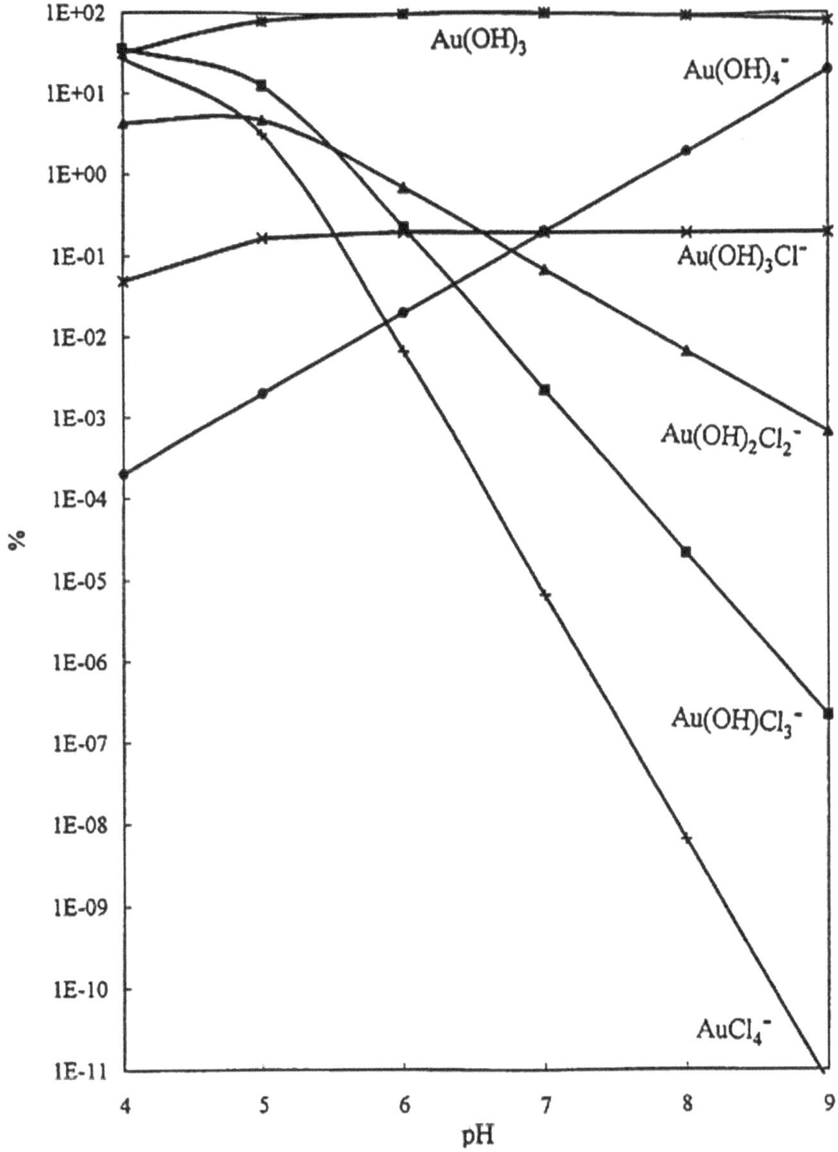

Figure 7 Calculated fractions of different gold species in solution as a function of the pH values at 300 K. *Reproduced from Ref. [79]. Copyright 1998, Elsevier.*

temperatures of preparation and washing, and the choice of bases, have been investigated in order to optimize the gold particle sizes [83,84]. This method works well with supports having a point of zero charge greater than 5, which includes ceria, titania, zinc oxide, iron oxide, and magnesia. Smaller gold particles were obtained on ceria when urea was included into the DP method [85]. This can be explained as the hydrolysis only occurs when the temperature rises above 80 °C. Thus a gradual and homogeneous release of hydroxyl ions takes place. Moreover, it was found that all the gold in solution was deposited onto the supports within the first hour.

It is noteworthy that the initially prepared structures will evolve according to the reaction conditions. Metastable structures may quickly change therefore the final structure is dictated by the subsequent treatment, i.e., gas composition and temperature. It is also challenging to disperse gold well on some "inert" supports, such as SiO_2 and Al_2O_3, because the interaction between gold and these supports is weaker compared with that of reducible oxides. When gold agglomeration takes place, a few reports of posttreatments have been demonstrated to redisperse the gold. For example, air treatment above 400 °C has been found to redisperse the gold [86]. In an interesting recent report, Sá et al. [87] have found that CH_3I treatment of gold particles on oxide supports at 240 °C redisperses them fully to the atomic state. The extent of the redispersion depends on the type of the oxide support used (TiO_2, SiO_2, Al_2O_3). Accordingly, the gold particle size was greatly decreased when CH_3I interacted with gold/alumina. However, the gold/silica displayed weaker response to this treatment. The addition of hydroxyls was reported by Veith et al. [88] to facilitate the formation of oxidized gold species in the range of 1.2–1.3 nm on the surfaces of titania. Strongly bound OH groups were suggested to structurally stabilize gold clusters. Absent the hydroxyls rapid agglomeration of gold took place even at room temperature and led to the formation of large gold particles on the titania surface.

Characterization. TEM and high-angle annular dark-field-scanning transmission electron microscopy have been used to image the gold/ceria system.

In situ TEM has identified the Ostwald ripening mechanism as dominant for the growth of Au particles on CeO_2 by simple heating [89]. Gold nanoparticles with crystalline structure, were found to shrink progressively to smaller and smaller particles to a final atomic layer on the ceria under the electron beam. This process reversed itself when the beam was cut off. A preferential orientation relationship between gold particles and the ceria was also identified; namely, $(111)\,[\bar{1}10]_{CeO_2}//(111)\,[\bar{1}10]_{Au}$ and $(111)\,[\bar{1}10]_{CeO_2}//(111)\,[\bar{1}10]_{Au}$. These observations revealed that changes depended on the oxidation state or the density of oxygen vacancies in the CeO_2 support, which essentially are determined by the type of oriented plane.

The Haruta group have applied aberration-corrected environmental transmission electronmicroscopy (ac-ETEM) to monitor the state of gold/ceria in situ during CO oxidation at room temperature [90]. They have shown that adsorbed carbon monoxide molecules caused the {100} facets of a gold nanoparticle to reconstruct during CO oxidation at room temperature. Theoretical calculations and image simulations confirmed the energetic favorability of this reconstructed structure. With the aid of ETEM, the Shen group reported that gold particles of 2–4 nm size which are strongly anchored onto ceria nanorods display high stability under CO oxidation and the WGS reaction conditions. Those strongly bound gold particles changed their shapes in response to alternating oxidizing and reducing atmosphere, but did not sinter at temperature up to 300 °C [91]. The interaction between gold and the defect oxygen sites of a support is important to stabilize gold atoms and clusters on oxides and inhibit their sintering. Allard et al. have shown that gold clusters on Au/FeO_x are stable under heating (up to 700 °C). Gold sintering did not begin below 500 °C, and the size of gold nanoparticles remained less than 2 nm in diameter even upon heating to 700 °C [92]. The strong interaction of gold adatoms on $Fe_3O_4(111)$ surfaces was first documented by scanning tunneling microscopy by Rim et al. [93], who showed that single atoms of gold are located atop the uncapped oxygen of the Fe-terminated Fe_3O_4 (111) surface and are stable in heating to 500 °C. CO adsorbs atop these (slightly oxidized) gold atoms.

Pioneering work done by the Gates group showed how in situ XANES and extended X-ray absorption fine structure (EXAFS) spectroscopy could be used to follow changes taking place on the structure of gold during reaction. For this measurement to be meaningful, the initial state of gold should be uniform; in the case of the Gates' work, the initial state was that of mononuclear gold carefully deposited on MgO [94,95] or zeolite supports [96] from Au(III) $(CH_3)_2$acetyl acetonate from the vapor and decomposed at mild temperatures. Isolated gold cations supported on MgO were identified as the exclusive active sites for the ethylene hydrogenation reaction [94], and for the CO oxidation reaction [95], the latter increasing activity as Au(III) was changing to Au(I). When this technique is applied to gold (or other) catalysts comprising a mixture of atoms, nanoclusters, and nanoparticles, averaging is unavoidable and a lot of inaccurate conclusions about the active sites are drawn. To avoid such effects, and to examine the evolution of isolated gold cations on ceria during the WGS reaction, Flytzani-Stephanopoulos and co-workers used cyanide solutions to leach gold nanoclusters and nanoparticles from a ceria surface, and reported for the first time that under the reaction, the surface gold changes to a mixture of oxidized and reduced gold species in amounts depending on the oxygen potential of the reaction mixture and temperature [86]. However, the initial state, i.e., isolated gold cations had the maximum activity. Figure 8 shows in situ XANES data collected on $0.75AuCeO_2$, a cyanide-leached catalyst, further calcined in air at 400 °C, and initially containing Au cations only, as verified by Au–Au coordination number = 0 by EXAFS analysis of the same data. Air regeneration at 400 °C was demonstrated to redisperse the gold and recover the initial activity [86]. In following years, other workers have employed similar approaches to follow the gold active site evolution in situ. For example, isolated gold cations were found as the active sites for 1,3-butadiene hydrogenation by the use of cyanide leached gold/ceria catalysts [97].

In situ FTIR spectroscopy is a useful technique to obtain mechanistic insight, by carefully evaluating the adsorbed species and their evolution during reaction, assuming well-defined and stable surface structures are used. The same averaging problem as explained above would affect the accuracy of the IR results when the structure is heterogeneous. Manzoli et al. [98] used FTIR to monitor the CO adsorption on a leached $Au/Ce(La)O_x$

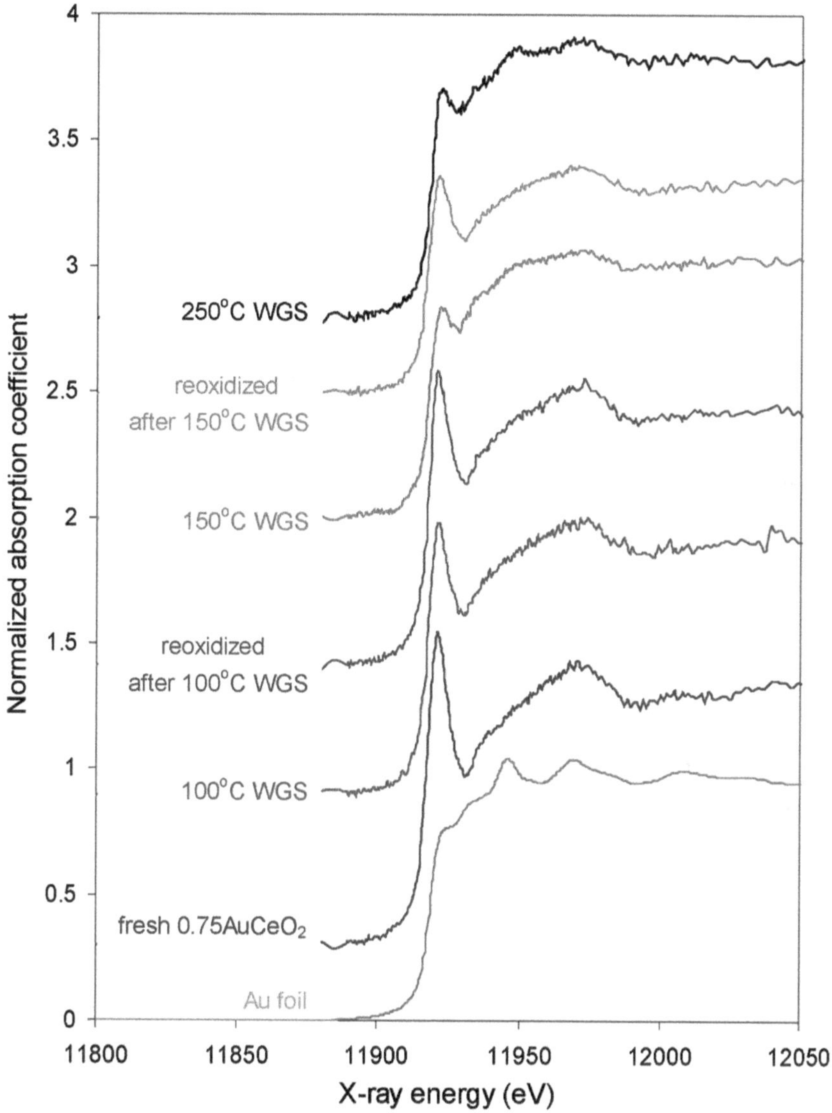

Figure 8 In situ XANES spectra of 0.75AuCeO_2, initially containing isolated gold cations, collected in WGS reaction conditions with 1%CO–3%H_2O-bal.He; and in re-oxidation in 5%CO_2/He at 150 °C, which partially restores the gold as Au–O_x species. *Reproduced from Ref. [86]. Copyright 2008, the American Chemical Society.*

surface containing only gold cations with FTIR. The gold cations, stable under the conditions employed, showed a maximum CO absorbance at 2130–2140/cm, which was more resistant to oxidation than the usual carbonyl band on Au^0 sites. Meunier et al. [99] applied combined SSITKA/DRIFTS to identify the reaction intermediates on the surface of the same Au/Ce(La)O_x catalyst. Under WGS reaction conditions (i.e., 2%CO + 7%H_2O), the rate of CO_2 formation was much faster than the rate of decomposition of surface formates, pointing to carbonates rather than formates as the reaction intermediates. The dynamic SSITKA/DRIFTS characterization should be used when one needs to discriminate the reaction intermediates from spectator species adsorbed on the surface [100].

Catalysis. Nanostructured gold/ceria has been examined as a catalyst for a variety of reactions, especially oxidation reactions, beginning with the CO oxidation [54,67,101], low-temperature methane combustion [54], the WGS reaction [7,35,47,68,86,98,99,102], the complete oxidation of volatile organic compounds [103], and the CO-PROX reaction [101].

Over the last five years, new reactions were found to be catalyzed most efficiently by the Au/ceria catalysts at low temperatures. An example is the steam reforming of methanol which was first reported by Yi et al. in 2010 as a very selective catalyst for H_2 production [104]. A wide operating temperature window (150–275 °C) is possible with this catalyst with H_2 and CO_2 as the only products, because the reaction is not limited by the WGS reaction equilibrium [104,105]. In these low-temperature applications, the Au/CeO_2 system is intrinsically more active than the commercial Cu/ZnO, and stable to air exposure, unlike the pyrophoric Cu/ZnO catalyst.

When hydrogen production is intended for PEM fuel cell use, the catalyst in the upstream WGS reactor must be stable to the operational requirements of the fuel cell system, i.e., frequent shutdowns to room temperature with water condensation and accidental exposures to air. The ceria supports suffer from instability due to the formation of surface Ce(III) hydroxycarbonate in shutdown in CO_2 and water mixtures. We have found that this problem is solved by including small (0.5–1%) amounts of oxygen in the feed gas which stabilizes both the Au/CeO_2 and the Pt/CeO_2 catalysts in cyclic shutdown/restart operation [106]. O_2−assisted WGS operation also improves the high-temperature (300 °C) stability of the Au/CeO_2 catalysts.

5. IDENTIFICATION OF THE ACTIVE SITE IN AU/CEO$_2$ CATALYSTS FOR THE WATER-GAS SHIFTS AND THE STEAM REFORMING OF METHANOL

Over the last 15 years, a plethora of papers have reported the interface of the Au and CeO_2 as the locus of active sites for oxidation reactions; some indirectly, and others directly like in the work of Zhou with precisely constructed nanotowers of alternating ceria and gold layers [61] described earlier, and by the use of inverse structures, as in the work of Rodriguez and coworkers [60] and Zhou et al. [61], also mentioned in Section 3. But the active site could not be identified in these materials due to the dominance of metallic gold. From the earliest work [35], it was clear that nanoscale ceria mattered. The surface oxygen on such ceria was higher in amount (higher OSC) than on annealed ceria and so was the activity of the final Au/ceria catalyst. Along that line, La-doped ceria was superior as La addition preserved the nanosize of ceria to high temperatures and increased its surface oxygen [35,55].

If nanoceria was important for the Au/CeO_2 catalyst, as was to be found later by others also for the CO oxidation reaction [107], what was the role of gold? Were particle size effects important? And if not, what was the active gold structure at the interface with ceria? Our initial WGS work [35] showed that the particle size of gold was not important for the activity. However, the gold loading was important. By the way of preparation (DP), the higher gold loadings ended up giving us more active sites (still unidentified) on the nanoscale Ce(La)O_2 used in that work. The latter apparently had a higher "capacity" for accommodating gold active sites. In 2003, Fu et al. [7] showed these sites to be Au–O_x sites, but not metallic gold. Indeed, all the metallic gold nanoparticles and nanoclusters were first removed by cyanide solutions; the derivative material was heated to 400 °C in air and characterized by XPS to show only oxidized gold on the surface. Only ~0.5 wt% Au stably anchored on ceria through –O linkages remained. Unexpectedly, this contained all the active sites of the parent material (5 wt% gold). The specific (per gram catalyst, not gold) reaction rates remained unchanged. Unchanged was also the surface oxygen of ceria which stabilized the gold. We thus had a technique that could identify the surface "capacity" of ceria for gold. We did not exactly understand it as such initially, but subsequent work proved the point.

To further examine how we could increase the number of atomically dispersed Au–O_x sites, we considered potential "shape effects" of ceria [47]. As discussed in Section 2, there were facile (hydrothermal) methods available to make single crystals of ceria at the nanoscale with enough surface area to be used as catalyst support. In a first demonstration of utilizing different single-crystal nanoshapes as supports of a metal (in our case, gold), we reported a strong shape effect for nanorods, nanooctahedra, and nanocubes of ceria, each supporting the same amount of gold (1 wt%) [47]. On the nanocube {100} surfaces, the gold was almost exclusively present as 3 nm-size particles, while on the nanorod {110} surfaces (and on the {111} surfaces of the nanooctahedra), the gold was fully dispersed, with 50–60% of it present as atoms, as leaching characterization showed. The gold from the cubes was almost fully removed (0.03% residual). The catalysts had very different activity because of the higher number of Au–O_x sites on the nanorods [47]. However, the intrinsic activity (TOF) of the leached catalysts (per gold atom) was the same and the apparent activation energy of the reaction was the same, ~41.0 kJ/mol [105]. Figure 9 shows the steady-state conversion of CO and the Arrhenius-type plots for the rates. We have thus identified the ceria effect as an "indirect" one. Ceria mattered but only as far as its number of anchoring sites for gold, a number that is different for each surface of ceria (different oxygen defect density). In other words, the active sites for the WGS reaction comprise Au–O_x sites anchored on ceria surfaces and displaying the same intrinsic activity.

Since 2008, several more papers have been published depicting shape effects for the Au/CeO_2 catalysts, both for the WGS [105,108], and for other reactions [91,108,109]. Other

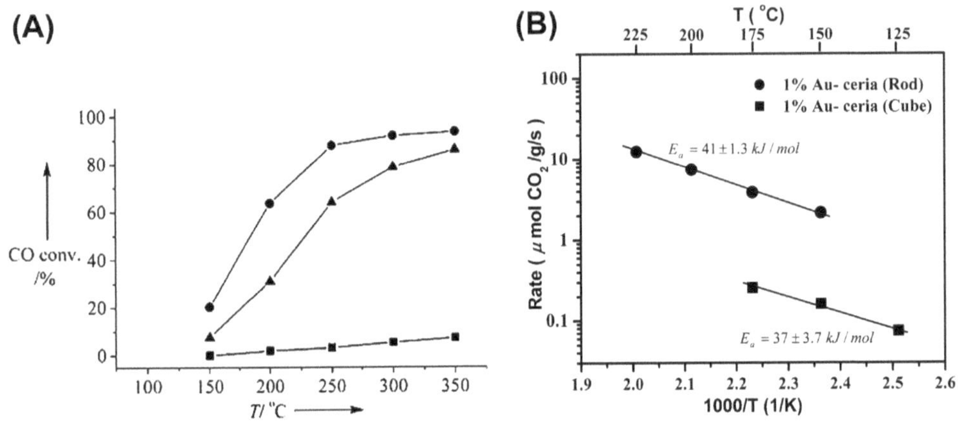

Figure 9 (A) WGS reaction "light-off" profiles for 1%Au on CeO_2 nanorods (●), cubes (■), and polyhedra (▲). *Reproduced from reference [47]. Copyright 2008, Wiley.* (B) Steady-state reaction rates for WGS reaction measured in 2%CO–10%H_2O–He. *Reproduced from Ref. [105]. Copyright 2010, the Royal Society of Chemistry.*

supports were also found to display indirect shape effects on gold, e.g., ZnO, Fe_3O_4, and TiO_2 for the WGS reaction [107,110]. In all cases, the apparent activation energies were the same, validating the Au–O_x site as the active site. Thus, the role of the support is to provide hydroxyl groups vicinal to the Au–O_x center to react with CO adsorbed on the site. However, the support does not participate in the reaction pathway. Figure 10 shows that the TOF of the WGS reaction on supported gold catalysts is invariant to the type of support used.

The atomically stabilized gold/ceria system can now be used to test other reactions occurring on oxidized gold sites. Yi et al. [104,105] first showed the Au/CeO_2 nanoshapes as a very active and highly selective catalyst for the steam reforming of methanol. This is briefly reviewed here.

While methanol steam reforming takes place on ceria, the "light-off" temperature is above 300 °C [104] and the selectivity to hydrogen poor. The addition of gold greatly improved the methanol conversion at lower temperatures (150–275 °C) with excellent selectivity (>98%) to CO-free hydrogen and CO_2. This is a wide temperature window for practical application of this new catalyst, vying that of Cu–ZnO catalysts but with much higher stability to air and water exposure. Figure 11 shows the key results obtained with this type of catalysts.

While this is an excellent catalyst for the WGS reaction, it is found that the WGS pathway is turned off completely, i.e., in the presence of methanol CO does not adsorb on the surface. From the TPRS of CH_3OH alone (in the absence of water vapor), we were able to identify the reaction intermediates by mass spectrometry. The pathway of methanol reactions is as follows:

$$2CH_3OH \rightarrow HCOOCH_3 + 2H_2$$

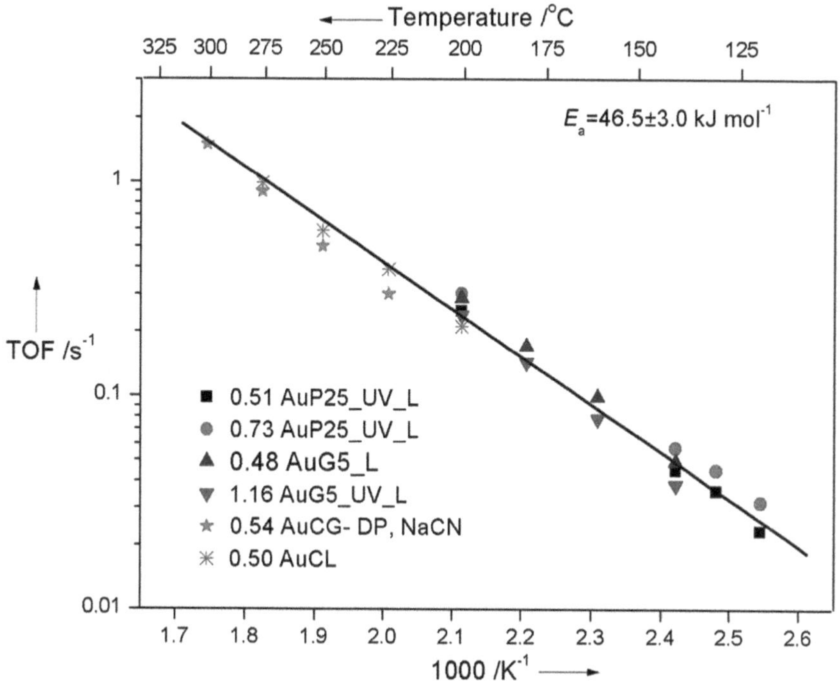

Figure 10 Arrhenius plot of TOFs for the WGS reaction over atomically dispersed gold catalysts retrieved after cyanide leaching and thermal treatment to 400 °C in air of the parent materials in a simulated reformate gas mixture (11% $CO/26\%H_2O/7\%CO_2/26\%H_2$ in He). Gold was supported on gadolinia-doped ceria (CG); lanthana-doped ceria (CL); P25 titania; and G5 titania (pure anatase). *Reproduced from Ref. [110]. Copyright 2013, the American Chemical Society.*

$$HCOOCH_3 + H_2O \rightarrow HCOOH + CH_3OH$$

$$HCOOH \rightarrow CO_2 + H_2$$

The reaction pathway was proposed as a cooperative mechanism for gold–ceria whereby reduced ceria adsorbs methanol as methoxy which then combines with the formaldehyde adsorbed on gold to produce methylformate. Adsorbed water helps to hydrolyze methylformate to formic acid, which is then decomposed to CO_2 and H_2. Because the WGS reaction was not involved in the reaction network, high CO_2 selectivity (>93%) over the whole range of reaction temperatures was achieved.

It was noted that the specific activity of parent and leached gold catalysts regardless of the type of ceria nanoshape was the same, as was the apparent activation energy of the reaction. Hence, the shape effect is indirect, simply manifesting the different ability of each ceria surface to bind gold, but ceria does not alter the mechanism of the steam reforming reaction.

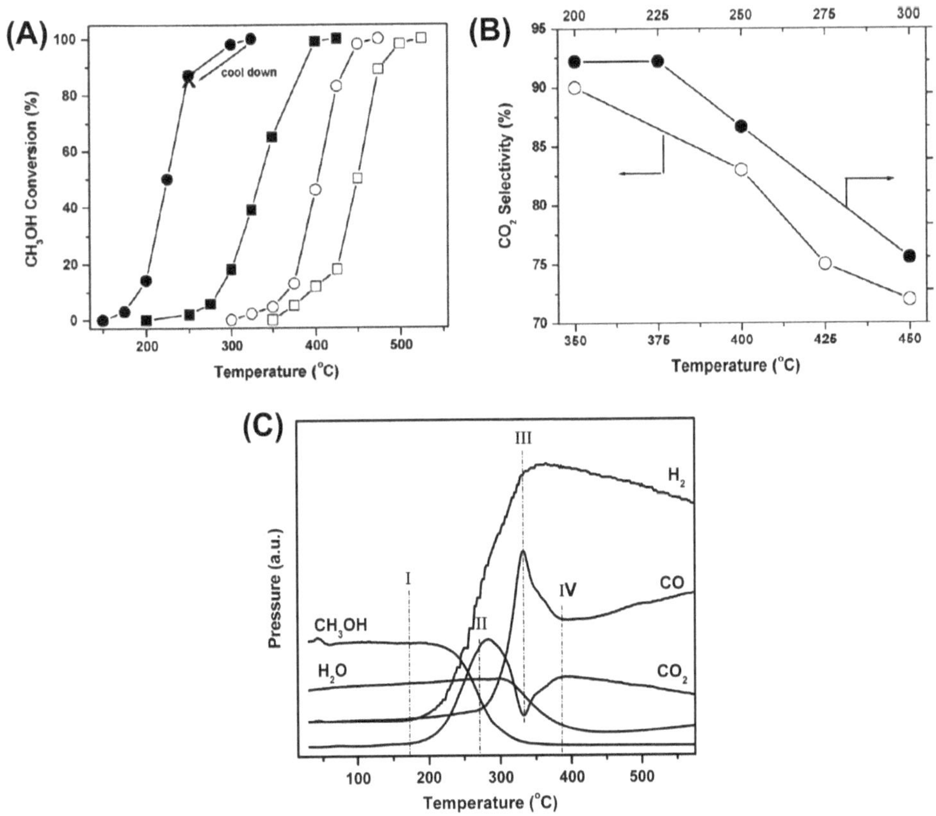

Figure 11 (A) CH_3OH conversion (%) in steady-state tests of steam reforming of methanol over nanoshapes of CeO_2 and Au–CeO_2; 1%Au–ceria (rod) (●); 1%Au–ceria (cube) (■); ceria (rod) (○); ceria (cube) (□); and 1%Au–ceria (rod) after cooling down to 250 °C (X); (B) CO_2 selectivity (%) in steady-state light off tests of steam reforming of methanol over nanoshapes of ceria (rod) (○) and 1%Au–ceria (rod) (●); 2%CH_3OH/2.6%H_2O/bal. He; GSHV = 42,000/h; and (C) (CH_3OH + H_2O)-TPSR over 1%Au–ceria (rod). *Reproduced from Ref. [104]. Copyright 2010, Elsevier.*

Working with Au on ZnO nanoshapes (rods and polyhedra), Boucher et al. [71] found the polar surface of ZnO(0001) to be the best support of the active gold sites. A similar apparent activation energy, ~110 kJ/mol, was observed on all Au/ZnO catalysts in methanol steam reforming. Moreover, the results from the temperature-programmed reaction suggested that Au/ZnO followed the same reaction pathway as that of Au/CeO_2. Thus, the active gold sites are of similar structure on both supports. From the practical viewpoint, these studies have identified gold catalysts, especially gold/ceria, as a suitable catalyst for hydrogen production through methanol steam reforming.

It is noteworthy that the mechanism discussed earlier agrees with the oxidized gold mechanism obtained from surface science studies of the oxidative self-coupling reaction of methanol to methylformate [70]. Work with single crystals of Au(110) [111] and

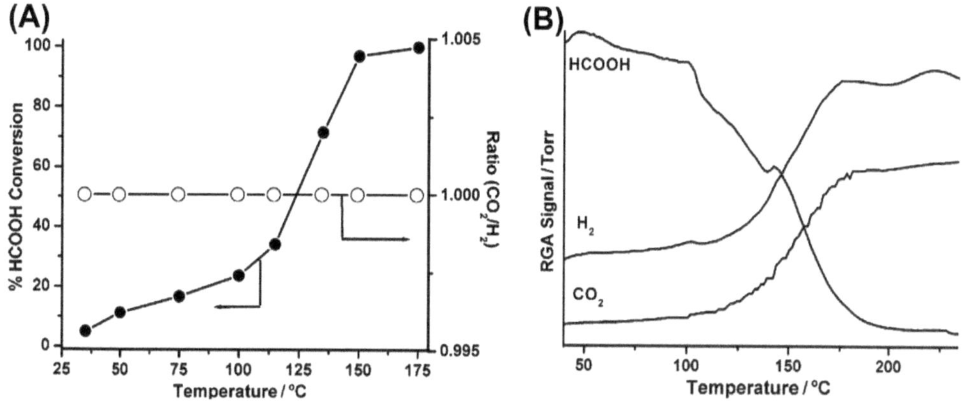

Figure 12 (A) Formic acid conversion (%) and product ratio (CO_2/H_2) at GHSV=42,000/h; (B) HCOOH-TPSR at heating rate 5 °C/min for decomposition of formic acid over Au/CeO_2(rod). *Reproduced from Ref. [113]. Copyright 2013, Wiley.*

nanoporous gold [112] produced the same results. Hence, the importance of Au–O_x species for this reaction irrespective of the choice of support is corroborated also from the (unsupported) nanoporous gold studies [112] and the investigations of single crystals of gold activated by atomic oxygen [70,111]. Thus, atomically stabilized Au–O_x species on various supports can be designed and new powerful catalysts developed for methanol oxidation and steam reforming, and methanol coupling reactions. The choice of the support will be dictated by the application at hand. Using Au/CeO_2 we can also demonstrate the dehydrogenation of formic acid as the exclusive pathway on gold catalysts [113]. Figure 12 shows the high activity and selectivity of this catalyst operating at near-ambient temperature [113].

6. OUTLOOK

Using various examples from the published literature, we have shown that the Au/CeO_2 system is a powerful oxidation catalyst and also an excellent choice for highly selective methanol steam reforming and formic acid decomposition to maximize hydrogen yields. The active sites do not reside on the gold particle surfaces, but at the interface with ceria. These are shown to be atomically stabilized Au–O_x sites anchored on ceria. The importance of ceria is demonstrated by studying its shape effect on the reactions. The shape effect is strong but indirect, i.e., the reaction occurs on the Au–O_x sites, but different shapes provide a different number of anchoring sites to gold, presumably related to the number of oxygen defects of the ceria surface, with {110} and {111} being superior to {100} surfaces. Different preparation protocols can be used to prepare more active catalysts. The stability of the active sites depends on the reaction gas used and the operating conditions. One approach to prevent deactivation even in cyclic operation with shutdown to ambient temperature was shown to be O_2–assisted operation. These findings may be used to guide the design of stable, atomically dispersed supported gold catalysts as well as other metals on ceria. They may also be extended to other supports either for fundamental study or to develop robust practical catalysts.

REFERENCES

[1] A.S.K. Hashmi, G.J. Hutchings, Angew. Chem. Int. Ed. 47 (2006) 7896.
[2] A.S.K. Hashmi, Chem. Rev. 107 (2007) 3180.
[3] A. Corma, H. Garcia, Chem. Soc. Rev. 37 (2008) 2096.
[4] M. Haruta, Faraday Discuss. 152 (2011) 11.
[5] M. Flytzani-Stephanopoulos, Acc. Chem. Res. 47 (2014) 783.
[6] J. Oliver-Meseguer, J.R. Cabrero-Antonino, I. Dominguez, A. Leyva-Perez, A. Corma, Science 338 (2012) 1452.
[7] Q. Fu, H. Saltsburg, M. Flytzani-Stephanopoulos, Science 301 (2003) 935.
[8] J. Lu, C. Aydin, N. Browning, B.C. Gates, Angew. Chem. Int. Ed. 51 (2012) 5842.
[9] M. Flytzani-Stephanopoulos, B.C. Gates, Annu. Rev. Chem. Biomol. Eng. 3 (2012) 545.
[10] J. Guzman, B.C. Gates, Angew. Chem. Int. Ed. 42 (2003) 690.
[11] M. Haruta, T. Kobayashi, H. Sano, N. Yamada, Chem. Lett. (1987) 405.
[12] P. Kracke, T. Haas, H. Saltsburg, M. Flytzani-Stephanopoulos, J. Phys. Chem. C 114 (2010) 16401.
[13] J.T. Calla, R.J. Davis, J. Catal. 241 (2006) 407.
[14] A. Trovarelli, P. Fornasiero, Catalysis by Ceria and Related Materials, second ed., Imperial College Press, 2013.
[15] W. Liu, A.F. Sarofim, M. Flytzani-Stephanopoulos, Appl. Catal. B 4 (1994) 167.
[16] T. Zhu, A. Dreher, M. Flytzani-Stephanopoulos, Appl. Catal. B 21 (1999) 103.
[17] L.F. Cordoba, M. Flytzani-Stephanopoulos, C.M. de Correa, Appl. Catal. B 33 (2001) 25.
[18] H. Yamashita, A. Tomita, H. Yamada, T. Kyotani, L. Radovic, Energy Fuels 7 (1993) 85.
[19] A. Trovarelli, Catal. Rev. 38 (1996) 439.
[20] R.J. Gorte, AIChE J. 56 (2010) 1126.
[21] Z.A. Qiao, Z. Wu, S. Dai, ChemSusChem 6 (2010) 1821.
[22] M.V. Ganduglia, A. Hofmann, J. Sauer, Surf. Sci. Rep. 62 (2007) 219.
[23] J. Paier, C. Penschke, J. Sauer, Chem. Rev. 113 (2013) 3949.
[24] D.S. Newsome, Catal. Rev. Sci. Eng. 21 (1980) 275.
[25] H.L. Tuller, S.R. Bishop, Ann. Rev. Mater. Res. 41 (2011) 369.
[26] W. Song, A.P.J. Jansen, E.J.M. Hensen, Faraday Discuss. 162 (2013) 281.
[27] H.C. Yao, Y.F. Yao, J. Catal. 86 (1984) 254.
[28] C. Zhang, A. Michaelides, S.J. Jenkins, Phys. Chem. Chem. Phys. 13 (2011) 22.
[29] M. Nolan, S.C. Parker, G.W. Watson, Surf. Sci. 595 (2005) 223.
[30] N. Baudin, M. Wojcik, K. hermansson, Surf. Sci. 468 (2000) 51.
[31] T.X.T. Sayle, S.C. Parker, C.R.A. Catlow, Surf. Sci. 316 (1994) 329.
[32] Y. Li, W.J. Shen, Chem. Soc. Rev. 43 (2014) 1543.
[33] W.X. Huang, Y.X. Gao, Catal. Sci. Technol. 4 (2014) 3772.
[34] H.X. Mai, L.D. Sun, Y.W. Zhang, R. Si, W. Feng, H.P. Zhang, H.C. Liu, C.H. Yan, J. Phys. Chem. B 109 (2005) 24380.
[35] Q. Fu, A. Weber, M. Flytzani-Stephanopoulos, Catal. Lett. 77 (2001) 87.
[36] W. Wang, J.Y. Howe, Y. Li, X. Qiu, D.C. Joy, M.P. Paranthaman, M.J. Doktycz, B. Gu, J. Mater. Chem. 20 (2010) 7776.
[37] M. Hirano, E. Kato, J. Am. Ceram. Soc. 79 (1996) 777.
[38] Y. Madier, C. Descorme, A.M.L. Govic, D.J. Duprez, J. Phys. Chem. B 103 (1999) 10999.
[39] F. Zhang, S.-W. Chan, J.E. Spanier, E. Apak, Q. Jin, R.D. Robinson, I.P. Herman, Appl. Phys. Lett. 80 (2002) 127.
[40] F. Zhang, C.H. Chen, J.M. Raitano, J.C. Hanson, W.A. Caliebe, S. Khalid, S.-W. Cham, J. Appl. Phys. 99 (2006) 084313.

[41] Z.L. Wu, M. Liu, J. Howe, H.M. Meyer III, S.H. Overbury, Langmuir 26 (2010) 16595.
[42] C. Li, K. Domen, K. Maruya, T. Onishi, J. Catal. 123 (1990) 436.
[43] M. Huang, S. Fabris, Phys. Rev. B 75 (2007) 081404.
[44] Z.L. Wu, M.J. Li, S.H. Overbury, J. Catal. 285 (2012) 61.
[45] R.M. Ferrizz, G.S. Wong, T. Egami, J.M. Vohs, Langmuir 17 (2001) 2464.
[46] D.R. Mullins, M.D. Robbins, J. Zhou, Surf. Sci. 600 (2006) 1547.
[47] R. Si, M. Flytzani-Stephanopoulos, Angew. Chem. Int. Ed. 47 (2008) 2884.
[48] Z.L. Wu, M.J. Li, D.R. Mullins, S.H. Overbury, ACS Catal. 2 (2012) 2224.
[49] C.H. Bartholomew, R.J. Farrauto, Fundamentals of Industrial Catalytic Processes, second ed., Wiley, 2005.
[50] A. Trovarelli, Catal. Rev. Sci. Eng. 38 (1996) 439.
[51] E. Aneggi, D. Wiater, C. de Leitenburg, J. Llorca, A. Trovarelli, ACS Catal. 4 (2014) 172.
[52] K.B. Zhou, X. Wang, X.M. Sun, Q. Peng, Y.D. Li, J. Catal. 229 (2005) 206.
[53] X.W. Liu, K.B. Zhou, L. Wang, B.Y. Wang, Y.D. Li, J. Am. Chem. Soc. 131 (2009) 3140.
[54] W. Liu, M. Flytzani-Stephanopoulos, J. Catal. 153 (2005) 304.
[55] L. Kundakovic, M. Flytzani-Stephanopoulos, J. Catal. 179 (1998) 203.
[56] Y. Li, Q. Fu, M. Flytzani-Stephanopoulos, Appl. Catal. B 27 (2000) 179.
[57] R. Si, J. Raitano, N. Yi, L.H. Zhang, S.-W. Chan, M. Flytzani-Stephanopoulos, Catal. Today 180 (2012) 68.
[58] W. Liu, M. Flytzani-Stephanopoulos, Catal. Today 28 (1996) 391.
[59] N. Yi, Y. Cao, Y.M. Liu, W.L. Dai, H.Y. He, K.N. Fan, Chem. Lett. (2005) 108.
[60] J.A. Rodriguez, S.D. Senanayake, D. Stacchiola, P. Liu, J. Hrbek, Acc. Chem. Res. 47 (2014) 773.
[61] Z. Zhou, M. Flytzani-Stephanopoulos, S. Kooi, H. Saltsburg, Adv. Funct. Mater. 18 (2008) 280.
[62] M. Cargnello, V.V.T. Doan-Nguyen, T.R. Gordon, R.E. Diaz, E.A. Stach, R.J. Gorte, P. Fornasiero, C.B. Murray, Science 341 (2013) 771.
[63] M. Hatanaka, N. Takahashi, T. Tanabe, Y. Nagai, K. Dohmae, Y. Aoki, T. Yoshida, H. Shinjoh, Appl. Catal. B 99 (2010) 336.
[64] S. Aranifard, S.C. Ammal, A. Heyden, J. Phys. Chem. C 116 (2012) 9029.
[65] Y.H. Tang, H. Zhang, L.X. Cui, C.Y. Ouyang, S.Q. Shi, W.H. Tang, H. Li, L.Q. Chen, J. Power Sources 197 (2012) 28.
[66] W.N. Delgass, M. Boudart, G. Parravano, J. Phys. Chem. 72 (1968) 3563.
[67] S.D. Gardner, G.B. Hoflund, D.R. Schryer, J. Schryer, B.T. Upchurch, E.J. Kielin, Langmuir 7 (1991) 2135.
[68] Q. Fu, W.L. Deng, H. Saltsburg, M. Flytzani-Stephanopoulos, Appl. Catal. B 56 (2005) 57.
[69] D. Pierre, W.L. Deng, M. Flytzani-Stephanopoulos, Top. Catal. 44 (2007) 199.
[70] B. Xu, R.J. Madix, C.M. Friend, J. Am. Chem. Soc. 132 (2010) 16571.
[71] M.B. Boucher, N. Yi, F. Gittleson, B. Zugic, H. Saltsburg, M. Flytzani-Stephanopoulos, J. Phys. Chem. C 115 (2011) 1261.
[72] P. Pyykkö, Angew. Chem. Int. Ed. 43 (2004) 4412.
[73] Z.P. Liu, S.J. Jenkins, D.A. King, Phys. Rev. Lett. 94 (2005) 196102.
[74] C. Zhang, A. Michaelides, D.A. King, S.J. Jenkins, J. Chem. Phys. 129 (2008) 194708.
[75] N.J. Castellani, M.M. Branda, K.M. Neyman, F. Illas, J. Phys. Chem. C 113 (2009) 4948.
[76] C. Zhang, A. Michaelides, D.A. King, S.J. Jenkins, J. Am. Soc. Chem. 132 (2010) 2175.
[77] G.C. Bond, C. Louis, D.T. Thompson, Catalysis by Gold, Imperial College Press, 2006.
[78] H.H. Kung, M.C. Kung, C.K. Costello, J. Catal. 216 (2003) 425.
[79] C.K. Chang, Y.J. Chen, C.T. Yeh, Appl. Catal. A 174 (1998) 13.
[80] J.A. Cusumano, Nature 247 (1974) 456.
[81] S. Tsubota, M. Haruta, T. Kobayashi, A. Ueda, Y. Nakahra, Stud. Surf. Sci. Catal. 72 (1991) 695.
[82] M. Haruta, Catal. Today 36 (1997) 153.
[83] A. Wolf, F. Schuth, Appl. Catal. A 226 (2002) 1.
[84] G.C. Bond, D.T. Thompson, Catal. Rev. Sci. Eng. 41 (1999) 319.
[85] R. Zanella, L. Delannoy, C. Louis, Appl. Catal. A 291 (2005) 62.
[86] W.L. Deng, A.I. Frenkel, R. Si, M. Flytzani-Stephanopoulos, J. Phys. Chem. C 112 (2008) 12834.
[87] J. Sá, S.F.R. Taylor, H. Daly, A. Goguet, R. Tiruvalam, Q. He, C.J. Kiely, G.J. Hutchings, C. Hardacre, ACS Catal. 2 (2012) 552.

[88] G.M. Veith, A.R. Lupini, N.J. Dudney, J. Phys. Chem. C 113 (2009) 269.
[89] T. Akita, M. Okumura, K. Tanaka, M. Kohyama, M. Haruta, J. Mater. Sci. 40 (2005) 3101.
[90] H. Yoshida, Y. Kuwauchi, J.R. Jinschek, K.J. Sun, S. Tanaka, M. Kohyama, S. Shimada, M. Haruta, S. Takeda, Science 335 (2012) 317.
[91] N. Ta, J. Liu, S. Chenna, P.A. Crozier, Y. Li, A. Chen, W.J. Shen, J. Am. Chem. Soc. 134 (2012) 20585.
[92] L.F. Allard, A. Borisevich, W.L. Deng, R. Si, M. Flytzani-Stephanopoulos, S.H. Overbury, J. Electron. Microsc. 58 (2009) 199.
[93] K.T. Rim, D. Eom, L. Liu, E. Stolyarova, J.M. Raitano, S.-W. Chan, M. Flytzani-Stephanopoulos, G.W. Flynn, J. Phys. Chem. C 113 (2009) 10198.
[94] J. Guzman, B.C. Gates, J. Phys. Chem. B 106 (2002) 7659.
[95] J. Guzman, B.C. Gates, J. Catal. 226 (2004) 111.
[96] J.C. Fierro-Gonzalez, B.C. Gates, Langmuir 21 (2005) 5693.
[97] Y. Guan, E.J.M. Hensen, Phys. Chem. Chem. Phys. 11 (2009) 9578.
[98] M. Manzoli, F. Boccuzzi, A. Chiorino, F. Vindigni, W.L. Deng, M. Flytzani-Stephanopoulos, J. Catal. 245 (2007) 306.
[99] F.C. Meunier, D. Reid, A. Goguet, S. Shekhtman, C. Hardacre, R. Burch, D. Deng, M. Flytzani-Stephanopoulos, J. Catal. 247 (2007) 277.
[100] F.C. Meunier, Chem. Soc. Rev. 39 (2010) 4602.
[101] J. Guzman, B.C. Gates, J. Am. Chem. Soc. 126 (2004) 2672.
[102] W. Deng, J. DeJesus, H. Saltsburg, M. Flytzani-Stephanopoulos, Appl. Catal. A 291 (2005) 126.
[103] S. Scire, S. Minico, C. Crisafulli, C. Satriano, A. Pistone, Appl. Catal. B 40 (2003) 43.
[104] N. Yi, R. Si, H. Saltsburg, M. Flytzani-Stephanopoulos, Appl. Catal. B 95 (2010) 87.
[105] N. Yi, R. Si, H. Saltsburg, M. Flytzani-Stephanopoulos, Energy Environ. Sci. 3 (2010) 831.
[106] W.L. Deng, M. Flytzani-Stephanopoulos, Angew. Chem. Int. Ed. 45 (2006) 2285.
[107] J. Guzman, S. Carrettin, A. Corma, J. Am. Chem. Soc. 127 (2005) 3286.
[108] M.B. Boucher, S. Goergen, N. Yi, M. Flytzani-Stephanopoulos, Phys. Chem. Chem. Phys. 13 (2011) 2517.
[109] X. Huang, H. Sun, L.C. Wang, Y.M. Liu, K.N. Fan, Y. Cao, Appl. Catal. B 90 (2009) 224.
[110] M. Yang, L.F. Allard, M. Flytzani-Stephanopoulos, J. Am. Chem. Soc. 135 (2013) 3768.
[111] D.A. Outka, R.J. Madix, J. Am. Chem. Soc. 109 (1987) 1708.
[112] A. Wittstock, V. Zielasek, J. Biener, C.M. Friend, M. Bäumer, Science 327 (2010) 319.
[113] N. Yi, H. Saltsburg, M. Flytzani-Stephanopoulos, ChemSusChem 6 (2013) 816.

CHAPTER 6

Surface Effects in the Reactivity of Ceria: A First Principles Perspective

Michael Nolan
Materials Modelling for Devices Group, Tyndall National Institute, University College Cork, Lee Maltings, Dyke Parade, Cork, Ireland
E-mail: michael.nolan@tyndall.ie

1. INTRODUCTION

In catalysis cerium dioxide, ceria, is one of the most widely studied materials [1–4], with a wide range of applications in CO oxidation, NO_x reduction, oxidative dehydrogenation, and methane oxidation, while other applications include solid oxide fuel cells, gate dielectrics in CMOS transistors, and photocatalysis. Ceria has traditionally been used as a support for noble metal particles in three-way automotive catalytic convertors, but is now being investigated as the material of interest in the catalyst since it does not merely function as a so-called innocent support for metal nanoparticles, such as gold, or a second oxide, as in the VO_x–CeO_2 system, but is instead intimately involved in the activity of these catalyst structures.

Key to the properties of ceria is its reducibility, characterized by the relatively facile change in oxidation state from Ce^{4+} to Ce^{3+} upon formation of oxygen vacancies or addition of electrons. CeO_2 is classed as a reducible metal oxide, similar to TiO_2, in that when neutral oxygen vacancies are formed, the two electrons given back to the oxide are localized on two Ce ions, reducing them from Ce^{4+} to Ce^{3+}, similar to the Ti^{4+} to Ti^{3+} reduction process in reduced TiO_2. This reduction process gives ceria great flexibility in its oxidation state and appears, from work over the last 10 years, to be the key aspect of its reactivity.

Crystallizing in the fluorite structure, CeO_2 has three low index surfaces that have been widely studied [5–30] and which form the second key aspect of the reactivity of ceria which we will discuss in this chapter. Other chapters in this book will discuss the experimental details of surface effects in the reactivity of ceria. There have been two important review articles in 2007 [4] and 2013 [1] that focused on discussing first principles modeling of ceria, with an emphasis on the issues in applying density functional theory (DFT) to modeling reducible oxides and surveying the literature on this fascinating material. The focus in first principles modeling of ceria has been on the (111) surface, as this is the most stable of the three low index surfaces [1,4,20] and is also experimentally well studied [5–18]. In this contribution, we will focus on discussing the low index

Catalysis by Materials with Well-Defined Structures
http://dx.doi.org/10.1016/B978-0-12-801217-8.00006-2

ceria surfaces, (111), (110), and (100), and how the surface structure impacts on (1) oxygen vacancy formation, (2) interaction with small molecules in redox reactions, (3) substitutional doping of ceria surfaces, and (4) the interaction with metal atoms and particles. We will discuss both our work on this topic over the last decade, and the important contributions to these topics by other research groups. I intend to provide the reader with insights into how the reactivity of ceria can be driven by the surface exposed and how this is a useful strategy in developing new catalysts based on ceria.

A ball and stick model of the (111), (110), and (100) surfaces is shown in Figure 1. Since ceria is an ionic material, we can follow the Tasker classification of these surfaces as follows. The (110) surface is a **Type I** surface in which each atomic layer contains a stoichiometric amount of Ce and O species. Since each atomic plane is charge neutral, the surface is nonpolar. The (111) surface is a **Type II** surface in which each atomic plane contains either Ce^{4+} cations or O^{2-} anions and is not charge neutral. However, a trilayer with composition O–Ce–O that repeats along the surface normal is charge neutral and hence this surface is also nonpolar. Finally, the (100) surface is a **Type III** surface. Such surfaces cannot be cut without leaving an excess of cations or anions, which creates a polar surface. Such surfaces are inherently unstable and will reconstruct in some way, e.g., through formation of vacancies or adsorption of hydrogen for example, to kill the polarity in the surface.

In order to facilitate the study of surface structure, thin films of ceria have been grown on a number of support materials including alumina, yttrium-stabilized zirconia (YSZ), platinum, and palladium [5–9]. We now briefly summarize the experimental findings. The (111) surface is by far the most well studied, and it is observed to be stable and undergoes little relaxation [10]. The (110) and (100) surfaces have been observed to

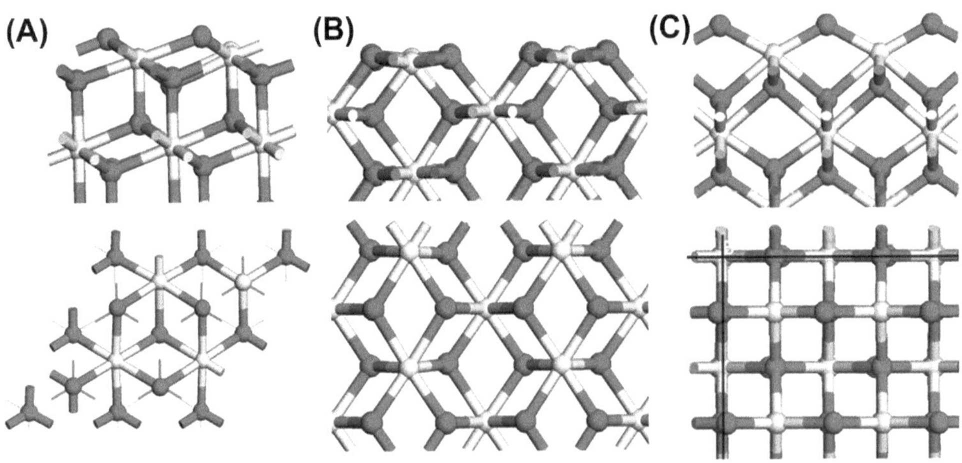

Figure 1 Atomic structure of the three low index surfaces of ceria. (A) (111), (B) (110), and (C) (100). In this, and in subsequent figures, Ce ions are denoted by gray spheres and O ions by red spheres.

undergo surface relaxations [10,11]. It has been determined through scanning tunneling microscopy (STM) [11], atomic force microscopy (AFM) [12], ion scattering spectroscopy [13], and low-energy electron diffraction [14] that the (111) surface is oxygen terminated. Later work has used state-of-the-art AFM [15,16] and STM [17,18] on ceria (111) films to examine the presence of point defects in this surface. Interesting features are observed that correspond to subsurface vacancy defects, cluster of defects in triangular or linear shapes and the location of reduced Ce cations that appear not to be nearest neighbors to the vacancy sites.

The (110) surface has been studied by Nörenberg and Briggs [19] using STM and electron diffraction, demonstrating that it is terminated with a stoichiometric layer. The least stable, and most studied, surface is (100). Cleaving this surface gives a dipole moment perpendicular to the surface and therefore requires a reconstruction since dipolar surfaces are unstable [20]. Generally, removal of 50% of the terminating oxygen species is observed [11,20], which Hermann demonstrated using angular resolved mass spectroscopy of recoiled ions [13]. Nörenberg and Harding [11] have presented an STM study of the pure and partially reduced (100) surface of ceria. These authors have found that surface relaxations take place in order to reduce the surface energy. While cation termination of the (100) surface is possible, the anion terminated (100) surface is found to have the lowest surface energy [11,20] and is taken as the observed termination. Evidence for the appearance of naturally occurring oxygen defects was derived from the nature of the bright spots in the STM image coupled to atomistic simulations.

Despite the fundamental importance of these species to automotive catalysis, an understanding of how oxygen vacancy defects modify the properties of ceria was lacking until the last decade and this understanding is crucial in allowing the development of more efficient ceria-based catalysts. The electronic structure of reduced ceria has received significant attention from experiments [10,17,18,21–27]. Two key techniques that probe the formation of oxygen vacancy defects and their nature have been ultraviolet photoemission spectroscopy (UPS) and STM.

In UPS, the resulting features in the spectrum are dependent on the occupation of the cerium 4*f* states [24]. In the work of Henderson et al. [21], it was demonstrated that upon reduction of the ceria (111) surface, a new occupied Ce 4*f* state appears in the gap between the valence band and the previously unoccupied Ce 4*f* states and lies 1.2 eV above the valence band. Mullins et al. [14] have observed the formation of this same peak for reduced ceria (111) thin films grown on an Ru(0001) support. This peak is due to formation of Ce^{3+} and the intensity of the XPS peak is related to the amount of Ce^{3+} present. The Ce 5*d* spectrum is also modified upon reduction, with a new peak appearing at a binding energy of 903.8–904.0 eV [11,21]. These observations are characteristic of the presence of Ce^{3+} species.

Matolin and co-workers [25–27] have shown in a number of papers that resonant valence band photoemission spectroscopy, RPES, is an important technique in studying reduced ceria. RPES is carried out at photon energies at which Ce^{4+} and Ce^{3+} states resonate most strongly; these energies are at 124.5 and 122 eV, respectively. In an oxidized (111) surface [25], the resonant enhancement of the Ce^{4+} signal is large, but it is very small for Ce^{3+}. If the sample is reduced by sputtering, then there is a significant resonant enhancement of the Ce^{3+} signal, which is significantly more sensitive to the presence of Ce^{3+} than the traditional XPS analysis of the Ce 3*d* feature. This method can determine the degree of reduction in a ceria film prepared under different conditions.

The defective (111) surface has also been studied by Nörenberg and Briggs [28] with STM. No significant lateral relaxation of the surface was observed and the surface was oxygen terminated. These authors claim that oxygen vacancy defects initially form in triangular clusters and upon further annealing form line defects, indicating that the clustering of the oxygen vacancy defects on ceria surfaces is energetically favorable. Namai et al. [12] have also observed these triangular and linear oxygen defects in the partially reduced CeO_2 (111) surface. These authors have also concluded that an oxygen defect density of greater than 1×10^{13} cm^{-2} (approximately 1% of top layer O^{2-}) is necessary for vacancy clustering to occur. Oxygen vacancy clustering is also predicted from atomistic simulation to be energetically more favorable than isolated surface oxygen vacancies [12,20].

There are more limited studies on the reduced (110) and (100) surfaces, although Mullins et al. [24] have studied the electronic structure of reduced CeO_2 (110) with UPS and have observed the appearance of the Ce 4*f* peak at 2.0 eV above the valence band. Thus all low index surfaces of ceria show the same behavior with respect to the electronic properties of the reduced surface, namely the clear appearance of an occupied Ce^{3+} state in the original energy gap between the valance band and the empty Ce 4*f* states.

The experimental work from Li et al. [29,30] on the interaction of CO with ceria powders indicated formation of many different species, namely CO, monodentate and bidentate adsorbed carbonate $(CO_3)^{2-}$ and carboxylates, that were assigned from the infrared (IR) vibrational spectrum, which showed many vibrational bands characteristic of these different CO-derived species. A weak adsorption mode, in which an intact CO molecule is oriented over a single Ce(IV) ion, was further characterized by observations that CO in this situation can be easily desorbed at temperatures above room temperature. However, in these studies, the ceria sample was a powder and thus no conclusions regarding surface sensitivity could be easily established.

Later work [31–36] on nanostructures of ceria has provided irrefutable evidence of the surface sensitivity of, e.g., CO oxidation on ceria [31,32] as well as the adsorption of oxygen [33] and methanol on reduced and oxidized structures [35]. This important advance has been achieved through synthesis of well-defined nanoshapes of ceria, primarily nanorods, nanoparticles, and nanocubes. Ceria nanoparticles expose the (111) surface, nanorods expose (110) and (100) faces, and the nanocubes expose the (100) surface. This feature of ceria nanostructures has allowed surface effects to be unraveled as discussed in other contributions to this volume.

Briefly, to set the scene for our discussion of surface effects, catalytic studies show that the interaction of CO with ceria nanorods is substantially stronger than with nanoparticles, with a greater conversion rate for CO to CO_2 at a given temperature [31]. In ref. [32], it was also demonstrated that as ceria polycrystallites are aged and heated, (111) faces disappear, while (100) faces were exposed. These aged samples have a greater activity for CO oxidation than the untreated samples. These two sample results are taken to be due to the presence of the more reactive (110) and (100) faces in the samples studied.

Wu and Overbury have applied Raman spectroscopy and TPD-QMS to examine the reoxidation of reduced ceria nanoshapes with O_2 [33] and to highlight the effect of the surface planes in the various ceria nanostructures. For example, the intensity ratio of the I_D/I_{F2g} Raman bands, which indicates the degree of reduction, show that nanooctahedra that expose the (111) surface are the least reducible; nanorods that expose the (110) surface are the most reducible; and nanocubes that expose (100) were in the middle of this sequence. Nanooctahedra showed no O_2 chemisorption either before or after reduction and the Raman bands corresponding to O_2 adsorption were very weak, again indicative of little reduction of this structure which results in little subsequent interaction with O_2. On the nanorods and nanocubes, Raman bands characteristic of superoxide or peroxide species were observed, with the latter being more stable. This is consistent with a higher degree of reduction on these structures compared to the nanooctahedra.

Given these interesting observations of surface effects in the reactivity of ceria, and the ability to synthesize in a controlled fashion structures that expose particular crystal planes, it is clear that this is an area where state-of-the-art first principles simulations are able to provide necessary insights into surface reactivity and make predictions regarding shape effects and highlighting these surface effects is the principal aim of this contribution.

2. FIRST PRINCIPLES MODELING OF CERIA

CeO_2 is a fluorite structured nonmagnetic insulating rare earth oxide. The cubic fluorite structure has four cerium and eight oxygen atoms per unit cell and the experimental lattice constant is 5.411 Å [37]. The vast majority of first principles studies of ceria have employed periodic simulation supercells using a plane wave basis set to describe the valence electrons, primarily using the VASP [38] and PWSCF codes [39]. DFT has been the most widely used approach in the literature [1,4]. We mention briefly here that a localized orbital basis set has been used in Hartree-Fock (HF) [40,41] and DFT [42] calculations of bulk ceria within the CRYSTAL code. Finite-embedded cluster approaches have been employed by Hermansson and co-workers [43], Paulus and co-workers [44], and Galea et al. [45] to study the properties of ceria surfaces and interactions with molecules. These have the advantage of allowing a broad range of theoretical approaches beyond DFT to be employed, namely hybrid DFT, MP2, and coupled cluster theory that are either not available or prohibitively expensive within a plane wave basis. However, since the majority of studies of ceria surfaces have used some flavor of DFT, we focus our discussion on these investigations.

Modeling of the two predominant forms of ceria, namely fluorite-structured CeO_2 and the sesquioxide Ce_2O_3, must take account of the fact that cerium is a rare earth element with 4f electrons. In CeO_2, with a formal 4+ oxidation state, the 4f electron configuration is $4f^0$, while in Ce_2O_3, with a formal 3+ oxidation state, the 4f electron configuration is $4f^1$. There are no particular pitfalls to modeling stoichiometric CeO_2 within what we term 'standard DFT approaches,' including the local density (LDA) and the different generalized gradient (GGA) approximations to the exact exchange-correlation functional, as demonstrated in the work on bulk ceria and its surfaces [46]. These calculations predict lattice constants consistent with experiment as well as giving a good description of the electronic structure. Regarding the latter, the valence band is derived from occupied O 2*p* electronic states, with a gap to the highly localized and empty Ce 4*f* states and another energy gap to the empty Ce 5*d* electronic states. Obviously, these energy gaps are underestimated with DFT (whether that be with LDA or GGA), but qualitative results are reasonable.

However, in Ce_2O_3 and in partially reduced ceria, which is formed when, e.g., oxygen is removed, the presence of occupied Ce 4*f* states in reduced Ce^{3+} is problematic for the standard LDA and GGA approximations of DFT, as discussed in refs [47–49]. This arises from the well-known difficulty these approximations have in describing partially occupied 4f states in rare earth metals; this is in fact a rather general problem that afflicts LDA and GGA approximations to DFT and is also observed for partially occupied 3*d* states in transition metals and partially filled oxygen 2*p* states in some doped metal oxides [1,4,50]. It arises from the fact that LDA and GGA, with their approximate description of exchange, cannot completely cancel the self-interaction of one electron, which in HF is correctly cancelled owing to the exact description of exchange in HF. The consequence of this is that the electron interacts with itself and it is delocalized over the 3*d* or 4*f* spin states on all cations in the material.

For ceria, we showed in ref. [48] that if one applies the GGA approach to the oxygen vacancy in the (100) ceria surface, then the occupied 4f states do not separate from the empty Ce 4*f* states and in fact cross the Fermi level, indicative of a metallic-like behavior. The electron density associated with the two electrons released by the removed oxygen is delocalized over the Ce atoms, as shown in Figure 2 for the example of an oxygen vacancy in the (100) surface from our first DFT + U study of reduced ceria [48].

In order to deal with this problem two approaches dominate. The first is the Hubbard + U correction to DFT, known as DFT + U [51,52], which has also been termed in the ceria literature as LDA + U, LSDA + U (local spin density approximation), GGA-DFT + U, PW91 + U, and PBE + U. The latter two indicate the precise GGA functional, PW91 or PBE, to which the +U correction is added. The second computational approach is the so-called hybrid DFT, which has come to dominate the world of molecular computational chemistry and involves a linear combination of a

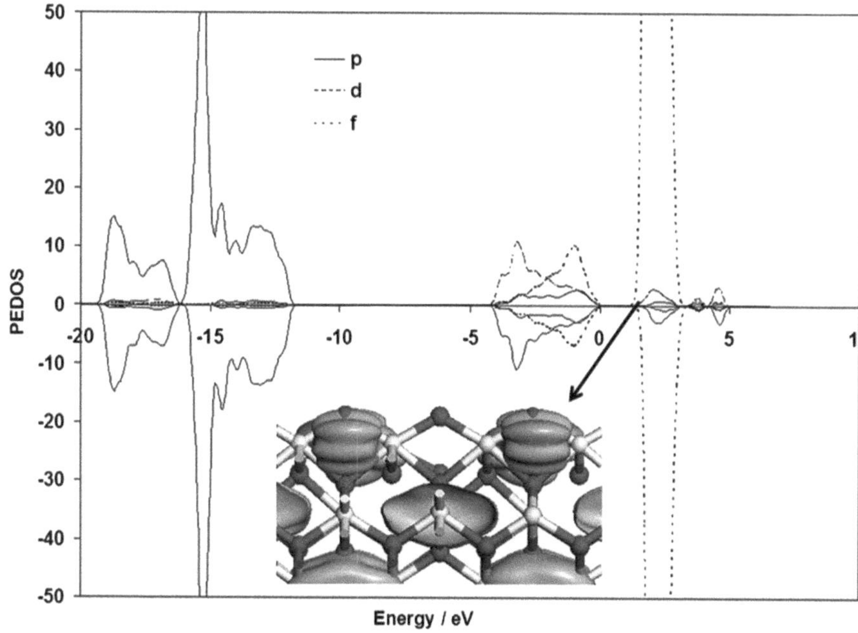

Figure 2 Electronic density of states (EDOS) projected onto Ce 5p, 5d and 4f electronic states for the CeO_2 (100) surface with one surface oxygen vacancy from DFT using the PW91 GGA functional. The zero of energy is the top of the valence band and the inset shows the partial charge density of the Ce 4f state indicated by the arrow. *Reprinted from Surface Science, Vol. 576, M. Nolan, S. Grigoleit, D. C. Sayle, S. C. Parker and G. W. Watson, Density functional theory studies of the structure and electronic structure of pure and defective low index surfaces of ceria, pp. 217-229. Copyright 2005, with permission from Elsevier.*

portion of exact (HF) exchange and the approximate DFT exchange-correlation functional to correct deficiencies in the approximate DFT description of exchange. If we set the exchange contribution to 25% ($a_{HF}=0.25$) then we get PBE0. The Heyd-Scuseria-Ernzerhof (HSE) functional relies on screening the nonlocal long-range exchange interaction to some distance which is determined by the screening length, which is generally 0.2/Å.

Extensive work in the last 10 years has shown that DFT + U can be successful in describing the formation of oxygen vacancies in ceria [48,49,53–55], the interactions with molecules [56–60], substitutional doping of ceria surfaces [61–69], and the formation of interfaces with metal atoms and nanoparticles [70–72] as well as with other metal oxides [73,74]. It provides a reasonable description of the localized Ce^{3+} states and has a computational cost similar to the DFT functional which it modifies. Important issues with DFT + U that must be considered include empiricism in the value of U, materials' properties can be very sensitive to U, it is not usually applied to all elements making up the material or to all electronic states and there can be many possible minima lying close in energy that can be dependent on the details of the implementation of DFT + U and the charge localization and structural distortions within the surface.

In recent years hybrid DFT, primarily within the HSE functional, has been applied to bulk ceria [75], ceria surfaces [76], and the formation of reduced ceria [54,76]. Despite having two parameters that are adjustable, the exchange contribution and the screening length, these tend to be universally applicable and the hybrid DFT description is applied to all elements in the material of interest, giving a significant advantage over DFT + U. A severe issue with hybrid DFT is that the computational cost tends to be 80–100 times that of DFT + U. For this reason, hybrid DFT tends to be used to provide accurate, benchmarked results for checking the consistency of DFT + U calculations. A consistent description of a system with hybrid DFT and DFT + U tends to give confidence in using DFT + U for larger-scale computations.

3. MODELING PERFECT SURFACES OF CERIA

We begin our discussion of surface effects in ceria by briefly considering modeling studies of perfect ceria surfaces. A variety of computational approaches have been applied, including interatomic potentials [77,78], Hatree-Fock within a localized atomic orbital basis set [41], DFT in the form of the LDA and GGA approximations [46–48], and hybrid DFT [76]. The focus in these studies has been on the surface stability and surface structure, i.e., relaxations in the surface after cleaving from bulk.

The stability of a series of surfaces with different Miller indices is generally assessed by computing the surface energy for these surfaces. The surface energy per unit area, γ, usually quoted in Jm^{-2}, is defined as

$$\gamma = \left(E^{surf} - nE^{bulk}\right) / 2A$$

where E^{surf} is the total energy of the surface slab, E^{bulk} is the total energy of bulk, n scales the bulk energy to the same number of CeO_2 units as the surface slab, and A is the surface area. The factor of 2 accounts for the fact that a periodic slab has two surfaces. The surface energy is computed as a function of slab thickness until it is converged, which means that the two faces in the slab are independent and the interior of the slab is bulk-like.

From shell model interatomic potential (IP) calculations on the (111) and (110) surfaces, Sayle et al. [77] showed that the most stable surface both before and after surface relaxation is (111). Baudin [79] used MD simulations with a shell model potential to study the (111), (110), and (100) surfaces of ceria; the polar (100) surface had every second surface oxygen removed in a checkerboard fashion to stabilize it. These authors found the following ordering of the surface energies for both unrelaxed and relaxed surfaces: (111) < (110) < (100). The largest change in surface energy upon relaxation was found for the (100) surface, consistent with it being the least stable of the three surfaces.

Since 2004, DFT calculations have been used to assess the stability of the (111), (110), and (100) surfaces of ceria [46–48,80]. Table 1 presents the computed surface energies of

Table 1 Computed surface energies, in Jm^{-2}, for ceria (111), (110), and (100) surfaces from different IP, HF, and DFT

Method	γ/Jm^{-2} (111) Surface	γ/Jm^{-2} (110) Surface	γ/Jm^{-2} (100) Surface	References
IP	1.35	2.10	–	[78]
IP	1.54	2.45	3.25	[20]
IP-MD	1.12	2.07	2.41	[79]
HF	1.31	2.11	–	[41]
PW91	0.60	1.01	1.43	[47]
PW91	0.68	1.01	1.41	[48]
LDA	1.04	1.35	–	[46]
GGA	0.68	1.05	1.41	[46]
LDA + U	1.02	1.40	–	[49]
GGA + U	0.72	1.06	–	[49]

these ceria surfaces from a variety of DFT calculations employing a plane wave basis set and different flavors of the exchange-correlation functional.

All DFT approaches show that the order of stability is (111) > (110) > (100), consistent with the IP results. The extent of surface relaxations is what one would expect—the defective type 3 (100) surface is unstable and thus undergoes significant rearrangements, which results in a greater energy gain due to the relaxation. The relative magnitudes of the surface energies are in agreement with experimental observations that the (110) and (100) surfaces are less stable than the (111) surface. The difference in surface energy between the (110) and (100) surfaces is smaller than the difference between these surfaces and the (111) surface.

While the ordering of the surface energies obtained with DFT is the same as that from IP studies, the magnitudes of the surface energies are significantly smaller than the values obtained using atomistic simulation [77–79] but are similar within DFT studies that use the same underlying exchange-correlation functionals [41].

As regards surface relaxations, since the (111) surface is the most closely packed and stable surface, any relaxations are rather small. The overall effect is an inward shift of the surface layer. The atomic displacements in this surface are smaller than those using HF [41] and the IPs [77–79]. The Ce–O distances around the surface layer contract by no more than 0.01 Å (0.5%). In the (110) surface, there is a rumpling in the surface layer similar to the (100) surface of rock salt structured oxides [81]. The inward displacement of the cerium ions is notably larger than the displacement of the oxygen ions. This is due to the charge on the cerium atom causing it to be pulled into the bulk. The (110) surface is the only surface which undergoes lateral displacement. The oxygen ions are shifted toward the nearest cerium ion in the same layer, so that the extension of the bond length due to the vertical relaxation is compensated. The result is that the Ce–O distances in the surface layer are 2.33 Å (a contraction of 1.7% compared to the bulk Ce–O distance of 2.37 Å). The displacements were similar to HF results [41] and smaller than those from the IPs [77–79].

The least stable (100) surface shows the strongest relaxations. The two coordinated surface oxygen shifts inward with a contraction of 7.60% in the Ce–O distance directly below the terminal oxygen. Each of the other oxygen layers is split up—one half of the oxygen ions are shifted out toward the surface, while the other half (directly below the surface oxygen) relaxes into the bulk. The contractions of the Ce–O distances reduced on going toward bulk, with a contraction of 5.50% in the second layer and the distances in the middle of the slab close to bulk values.

4. MODELING OF SURFACE DEFECTS IN CERIA

Given the importance of surface reduction and the formation of oxygen vacancy defects in ceria, this topic has received a significant level of attention. Sayle et al. [77] used a shell model interatomic potential to examine the energetics of formation of oxygen vacancy defects in the (111) and (110) surfaces of ceria. They showed that oxygen vacancy formation is more favorable on both surfaces than in bulk ceria. Furthermore, they showed that the energy of reduction of the (110) surface is smaller than on the (111) surface, which was reasonably linked to the lower stability of the (110) surface. A similar result was found for the energetics of CO oxidation over the (111) and (110) surfaces.

Yang et al. [47] presented the first GGA-DFT study of ceria (111) and (110) surfaces with oxygen vacancies, and they treated the Ce 4*f* states as valence, allowing reduction of Ce from Ce^{4+} to Ce^{3+}. Their computed oxygen vacancy formation energies are 3.39 and 3.98 eV for a surface oxygen in the (111) surface; the larger value is for a higher concentration of oxygen vacancies in the surface. A first sublayer oxygen vacancy formation energy of 3.21 eV was computed (3.80 eV for the larger vacancy concentration), indicating that surface and subsurface oxygen vacancies are quite close in energy. In the (110) surface, the formation energy is 2.48 eV for a surface oxygen, which is consistent with the lower stability of this surface relative to the (111) surface. They found strong surface modifications after structural relaxation upon formation of the oxygen vacancy. In the (110) surface, a surface oxygen moves to bridge two surface Ce ions. However, the description of the electronic structure of the reduced surfaces from GGA-DFT was inconsistent with experiment [21], in that GGA predicts delocalization of the charge density of the electrons released by the oxygen vacancy and no gap state was present between the valence band and the unoccupied Ce 4*f* states in the electronic density of states (EDOS), inconsistent with the UPS data.

In our first study of oxygen vacancies in the (100) surface [48], we showed that the PW91 GGA functional delocalizes the two electrons released by the vacancy over all Ce ions in the surface and the EDOS shows that there is no gap state arising from formation of localized Ce^{3+} states, as shown in Figure 2.

In 2005, we [48,80] and Fabris and co-workers [49] presented the first DFT + U studies of the formation of oxygen vacancies in the (111) and (110) surfaces; we also studied the (100) surface. Our computed oxygen vacancy formation energies in the (111), (110), and (100) surfaces are given in Table 2, indicated as PW91 + U for a 25%

Table 2 Computed oxygen vacancy formation energies in eV for low index ceria surfaces from different DFT approaches

Method	O vacancy concentration	E^{vac}/eV (111) Surface	E^{vac}/eV (110) Surface	E^{vac}/eV (100) Surface	References
GGA	25%	3.39 (Surface) 3.21 (Subsurface)	2.48	–	[47]
PW91 + U	25%	2.60	1.99	2.27	[48, 80]
LDA + U	25%	2.29	2.33	–	[49]
GGA + U	25%	2.15	1.57	–	[49]
GGA + U	25%	2.76	2.10	2.26	[53]
PBE + U	–	1.76	–	–	[1]
PBE + U	8.3%	2.13	–	–	[82]
PBE + U	25%	2.23	–	–	[54]
PBE + U	12.5%	–	1.61	–	[55]
HSE06	25%	2.65	–	–	[54]
HSE06	12.5%	–	2.00	2.60	[76]

vacancy concentration in the supercells we used. Fabris et al. used LDA + U and GGA + U. Very few papers have considered surfaces other than (111) and (110), with the majority focusing on (111), and some of the oxygen vacancy formation energies from these studies are also presented in Table 2.

All studies to date indicate that the (111) surface has the highest oxygen vacancy formation energy of the three low index ceria surfaces, which arises from it being the most stable surface of the three. Due to its stability that makes it amenable to study with experimental methods such as STM, the focus in recent years has tended to be on studying oxygen vacancy defects in the (111) surface. Our first study considered only oxygen vacancies in the surface layer and we found that upon relaxation the symmetry around the oxygen site is broken due to localization of two electrons onto two Ce species that neighbor the oxygen vacancy site. This is shown in the spin density plots in Figure 3(A)–(C). The same symmetry breaking was found for the surface oxygen vacancy by Fabris et al., who also found, similar to Yang et al. [47], that the subsurface and surface oxygen vacancies were very close in energy.

In the (110) and (100) surfaces, we found that oxygen vacancy formation is more favorable compared to the (111) surface [48,80]. Fabris et al. [49] also found that oxygen vacancies are more easily formed in to the (110) surface, but did not present any results for the (100) surface. One interesting finding from our DFT + U work was that the (110) surface showed the lowest oxygen vacancy formation energy of the (111), (110), and (100) surfaces [80], which is not what one would predict from a simple comparison of surface energies in the stoichiometric surfaces. This ordering of stability was also found in the later work of Meyernick and Janik [53]. Analysis of unrelaxed and relaxed oxygen vacancy structures showed that for unrelaxed vacancies the order of stability follows that expected from the surface energies, with the (100) surface having the smallest formation

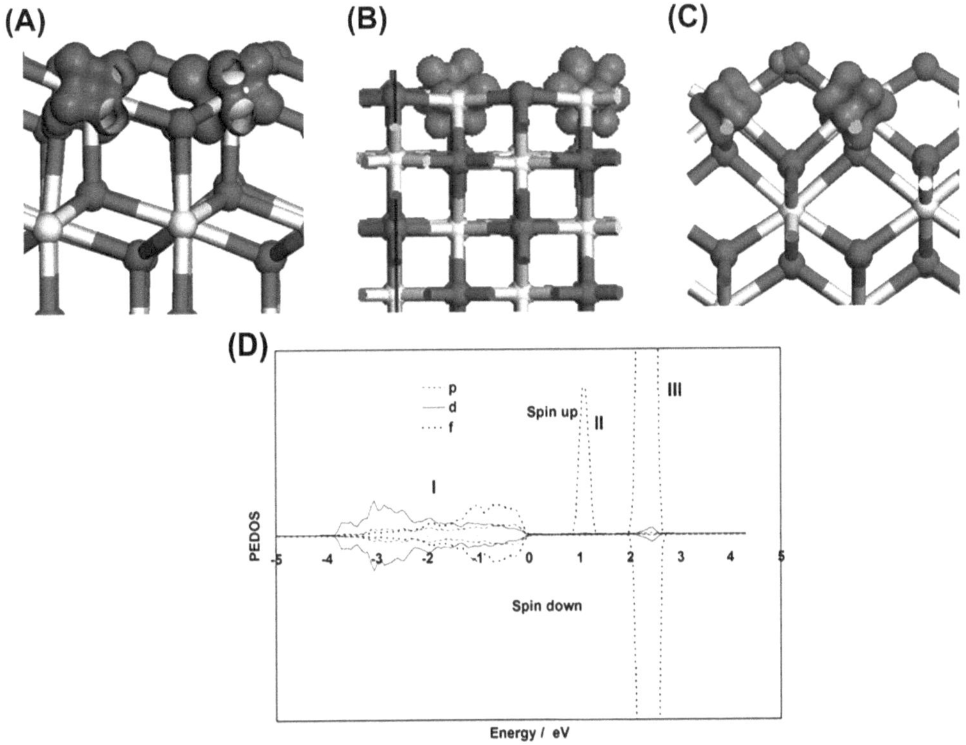

Figure 3 Isosurfaces of excess spin density (the difference between up spin and down spin electron density) for a surface oxygen vacancy in the (a) (111), (b) (110) and (c) (100) surfaces of CeO_2. The grey spin density isosurfaces enclose values up to 0.02 electrons / Å^3 and the colour scheme is the same as figure 1. Part (d) shows the EDOS projected onto Ce 5p, 5d and 4f electronic states for the example of the surface oxygen vacancy in the (111) surface; the EDOS for the (110) and (100) surfaces are similar. *Part (d) Reprinted from Surface Science, Vol. 595, M. Nolan, S. C. Parker and G. W. Watson, The electronic structure of oxygen vacancy defects at the low index surfaces of ceria, pp. 223-232. Copyright 2005, with permission from Elsevier.*

energy, but allowing for relaxation, changes the order of stability [80]. We traced this to the strong surface relaxations in the (110) surface upon relaxation.

The Ce sites neighboring the vacancy were reduced from Ce^{4+} to Ce^{3+} in both the (110) and (100) surfaces, Figure 3, and these sites show a strong elongation of their respective Ce–O distances consistent with formation of a Ce^{3+} cation with a larger ionic radius than the Ce^{4+} cation [83]. The EDOS showed the presence of a new defect state in the gap between the valence band and the empty Ce 4*f* states, indicated for the (111) surface in Figure 3(D), which arises from formation of the reduced Ce^{3+} cations [48,49,80]. These results, along with those for the (111) surface, showed that a DFT + U approach is suitable to describe the key features of reduced ceria surfaces.

In a comprehensive study of oxygen vacancies in the (111) surface using hybrid DFT and PBE + U (U = 4.5 eV), Ganduglia-Pirovano et al. [54] found the following

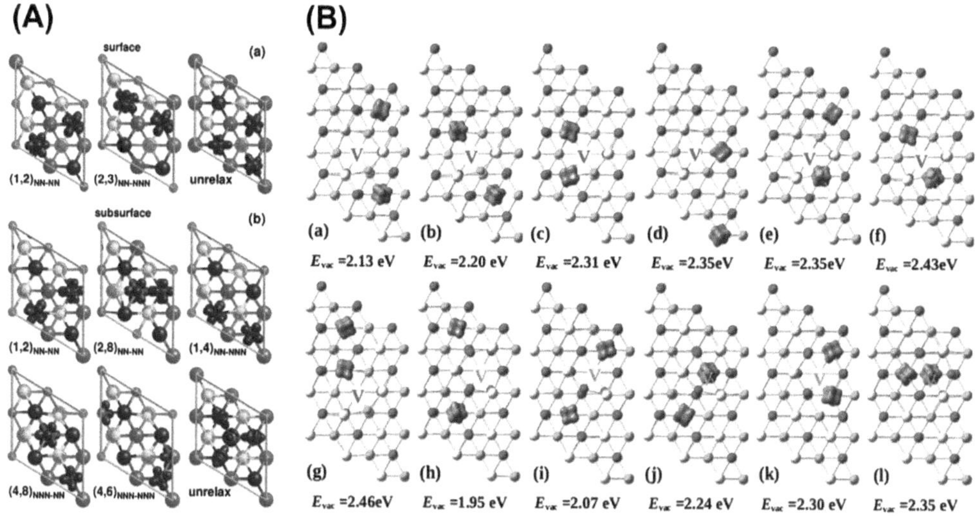

Figure 4 Ce^{3}+ charge localisation patterns in the CeO_2 (111) surface with one oxygen vacancy for different oxygen vacancy sites. (A): results from ref 54 showing different Ce^{3+} localisation sites. (B): results from ref 82 showing both different Ce^{3+} localisation sites and oxygen vacancy formation energies that highlights the spread of vacancy formation energies. *Part (A) M. Veronica Ganduglia-Pirovano, J. L. F. Da Silva and J. Sauer, Physical Review Letters, 102, 026101, 2009. Copyright (2009) by the American Physical Society. Part (B) H. Y. Li, H-F Wang, X-Q Gong, Y-L Guo, Y. Guo, G. Lu and P. Hu, Physical Review B, 79, 193401, 2009. Copyright (2009) by the American Physical Society.*

important results: (1) confirmation that the subsurface oxygen vacancy is the most stable vacancy site in the (111) surface; (2) with large enough supercells, the preferred site for Ce^{3+} localization is not directly neighboring the vacancy site, but is instead on next nearest neighbor Ce sites to the vacancy; and (3) there is a very strong link between local structural distortions and the Ce^{3+} localization, in that the Ce–O distances around the Ce^{3+} site are strongly modified from the stoichiometric surface. Different Ce^{3+} localization patterns are found to induce different local distortions, with correspondingly different formation energies in a range of a few tenths of an eV, as indicated in Figure 4(A). A similar result for the spread of energies with different Ce^{3+} charge localization patterns was also presented in ref. [82], Figure 4(B). The modeling results in ref. [54] were compared with STM studies from the group of Freund where the preference for Ce^{3+} localization away from the oxygen vacancy site was further established [84].

In 2009, Scanlon et al. [55] re-examined the formation of oxygen vacancies in a larger surface supercell model of the (110) surface and they found a second Ce^{3+} distribution in addition to that found in our original work. In this structure, which they termed the 'split oxygen vacancy,' the surface oxygen nearest to the removed oxygen migrates to bridge two surface Ce cations, and thus increase their coordination relative to the simpler oxygen vacancy structure in which this oxygen does not migrate. They found the split oxygen vacancy to be more stable than the simple oxygen vacancy. In addition, the Ce^{3+} cations were distributed so that one was present in the surface layer and one in the subsurface layer, so that similar to the (111) surface there is a preference to localize electrons on Ce^{3+} sites that do not directly neighbor the vacancy site.

We published an HSE06 and DFT + U study of oxygen vacancies in the (110) and (100) surfaces using surface supercells with a 12.5% vacancy concentration in the surface to compare HSE06 and DFT + U for these surfaces and examine the important question of different charge localization sites in these ceria surfaces [76]. For both (110) and (100), we found more than one stable distribution of the Ce^{3+} sites for the same oxygen vacancy. In the (110) surface, the split oxygen vacancy with one Ce^{3+} cation in the surface layer and one in the first subsurface layer was found to be more stable, Figure 5, which was also found with DFT + U. In the (100) surface, the more

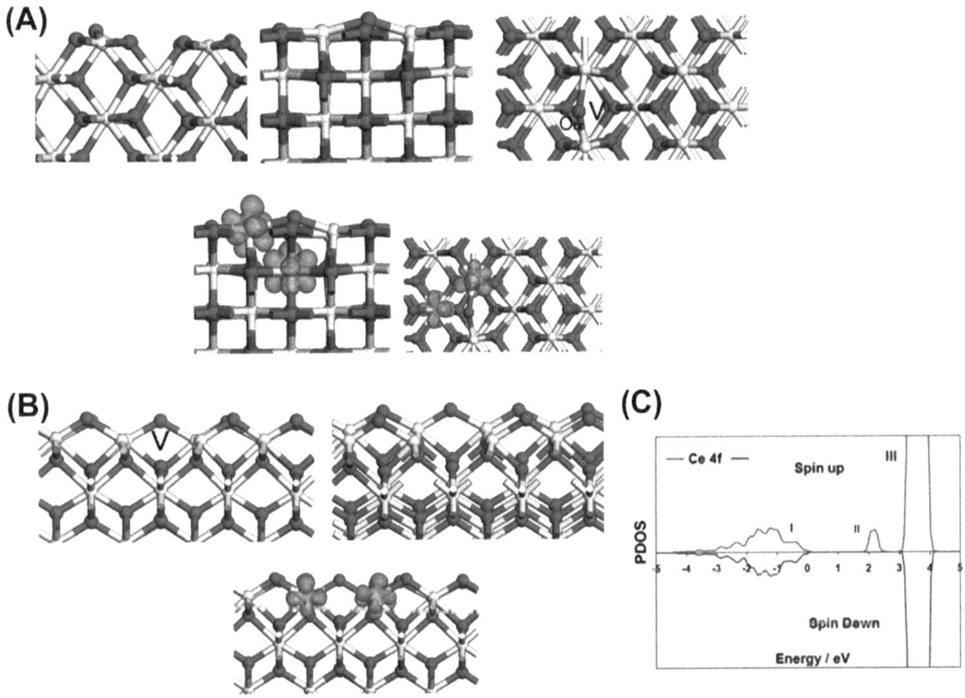

Figure 5 Atomic structure and excess spin density for the most stable oxygen vacancy site in (a) the (110) and (b) the (100) surfaces of CeO_2 from the HSE06 functional. Part (c) shows the Ce 4f EDOS for the oxygen vacancy in the (110) surface from HSE06 in which the zero of energy is set to the top of the valence band. The spin density isosurfaces enclose values up to 0.02 electrons / Å^3 and the vacancy site is indicated by a "V". The bridging oxygen that moves across the (110) surface is indicated by O^{br}. *Reprinted from Chemical Physics* Letters , *Vol. 499, M. Nolan, Hybrid density functional theory description of oxygen vacancies in the CeO_2 (110) and (100) surfaces, pp. 126-130. Copyright 2010, with permission from Elsevier.*

stable solution is that in which the two Ce^{3+} cations are neighbors to the oxygen vacancy site, with the next most stable site being only 0.30 eV higher in energy. This is in contrast to the (111) and (110) surfaces where Ce^{3+} cations are not neighbors to the vacancy site. Figure 5 shows the most stable oxygen vacancy structure and the resulting spin density for these structures from HSE06, showing the location of the reduced Ce cations.

In our work, oxygen vacancy formation energies computed with HSE06 are around 0.3–0.4 eV higher in energy than the same energies from PW91 + U (U = 5 eV), but the relative stability of the oxygen vacancies, the lower cost of vacancy formation in the (110) surface, and the nature of the Ce^{3+} distribution in both surfaces are the same with both DFT approaches.

Thus, DFT-based simulations have established clearly the order of reactivity of the dominant low index surfaces of ceria: (110) is the most reactive and (111) the least reactive, which nicely explains the experimental findings regarding the activity of ceria nanostructures, with (111)-dominated nanoparticles being least reactive and (110)-dominated nanorods being the most reactive. However, despite this agreement in trends with different DFT-based approaches, it is clear that the choice of exchange-correlation functional and DFT + U parameters can give different energies for the cost to form the oxygen vacancy in a given surface, with differences between, e.g., HSE06 and PBE + U (U = 4.5 eV) of up to 0.60 eV. Part of the issue with the energies computed using DFT + U is that the well-known overbinding of O_2 with GGA-DFT stabilizes the oxygen vacancy relative to the perfect surface, giving formation energies that are too small. This arises from the fact that DFT + U is usually only applied to the Ce 4*f* states and the magnitude of the error in the O_2 bonding energy is *ca.* 0.7 eV. Adding this to the computed oxygen vacancy formation energy will increase the DFT + U oxygen vacancy formation energy by 0.7 eV. However, since the error is constant, it will not change any trends in oxygen vacancy formation energy, and this is the pragmatic approach taken in the majority of studies. Since it is difficult to measure the oxygen vacancy formation energy in a film, there is still work to be done on assessing the reliability of the computed energetics of oxygen vacancy formation in ceria surfaces using different DFT-based approaches.

However, regardless of the DFT approach used, there are many possible local energy minima for the distribution of the Ce^{3+} cations and the associated structural distortions, so one must be careful in searching the space of possible structures for oxygen vacancy defects in ceria surfaces [54].

In spite of these difficulties, the general picture of oxygen vacancy formation in ceria surfaces is by now well established as is the order to reactivity of the low index surfaces. Thus, using the oxygen vacancy formation energy as a descriptor allows one to understand other aspects of the reactivity of ceria surfaces, some of which we will discuss in the following sections.

5. INTERACTION OF OXYGEN WITH CERIA SURFACES

Although the question of oxygen vacancy formation in ceria has received a significant amount of attention, the important questions of vacancy healing by oxygen and the interaction of the vacancy with molecular oxygen, which are important in redox reactions, have been rather less studied. Two important DFT studies of the interaction of oxygen with vacancies in ceria that are of interest are from Huang and Fabris [85] and Choi et al. [86], both of whom studied oxygen adsorption at the undoped stoichiometric and reduced (111) surface, while Huang and Fabris studied the stoichiometric (110) surface. Both papers used DFT + U, while Choi et al. also used Raman spectroscopy. Both studies find that atomic and molecular oxygen interact weakly with nondefective surfaces, forming a tilted peroxo O_2^{2-} unit, with an elongated O–O bond. Of more interest is the interaction of oxygen with the defective (111) surface. On this surface, atomic oxygen exothermically reoxidizes the vacancy site and there is a strong interaction of molecular O_2 at the oxygen vacancy site. Experimentally two signals were observed in the Raman spectrum given in ref. [86] at 825 and 1131 cm^{-1} that are characteristic of peroxo and superoxo O^{2-} species. In the computations, the most stable adsorption site is that in which one oxygen atom sits in the vacancy site, forming a peroxo species, with an O–O bond length of 1.43 Å and a stretching frequency of 972 cm^{-1}. These findings were consistent with experiment, although the stretching frequency was overestimated. The superoxo adsorption structure was only 0.05 eV less stable.

In ref. [87] we studied the interaction of undoped reduced ceria (110) and (100) surfaces, in which a preexisting oxygen vacancy is present, with atomic and molecular oxygen. Taken together with the work from Huang and Fabris and Choi et al., these works provide a good picture of both the vacancy healing process, in which atomic oxygen heals the vacancy, the interaction with O_2 and charge transfer from the reduced surfaces to adsorbed O_2. At all reduced surfaces, atomic oxygen resides in the vacancy site, healing the vacancy with an energy gain that is the same magnitude as the energy cost to form the vacancy, where the energy of oxygen is referenced to half that of the O_2 molecule. This is intuitively reasonable as vacancy healing is the reverse process of formation.

For the interaction of the O_2 molecule with reduced ceria, O_2 adsorbs with one oxygen atom in the vacancy site, shown in Figure 6(A) and (B) for the (110) surface and Figure 6(C) and (D) for the (100) surface. The adsorption energy is practically the same on both surfaces, being −2.05 eV on (110) and −2.02 eV (100). The O–O bond elongates from 1.20 Å in the gas phase to 1.44 Å on (110) and 1.42 Å on (100). The computed O–O stretching frequencies are 900 cm^{-1} on (110) and 885 cm^{-1} on (100). These data are consistent with formation of a peroxo species upon O_2 adsorption at all reduced ceria surfaces, with the two electrons in the peroxo supplied by the reoxidation of Ce^{3+} to Ce^{4+}. In experimental work on O_2 interaction with reduced ceria nanoshapes, the peroxo adsorption mode of O_2 was the more stable, consistent with our results.

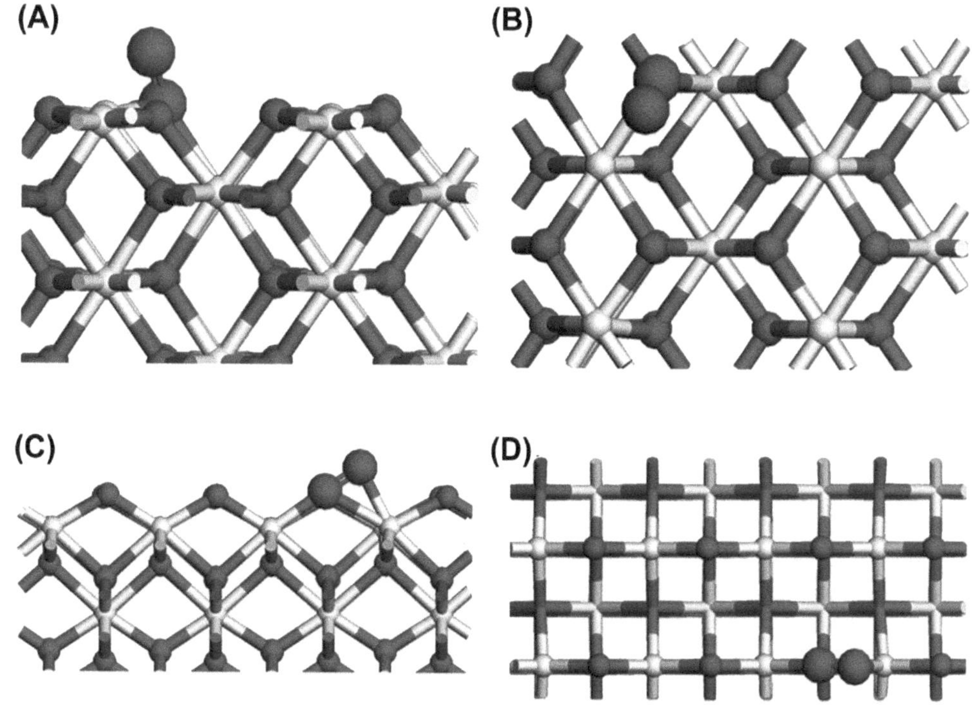

Figure 6 Atomic structure for O_2 adsorbed at CeO_2 surfaces with an existing oxygen vacancy. (a), (b): (110), (c), (d): (100). The colour scheme is the same as figure 1 and the oxygen atoms making up adsorbed O_2 are indicated by the large red spheres. *Reprinted with permission from M. Nolan, Journal of Chemical Physics, vol. 130, 144702 (2009). Copyright 2009, American Institute of Physics.*

It is interesting to see that both oxygen vacancy healing and the interaction of O_2 at reduced ceria surfaces appear to be independent of the surface structure in the sense that the interaction of atomic oxygen and molecular O_2 with the reduced surfaces are characterized by favorable adsorption energies, with vacancy healing for atomic oxygen and adsorption of O_2 at the vacancy site as a peroxo species.

6. MODELING OF MOLECULAR ADSORPTION AT SURFACES

While we have seen some differences in the low index ceria surfaces in terms of their oxygen vacancy formation energies, it is in their interaction with molecules whereby notable differences between the ceria surfaces can be most obviously observed. This is particularly the case for the CO–ceria interaction, the discussion of which forms the bulk of this section.

In their interatomic potential study in refs. [77,78], Sayle et al., also presented the energetics of CO oxidation to CO_2 on the (111) and (110) ceria surfaces. These authors showed that CO oxidation at the (110) surface is more exothermic than on the (111) surface; CO oxidation with bulk ceria was endothermic. Again, the origin of the differences between the two ceria surfaces was traced to their stability. However, such IP studies cannot study the atomic-level details of the interaction between CO and ceria surfaces.

In 2004, Yang et al. [88] presented the first DFT study of the interaction of CO with ceria (111) and (110) surfaces. The authors noted the spread of results from experimental studies of CO adsorption on ceria powders using IR spectroscopy: CO can adsorb weakly in an upright configuration or it can adsorb strongly forming surface-bound carbonate or carboxyl species. Given these were powder samples, this suggested that effects due to the nature of the surfaces in the powders must be playing a role. This important work predates the publication of experimental works on shape-controlled ceria synthesis that preferentially exposes particular surfaces, which were discussed in the introduction to this chapter.

Yang et al., did not apply the DFT + U correction to the Ce 4*f* states, instead they used the PW91 GGA functional. Their principal results regarding surface effects are as follows. The CO molecule interacts only weakly with the (111) surface, in a vertical orientation, with the carbon atom lying atop the outermost Ce atom in the surface. The adsorption energy was computed to be 0.17 eV, indicating no interaction between CO and the (111) surface.

The interaction of CO with the (110) surface is more complex, with many possible structures. There are two possible bonding motifs: the CO molecule can bind with two surface oxygens and a Ce atom lies under the CO (Ce-bridge) or the CO molecule can bridge two surface oxygens with no Ce atom lying under the CO molecule (O-bridge). The O bridging adsorption structures were found to be the most stable, with adsorption energies relative to a free CO molecule and a bare surface of −1.96 eV, compared to −0.76 eV for the Ce-bridging adsorption mode.

In the strongly binding adsorption mode, the CO molecule undergoes an elongation of up to 0.13 Å in the C–O distance and two surface oxygen atoms are pulled out of the surface layer to bind to carbon, making a carbonate-like adsorption species. The adsorption species were further characterized by their vibrational frequencies, in particular the C–O stretching mode which has a characteristic wavenumber for different C–O bonds in CO, CO_2, or carbonate. A weakly bound CO on (111) and (110) had a computed blue shift of 6–15 cm^{-1}, while strongly bound CO on the (110) surface had a computed red shift of 266 cm^{-1} relative to free CO. Compared to experiment, the blue shift for weakly bound CO and the red shift for strongly bound CO were consistent.

Finally, the electronic structure was presented. For weakly bound CO on both surfaces, there was little modification to the electronic structure of ceria. But for the Ce-bridging and O-bridging adsorption modes on the (110) surface, there was a significant change in the electronic structure of ceria. This was a partial occupation of the Ce 4*f* states in the EDOS, which indicates reduction of the surface upon interaction with CO. However, the resulting electronic structure is not well described with the GGA approach used in this study.

In 2006, we published two papers [57,89] studying the adsorption structures and vibrational modes of CO on the (111), (110), and (100) surfaces of ceria. Given the experimental work at the time in refs [31] and [32] that clearly showed the effect of surface structure on CO oxidation, our work aimed to understand these experimental findings. An important advance on previous work was that we applied the DFT + U approach in modeling CO adsorption at the ceria surfaces, which was particularly important given the likelihood that the surfaces would be reduced upon interaction with CO.

We determined the adsorption structures of CO at different sites on the (111), (110), and (100) surfaces of ceria using surface models from our earlier work on oxygen vacancy defects in ceria, see Figure 7. At the (111) surface, the most stable adsorption mode was similar to that found by Yang et al. [88] in which CO adsorbed perpendicular to the (111) surface, Figure 7(A), with a very small adsorption energy of 0.26 eV and little change to the atomic or electronic structure of the surface or the CO molecule. The change in the C–O stretching frequency was +25 cm^{-1}, in very good agreement with experiment.

At the (110) surface, we found a number of adsorption modes. The most stable CO adsorption mode had an adsorption energy of −1.95 eV. This adsorption mode is the symmetric O-bridging mode, in which the carbon atom of CO interacts with two surface oxygen in the (110) surface and pulls them out of the surface layer to make two new C–O bonds, leading to formation of a CO_3 carbonate-like species. The C–O distance

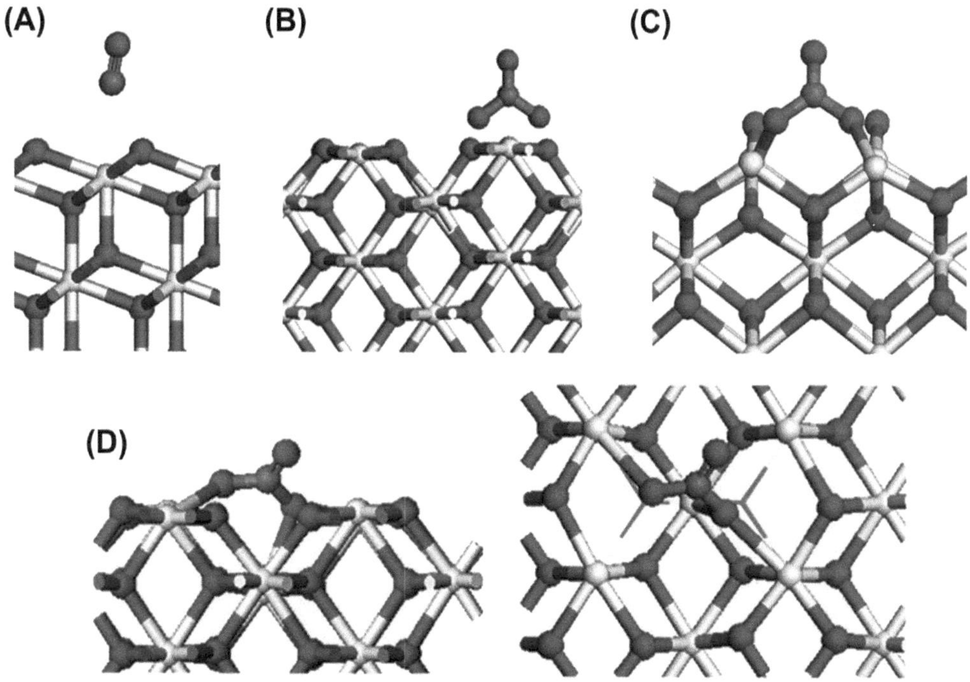

Figure 7 Adsorption structures of CO on ceria surfaces. (a): (111), (b): (110) and (c): (100) surfaces. The colour scheme is the same as figure 1 and carbon of CO is indicated by the grey sphere. In part (d): we show the adsorption structure of CO at the CeO_2 (110) surface from HSE06. *Parts (a) – (c) Adapted with permission from M. Nolan and G. W. Watson, J. Phys.* Chem. *C 110 (2006) 16600. Copyright 2006 American Chemical Society.*

the adsorbate is elongated to is 1.23 Å from 1.14 Å, while the other two C–O distances are 1.37 Å. We proposed that the geometry of the adsorbed CO is more consistent with a carbonate than other possible CO-derived species such as CO_2. A red shift of 393 cm^{-1} was found for the C–O stretch in our most stable CO-binding mode [89], in reasonable agreement with the experimental shift of 420 cm^{-1} [29,30], which was assigned to a carbonate. In addition, two further vibrational modes at 1108 and 1271 cm^{-1} were found [89], corresponding to the experimental modes of 1020 and 1260 cm^{-1} [29,30].

At the (100) surface, we found a stable adsorption mode of CO in which the CO species bridges the two rows of terminal oxygen atoms, as shown in Figure 7, pulling two surface oxygen atoms off their initial lattice sites to make two new C–O bonds. The adsorption energy was −3.21 eV and the adsorption structure was also characteristic of a carbonate species, with a C–O distance of 1.21 and 1.37 Å involving the oxygen from CO and oxygen from the (100) surface, respectively. The shift in the C–O stretching mode was 345 cm^{-1}, consistent with the experimental red shift [29,30].

On both the (110) and (100) surfaces, CO adsorption resulted in two surface oxygen atoms being pulled out of the surface to make the new C–O bonds in the surface-adsorbed carbonate. This can be considered an oxygen vacancy formation process which leads to surface reduction. We examined the EDOS of the CO–ceria systems and the excess spin density to characterize any surface reduction processes that may have taken place. The EDOS and spin densities are shown in Figure 8 for the (110) and (111) surfaces. The spin density isosurfaces show that two Ce ions neighboring the CO adsorption sites are reduced from Ce^{4+} to Ce^{3+}, which is consistent with removal of oxygen from the ceria surfaces. The DOS also clearly shows the formation of a Ce 4*f* derived gap state that results from reduction of Ce^{4+} to Ce^{3+}, with localization of the electrons on two Ce ions.

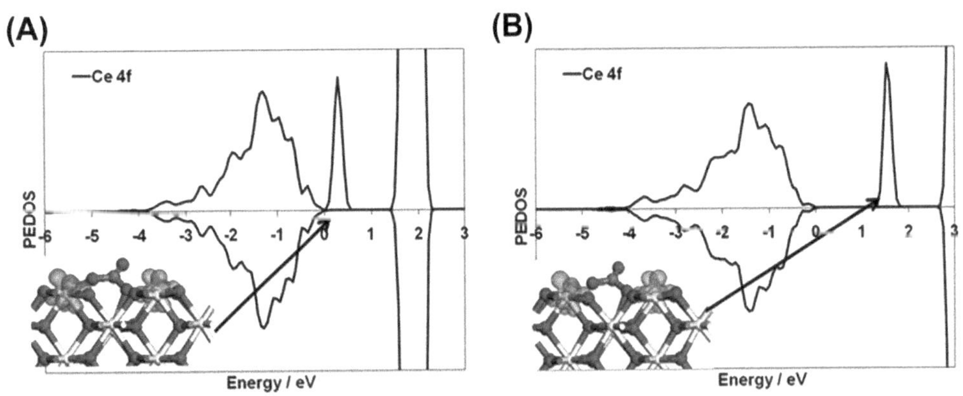

Figure 8 Electronic density of states projected onto Ce 4*f* states for CO adsorbed at the CeO_2 (110) surface from (A) DFT+U (U = 5 eV) and (B) HSE06. The zero of energy is set to the top of the valence band. The insets show the excess spin density in each structure showing the reduction of two Ce ions in the surface to Ce^{3+}. The color scheme is the same as Figure 7.

With two oxygens removed from the lattice to form the carbonate species, four electrons are released. We accounted for the remaining two electrons by examining the DOS for the surface-bound carbonates. The EDOS for the CO_3 adsorbate is consistent with that of carbonate and the integrated EDOS for the surface-bound CO_3 was computed to be 24 electrons, which is the same number of electrons in $(CO_3)^{2-}$. Thus, the formation of a strongly bound carbonate at the ceria (110) and (100) surfaces is clear. Differences in the precise atomic geometry of the surface-bound carbonate species on (110) and (100) can be attributed to the differences in the surface structures: the more closely packed (110) surface has oxygen–oxygen distances that are smaller than on the open (100) surface. It is therefore more favorable for carbon of the CO molecule to interact with two oxygens on the (110) surface than on the (100) surface, which leads to the shorter C–O distances involving surface oxygen on the (110) surface.

It is interesting to note that despite the fundamental differences between the DFT and DFT + U approaches in treating the reduced Ce^{3+} species, the same adsorption structures are found with both DFT approaches, that is a weakly bound CO at the (111) surface and a strongly bound carbonate at the (110) surface. Differences between the two approaches arise in the nature of the reduced surface and the shifts in the C–O vibrational modes, both of which appear to be sensitive to the details of the underlying electronic structure in the ceria surface.

In 2008, Huang and Fabris [90] studied CO adsorption and oxidation on the (111) and (110) surfaces of ceria using DFT + U. A different value of U for the Ce 4*f* states was used, namely 4.5 eV as against 5 eV in, e.g., ref. [89]. This work examined more CO-binding sites at the (110) surface and the effect of the +U correction on the binding energies of CO at the (110) surface. Two CO adsorption sites were studied at the (110) surface: the Ce-bridging and O-bridging sites. The latter site was the most stable, with an adsorption energy of −3.71 eV, as against −2.04 eV for the Ce-bridging site that we studied. In both adsorption modes, a carbonate was formed and the surface was reduced, with formation of two Ce^{3+} species in the surface.

Scanlon et al. [55] reexamined CO oxidation at the ceria (110) surface in a larger (2 × 2) surface supercell, compared to (2 × 1) surface supercells in earlier studies, and found that the O-bridging site is more stable than the Ce-bridging site and also found a tilted CO_3 adsorption structure that had not been published before; this may arise from the larger surface supercell used that would permit such an adsorption mode compared to the smaller surface supercells used by Nolan and Fabris.

An interesting analysis in Huang and Fabris' work is the interplay between the reaction energetics and the size of the +U correction. They showed that (1) values of U that correctly describe the electronic structure of reduced ceria (4–6 eV) will strongly underestimate the reduction energy of CeO_2 to Ce_2O_3, (2) the oxygen vacancy formation energy depends linearly on the value of U (reducing with an increase in the value of U), and (3) the CO adsorption energy increases (becoming more negative) with an increase in the value of U. Taking the experimental adsorption energy of CO at the (110) surface to be 2.27 eV from microcalorimetry, they showed that the values of U used to correct the electronic structure of reduced ceria strongly *overbinds* CO at the (110) surface and a smaller value of U, namely 2 eV, is in better agreement with the experimental binding energy. However, such a small value of U is unable to describe electron localization in reduced CeO_2. A hybrid DFT, B3-LYP, based cluster model study of CO adsorption at the (110) surface [43] gave a CO binding energy of 2.12 eV, which is in better agreement with experiment. However, the energetics from such models can be dependent on the number of atoms in the quantum regimen and the details of the embedding procedure. Shapovalov et al. [91] also showed that if one uses reduction energies to determine a value of U for DFT+U calculations, then U = 2 eV would be the best choice. However, we need to bear in mind that the energetics of oxidation and reduction reactions will be affected by the details of the electronic structure, that is if the resulting Ce^{3+} species are localized (with U = 4–6 eV) or delocalized (U < 4 eV), so that the results for reduction energies obtained with small U may be fortuitous. In addition the results may also be affected by the fact that the +U correction is applied only to the Ce 4*f* states and there is no correction to the remaining Ce states or any oxygen states. A similar issue also pertains to TiO_2, where the value of U that describes the reduction energy of $TiO_2 \rightarrow Ti_2O_3$ is smaller than the value of U needed to describe the localization of Ti^{3+} states [91,92].

It would appear that hybrid DFT may be a suitable approach to deal with this conflict that plagues DFT+U, since the correction in hybrid DFT is applied to all atoms and all electronic states equally and previous work has shown that hybrid DFT describes defect formation in metal oxides very well. In addition, we recall the B3-LYP study of CO adsorption at a cluster model of the (110) ceria surface, for which an adsorption energy consistent with experiment was obtained. We undertook a DFT+U (U = 5 eV) and a HSE06 study of CO adsorption at the (110) surface of ceria with a (4 × 2) surface supercell and found the adsorption mode in Figure 7, namely carbonate formation at the O-bridging site, with a tilting of the carbonate and surface reduction. The computed binding energy was −4.60 eV with DFT+U and −3.60 eV with HSE06. Both energies display overbinding compared to experiment and the embedded cluster B3-LYP result. Understanding the origin of this difference between B3-LYP and HSE06 is an interesting challenge for workers in this field. On this basis, the extra expense involved in a hybrid DFT calculation compared to DFT+U appears to lead to no gain in the accuracy of the energetics. Similar to Huang and Fabris, we find that smaller values of U give better agreement with the experimental adsorption energies. Given the number of studies of redox reactions in which ceria predominates, this question of obtaining accurate reaction energies and electronic structures is a topic that requires further investigation.

Ceria can also take part in reduction reactions, e.g., the reduction of NO_2 or NO to N_2. Experimental work [93,94] showed that NO_x reduction only takes place on a previously reduced ceria surface, with no reaction on an oxidized ceria surface. We presented the first DFT study of NO_2 reduction at the reduced (111), (110), and (100) ceria surfaces, applying the DFT + U approach, using the same set up as our work on oxygen vacancy defects in the oxidized surfaces. We found that as long as the ceria surface has oxygen vacancies present, then the interaction with NO_2 shows no notable effects due to the surface. The only small difference arises from the magnitude of the adsorption energies, being −2.40, −2.25, and −2.23 eV on the (111), (110), and (100) surfaces, respectively. In all cases, an oxygen from NO_2 interacts with the oxygen vacancy site, lengthening this N–O bond by 10%, which means that upon adsorption of NO_2 at reduced ceria, the molecule is dissociating into NO and an oxygen that heals the vacancy

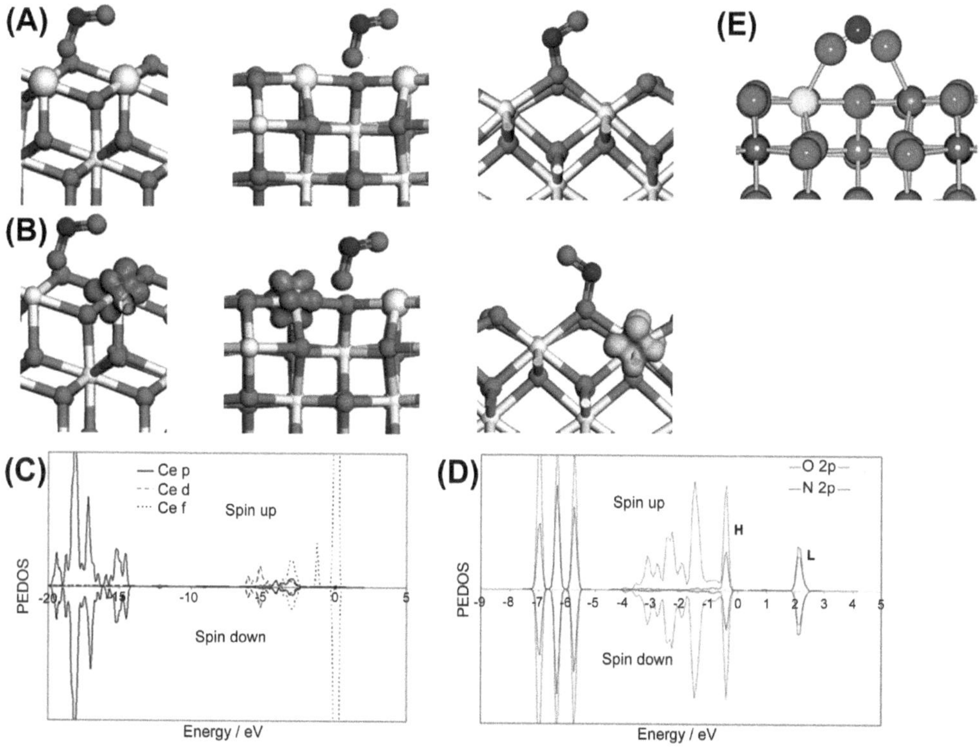

Figure 9 Adsorption structures of NO_2 on reduced ceria surfaces with an existing oxygen vacancy. (a): (111), (110) and (100) surfaces. The colour scheme is the same as figure 1 and N of NO_2 is indicated by the blue sphere. In part (b): we show the corresponding excess spin densities that indicate re-oxidation of one surface Ce^{3+} ion to Ce^{4+} with isosurfaces enclosing spin densities up to 0.02 electrons / Å^3. Parts (c) and (d) show the Ce EDOS and the adsorbed NO_2 EDOS for NO_2 adsorbed at the reduced (111) surface; similar EDOS plots are found for the (110) and (100) surfaces. Part (e) shows a structure found in ref. 58 for NO_2 adsorbed at the reduced (110) surface where the two oxygen atoms in NO_2 bind to the CeO_2 surface. *Parts (c) and (d) Adapted with permission from M. Nolan, S. C. Parker and G. W. Watson, J. Phys.* Chem. *C 110 (2006) 2256.* Copyright 2006 American Chemical Society. *Part (e) Adapted with permission from D. O. Scanlon, N. M.* Galea, *B. J. Morgan and G. W. Watson, J. Phys.* Chem. *C 113 (2009) 11095.* Copyright 2009 American Chemical Society.

site, Figure 9. A transfer of an electron from a Ce^{3+} site to NO_2 results in formation of an NO_2 anion at the surface and partial reoxidation of the ceria surface. In a later work, Scanlon et al. [55] revisited the interaction of NO_2 with the (110) surface and found a similar result to our earlier work, but identified a new binding mode of NO_2, shown in Figure 9(E).

In general, the work on oxygen adsorption and NO_x reduction shows clearly that the reoxidation of ceria by oxygen-containing species is not as sensitive to the surface structure as reactions that involve removal of surface oxygen. Reduced ceria surfaces can be easily reoxidized, which is, of course, important for their use in redox reactions and provides a basis from which to justify the focus on reduction of ceria surfaces since this is the difficult step in redox reactions catalyzed by ceria.

7. ADSORPTION OF METAL ATOMS AT CERIA SURFACES

The adsorption of metal atoms and metal nanoclusters or particles at CeO_2 surfaces is dominated by work on gold adsorption at the (111) surface of CeO_2 [95–99]. This topic is of great interest since ceria is widely used as a so-called 'noninnocent' support for metal particles in catalysis, in particular for gold. First principles studies of this topic have focused on understanding adsorption of *single* gold, copper, and silver atoms at the CeO_2 (111) surface. There is also some recent work on Au, Cu, and Ag adsorption at the (110) and (100) surfaces [100–102], which has allowed an understanding of surface effects related to metal atom adsorption. There has also been work on small gold nanoclusters supported on the (111) surface [70,71]. However, while that topic is obviously of great interest, it lies beyond the scope of this discussion.

The question of adsorption of atomic gold at the ceria (111) surface is one that appears rather simple but has been the subject of much debate. A gold atom adsorbed at an oxide can either accept one electron to become anionic, filling the 6*s* shell, or donate an electron, becoming cationic gold and emptying the 6*s* shell. In fact, a comprehensive study of different DFT + U and hybrid DFT set-ups by Branda et al. [97] showed that the precise nature of gold adsorbed at the most stable atop oxygen site depends very sensitively on the DFT approach employed. The lattice constant, the choice of U value, and atomic relaxations all play a role in determining if gold is oxidized, thus reducing a Ce^{4+} ion to Ce^{3+}, or remains neutral. The largest energy difference between the two scenarios was only 153 meV, suggesting that both could be present and that seemingly conflicting findings of oxidized [96,98] or neutral gold [98] in fact highlight the sensitive dependence on the approach employed. Upon gold oxidation, the reduced Ce^{3+} ion is not a nearest neighbor to the adsorption site, but is instead on a third nearest neighbor surface Ce ion to the adsorption site.

It is interesting to see that this problem with gold adsorption at the (111) surface is not seen for adsorption of Ag and Cu [99]. These metal atoms adsorb on top of an oxygen in the surface, denoted O_d (3-fold) and are oxidized forming a reduced Ce^{3+} species. Cu has the largest adsorption energy, followed by Ag and Au.

Effects on metal atom adsorption due to surface structure, comparing the different ceria surfaces, are more limited. Of note are some recent works in which adsorption of Au, Ag, and Cu atoms [100] and additional transition metal atoms [101] at the (110) surface are presented. In addition, Chen et al. [102] studied Au, Ag, and Cu adsorption at the (100) surface of ceria. On these surfaces, Cu adsorbs the most strongly followed by Au and then Ag, which is different to the (111) surface, where Ag binds more strongly. In addition, the computed adsorption energies at the (110) and (111) surfaces are larger than on the (111) surface, presumably arising from the higher reactivity of these surfaces compared to (111).

8. TUNING REACTIVITY: DOPING OF CERIA SURFACES

Ceria surfaces in which the Ce cation is substituted by another metal species, which is generally termed *substitutional doping*, in analogy to substitutional doping in silicon, is widely studied, particularly with various DFT methods, e.g., [68,69]. While not having as many publications as TiO_2 doping, there is still a significant body of literature on this interesting topic. We limit our discussion in this section to work on the doping of ceria surfaces, in which Ce atoms in or near to the surface layer are substituted with foreign metal atoms; there is also the possibility of bulk doping, which is of interest in solid oxide fuel cells [103–105] but we are not concerned with bulk doping in this contribution.

Again, the (111) surface predominates, although we have also presented a number of studies in which the (110) surface is doped. An early work is that of Shapovalov and Metiu, which they doped the (111) surface of CeO_2 with Au [61] and examined how this improves the activity of the (111) surface to CO oxidation. Essentially, these authors found that the presence of Au doped into the ceria surface enhances the formation of oxygen vacancies compared to the undoped surface and this makes it easier to form a CO_2 species from CO and lattice oxygen. In their reaction scheme, CO interacts with Au-doped CeO_2 to give adsorbed carbonate with an energy gain of 4.43 eV, while on the undoped (111) surface the same interaction is endothermic with a cost of 0.41 eV. These authors point out that the oxygen vacancy formation step is the most costly aspect of formation of carbonate and therefore any modification of the surface that lowers this energy will enhance CO oxidation.

Together with Metiu we extended this system to a study of Au doping of the (110) and (100) surfaces of ceria [62]. There we found that the Au-doped (110) surface will have oxygen vacancies present as a result of their spontaneous formation arising from negative formation energies and that both surfaces will show enhanced oxygen vacancy formation over the undoped surfaces. We furthermore found that different CO oxidation products can be obtained, as shown in Figure 10: at Au-doped (110) a strongly bound carbonate species is formed (adsorption energy of −4.70 eV) while CO_2 forms directly on the Au-doped (100) surface, with an adsorption energy of −3.90 eV. Similar to our discussion on undoped (110) and (100), the differences between the two surfaces arise from their surface geometry. The shorter O–O distance at the (110) surface facilitates the interaction of carbon in CO with two surface oxygens which are pulled out of

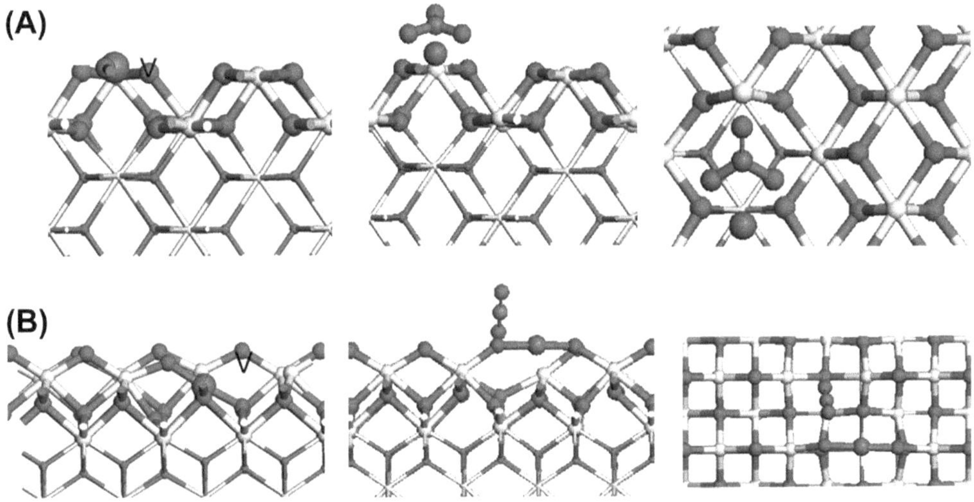

Figure 10 Atomic structures for Au doped (110), part (a), and (100), part (b), surfaces of CeO_2 which also show the adsorption modes of CO adsorbed at Au-doped (110) and (100) surfaces. The colour scheme is the same as figure 1, carbon is indicated by a small grey sphere and gold by a large grey sphere. The O vacancy site is indicated with a "V". *Reprinted from Surface Science, Vol. 602, M. Nolan, V. Soto* Verdugo *and H.* Metiu, *Vacancy formation and CO adsorption on gold-doped ceria surfaces, pp. 2734-2742. Copyright 2008, with permission from Elsevier.*

the surface by CO to form the carbonate, while the longer O–O distance at the (100) surface makes it more difficult to remove two oxygens from the outermost layer and instead one surface oxygen interacts with CO.

Mayernick and Janik studied the (111), (110), and (100) surfaces doped with Zr and Pd, with an emphasis on oxygen vacancy formation and methane activation [53]. These authors find that incorporation of the dopant into the ceria surfaces lowers the oxygen vacancy formation energy in each surface compared to having no dopant. The order of stability is unchanged, with (110) still the most reactive. Interestingly with Pd as the dopant, the formation energy of the oxygen vacancy is slightly negative, indicative of spontaneous formation of the vacancy; a plausible origin for this is that the oxygen vacancy charge compensates the lower valence of Pd (2+) compared to Ce (4+), see below. Methane adsorption also correlates with surface activity—with a larger gain in energy on the (110) and (100) surfaces compared to the (111) surface; this finding pertains irrespective of the surfaces being undoped or doped.

We used Ti, Zr, and Hf as dopants in the (110) surface [65], Figure 11, with a focus on the effect of dopant on the oxygen vacancy formation energies and the interaction with CO, comparing with our results on the undoped (110) surface. We find that doping reduces the oxygen vacancy formation energy compared to the undoped surface and a carbonate species forms upon interaction with CO. However, this carbonate is bound in a *monodentate* fashion, shown in Figure 11(E), through one oxygen atom, to the Ti- and Zr-doped surface, compared to the very strong bidentate adsorption mode found on the undoped surface. In addition, we estimate a small barrier of 0.40 eV to dissociation into CO_2 and a reduced surface.

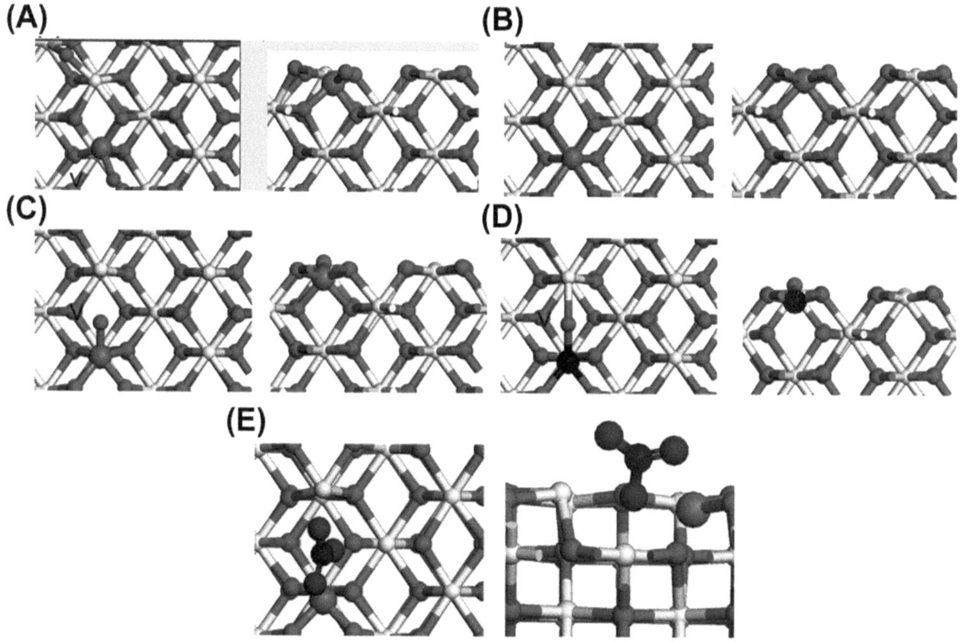

Figure 11 Atomic structure for ceria (110) surface doped with (a) Ti and (b) Zr. The relaxed oxygen vacancy structures are shown in parts (c) and (d) for Ti and Zr, respectively. In all parts, the colour scheme of figure 1 is used and the dopant is indicated by the large sphere. In part (e) we show the monodentate adsorption mode of CO at the Ti-doped (110) surface., with carbon from CO indicated by a purple sphere and oxygen making up the adsorbed carbonate species indicated by the blue spheres. *Adapted with permission from M. Nolan, J. Phys.* Chem. *C 113 (2009) 2425. Copyright 2009 American Chemical Society.*

This work shows that doping of ceria can introduce a flexibility into the surface structure, by distortion of the local atomic structure around that dopant that allows the adsorption structures of intermediates to be modified compared to the undoped surface.

Further work on doping of the (110) surface includes trivalent Al, Y, Sc, In [64], and La [103,104] as dopants. These dopants lead to formation of what is commonly termed a charge compensating oxygen vacancy which is a defect complex composed of two dopants and an oxygen vacancy, as is well known in YSZ and Al-doped TiO_2. The two dopants introduce two oxygen holes into the system as a result of the 3+ valance of the dopant compared to Ce which are compensated by the removal of a neutral oxygen. Interestingly, our DFT + U calculations showed that the defect complex for, e.g., La, was a Ce^{3+} cation and an oxygen hole, rather than two compensated oxygen holes [106,107]. We attributed this to the reducibility of Ce^{4+}. Calculation of the oxygen vacancy formation energy showed a notable reduction compared to the undoped (111) surface, while of note was the finding that the interaction of CO with the doped (111) surface was more favorable than with the undoped (111) surface [106].

Finally, we showed that pentavalent dopants, Ta and Nb, in the (110) surface will form reduced Ce^{3+} by charge transfer from the dopant to a Ce ion in the surface [108].

The resulting structure is then more reactive to O_2 and NO_2 than the stoichiometric undoped surface, transferring an electron to the adsorbed molecule but in a surface that has no oxygen vacancies. It is possible that this way of creating reduced Ce^{3+} species in ceria may be an alternative to surface reduction by formation of oxygen vacancies.

We have compared the (111) and (110) surfaces of ceria doped with the divalent dopants, Pd and Ni [63], with atomic structures shown in Figure 12 for the Ni-doped surfaces.

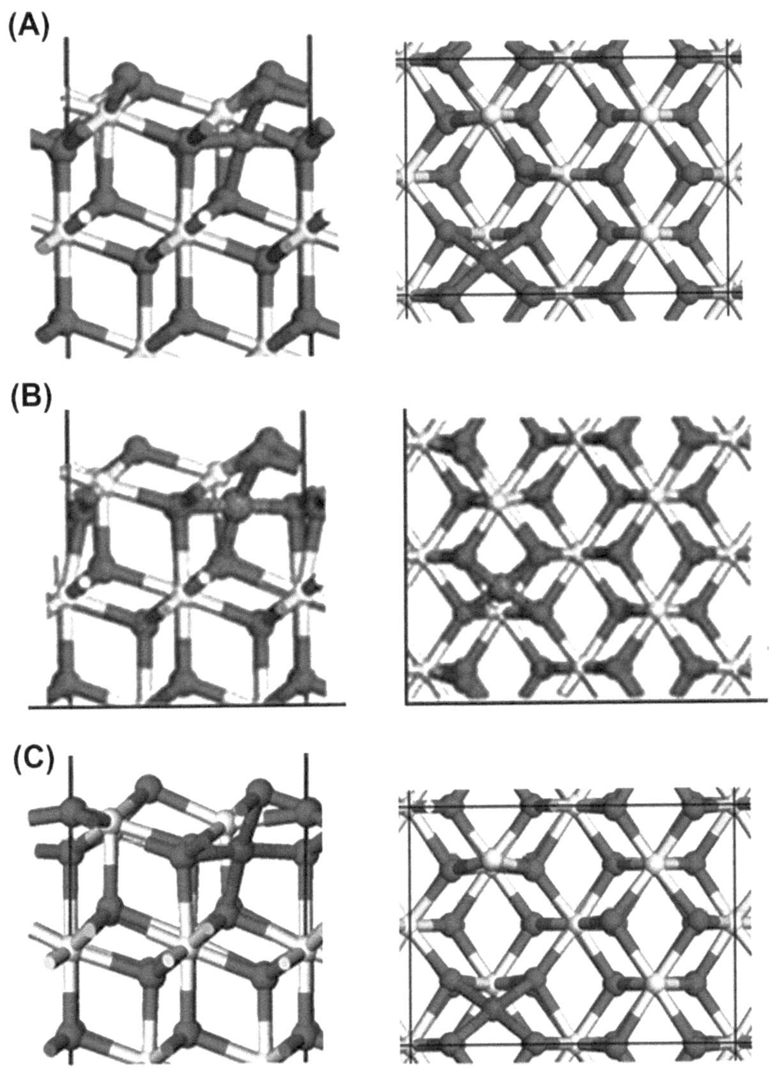

Figure 12 Atomic structure for ceria (111) and (110) surfaces doped with Ni from HSE06; DFT+U structures are similar. (a) Relaxed atomic structures for Ni-doped (111) and (110) surfaces. The colour scheme of figure 1 is used and the dopant is indicated by the blue sphere. Part (b) shows the relaxed structures with a compensating oxygen vacancy in the surface. Part (c) shows the atomic structure for the active oxygen vacancy. *M. Nolan, J. Mat. Chem. 21 (2011), 9160 – Reproduced by permission of The Royal Society of Chemistry.*

Here we find that the formation energy of the first oxygen vacancy is negative in both ceria surfaces doped with the cations. This means that an oxygen vacancy forms spontaneously upon incorporation with Pd and Ni as a result of charge compensation arising from the 2+ oxidation state in Pd and Ni compared to the 4+ oxidation state in Ce. We also find that the dopant takes on a local 4-coordinated square planar geometry that is a particularly stable coordination environment [63,66] and this also induces the formation of the charge compensating oxygen vacancy in the doped ceria surfaces. The next oxygen vacancy is therefore the *active* oxygen vacancy, in the sense that this vacancy reduces the ceria surface. In both surfaces with either dopant, the formation energy of the *active* oxygen vacancy is reduced over the corresponding undoped surfaces: 1.32 and 1.45 eV for Pd and Ni-doped CeO_2 (111) and 1.36 and 1.30 eV for Pd- and Ni-doped CeO_2 (110). In both surfaces, the reduction in the cost to form an oxygen vacancy is significant.

It is interesting to observe that studies of doping of ceria surfaces show that doping lowers the cost of forming oxygen vacancies and enhances the interaction with molecules such as methane [53] or CO [94]. The substantial difference in reactivity seen for the undoped surfaces is practically wiped out when the surfaces are doped with other metals, e.g., the formation energy of the active oxygen vacancy in Pd and Ni doped (111) and (110) surfaces from DFT calculations [63] is practically the same.

9. PERSPECTIVE AND CONCLUSIONS

Despite a significant amount of work on ceria surfaces over the last 10 years or so, the properties of its low index surfaces continue to fascinate and drive further innovations in our understanding and in applications. In particular, the twin advances of controlled synthesis of nanoshapes exposing particular surfaces and the advent of DFT-based simulations applied to ceria have allowed the effect of surface structure on stability, reactivity, and interaction with molecules to be unraveled. In this contribution we discussed our own work and that of other groups on DFT simulation of ceria surfaces to highlight these differences between ceria surfaces and how they can be rationalized and exploited.

Taking the oxygen vacancy formation energy as a reasonable descriptor for the reactivity of ceria in redox reactions, the order of reactivity is not what one would expect from considering the stability of the clean surfaces. DFT + U and hybrid DFT calculations show clearly that the (110) surface has the lowest oxygen vacancy formation energy, which arises from the strong relaxations in the surface after formation of the oxygen vacancy. The (111) surface is the least reactive as measured by oxygen vacancy formation energy. This finding is robust to the precise details of the calculation and is found with DFT + U and hybrid DFT. While the trends in the vacancy formation energies are clear, there still remains an issue with the quantitative formation energies between DFT + U and hybrid DFT which is a topic worthy of further investigation.

The reverse process of oxygen vacancy healing of a reduced surface with oxygen by contrast shows no obvious surface effects. Adsorption of molecular oxygen at each surface is exothermic with adsorption energies around 2 eV indicating a strong interaction. The O–O bond lengthens and a peroxo adsorbate is formed at all surfaces. Doping (110) and (100) with Au can modify this picture so that superoxo is also formed, thus showing a notable effect due to the surface. Experimentally, the work on nanoshapes indicates that nanooctahedra, exposing (111) surfaces, are hardest to reduce and correspondingly interact least strongly with oxygen, while nanorods and nanocubes that expose (110) and (100) are more easily reduced and interact more strongly with oxygen-forming peroxo species.

This difference between the surfaces manifests itself most clearly in the interaction with CO, which is a surface reduction process where the CO molecule removes oxygen from the ceria surface. Only physisorption is found at the (111) surface and chemisorption at the (110) and (100) surfaces, consistent with the reactivity pattern established from oxygen vacancy formation energies. Differences between the (110) and (100) surfaces are found in the geometry of adsorbed carbonate where the shorter O–O distance on (110) facilitates interaction of oxygen with carbon of CO and formation of the carbonate by displacing oxygen of its lattice site. The longer O–O distance in the (100) surface does not facilitate this oxygen displacement and the C–O interaction with lattice oxygen is less strong.

The reoxidation of reduced ceria surfaces with NO_2 is similar to reoxidation with O_2 in that NO_2 interacts exothermically at all reduced surfaces with a lengthening of one N–O distance and partial reoxidation of the surface by formation of an $[NO_2]^-$ anion. We can understand clearly from these examples the focus placed on investigating reduction of ceria by oxygen vacancy formation, since ceria reoxidation is facile on all surfaces.

The interaction of metals on ceria surfaces is beginning to deliver some interesting results with recent work highlighting differences between the three low index surfaces. At (111) and (110), metals adsorb and are oxidized with the reduced Ce^{3+} cations being found at second or third neighbors to the adsorption site. At (100), the reduced Ce^{3+} cations are found as nearest neighbors to the metal adsorption site. The magnitude of the adsorption energies correlates with the oxygen vacancy formation energy, in that adsorption at the (111) surface is the least stable, but this is still exothermic. The special case of a gold atom on the (111) surface which is a delicate balance between lattice constant, surface geometry, and theoretical approach demonstrates how the choice of simulation model and method can lead to very different results.

For doping of ceria, the energy of oxygen vacancy formation or the energy of methane adsorption (which reduces the ceria surface) can be enhanced over the corresponding undoped surface. This manifests as a predicted spontaneous vacancy formation process, e.g., in Pd [53,63] and Ni [63] doped ceria surfaces and as a reduction in the formation energy of the active oxygen vacancy [53,61–69,103–108]. In addition, for methane adsorption at Zr-doped and Pd-doped ceria surfaces [53], the doped surfaces are clearly more reactive than the undoped surfaces. These modifications of surface reactivity due to doping are a very interesting and active area of research [68,69].

While we have made good progress in understanding how ceria surface structure can be a determining factor in key properties such as oxygen vacancy formation or methane adsorption, this important question still has to be explored in other situations. These include key reactions or materials such as (1) interaction with H_2S as a model for sulfur poisoning, (2) structures and activities of metal particles on (110) and (100) surfaces of ceria, (3) interaction of water and hydrogen with the (110) and (100) surfaces, and (4) surface modification with metal oxide clusters, e.g., VO_n, which to date has only been widely explored for VO_n–CeO_2 (111) and where the enhanced reactivity of the (110) and (100) surfaces may be important in determining the activity of the VO_n-modified surfaces. It is our hope that this contribution will provide the interested reader with useful insights into and inspiration to study two areas: (1) the power of DFT-based approaches in unraveling the fundamentals of ceria's fascinating properties and (2) how surface properties can be a determining factor in the potential reactivity of ceria. Coupled with advances in controlled nanoshape synthesis and characterization, this will allow further exciting properties of ceria to be unveiled and used in developing new catalysts.

ACKNOWLEDGMENTS

Ceria remains a metal oxide that even after 10 years of study can still surprise and fascinate, and my sincere thanks are due to colleagues Prof. Graeme Watson, Prof. Steve Parker, and Dr. Dean Sayle for introducing me to this fascinating material. I also gratefully acknowledge many stimulating and inspiring interactions with colleagues in both modeling and experiment, Prof. Horia Metiu, Prof. Joerg Libuda, Prof. Vladimir Matolin, Prof. Javier Sanz, Dr. Stefano Fabris, Dr. Steve Overbury, Dr. Zili Wu, Dr. David Scanlon, Dr. Veronica Ganduglia-Pirovano, Prof. Konstantin Neyman, and Prof. Francesc Illas. I gratefully acknowledge the American Chemical Society Petroleum Research Fund, the Science Foundation Ireland Basic Research Grant Program and the SFI Starting Investigator Research Grant Program (project 'EMOIN' grant number SFI 09/SIRG/I1620), and the European Commission 6th and 7th Framework Programs for funding my research. I also acknowledge computing resources provided by SFI to the Tyndall National Institute and by the SFI and Higher Education Authority Funded Irish Centre for High End Computing, the Trinity High Performance Computing Centre, and the European Commission Research Infrastructures PRACE program. I acknowledge support from the European Union through the COST Action CM1104 'Reducible Oxide Chemistry, Structure and Functions.'

REFERENCES

[1] J. Paier, C. Penschke, J. Sauer, Chem. Rev. 113 (2013) 3949.
[2] A. Trovarelli, Catalysis by Ceria and Related Materials, Imperial College Press, London, 2002.
[3] A. Gorte, AIChE J. 56 (2010) 1126.
[4] M.V. Ganduglia-Pirovano, A. Hofmann, J. Sauer, Surf. Sci. Rep. 62 (2007) 219.
[5] T. Bunleusin, H. Cordatos, R.J. Gorte, J. Catal. 157 (1995) 222.

[6] C. Hardacre, G.M. Roe, R.M. Lambert, Surf. Sci. 326 (1995) 1.
[7] M. Alexandrou, R.M. Nix, Surf. Sci. 321 (1994) 47.
[8] F. Wu, A. Pavlovska, D.J. Smith, R.J. Culbertson, B.J. Wilkens, E. Bauer, Thin Solid Films 516 (2008) 4908.
[9] T. Staudt, Y. Lykhach, L. Hammer, M.A. Schenider, V. Matolin, J. Libuda, Surf. Sci. 603 (2009) 3382.
[10] H. Nörenberg, G.A.D. Briggs, Surf. Sci. 424 (1999) L352.
[11] H. Nörenberg, J.H. Harding, Surf. Sci. 477 (2001) 17.
[12] Y. Namai, K. Fukui, Y. Iwasawa, J. Phys. Chem. B 107 (2003) 11666.
[13] G.S. Herman, Phys. Rev. B 59 (1999) 14899.
[14] D.R. Mullins, P.V. Radanovich, S.H. Overbury, Surf. Sci. 429 (1999) 186.
[15] S. Torbrügge, M. Reichling, A. Ishiyama, S. Morita, O. Custance, Phys. Rev. Lett. 99 (2007) 056101.
[16] S. Gritschneder, Y. Namai, Y. Iwasawa, M. Reichling, Nanotechnology 16 (2005) S41.
[17] F. Esch, S. Fabris, L. Zhou, T. Montini, C. Africh, P. Fornasiero, G. Comelli, R. Rosei, Science 309 (2005) 752.
[18] X. Shao, J-F. Jerratsch, N. Nilius, H.-J. Freund, Phys. Chem. Chem. Phys. 13 (2011) 12646.
[19] H. Nörenberg, G.A.D. Briggs, Surf. Sci. 433–435 (1999) 127.
[20] J.C. Conesa, Surf. Sci. 339 (1995) 337.
[21] M.A. Henderson, C.L. Perkins, M.H. Engelhard, S. Thevuthasan, C.H.F. Peden, Surf. Sci. 526 (2003) 1.
[22] V. Matolín, I. Matolínová, F. Dvořák, V. Johánek, J. Mysliveček, K.C. Prince, T. Skála, O. Stetsovych, N. Tsud, M. Václavů, B. Šmíd, Catal. Today 181 (2012) 124.
[23] S.C. DeCaluwe, M.E. Grass, C. Zhang, F. El Gabaly, H. Bluhm, Z. Liu, G.S. Jackson, A.H. McDaniel, K.F. McCarty, R.L. Farrow, M.A. Linne, Z. Hussain, B.W. Eichhorn, J. Phys. Chem. C 114 (2010) 19853.
[24] D.R. Mullins, S.H. Overbury, D.R. Huntley, Surf. Sci. 409 (1998) 307.
[25] V. Matolín, M. Cabala, V. Cháb, I. Matolínová, K.C. Prince, M. Škoda, F. Šutara, T. Skála, K. Veltruská, Surf. Interface Anal. 40 (2008) 225.
[26] F. Dvorak, O. Stetsovych, M. Steger, E. Cherradi, I. Matolínová, N. Tsud, M. Škoda, T. Skála, J. Mysliveček, V. Matolín, J. Phys. Chem. C 115 (2011) 7496.
[27] T. Duchon, F. Dvořák, M. Aulická, V. Stetsovych, M. Vorokhta, D. Mazur, K. Veltruská, T. Skála, J. Mysliveček, I. Matolínová, V. Matolín, J. Phys. Chem. C 118 (2014) 357.
[28] H. Nörenberg, G.A.D. Briggs, Phys. Rev. Lett. 79 (1997) 4222.
[29] C. Li, Y. Sakata, T. Arai, K. Domen, H. Maruya, T. Onishi, J. Chem. Soc. Faraday Trans. 1 85 (1989) 929.
[30] C. Li, Y. Sakata, T. Arai, K. Domen, K. Maruya, T. Onishi, J. Chem. Soc. Faraday Trans. 1 85 (1989) 1451.
[31] E. Aneggi, J. Llorca, M. Boaro, A. Trovarelli, J. Catal. 234 (2005) 88.
[32] K. Zhou, X. Wang, X. Sun, Q. Ping, Y. Li, J. Catal. 229 (2005) 206.
[33] Z. Wu, M. Li, J. Howe, H.M. Meyer III, S.H. Overbury, Langmuir 26 (2010) 16595.
[34] Z. Wu, A.J. Rondinone, I.N. Ivanov, S.H. Overbury, J. Phys. Chem. C 115 (2011) 25368.
[35] Z. Wu, M. Li, D.R. Mullins, S.H. Overbury, ACS Catal. 2 (2012) 2224.
[36] M.B. Boucher, S. Goergen, N. Yi, M. Flytzani-Stephanopolous, Phys. Chem. Chem. Phys. 13 (2011) 2517.
[37] E.A. Kuemmerle, G. Heger, J. Solid State Chem. 485 (1999) 485.
[38] G. Kresse, J. Hafner, Phys. Rev. B 49 (1994) 14251 G. Kresse, and J. Furthmüller., Comp. Mat. Sci. 6 (1996) 15.
[39] P. Giannozzi, S. Baroni, N. Bonini, M. Calandra, R. Car, C. Cavazzoni, D. Ceresoli, G.L. Chiarotti, M. Cococcioni, I. Dabo, A. Dal Corso, S. Fabris, G. Fratesi, S. de Gironcoli, R. Gebauer, U. Gerstmann, C. Gougoussis, A. Kokalj, M. Lazzeri, L. Martin-Samos, N. Marzari, F. Mauri, R. Mazzarello, S. Paolini, A. Pasquarello, L. Paulatto, C. Sbraccia, S. Scandolo, G. Sclauzero, A.P. Seitsonen, A. Smogunov, P. Umari, R.M. Wentzcovitch, J. Phys. Condens. Matter 21 (2009) 395502.
[40] S.E. Hill, C.R.A. Catlow, J. Phys. Chem. Solids 54 (1993) 411.
[41] S. Gennard, F. Cora, C.R.A. Catlow, J. Phys. Chem. B 103 (1999) 10158.
[42] J. Graciani, A.M. Márquez, J.J. Plata, Y. Ortega, N.C. Hernández, A. Meyer, C.M. Zicovich-Wilson, J.F. Sanz, J. Chem. Theory Comput. 7 (2011) 56.

[43] B. Herschend, M. Baudin, K. Hermansson, Chem. Phys. 328 (2006) 345.
[44] C. Mueller, B. Herschend, K. Hermansson, B. Paulus, J. Chem. Phys. 128 (2008) 214701.
[45] N.M. Galea, D.O. Scanlon, P. Martin, P. Sherwood, G.W. Watson, G.W. e-J, Surf. Sci. Nanotech 7 (2009) 413.
[46] N.V. Skorodumova, M. Baudin, K. Hermansson, Phys. Rev. B 69 (2004) 075401.
[47] X.Z. Yang, T.K. Woo, K. Hermansson, J. Chem. Phys. 120 (2004) 7741.
[48] M. Nolan, S. Grigoleit, D.C. Sayle, S.C. Parker, G.W. Watson, Surf. Sci. 576 (2005) 217.
[49] S. Fabris, G. Vicario, G. Balducci, S. de Gironcoli, S. Baroni, J. Phys. Chem. B 109 (2005) 22860.
[50] G. Pacchioni, J. Chem. Phys. 128 (2008) 182505.
[51] V.I. Anisimov, J. Zaanen, O.K. Andersen, Phys. Rev. B 44 (1991) 943.
[52] S.L. Dudarev, G.A. Botton, S.Y. Savrasov, C.J. Humphreys, A.P. Sutton, Phys. Rev. B 57 (1998) 1505.
[53] A.D. Mayernick, M.J. Janik, J. Phys. Chem. C 112 (2009) 14955.
[54] M.V. Ganduglia-Pirovano, J.L.F. Da Silva, J. Sauer, Phys. Rev. Lett. 102 (2009) 026101.
[55] N.M. Galea, D.O. Scanlon, B.J. Morgan, G.W. Watson, Mol. Simul 35 (2009) 577.
[56] M. Nolan, S.C. Parker, G.W. Watson, J. Phys. Chem. B 110 (2006) 2256.
[57] M. Nolan, G.W. Watson, J. Phys. Chem. B 110 (2006) 16600.
[58] D.O. Scanlon, N.M. Galea, B.J. Morgan, G.W. Watson, J. Phys. Chem. C 113 (2009) 11095.
[59] M.B. Watkins, A.S. Foster, A.L. Shluger, J. Phys. Chem. C 111 (2007) 15337.
[60] D. Marrocchelli, B. Yildiz, J. Phys. Chem. C 116 (2012) 2411.
[61] V. Shapovalov, H. Metiu, J. Catal. 245 (2007) 205.
[62] M. Nolan, V.S. Verdugo, H. Metiu, Surf. Sci. 602 (2008) 273.
[63] M. Nolan, J. Mat. Chem. 21 (2011) 9160.
[64] M. Nolan, J. Phys. Chem. C 115 (2011) 6671.
[65] M. Nolan, J. Phys. Chem. C 113 (2009) 2425.
[66] A.B. Kehoe, D.O. Scanlon, G.W. Watson, Chem. Mat 23 (2010) 4464.
[67] P.R.L. Keating, D.O. Scanlon, G.W. Watson, J. Mat. Chem. C 1 (2013) 1093.
[68] E.W. McFarland, H. Metiu, Chem. Rev. 113 (2013) 4391.
[69] H. Metiu, S. Chrétien, Z. Hu, B. Li, X.Y. Sun, J. Phys. Chem. C 116 (2012) 10439.
[70] C. Zhang, A. Michaelides, D.A. King, S.J. Jenkins, J. Am. Chem. Soc. 132 (2010) 2175.
[71] H.Y. Kim, G. Henkelman, J. Phys. Chem. Lett. 3 (2012) 2194.
[72] J. Carrasco, L. Barrio, P. Liu, J.A. Rodriguez, M.V. Ganduglia-Pirovano, J. Phys. Chem. C 117 (2013) 8241.
[73] J. Graciani, J.J. Plata, J.F. Sanz, P. Liu, J.A. Rodriguez, J. Chem. Phys. 132 (2010) 104703.
[74] A.C. Johnston-Peck, S.D. Senanayake, J.J. Plata, S. Kundu, W. Xu, L. Barrio, J. Graciani, J.F. Sanz, R.M. Navarro, J.L.G. Fierro, E.A. Stach, J.A. Rodriguez, J. Phys. Chem. C 117 (2013) 14463.
[75] J.L.F. Da Silva, M.V. Ganduglia-Pirovano, J. Sauer, V. Bayer, G. Kresse, Phys. Rev. B 75 (2007) 045121.
[76] M. Nolan, Chem. Phys. Lett. 499 (2010) 126.
[77] T.X.T. Sayle, S.C. Parker, C.R.A. Catlow, Chem. Commun. (1992) 977.
[78] T.X.T. Sayle, S.C. Parker, C.R.A. Catlow, Surf. Sci. 316 (1994) 329.
[79] M. Baudin, M. Wojcik, K. Hermansson, Surf. Sci. 468 (2000) 51.
[80] M. Nolan, S.C. Parker, G.W. Watson, Surf. Sci. 595 (2005) 223.
[81] P. Broqvist, H. Groenbeck, I. Panas, Surf. Sci. 554 (2004) 262 D. R. Alfonso, J. A. Snyder, J. E. Jaffe, A. C. Hess and M. Gutowski, Phys. Rev. B, 63 (2000) 8318.
[82] H.Y. Li, H.-F. Wang, X.-Q. Gong, Y.-L. Guo, Y. Guo, G. Lu, P. Hu, Phys. Rev. B 79 (2009) 193401.
[83] see for example: www.mrl.ucsb.edu/~seshadri/Periodic/(accessed April 22 2014).
[84] J.-F. Jerratsch, X. Shao, N. Nilius, H.-J. Freund, C. Popa, M.V. Ganduglia-Pirovano, A.M. Burow, J. Sauer, Phys. Rev. Lett. 106 (2011) 246801.
[85] M.H. Huang, S. Fabris, Phys. Rev. B 75 (2007) 081404.
[86] Y.M. Choi, H. Abernathy, H.T. Chen, M.C. Lin, M.L. Liu, ChemPhysChem 7 (2006) 1957.
[87] M. Nolan, J. Chem. Phys. 130 (2009) 144702.
[88] X.Z. Yang, T.K. Woo, K. Hermansson, Chem. Phys. Lett. 396 (2005) 384.
[89] M. Nolan, S.C. Parker, G.W. Watson, Surf. Sci Lett 600 (2006) 175.
[90] M.H. Huang, S. Fabris, J. Phys. Chem. C 112 (2008) 8643.
[91] S. Lutfalla, V. Shapovalov, A.T. Bell, J. Chem. Theory Comput. 7 (2011) 2218.
[92] Z. Hu, H. Metiu, J. Phys. Chem. C 115 (2011) 5841.

[93] U. Berner, K. Schierbaum, G. Jones, P. Wincott, S. Haq, G. Thornton, Surf. Sci. 467 (2000) 201.
[94] J.A. Rodriguez, T. Jirsak, S. Sambasivan, D. Fischer, A. Maiti, J. Chem. Phys. 112 (2000) 9929.
[95] Z.-P. Liu, S.J. Jenkins, D.A. King, Phys. Rev. Lett. 94 (2005) 196102.
[96] C. Zhang, A. Michaelides, D.A. King, S.J. Jenkins, J. Chem. Phys. 129 (2008) 194708.
[97] M.M. Branda, N.J. Castellani, R. Grau-Crespo, N.H. de Leuw, N. Cruz Hernandez, J. Fdez Sanz, K.M. Neyman, F. Illas, J. Chem. Phys. 131 (2009) 094702.
[98] C. Zhang, A. Michaelides, S.J. Jenkins, Phys. Chem. Chem. Phys. 13 (2011) 22.
[99] M.M. Branda, N.C. Hernandez, J.F. Sanz, F. Illas, J. Phys. Chem. C 114 (2010) 1934.
[100] L. Cui, Y. Tang, H. Zhang, L.G. Hector Jr, C. Ouyang, S. Shi, H. Lib, L. Chen, Phys. Chem. Chem. Phys. 14 (2012) 1923.
[101] M. Nolan, J. Chem. Phys. 136 (2012) 134703.
[102] L.-J. Chen, Y. Tang, L. Cui, C. Ouyang, S. Shi, J. Power Sources 234 (2013) 69.
[103] D.A. Andersson, S.I. Simak, N.V. Skorodumova, I.A. Abrikosov, B. Johansson, Proc. Natl. Acad. Sci. 103 (2006) 3518.
[104] D.A. Andersson, S.I. Simak, N.V. Skorodumova, I.A. Abrikosov, B. Johansson, Appl. Phys. Lett. 90 (2007) 031909.
[105] M. Burbano, S. Nadin, D. Marrocchelli, M. Salanne, G.W. Watson, Phys. Chem. Chem. Phys. 16 (2014) 8320.
[106] I. Yeriskin, M. Nolan, J. Chem. Phys. 131 (2009) 244702.
[107] I. Yeriskin, M. Nolan, J. Phys. Condens. Matt. 22 (2010) 135004.
[108] M. Nolan, Chem. Phys. Lett. 492 (2010) 115.

CHAPTER 7

Applications of Electron Microscopy in Heterogeneous Catalysis

Yuyuan Lin*, James A. McCarthy, Kenneth R. Poeppelmeier, Laurence D. Marks
Northwestern University, Evanston, IL, USA
*Corresponding author: E-mail: linyuyuan630@gmail.com

LIST OF ABBREVIATIONS

ABF Annular bright field
ADF Annular dark field
AFM Atomic force microscopy
BF Bright field
CCD Charge coupled device
DF Dark field
DFT Density functional theory
EELS Electron energy loss spectroscopy
EDS Energy-dispersive X-ray spectroscopy
ETEM Environmental transmission electron microscopy
FT-IR Fourier transform infrared
GNP Gold nanoparticle
HAADF High angle annular dark field
HOR Hydrogen oxidation reaction
HREM High-resolution electron microscopy
IR Infrared
LB Langmuir–Boldgett
NPG Nanoporous gold
ORR Oxygen reduction reaction
PEEM Photoelectron electron microscopy
PEM Proton exchange membrane
PROX Preferential CO oxidation in excess hydrogen
REM Reflection electron microscopy
SACs Single atom catalysts
SEM Scanning electron microscopy
SOFC Solid oxide fuel cell
STEM Scanning transmission electron microscopy
STM Scanning tunneling microscopy
TEM Transmission electron microscopy
UHV Ultra-high vacuum
WGS Water gas shift
XRD X-ray diffraction

Catalysis by Materials with Well-Defined Structures
http://dx.doi.org/10.1016/B978-0-12-801217-8.00007-4

1. INTRODUCTION

A good catalyst should have high activity, selectivity, and stability as well as low economic and environmental costs. Owing to the development of controlled synthetic methods, catalysts with novel chemical and morphological designs have been found to enhance catalytic properties. For example, catalytic systems with well-defined shapes, bimetallic catalysts, core–shell nanoparticles, and single atom catalysts have been obtained by a variety of chemical and physical preparation methods. Those systems offer unique advantages for optimizing catalytic performance. On the other hand, direct characterization of the catalysts at the nanoscale or atomic scale can help to verify the rational design and provide a mechanistic understanding. A clear mechanism in turn can aid in the development of future catalysts.

Transmission electron microscopy (TEM) is one of the most often used techniques for characterization of catalysts, and can provide rich information about the morphology, structure, and chemical composition. TEM has undergone great improvements in resolution and functionality since the first one was built by Max Knoll and Ernst Ruska in 1931. Figure 1 shows the historical improvement of the best point-to-point resolution in TEM. The development of the TEM design, electron gun, vacuum, aberration correctors, and other components has led to some microscopes with sub-Å resolution.

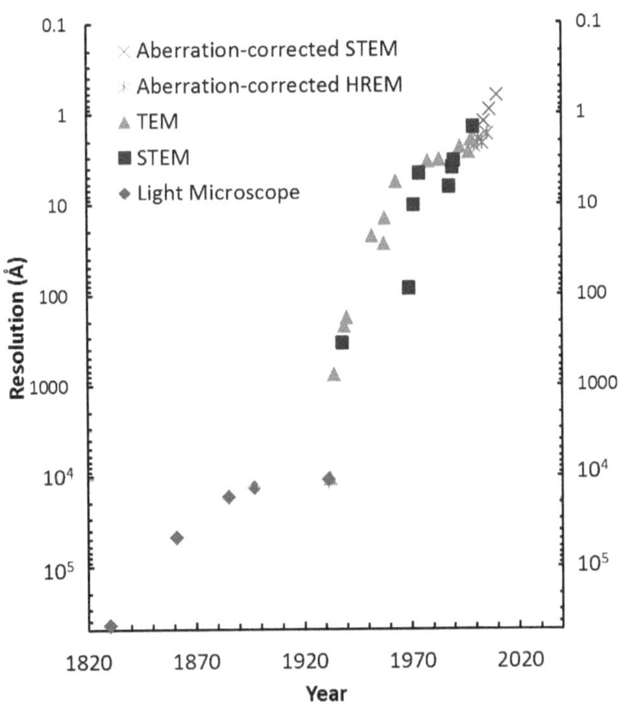

Figure 1 Chronological advancement of microscope resolution. *Reproduced from S.J. Pennycook et al. in: A.V. Narlikar, Y.Y. Fu, (Eds.), The Oxford Handbook of Nanoscience and Nanotechnology, Oxford University Press, Oxford, United Kingdom (2010) p. 205.*

Understanding catalysts at the atomic level is no longer limited to the model extended surfaces on single crystals using conventional surface-sensitive techniques. Imaging the active surface sites of real catalytic nanoparticles can now be achieved by operating TEM in high resolution electron microscopy (HREM) and scanning transmission electron microscopy (STEM) modes. Hence the materials gap in the characterization of catalysts can be bridged by performing TEM characterization. Recently it has been reported that the secondary electrons signals generated by the STEM electron probe can achieve atomic resolution [1]. This is essentially the same as performing scanning electron microscopy (SEM) in a TEM, and it has a promising future for studying the surfaces of materials. In addition to advanced imaging techniques, modern TEMs are often equipped with energy dispersive X-ray spectroscopy (EDS) and/or electron energy loss spectroscopy (EELS), which can help to analyze the local chemical composition of materials. Advances in TEM have also increased the spatial resolution and energy resolution of EDS and EELS. The combination of imaging and spectroscopic studies often results in a more conclusive interpretation. Furthermore, a functional TEM holder or an environmental TEM (ETEM) enables the in situ characterization of materials under different gaseous environments and/or at elevated temperatures. The pressure in a typical TEM column is ~10^{-7} Torr while ETEM can be operated under a pressure of up to ~20 Torr [2]. Therefore, performing direct TEM studies can bridge the pressure gap between ultra-high vacuum (UHV) surface science and realistic gaseous conditions.

Rather than a comprehensive introduction to TEM, the purpose of the TEM part of this chapter is to provide an understanding of image contrast in HREM and STEM. Furthermore, an additional aim of this part is to illustrate how to correctly interpret atomic resolution TEM images. Several recent applications of TEM in studies of heterogeneous catalysis are discussed in the case study section. In addition to illustrating the structural and chemical information included in the images, topics including oxide surface stabilization mechanisms, catalytic reaction mechanisms, nanostructure syntheses, and equilibrium metal shapes will be briefly discussed in the case studies. For the readers who are interested in a more detailed knowledge of TEM, there are several excellent textbooks available [3–5]. As stressed by many researchers within the field, a better understating of catalysis should always combine the results of different characterization methods. Other techniques including Fourier transform infrared spectroscopy (FT-IR), Raman spectroscopy, extended X-ray absorption fine structure (EXAFS), and density functional theory (DFT) should often be used to confirm the structural or chemical information provided by TEM images.

2. BASICS OF TEM

The wealth of information TEM can provide is due to the different electron-specimen interactions and versatility of TEM operation methods. Figure 2(A) shows a cross-sectional view of a simplified TEM model. Changing the convergence of the illuminating beam

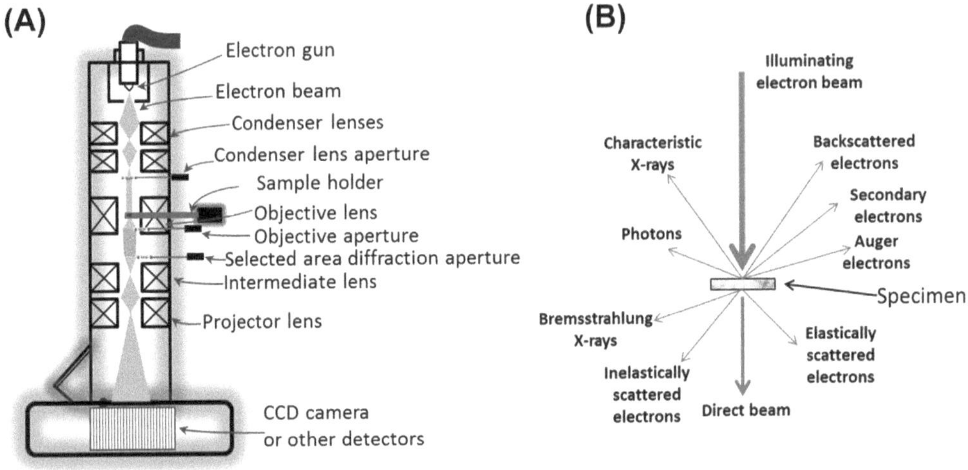

Figure 2 Basics of transmission electron microscopy (TEM): (A) Schematic setup of TEM, and (B) signals generated when high-energy electrons interact with a specimen.

and using different apertures or detectors can lead to different operating modes for specific applications. Figure 2(B) shows different types of signals generated by a high energy electron beam (usually 80–300 KeV) passing through a thin specimen. Images are formed by the detection of transmitted beams. The objective lens aperture can be used to selectively choose the direct beam to form images, which are known as bright field (BF) images. If the scattered electrons are chosen, the corresponding images are dark field (DF) images. The BF and DF images sometimes are very useful to deliberately enhance or suppress signals of a particular material, crystal plane, or defects. Analyzing the energy loss of the inelastically scattered beam can help to determine the elemental composition and bonding configurations of materials, which is the principle of EELS. Above the specimen, the characteristic X-ray signals can be utilized for EDS analysis. Collecting the signals of the secondary electrons and backscattered electrons can form surface-sensitive SEM-like images.

In imaging techniques, an important issue is the image contrast. The contrast can be understood as the differentiation of the object of interest from the background due to differences in intensity (brightness or darkness). Whether the object is brighter or darker than the background depends on the TEM operation and the materials themselves. A common feature of BF images is that the vacuum is whiter than the specimen. In contrast, the vacuum is black in DF images as the direct beam is blocked by the apertures.

The two most often used operation modes for atomic resolution imaging are HREM and STEM. Figure 3 shows a comparison between STEM mode and HREM mode. In STEM mode, the electron beam converges to a probe (typically close to 1 Å to obtain atomic resolution) which scans across the sample, while HREM makes use of static

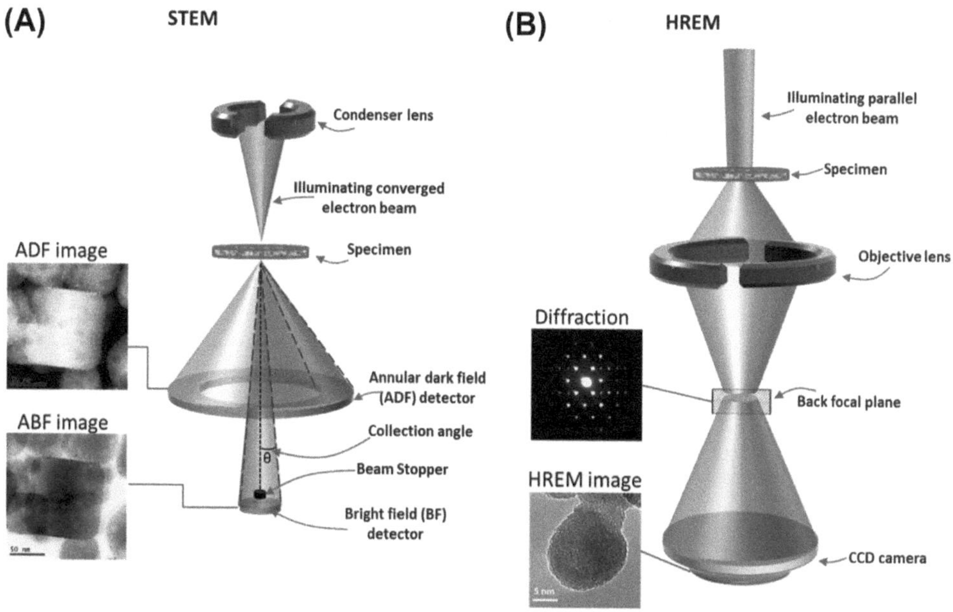

Figure 3 Comparison of image formation in scanning transmission electron microscopy (STEM) and high resolution electron microscopy (HREM). (A) Schematic view of image formation in STEM mode. (B) Schematic showing image formation in HREM mode.

parallel beam imaging. In STEM mode, after the electrons exit the sample, the scattered beams and direct beam can be collected in the diffraction plane by annular detectors. The analog signal on each detector is digitalized and displayed on the computer screen. The electrons collected by the annular dark field detectors (ADF) can form images in which the contrast is related to the atomic number of the materials. Usually the larger Z is, the brighter the material appears in the Z-contrast ADF images while the vacuum is displayed as a black background. The Z-dependency can be tuned by the collection angle of annular detectors. For example, increasing the collection angle (θ) of ADF detectors increases the Z-dependency of the image contrast. The detector with collection angle larger than 50 mrad is considered to be a high angle ADF (HAADF) detector. For HAADF images, the contrast is proportional to Z^n (n can be ~1.7), and thus the HAADF images are often called Z-contrast images. The direct beam and scattered beams within a very small angle can be collected by the bright field (BF) detector. Usually the collection angle for BF detector is less than 10 mrad. If a beam stop is used with the BF detector to block the direct electrons, the resulting images are called annular bright field (ABF) images. Typically in HREM mode, all of the signals in the diffraction plane pass through additional TEM lenses and form an HREM image on a CCD camera. Selectively allowing the direct or diffracted spots in the diffraction plane to pass through the remaining lens optics can form BF or DF images, respectively. The lattice

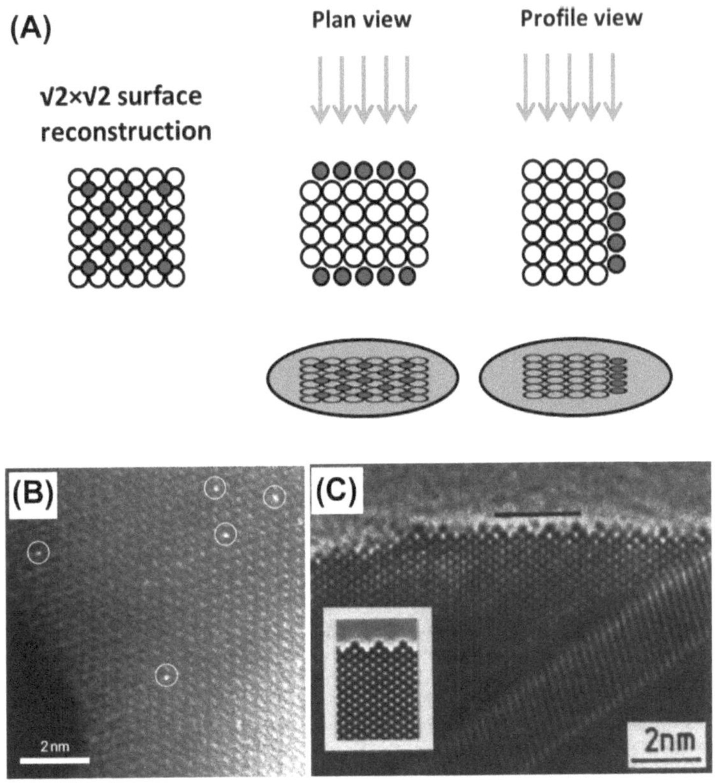

Figure 4 Illustration of plan-view and profile-view surface imaging modes in transmission electron microscopy. (A) Schematic view of the plan-view and profile-view imaging. (B) A high angle annular dark field image in plan-view mode showing single Pt atoms dispersed on a FeO_x support. (C) A profile-view high resolution electron microscopy image showing a (2×1) surface reconstruction on an Au (110) surface. The simulated image is shown in the inset. *Panel (B) was reprinted by permission from Macmillan Publishers Ltd: Qiao et al. Nat. Chem. 3 (2011) 634–641, copyright 2011.*

fringes in HREM images are a result of the interference between the direct and diffracted spots. In STEM, as the collection angle of the BF detector is small enough so that the beam can be considered a parallel beam, BF-STEM and HREM are related by reciprocity and thus the mechanisms of image contrast are identical if inelastic scattering is minimal [6].

Most TEM images can be considered as projections of a 3D object. Thus TEM images contain both bulk and surface information about materials. In heterogeneous catalysis, the surface structures of catalysts and supports are of particular interest. By tilting the specimen to a direction (plane normal) parallel or perpendicular to the beam direction, the surface information can be visualized in plan view and profile view, as shown in Figure 4(A). The plan-view images in TEM of 3D materials have signal overlap between the bulk and surface. Plan-view imaging is useful when the surface has a different structure/material from the bulk structures. For example, a supported single atom catalyst site can be readily distinguished from the support, as shown in Figure 4(B) [7]. If a surface reconstruction is present, the reconstructed surface structure can be abstracted by filtering out the bulk information using imaging processing [8,9]. In the case when the supported metal particles and supports are crystalline, the metal-support interface structure can be analyzed by the Moiré patterns [10]. A Moiré pattern is generated when two periodic patterns overlap.

In profile-view mode, the surface and bulk are separated. Sometimes the surface structures as well as the registry to the bulk can be conveniently imaged. The first profile imaging was reported by Marks and Smith in 1983 using conventional HREM [11] (Figure 4(C)), and they found a (2×1) reconstructed (110) surface of Au nanoparticles. Particularly after the implementation of aberration correctors for HREM images, light elements such as O atoms were able to be imaged clearly. The atomic surface structures of Co_3O_4 [12], CeO_2 [13], $SrTiO_3$ [14] nanoparticles as well as other materials [15–18] have been solved using profile-view mode in the past few years. A drawback is that the 2D surface structure is displayed as a 1D edge-on contrast, as shown in Figure 4(A). It is very difficult to image the 2D surface defect structures as conventional scanning tunneling microscopy (STM) or atomic force microscopy (AFM) does. The reduction of dimensionality can also result in misinterpretations. For example, a reconstructed $(\sqrt{2} \times \sqrt{2})$ surface structure can show up as a (1×1) surface. Therefore, image simulation as well as imaging in a different crystal zone axis can often differentiate the correct surface structures from the other possible ones.

3. ATOMIC RESOLUTION CHARACTERIZATION USING TEM

3.1 Aberration Correction

Similar to optical lenses, the magnetic lens in TEM is not perfect. Lens artifacts such as astigmatisms, coma, spherical aberration (Cs), and chromatic aberration which exist in glass lenses also exist in the electromagnetic lens of a TEM. For STEM mode, the aberrations in the condenser lens are the most important, while in HREM mode, the aberrations of the objective lens are the most important. The aberrations can reduce the image resolution and increase the difficulty of interpreting image contrast. A significant obstacle that limits the resolution of TEM images is Cs. Cs can be understood as the edge of a converging lens always bending the transmitting beam too much such that the beam cannot be brought to a focal spot, as shown in Figure 5. The nonuniform focal plane causes blurriness and increases the complexity. The concept Cs correction was first suggested by Scherzer [19] and further developed Crewe, Rose, Haider, Krivanek, et al. [20,21]. For a glass lens, the Cs can be corrected by using a diverging lens. An "effective" diverging lens can be created in TEM by using non-cylindrical lenses such as octapoles, which are used for Cs correction.

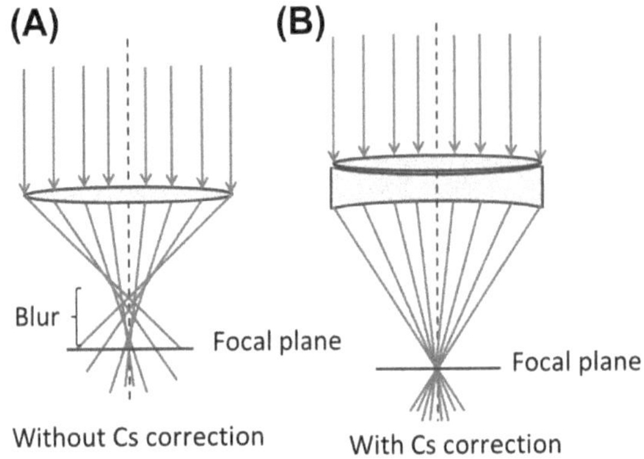

Figure 5 Schematic illustration of Cs and Cs correction. (A) A lens with Cs. (B) A lens with Cs corrected by a diverging lens.

Correction of chromatic aberration (Cc) can further increase the resolution and interpretability of TEM images. The unstable electron gun source can generate electrons with different energies, as can instabilities in the high voltage source and inelastic scattering in the sample. The lens in TEM bends the electrons differently due to the different energies. The resulting nonuniform focal plane for the electrons is what limits the resolution. The principle of Cc correction is similar to Cs correction. An "effective" diverging lens can be used to bend the electron beams to the same focal plane. With both Cc and Cs correction, the best point-to-point resolution that has been achieved is close to 0.5 Å [22].

3.2 Resolution and Precision

The point-to-point resolution (ρ) in TEM can be defined as the smallest observable distance between two objects. Modern advances in TEM design have improved the point-to-point resolution to 0.5 Å. For simplicity, consider the objects as Gaussian functions as shown in Figure 6. If the two Gaussian functions start to overlap, the resolution of the functions worsens. At the point when the total intensity of the two overlapping functions is ~25% less than the peak intensity of the individual functions, the distance between the two peaks meets the Rayleigh criterion for resolution (best point-to-point resolution). We can see that resolution has a subjective definition. There is no strict requirement for the intensity of the overlapping area. For aberration-corrected microscopes the resolution can be below 1 Å, which allows for visualization of separated atom columns of most crystals in low index zone axes.

Precision (σ) is related to resolution by $\sigma = \frac{\rho}{\sqrt{N}}$, where N is the total electron dose for the object. For a typical HREM image taken with a modern aberration-corrected electron microscope, the electron dose is 500–40,000 e/Å^{2}s. Assuming a resolution of 0.8 Å and a Gaussian function width of 1 Å, the precision can be 0.036–0.004 Å. That is

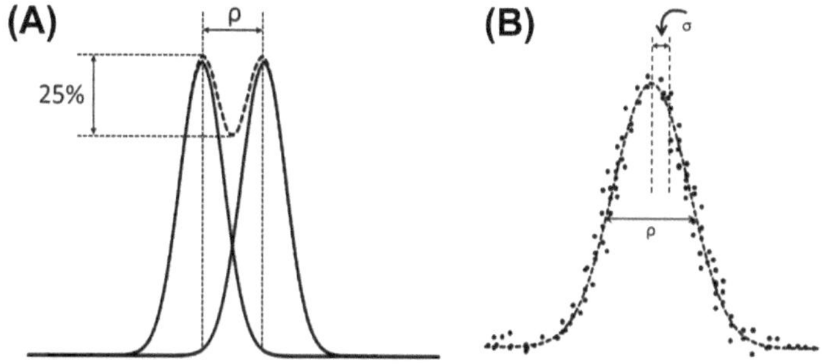

Figure 6 Comparison of resolution and precision. (A) The point-to-point resolution defined as the Rayleigh criterion. (B) Illustration of the definition of precision. The precision of peak position (σ) is related to both resolution and signal-to-noise ratio.

why some HREM images can predict atomic locations at the picometer scale, which is very useful for measuring lattice relaxations. For example, Yu et al. showed that the surface plane of Co_3O_4 (111) has an inward contraction of 20 pm [12]. When comparing two images, it is possible that one has a worse resolution but a better precision if it has a high signal-to-noise ratio. Hence resolution is not the only concern in microscopic studies.

3.3 Atom Locations STEM and HREM Images

In both STEM and HREM images, the contrast is a combined result of the intrinsic electron-specimen interaction and the electron-lens interaction. We can "see" atoms in "good" HREM or STEM images because the electron-specimen interaction is localized around the atoms, that is, good localization. In a crystalline specimen, the electron signals are mostly confined around the atom columns instead of undergoing diffuse scattering. The well-defined atom columns can be observed by tilting the crystalline specimen to a low index zone axis. We generally want the TEM lens to magnify the images and enhance the contrast, but to not influence the relative localization of the projected atom positions. The following content intuitively illustrates the atom locations in HREM and STEM images.

3.3.1 High Resolution Electron Microscopy

As long as TEM has a coherent illuminating electron source, obtaining an HREM image with lattice fringes is not difficult. The spacing of the lattice fringes indicates the distance between crystallographic lattice planes (d-spacing), which is very useful for characterizing the structure of a specimen. However, interpretation of HREM contrast beyond d-spacing is not trivial. The spots (either bright or dark spots) can represent real atoms or may just be artifacts. In HREM, both the intrinsic electron-specimen interaction and the electron-lens interaction can cause the atoms to appear as white or black spots in the images. The electron-specimen interaction and electron-lens interaction can be

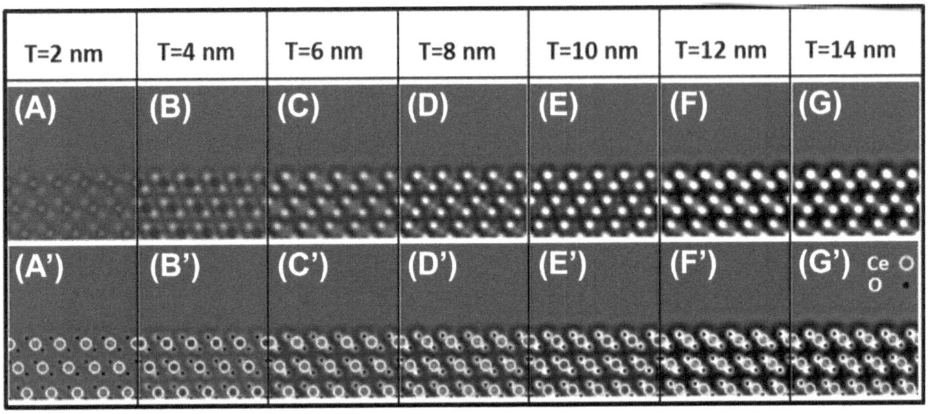

Figure 7 (A)–(G) Simulated high resolution electron microscopy images of the CeO_2 (111) surface with varying thicknesses at a fixed defocus (2 nm). The atom locations of each image are shown in (A′)–(G′), respectively.

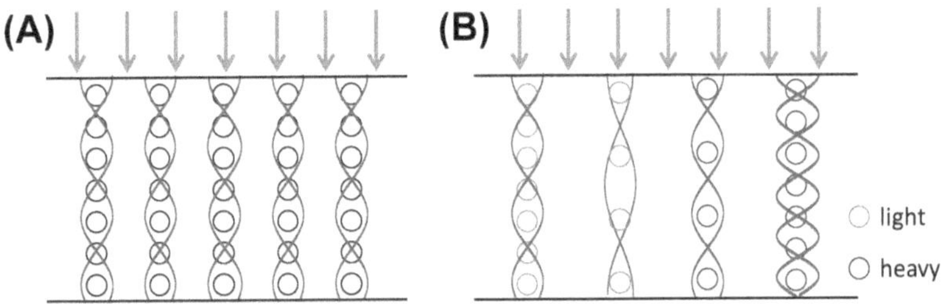

Figure 8 Schematic illustration of electron channeling theory in high resolution electron microscopy. (A) Channeling in a crystal. (B) Channeling along different atom columns. *Reproduced from D. Van Dyck, M. Op de Beeck, Ultramicroscopy 64 (1996) 99–107.*

simplified to the thickness effect and lens defocus effect, respectively. Figure 7 shows simulated HREM images for CeO_2 (111) surfaces with varying thicknesses and a fixed defocus. When the specimen thickness is 2 nm (T = 2 nm), both Ce and O atoms appear as white spots (see Figure 7(A) and (A′)). As the thickness increases, the intensity of Ce positions oscillates from white to black. The intensity of the O positions increases but the contrast is maintained as white spots.

The thickness dependence of HREM contrast arises from the complex interaction between the direct transmitting beam and diffracted beam in the specimen, which can be intuitively understood by the s-state electron channeling model [23]. When a crystal is tilted to a zone axis, the atoms are aligned into columns along the electron beams, as shown in Figure 8(A). After the electrons reach the sample, the dominant electron scattering is confined to the channels provided by the atoms in the material owing to the electron-atom interactions. The intensity of the electron wave along each atom column oscillates with a certain periodicity. Thus the intensity of the wave as it exits the material is dependent on the material thickness. Furthermore, the intensity of the exit wave is

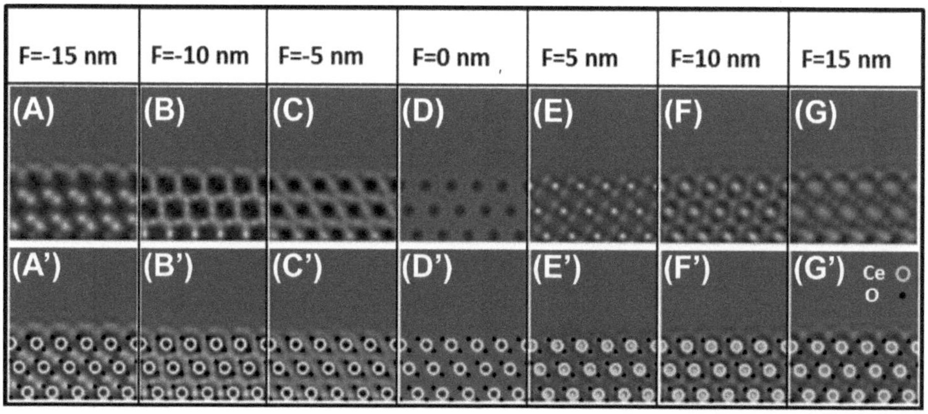

Figure 9 Simulated high resolution electron microscopy images of the CeO_2 (111) surface with varying degrees of defocus at a fixed thickness (2 nm). The atom locations for each image are shown in (A′)–(G′), respectively.

zone-axis dependent and element dependent, as shown in Figure 8(B). For a zone axis along which the atoms are packed sparsely, the oscillation is slower than in the columns where atoms are packed closely. With the same thickness and same atom density within atom columns, an electron wave traveling along a "heavy" element column will oscillate faster than one traveling along a column containing lighter elements. Therefore, a significant advantage of HREM is that even in the presence of heavy elements, light elements can still be visualized for a specimen with a certain thickness.

To understand the defocus effect on HREM images, we can again use the CeO_2 (111) surface as an example. Figure 9 shows a 2 nm thick CeO_2 sample with varying defocus values ranging from −15 to 15 nm. Several impacts of defocus on the images can be summarized based on Figure 9. (1) Small defocus results in good image localization. At small defocus such as F = 5 nm, the spots in the images directly correspond to atom locations. At a defocus of 15 nm, it is hard to relate the image contrast to atom positions. Particularly for the surface, additional fringes (Fresnel fringes) other than those representing real atom contrast (visible at the surface areas of Figure 9(A) and (G)) can make the interpretation of surface contrast very difficult. (2) Positive defocus and negative defocus can reverse the image contrast. For F = −5 nm, the black spots are atoms. However, at F = 5 nm, the O and Ce atoms appears as white spots. (3) Zero defocus is not necessarily the best imaging condition. The image with F = 5 nm has a better contrast than the image with F = 0 nm. According to HREM theory, zero aberration results in an image with minimal contrast for a thin specimen although the image localization is optimized. The "best" image is a balance between localization and image contrast. Sometimes a small defocus or Cs is intentionally applied to obtain images with good contrast [24]. Nevertheless, a minimal contrast image is very useful in practice, as it is a way to identify the zero defocus condition. The absolute values of defocus can then be approximately determined based on the zero defocus reference.

Defocus is not the only aberration. There are other aberrations such as astigmatisms, coma, and Cs, all of which can induce delocalization and render the HREM images difficult to interpret. Fortunately, for HREM imaging there are many simulation programs [25–30] which work quite well, and the computational time for each simulation is very fast (several seconds). In a typical simulation program, besides the sample structure, imaging parameter inputs such as the defocus, Cs, sample thickness, astigmatism, and sample tilt are all required as input parameters to generate simulated HREM images. Better control of imaging parameters in an HREM experiment facilitates the simulation processes, although sometimes it is hard to know the correct values to use.

3.3.2 Scanning Transmission Electron Microscopy

When the electron beam converges to a probe with a size on the atomic scale, atomic resolution imaging is possible. However, the contrast differs considerably between BF, ABF, ADF, and HAADF images in STEM. A BF-STEM image should be understood (by reciprocity) as an HREM image. The changing of specimen thickness and defocus can cause contrast reversal in the atom positions. In ADF images, good image localization at atom positions is also due to s-state channeling although changes in orientation (strain) can also lead to significant contrast. In general it is better to use larger collection angles, that is, HAADF, in order to minimize this, although it can never be completely eliminated. However, the behavior of the electron channeling differs between ADF and particularly HAADF and HREM. In HAADF, the intensity of the spots only slightly changes with the specimen thickness. There is no contrast reversal with respect to the thickness change provided the thickness is not too large. The atoms in HAADF images appear as white spots in most cases. The aberrations such as defocus can also induce significant delocalization in HAADF images. However, the contrast will not reverse as in HREM. The atoms remain white as the defocus changes in HAADF images.

The significant difference between ADF/HAADF and BF images arises from the direct beam which is collected by the BF detector. In principle, if the direct beam is blocked, the interference between the direct and diffracted beams in the specimen can be neglected. In practice, a beam stopper can be put on top of the BF detector to block the direct beam. The resulting images collected by the BF detector are ABF images. The ABF technique has the same advantage as HAADF in that it has relatively low sensitivity to specimen thickness and defocus compared to HREM. In addition, it has good sensitivity to both heavy and light elements, similar to BF imaging. Intuitively, as the large angle scattered beam (the electrons collected by HAADF) is not collected, the atoms appear black in ABF. Experimental and simulated ABF images on several materials all demonstrated that the ABF technique is less sensitive to thickness and defocus [31,32]. Examples of HAADF and ABF images of four different materials [31] are shown in Figure 10.

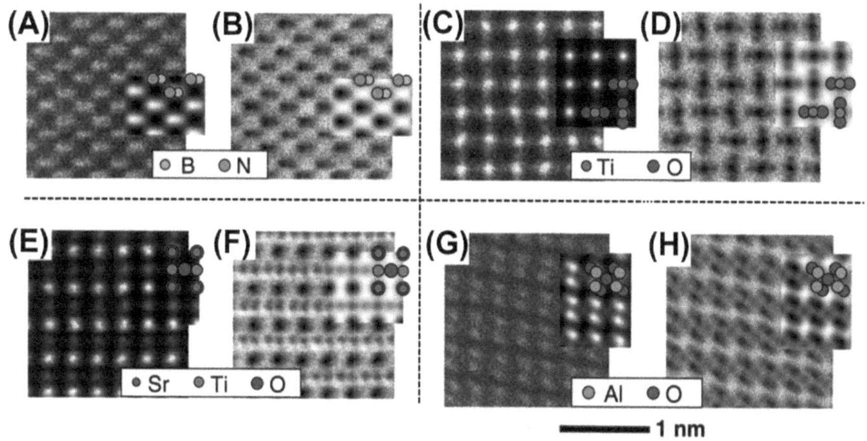

Figure 10 Comparison of high angle annular dark field and annular bright field images of (A), (B) cubic BN [110], (C), (D) TiO_2 [001], (E), (F) $SrTiO_3$ [011], and (G), (H) Al_2O_3 [$\bar{1}2\bar{1}0$]. The simulated images are shown in the insets with the structural models overlaid. *Reprinted from Findlay et al. Dynamics of annular bright field imaging in scanning transmission electron microscopy, Ultramicroscopy 110 903–923, Copyright (2010), with permission from Elsevier.*

Although the interpretation of HAADF and ABF images is relatively easy, sometimes image simulations are necessary, particularly for two materials with different thickness. In HAADF images, the "Z-contrast" is only for specimen with same imaging condition (such as thickness and defocus). For two species in a specimen with different thicknesses, the intensity is not necessarily directly related to Z. The popular HREM simulation programs usually contain a package for STEM simulation. The simulation of STEM images usually requires more computational time than HREM simulation.

3.3.3 HREM versus STEM

HREM, ADF-, ABF-, and BF-STEM techniques are all capable of imaging atoms. Operation in STEM mode allows users to obtain images with a desired contrast. Most often the HAADF-STEM images are Z-contrast images, which can significantly facilitate image interpretation. However, HAADF-STEM has relatively low sensitivity to light elements when heavy elements are present. ABF- and BF-STEM can image the light and heavy atoms simultaneously. The contrast in BF-STEM and HREM images can reverse with respect to thickness and defocus. Performing image simulation is very important to correctly interpret the images.

In practice, the HAADF and BF signals can often be obtained simultaneously. Or for a microscope equipped with EDS or EELS, operating in STEM mode significantly enhances the spatial resolution of elemental analysis. This is very helpful for visualizing some systems such as bimetallic catalysts in which each element cannot be identified easily using imaging techniques. In some other aspects HREM has advantages over

HAADF. Usually HREM utilizes all the electrons passing through the specimen, so the signal-to-noise ratio is higher than in HAADF. The high signal-to-noise ratio in some cases is very important for imaging the surface structures. Specimen drift should also be considered in order to choose the right imaging mode. For example, specimen drift during the scanning of the electron probe results in distorted atom column positions in an STEM image for both BF and HAADF. Hence the measurement of surface relaxation typically involves more noise or errors in STEM than in HREM.

4. CASE STUDIES

4.1 Atomic Surface Structures of Well-Defined Oxide Supports

4.1.1 CeO_2 Nanocubes

Owing to its remarkable redox ability, oxygen storage capability, and high ion conductivity, ceria (CeO_2) is a popular material in chemical catalysis [33–36] and solid oxide fuel cells (SOFCs) [37–39]. CeO_2-based oxides can be used in three-way catalysts for the conversion of toxic gases in automobile exhaust [40]. As a reducible oxide, the surface oxygen atoms can be easily released to form O vacancies. The O vacancies at the surface of CeO_2 are believed to enhance the stability, dispersion, and catalytic activity of supported metals [41–43]. To increase the oxygen vacancy formation and catalytic properties of CeO_2, generally two approaches have been explored. One is the development of a solid solution with other oxides such as ZrO_2 [33]. The oxygen vacancy formation energy was found to be lower in CeO_2–ZrO_2, which is possibly due to the lattice strain induced by incorporating Zr atoms [44,45]. For the substitution of some other metals such as Mn, Pr, and Sn, the decrease of vacancy formation energy is due to both electronic modification and structural distortion [46]. The other approach is to synthesize metastable CeO_2 nanoparticles with high energy facets. DFT studies have shown that the vacancy formation energy is lower on the (100) and (110) surfaces than on the (111) surface. It has been demonstrated that CeO_2 nanocubes with predominately {100} facets exposed are more active than CeO_2 nanooctahedra with primarily {111} facets exposed for water gas shift (WGS) and CO oxidation [47]. The ease of O vacancy formation on the (100) surface originates from the surface stability. Atomic or electronic reconstructions usually occur to lower the surface free energy and stabilize the material. One of the possible ways is the formation of surface O vacancies.

The models of bulk and surface structures are shown in Figure 11. Tasker classified the surfaces of ionic crystals into three types purely based on electrostatic considerations [48]. (This is a useful first approach to understanding the surfaces of oxides, but since this assumes that only ions exist it can in some cases be misleading, and a better strategy is a bond valence approach [49].) The crystal with type I surface consists of valance neutral layers. Along the [110] direction, CeO_2 with a fluorite structure consists of repeating valance neutral CeO_2 layers, as shown in Figure 11(B). The bulk terminated CeO_2 (110)

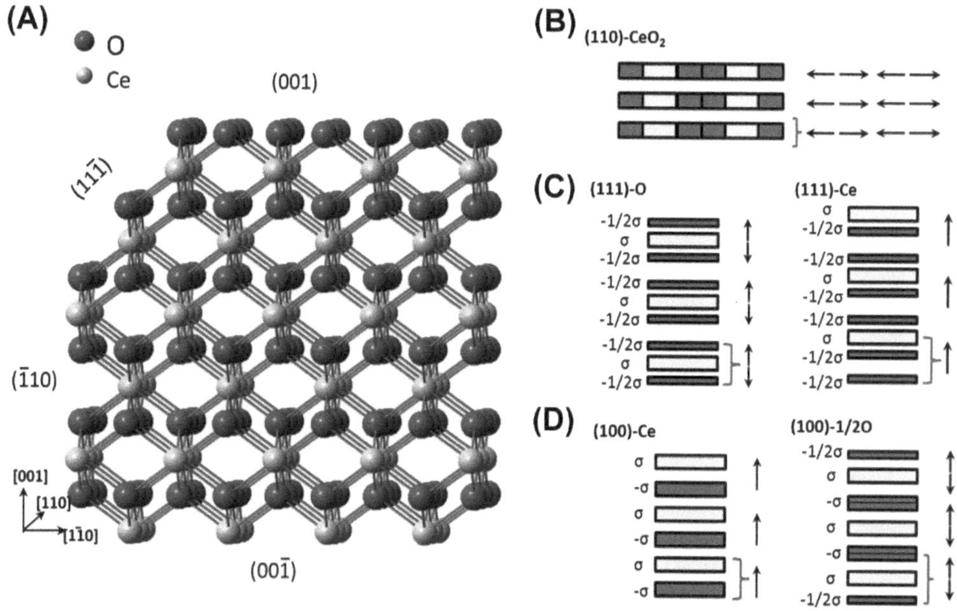

Figure 11 Schematic drawings of CeO_2 bulk structure (fluorite), surface structures, and associated dipole moments. (A) A CeO_2 structural model with truncated {100}, {110}, and {111} surfaces. (B) Schematic view of the stacking of dipole moments along the [110] direction. The red (black in print versions) and yellow (gray in print versions) boxes represent O and Ce atoms and each arrow represents a dipole moment. The curly brackets represent a repeating unit. (C) Schematic view of the stacking of dipole moments along the [111] direction. The O-terminated (left) and Ce-terminated (111) (right) surfaces result in nonpolar and polar surfaces, respectively. (D) Schematic view of the stacking of dipole moments along the [100] direction. The Ce-terminated (100) is a type III polar surface (left). The dipole moment can be compensated by transferring ½O atoms from an O-terminated side to a Ce-terminated side (right).

is a type I surface and is considered to be stable in most gas environments. The type II surface comprises repeating units of a tri-layer system in which the dipole moments are in opposite directions and cancel each other out. The type III surface consists of a stacking of repeated bilayers. Each repeating unit is associated with a dipole moment. The net dipole moment stacks up as the thickness increases and the resulting surface is polar and unstable except under highly reducing or oxidizing conditions. The CeO_2 (111) surface can be a type II or type III surface depending on the cation or anion surface termination. If the (111) surface is O terminated, the repeating unit is a sandwich-like O^{2-}–Ce^{4+}–O^{2-} along the [111] direction, as shown in Figure 11(C). In each repeating unit, the dipole moments with opposite directions cancel each other out. Therefore, the net dipole moment of each repeating unit is zero and the O-terminated CeO_2 (111) is a type II stable surface for most gas environments. If the (111) surface is Ce terminated, the (111) surface is then a type III unstable polar surface, as shown in Figure 11(C). In this case, the repeating unit in the [111] direction should be chosen as Ce^{4+}–O^{2-}–O^{2-}. The net dipole moments from each unit stack up and the electrostatic energy of the crystal diverges as the thickness increases. Therefore, the Ce-terminated CeO_2 (111) is expected to have high surface energy and the O-terminated CeO_2 (111) is energetically preferred.

The (100) surface is classified as a "type-III" polar surface. This type of surface is unstable in most cases and surface reconstructions or other stabilization mechanisms are expected to take place. Along the [100] direction, CeO_2 consists of repeating units containing a Ce^{4+} and a O^{2-} plane, as shown in Figure 11(D). For a stoichiometric CeO_2 nanocube (i.e., Ce and O layers on the opposite sides), a total dipole moment perpendicular to the (100) surface is induced. The macroscopic dipole moment can be quenched if the Ce-side surface is less positive and O-side surface is less negative by a factor of ½. This value can be calculated based on the spacing between the cation and anion planes [50,51]. The most often assumed surface termination of CeO_2 (100) is O termination with half of the oxygen atoms removed on both sides. In this case, $\frac{1}{2}O^{2-}$–Ce–$\frac{1}{2}O^{2-}$ can be chosen as the repeating unit, as shown in Figure 11(D). The net dipole moment in each unit is zero and the surfaces become type-II nonpolar surfaces. For completeness, we will mention that under highly oxidizing conditions both the (100) and $(\bar{1}00)$ could be oxygen terminated; similarly Ce terminated under highly reducing conditions. However, the relatively large oxidizing or reducing conditions required for this will rarely be encountered in practice.

CeO_2 nanocubes have the {100} facet predominantly exposed, but the edges and corners are truncated by {111} and {110} surfaces, as shown in Figure 12(A). The surface structures of the three facets can be imaged simultaneously when the nanocube is tilted to the [110] zone axis, as shown in Figure 12(B). The Cc and Cs corrected HREM study performed by our group shows that the {111} surface is O terminated [13], which agrees well with electrostatic arguments. The experimental HREM image is shown in Figure 12(D) with the simulated image overlaid. Although the bulk terminated (110) surface is considered a stable surface, HREM experiments show that the (110) surface consists of (111) nanofacets and flat CeO_{2-x} layers. The simulated images with the Ce terminated (110) surface are shown in Figure 12(C). DFT studies indicate that the (111) surface has a smaller surface energy than the (110) surface. Thus (111) faceting is energetically plausible. However, inducing additional oxygen vacancies will eliminate the (111) nanofacets. Crozier et al. performed an environmental HREM study and showed that the (110) surface, which initially contains sawtooth-like features, becomes flat after annealing the nanoparticles at 600 °C in 0.5 Torr H_2 [52].

In contrast to the well-ordered reconstruction with ½O removed, the (100) surface appears to have multiple terminations [13]. Figure 12(E) and (F) are the experimental images of the top and bottom (100) surfaces. The two surfaces show the coexistence of O and Ce terminations. In Figure 12(E), the left part shows a Ce layer termination while the right part is O terminated. The structure with ½O removed seems to exist, as shown in region I of Figure 12(F), but it is not present over extended regions. Instead, the (100)

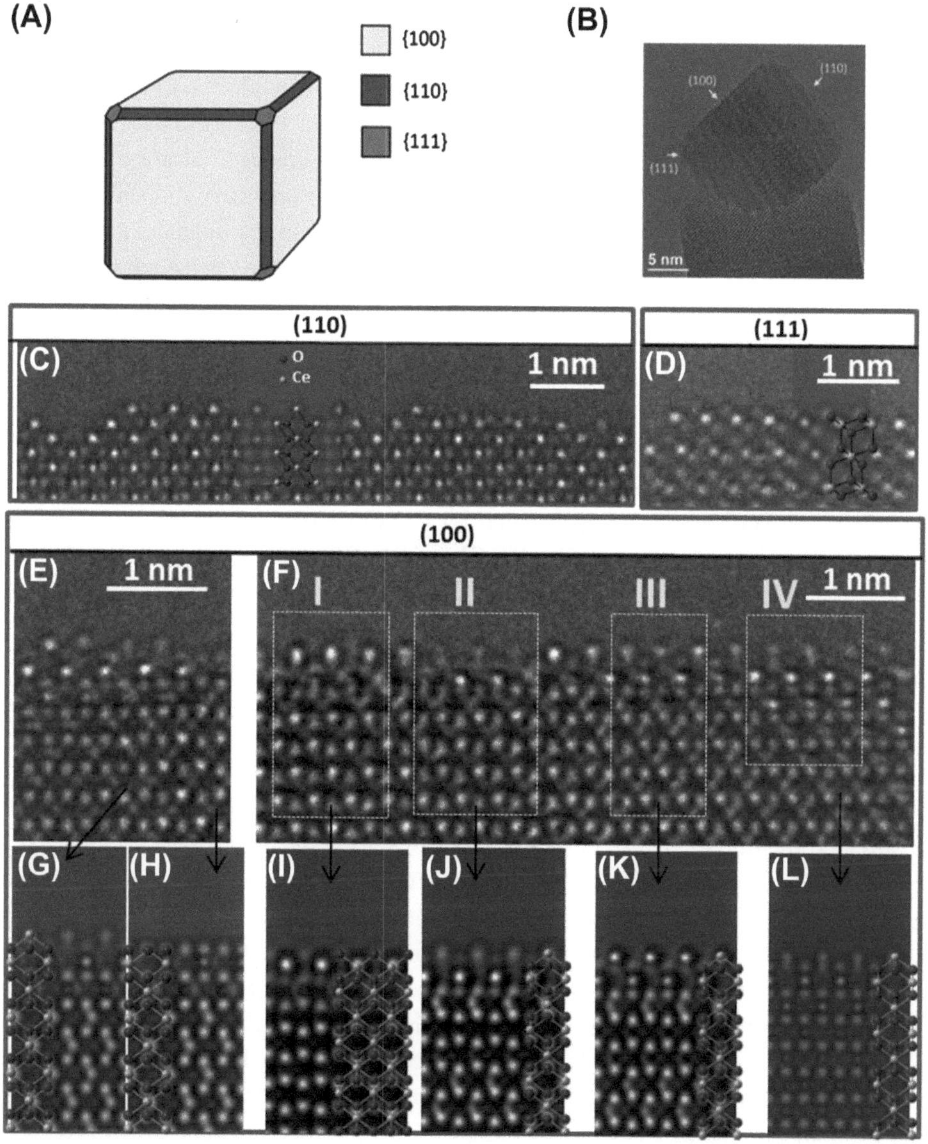

Figure 12 High resolution electron microscopy (HREM) studies of CeO_2 nanocubes. (A) A schematic drawing of the general shape of CeO_2 nanocubes. The {111} and {100} surfaces present on the edges of CeO_2 nanocubes with the {100} surface predominantly exposed. (B) A transmission electron microscopy image of a CeO_2 nanocube viewing in [110] direction. The (100), (111), and (110) facets can be observed simultaneously. (C) Experimental and simulated HREM profile-view images of CeO_2 (110) surfaces. The atom model is directly overlaid on the simulated image. (D) Experimental and simulated HREM profile-view images of CeO_2 (111) surfaces. (E), (F) Experimental HREM profile-view images of the (100) and $(\bar{1}00)$ facets. (G)–(L) Simulated HREM images of each area in the experimental images as indicated by the arrows.

surface contains a large number of cation and anion vacancies, which is evidenced by the significant contrast difference of the surface and bulk Ce atoms in the HREM image. As the thickness increases, the Ce columns become black spots while the O columns maintain white contrast. However, at the near-surface in regions I–III, both Ce and O atoms show white contrast. This is mainly due to the low occupancies of atoms at the surface region. The low occupancy lowers the average density of the atoms, and thus the intensity oscillation is slower compared to the fully occupied regions. The atom positions in the surface area (~1 nm) are partially occupied, which implies the presence of Ce and O vacancies. In addition, the vacancy concentrations are inhomogeneous on the surface, as the contrast of both surface Ce and O atoms changes randomly. DFT studies show that the formation of O vacancies is accompanied by the reduction of neighboring Ce^{4+} to Ce^{3+} [53], as expected from valence arguments, and this reduction was observed in the near surface region (>1 nm) in EELS studies [54,55]. Therefore, (100) surfaces of as-prepared CeO_2 nanocubes contain reasonable concentrations of Ce and O vacancies. The outermost surface can be terminated by Ce, O, or reconstructed CeO. The TEM characterization is in agreement with FT-IR studies [56] using methanol adsorption on CeO_2 nanocubes, nanorods (primarily {110} and {100} facets exposed), and nanooctahedra (predominantly {111} facets exposed). Only on-top methoxy adsorption was observed on the nanooctahedra. In contrast, on-top and several bridging adsorptions were found on CeO_2 nanocubes and nanorods. The IR studies imply the (100) surface consists of a large amount of disordered O vacancies as well as some amount of Ce termination.

Clearly the stabilization mechanisms affect the surface structures of CeO_2. However, the surfaces are not solely dependent on electrostatic effects. For the nonpolar (110) surfaces, (111) faceting can further reduce the surface energy. For the polar (100) surface, a reduced surface region with multiple surface terminations is present instead of a single well-ordered surface termination. In this case, strictly speaking, the (100) surface is not a well-defined surface, which helps to compensate the dipole moment. The formation of the "partially disordered" (100) surface is possibly due to the relatively small energy difference between the different orderings of surface O vacancies. It is also possible that the ordering of O vacancies is related to O partial pressures. The phase diagram of the Ce–O binary system shows a variety of intermediate CeO_{2-x} ($x \leq 0.5$) structures with respect to O partial pressure [57]. The reduction process typically starts at the surface owing to the reducing environment. With diffusion of the O atoms to the surface, the O vacancies diffuse into the crystal simultaneously [58]. A certain type of O vacancy ordering is energetically preferred if the surface is reduced to a certain extent. Indeed, well-ordered surface reconstructions are resolved by STM studies on single crystal CeO_2 at elevated temperatures under UHV conditions [59] (highly reducing). It is worth noting that CeO_2 has a centrosymmetric fluorite structure, hence having similar features on the two opposite surfaces (e.g., the (001) and $(\bar{1}00)$ surfaces) is reasonable. For the polar surfaces of some non-centrosymmetric material, the chemistry of two opposite surfaces can be different. For example, the ZnO (0001)-Zn and $(000\bar{1})$-O surfaces possess distinct structures according to DFT and STM studies [60].

Although HREM profile-view imaging can show occupancies of atom columns, it is not sensitive to 2D defect surface structures. Further detailed atomic structures can be obtained with the help of single crystal studies on CeO_2 surfaces, particularly for the (111) surface. For example, Esch et al. used DFT and STM studies to show that subsurface oxygen vacancies are critical for the formation of linear vacancy clusters on the (111) surface [61]. The study indicates that subsurface vacancies can enhance further surface oxygen release in catalytic applications. Here the subsurface refers to the third surface layer (O layer). Recently a (2×2) ordering of subsurface O vacancies was reported by Torbrügge et al. using AFM [62], which is also in agreement with later STM and DFT studies [63,64]. The structural characterization of the surfaces of nanoparticles and single crystals using different techniques can provide us with a more comprehensive understanding.

4.1.2 $SrTiO_3$ Nanocuboids

$SrTiO_3$ has a wide range of applications from thin film growth [65–67] to catalysis [68]. Recently it was discovered that $SrTiO_3$-based catalysts have high catalytic activity in soot combustion [69]. The synthesis of well-defined $SrTiO_3$ nanocuboids has provided new catalytic insights that are unobtainable from standard bulk $SrTiO_3$ [70–72]. The nanocuboids have the {100} facet exposed, which is close to a thermodynamically stable Wulff shape but more accurately is a growth shape (kinetic Wulff shape, e.g., [73] and references therein). As $SrTiO_3$ has a lattice constant of 3.905 Å, which is very close to the lattice parameters of Pt and some other metals, epitaxial growth of these metals on $SrTiO_3$ nanocuboid supports is expected. The stable support and good epitaxy between the support and metal can improve the stability of catalysts [74].

The surface structures of $SrTiO_3$ single crystals are rather complex and have been extensively investigated. $SrTiO_3$ has a standard perovskite structure consisting of a repeated stacking of strontium oxide (SrO) layers and titanium dioxide (TiO_2) layers in the [100] direction, as shown in Figure 13. If a $SrTiO_3$ surface is truncated in the [100] direction, the question of simple surface termination inevitably arises: Is the (100) surface terminated with SrO or TiO_2? In principle, the SrO and TiO_2 terminations can coexist for a vacuum-cleaved surface. A variety of surface reconstructions other than the SrO and TiO_2 terminations were observed after high temperature annealing (typically >800 °C). A number of surface reconstructions have also been reported on $SiTiO_3$ (100) single crystals prepared under different conditions [75–83]. Among them, the well-ordered (2×1) [75], c(4×2) [81], and (2×2) [84] surface reconstructions have been structurally solved by using a variety of techniques and there are additional, plausible structures such as locally ordered (3×3), $(\sqrt{5}\times\sqrt{5})$ R26.6° and $(\sqrt{13}\times\sqrt{13})$ R37.7°

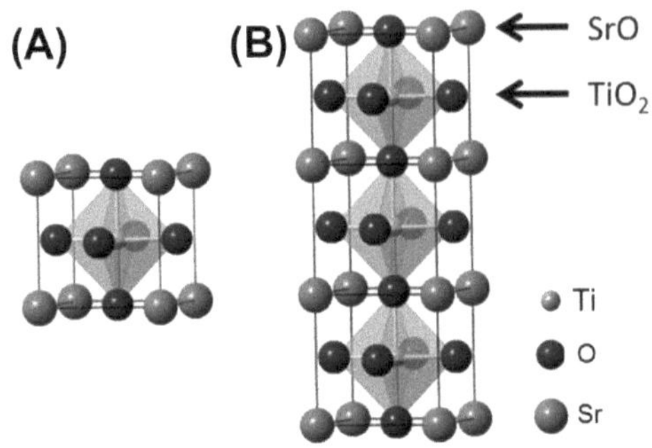

Figure 13 Bulk and surface structures of $SrTiO_3$. (A) Schematic view of a $SrTiO_3$ unit cell. (B) Illustration of two possible bulk truncations of $SrTiO_3$ (100): SrO and TiO_2 surface terminations.

surfaces [85]. Interestingly, all of the reconstructions have TiO_2 double-layer termination features, in which both the surface and subsurface are TiO_2 layers. The TiO_2-rich (100) surface can be obtained by etching a single crystal in a buffered NH_4F–HF solution [86], which is a standard technique in thin film processing. This is probably due the greater basicity of surface SrO compared to the TiO_2 layer. Prolonged annealing can also result in a variety of surface features such as islands [87], step-terrace structures [88,89], ridges [90], and canyon-like features [91]. The non-flat morphological features can either be Sr-rich or Ti-rich depending on the O partial pressure during annealing [92–94], which suggests that O and Sr vacancies may play an important role in the surface chemistry of $SrTiO_3$.

However, most of the reconstructed surface structures on $SrTiO_3$ single crystals are obtained after high-temperature annealing. The sample preparation is very different from the $SrTiO_3$ nanocuboid growth condition or catalytic reaction environments. The temperatures for some wet syntheses of $SrTiO_3$ nanocuboids are around 200 °C. Some surface reconstructions may not yet be activated in these conditions owing to the low temperature, as the formation of some stable reconstructions requires the overcoming of energy barriers. In addition to the temperature difference, solution-based synthesis involves the addition of different precursors with or without organic ligands. Whether or not the surface structures of $SrTiO_3$ single crystals are the same as the nanocrystals is a question of interest. Additionally, the impact of synthetic method on nanocuboid surface structure is an intriguing research topic.

To address these problems, our group studied the surface structures of $SrTiO_3$ nanocuboids prepared by three different hydrothermal syntheses. The first type of synthesis uses oleic acid as a capping agent as proposed by Hu et al. [72]. The end products are nanocuboids with a size of ~20 nm. The second hydrothermal route does not use oleic acid and results in nanocuboids with a size of ~65 nm [70]. During this synthesis, acetic acid was used to dissolve the $Sr(OH)_2$, which was used as the Sr precursor. The third method is the same as the second one but uses a microwave oven instead of a

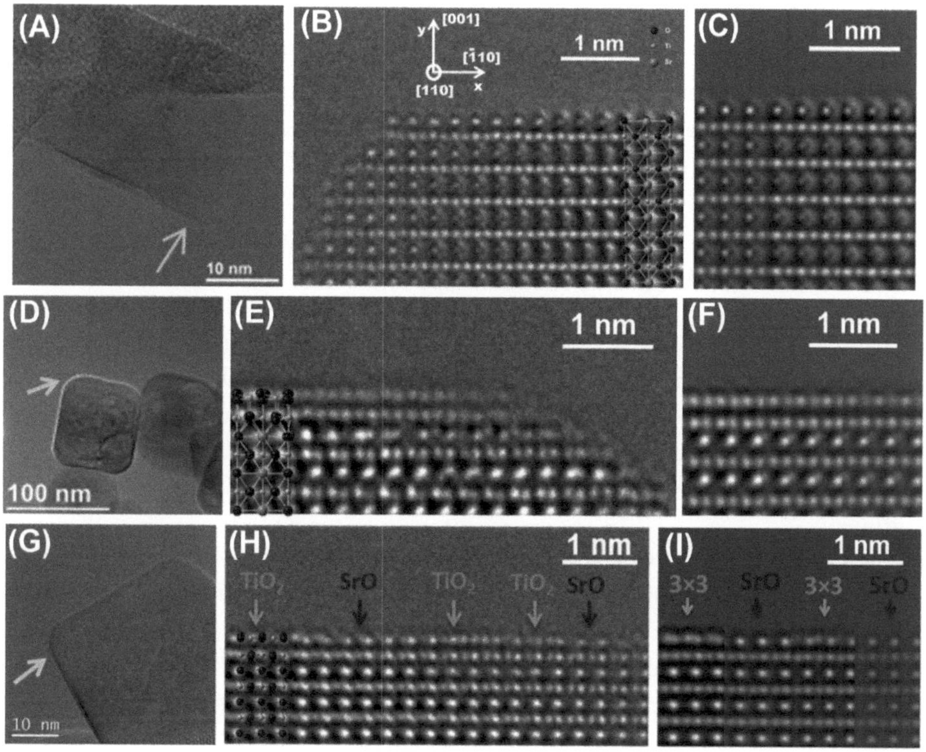

Figure 14 Comparison between the (100) surfaces of $SrTiO_3$ nanocuboids obtained by three hydrothermal processes. The atomic resolution surface structures are imaged by high resolution electron microscopy (HREM) profile-view studies. (A)–(C) Low magnification transmission electron microscopy, experimental HREM, and simulated HREM images of the $SrTiO_3$ (100) surface prepared by oleic acid synthesis. (D)–(F) Low magnification, experimental HREM, and simulated HREM images of the $SrTiO_3$ (100) surface prepared by acetic acid synthesis. (G)–(I) Low magnification, experimental HREM, and simulated HREM images of the $SrTiO_3$ (100) surface prepared by microwave synthesis.

conventional oven [14]. The apparent "benefit" of microwave synthesis is that it significantly reduces the hydrothermal treatment time from ~20 h to ~20 min. The microwave synthesis results in nanocuboids with a size of ~35 nm. Typical TEM images are shown in Figure 14 to demonstrate the general shapes of $SrTiO_3$ nanocuboids obtained by different preparation methods. The three methods will be referred to as oleic acid synthesis, acetic acid synthesis, and microwave synthesis, respectively.

The atomic surface structures of the nanocuboids obtained by the three different methods are obviously different, as shown in Figure 14. The nanocuboids obtained by oleic acid synthesis are SrO terminated. Figure 14(B) shows an experimental HREM image along the [110] viewing direction. The simulated HREM image using a SrO terminated surface matches with the experimental image well, as shown in Figure 14(C). In addition, HREM simulation reveals that some of reconstructed surface structures show (1 × 1) periodicity along the [110] zone axis. The simulated surface contrast is significantly different from the experiment [14]. The simulated image of the SrO surface matches the experimental image best among the tested structures. The acetic acid synthesis results in a double-TiO_2 layer reconstructed surface, as shown in Figure 14(E). The simulated HREM images using the locally ordered ($\sqrt{13}\times\sqrt{13}$) R37.7°, ($\sqrt{2}\times\sqrt{2}$) R45°, and ($\sqrt{5}\times\sqrt{5}$) R26.6° surface structures match better than the other surface reconstructions. As the difference between the three locally ordered reconstructions is small, it is possible that these structures can coexist on the nanocuboids. A simulated HREM image using a ($\sqrt{13}\times\sqrt{13}$) R37.7° surface model is shown in Figure 14(F). Interestingly the microwave synthesis not only reduces the required time of hydrothermal treatment, but also changes the atomic surface structures, as shown in Figure 14(H). The surface has a mixed termination of both double-TiO_2 layer reconstructions and SrO terminations. The simulated HREM image is shown in Figure 14(I), in which the (3 × 3) reconstruction [85] was used to simulate the double-TiO_2 layer feature in the experiment.

The presence of the SrO termination on the as-prepared nanoparticles is attributed to the capping oleate from the oleic acid synthesis. Hu et al. carried out in situ SAXS and WAXS and found that a lamellar microemulsion was formed during the hydrothermal synthesis [72]. The oleic acid in the microemulsion confines the nanocuboids and limits their size. The end products are nanocuboids with surfaces covered by a layer of oleate, which was confirmed by FT-IR studies [14]. As the oleic acid is expected to bond preferentially with the more basic SrO termination, the relatively unstable SrO-terminated surface is protected by the oleate molecules. For the acetic acid synthesis, no organic ligands are present, and the surface structures should be mostly controlled by thermodynamics, with the lowest energy structure dominating. The observed surface reconstructions are consistent with DFT predictions under vacuum conditions. The ($\sqrt{13}\times\sqrt{13}$) R37.7°, ($\sqrt{2}\times\sqrt{2}$) R45°, and ($\sqrt{5}\times\sqrt{5}$) R26.6° surface structures are located in the energy convex hull [85]. The microwave synthesis is a relatively new technique and its complete influence on the synthesis is not yet well understood. The effect of the microwave synthesis on the surface structure can either be simply due to the shorter reaction time or more complex mechanisms. More future studies are required for a comprehensive understanding.

The effects of ligands on the properties of catalysts have been reported in the past. Mostly the ligand was assumed to block reactant access to the catalyst surface [95,96] or affect the electronic structure of catalyst [97–100]. This study shows that the atomic surface structures of mixed metal oxides may in fact be changed by the ligands. The results may be extended to other similar materials, particularly for mixed metal oxides with different surface acidities. In addition, the SrO- and TiO_2-rich surfaces can result in the morphological change of the supported Pt particles [101]. It has been demonstrated that the oxygen reduction reaction (ORR, O_2 to H_2O and O_2 to H_2O_2) is sensitive to the shape of Pt on $SrTiO_3$ substrates [102]. And for some reactions, the selectivity can be tuned by the facets of Pt [103]. In these studies, either Pt single crystals or relatively large Pt nanoparticles are used. Controlling the morphology of Pt through controlling the atomic surface structures of the supports may be a more viable method in catalytic applications.

4.2 Surface Relaxations: Adsorbate Induced or Intrinsic

In aberration-corrected HREM or STEM images, picometer precision can be achieved [12,104]. At such a level of precision, small surface relaxations can be clearly imaged. Lattice relaxation typically happens at the low-coordinate atoms except for those on very close-packed surfaces. Together with DFT calculations, the measurement of surface relaxations by HREM or STEM can provide additional information about the surface atomic and electronic structures of catalysts. However, the relaxation measurements should be carefully conducted as the exact peak positions of the spots are not necessarily the real atom positions. The peak positions can deviate from the real atom positions due to residual aberrations, specimen tilt, or even the intrinsic phenomena of the electron-specimen interaction (in the case when column approximation does not hold). In STEM imaging, the scanning noise can also cause mismatch between the spot and atom positions. Performing DFT calculations and quantitative image simulations can identify whether the surface relaxations are real or simply the effects of HREM/STEM imaging.

Our group investigated the surface relaxation of $SrTiO_3$ nanocuboids with a SrO termination [14]. The surface layer has an inward contraction while the second surface layer has an outward expansion. The interlayer spacing measured from the experimental image was compared to the measurement from simulated HREM images with DFT relaxed surface structures. The experimental surface relaxation matches with the DFT results well, which implies a clean SrO-terminated surface without adsorption. Interestingly the lattice spacing shows oscillations of ~50 pm in the bulk region while in principle these should not exist, as the interlayer spacing between the SrO and TiO_2 layers should be the same if there is no relaxation. This implies that the spot peak positions in an HREM image are not always the exact atom positions. We found that residual aberrations and specimen tilt cause deviation of the peak positions from the exact atom column positions. More specifically, the peak intensity of the Ti column is higher than the SrO column at a certain imaging condition. A distortion of the HREM image will distort the low intensity spots (SrO) more than the high intensity spots (Ti). As a result, the distances between the alternating Ti and SrO peak positions appear to oscillate.

As one example, surface relaxation measurements were conducted by Yu et al. in studying the (111) surface of Co_3O_4 nanocubes using aberration-corrected HREM (Figure 15) [12]. Co_3O_4 has attracted great interest for its applications in lithium battery [105,106] and catalysis [107–110]. It has shown that the Co_3O_4 nanorods

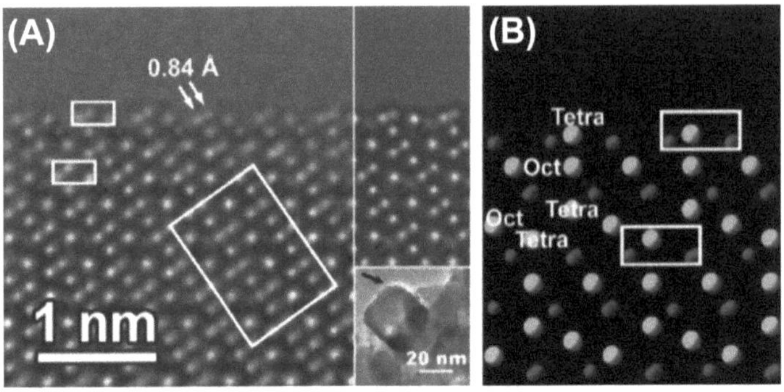

Figure 15 The (111) surface of Co_3O_4 nanocubes. (A) High resolution electron microscopy (HREM) profile-view image shows the (111) surface is terminated by the tetragonal Co atoms. The simulated HREM image is shown in the right side with the low magnification transmission electron microscopy image shown on the lower right corner. (B) The structural model of tetragonal Co-terminated (111) Co_3O_4 surface. *Reprinted figure with permission from Yu et al. Phys. Rev. Lett. 105 226101 (2010). Copyright (2010) by the American Physical Society.*

have high activity for CO oxidation even at −77 °C [109]. Co_3O_4 has a spinel structure with the AB_2O_4 general formula. For Co_3O_4, A and B are, respectively, the tetrahedral and octahedral sites of Co. Along the [111] direction, the sequential stacking of B3-O4-A-B-A-O4 results in six possible terminations of the (111) surface. Yu et al. performed aberration-corrected HREM and found that the (111) surface of Co_3O_4 is a tetrahedral Co layer (Co^{2+}) with an O layer as the subsurface. Beyond the surface termination determination, the lattice spacing measurement shows a significant inward relaxation (~0.35 Å) of the outmost surface and a moderate inward relaxation (~0.1 Å) of the subsurface. The relaxations of the first three layers of the (111) surface measured from HREM images match well with the surface relaxations predicted by DFT calculations. Based on electrostatic arguments, the (111) surface is a polar surface. The surface Co^{2+} should be more positive in order to compensate for the surface polarity. As only a stoichiometric and clean (111) surface were considered in the DFT calculations, a good match of DFT and experiment measurement of surface relaxation can almost rule out the possibility of surface polarity compensation by adsorbing charged external species. The authors assigned the polarity compensation to the electron redistribution associated with surface relaxation, which was further supported by their calculation of the Bader charge of surface O and Co.

4.3 Supported Metal Nanoparticles

Supported metal catalysts are used in many catalytic reactions. Catalytic supports can improve the dispersion and stability of the metal. In addition, many studies have demonstrated that supports can have strong interaction with the metal, which can significantly affect reactivity and selectivity. Designing a supported metal system with high activity, selectivity, and stability is of great interest in heterogeneous catalysis.

4.3.1 The Equilibrium Shapes of Freestanding and Supported Metal Particles

Under thermodynamic considerations, the equilibrium shape of a freestanding metal particle can be derived from the Wulff construction. The exposed facets of metal particles are typically the low-index surfaces with relatively small surface free energies (the surface energy per unit area). The distance from the center of the particle to an exposed facet (h_{hkl}) is proportional to the surface free energy (γ_{hkl}) of that facet, which can be expressed as:

$$\gamma_{hkl}/h_{hkl} = const. \tag{1}$$

Therefore the smaller the surface free energy is for a given facet, the larger the exposed area is for that facet. The calculated Wulff shapes of some popular catalytic metals can be found in Ref. [111]. The close-packed surfaces usually have low surface free energies, and they are therefore predominately exposed for equilibrium shapes. Examples are the (111) surface for fcc metals and the (110) surface for bcc metals. For a Pt particle, the Wulff shape is a truncated octahedron with {111} and {100} facets exposed, as shown in Figure 16(A).

If the freestanding metal is synthesized on a substrate, the interface should be accounted for in the shape determination. Qualitatively we can expect an orientation of the metal particle on the support should create a low energy interface. In materials science, low energy grain boundaries usually have a high lattice match (coincidence sites), which is why the epitaxial growth in a specific direction is usually favored. The total system energy can be further reduced by tuning the ratio of surface and interface areas. It is expected that a metal layer will completely wet the surface of the substrate or even form a solid solution with the substrate if the metal-substrate interface is extremely favorable. An energetically unfavorable interface has a strong driving force to minimize the interfacial contact, in which case the metal particles are very hard to grow on the substrate and forming freestanding particle is favorable. A more detailed investigation of supported metal shapes was developed by Kaichew [112] and Winterbottom [113]. According to the Winterbottom shape, the wetting of the metal particle on the substrate is determined by the energy of the interface as well as the energies of the surfaces of substrate and metal particles. The interfacial free energy (γ_{int}) is defined as:

$$\gamma_{int} = \gamma_{metal} + \gamma_{substrate} - \gamma_{bond} \tag{2}$$

where γ_{metal} is the surface free energy of the metal, $\gamma_{subtrate}$ is the free energy of the substrate, and γ_{bond} is the free energy due to the bonding of the metal and substrate. If $\gamma_{int} - \gamma_{substrate} = 0$, exactly half of the metal will be exposed. Otherwise the exposed metal shape will deviate from the half Wulff shape. With an increase in interface energy, the

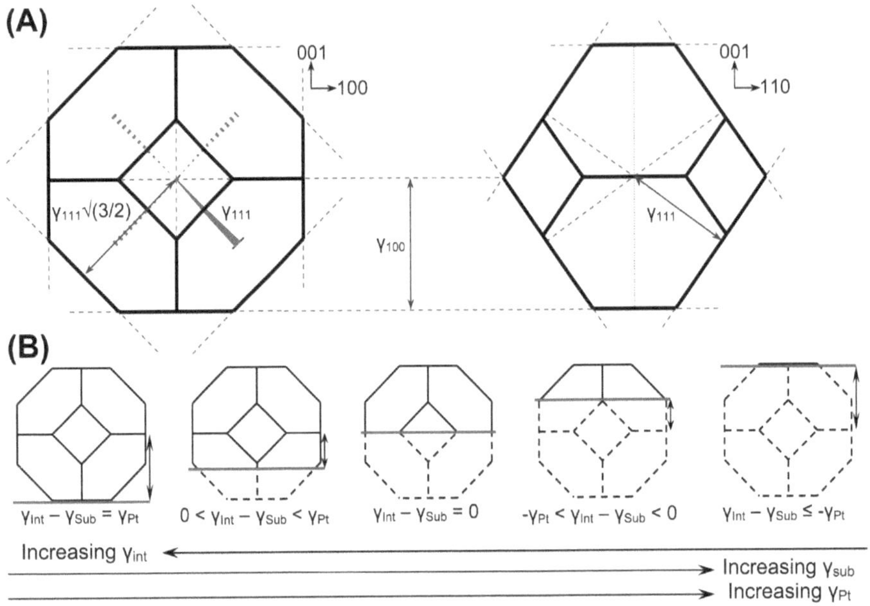

Figure 16 Equilibrium shapes of freestanding and supported Pt particles. (A) Equilibrium shape of a freestanding Pt particle using the γ{111}:γ{100} ratio of 0.84 from Vitos et al. [111] (B) The Winterbottom shape of Pt particles on supports with different interfacial and surface energies. A strong bonding between Pt and substrate results in larger extent of wetting. *Reprinted with permission from Enterkin et al. Nano Lett. 11 (2011) 993–997. Copyright (2011) by the American Chemical Society.*

exposed area of the Wulff shape can increase up to the limit of a "fully" exposed metal particle ($\gamma_{int} - \gamma_{substrate} = \gamma_{metal}$). For the opposite scenario in which $\gamma_{int} - \gamma_{substrate} \leq -\gamma_{metal}$, the substrate will be completely wetted by the metal and no Wulff shape will be exposed. Figure 16(B) shows the Winterbottom shape of Pt with the change of interfacial energy.

Therefore, the shape of metal catalyst can be controlled thermodynamically if the substrate is homogeneous. Enterkin et al. demonstrated an example of the equilibrium Pt metal shapes on $SrTiO_3$ nanocuboids, which have the {100} surface primarily exposed [114]. The Pt metal particles were synthesized on the $SrTiO_3$ nanocuboids by atomic layer deposition (ALD). The $SrTiO_3$ nanocuboids were prepared by the acetic acid synthesis [70], which resulted in TiO_2-rich surfaces, as described in the previous section. $SrTiO_3$ has a lattice parameter 3.905 Å which is very close to that of Pt (3.920 Å). Previous experimental and theoretical studies show that there is a low energy interface between Pt (100) and $SrTiO_3$ (100) surfaces [101]. As a result, epitaxial growth occurs in the [100] direction (cube-on-cube). DFT calculations show that the TiO_2-rich $SrTiO_3$ surface has a surface free energy similar to the {100} surface of Pt. Therefore the expected shape of Pt particles deposited on $SrTiO_3$ nanocuboids is close to a half Wulff shape with a cube-on-cube epitaxial direction. This expectation was confirmed by the HREM study, as shown in Figure 17.

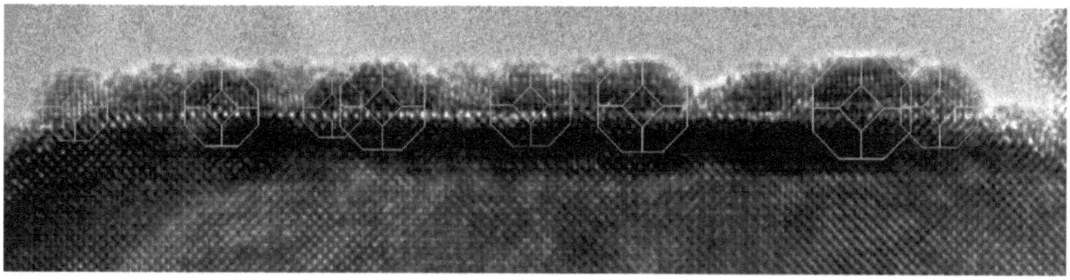

Figure 17 An high resolution electron microscopy image demonstrating the shape and interfacial structures of Pt grown on $SrTiO_3$ nanocuboids by atomic layer deposition. *Reprinted with permission from Enterkin et al. Nano Lett. 11 (2011) 993–997. Copyright (2011) by the American Chemical Society.*

By having a stable metal-support interface as well as equilibrium metal shapes, the sintering probability of the metal particles is reduced. Hence the stability of catalysts during catalytic cycles can be improved. A study of propane oxidation shows that the Pt-$SrTiO_3$ system has better stability compared to the Pt-γAl_2O_3 system [74]. It is also possible that by tuning the lattice parameter of the substrate, the wetting of the metals on the support can be controlled. For the $SrTiO_3$ case, this can be achieved by substituting Ba for Sr (for $BaTiO_3$ $a=b=3.992$ Å, $c=4.036$ Å), which changes the lattice parameters slightly and can therefore alter the ratios of exposed Pt {100} and {111} facets [115]. As Pt {111} and {100} facets are known to have different selectivities for some reactions [103], it is possible to control the catalysis in a thermodynamic way [115].

4.3.2 The Shapes of Supported Metal Catalysts under Reaction Conditions

The DFT calculated energies are for the ground state configuration, and are often for clean surfaces. It is possible that the equilibrium shape or interface structure changes during the catalytic reaction. This is primarily due to chemisorption on the particle or a change of the interface between the particle and support. The changes usually depend on specific gases and catalytic systems. In this review, we focus on some experimental studies to obtain insight into reaction-induced morphological changes. The dynamic process can be imaged by ETEM.

Hansen et al. provided clear evidence of shape changes in supported copper nanocrystals using ETEM [116]. The authors conducted a comparative study on the shapes of Cu nanoparticles supported on ZnO and SiO_2 under different gas environments at 220 °C. They observed shape changes of Cu particles after adding H_2O to pure H_2 on both the ZnO and SiO_2 supports. They also demonstrated that the shape change can be reversed by removing the H_2O gas from the systems. Thus the authors deduced that H_2O is adsorbed on the Cu surface and induces the shape change. In contrast, if CO was added to the pure H_2 environment instead of H_2O, a shape change was only observed on the ZnO-supported Cu. This indicates that the shape change is due to the reduction of the Cu–ZnO interface by CO. Therefore, both metal surfaces and the metal-support interfaces can be affected by the surrounding gases.

Systematic studies of the dynamic behavior of supported particles provide insights into their catalytic properties. Several elegant ETEM experiments have been performed to understand the reaction mechanism for CO oxidation on supported Au nanoparticles (GNPs) [10,117–119]. Au shows remarkable catalytic activity for CO oxidation even at room temperature. Unlike other metal catalysts, such as Pt, Pd, and Ru, bulk Au is almost completely inert to chemisorption [120–122], and adsorption is essential for the surface reactions on catalysts. The high activity of CO oxidation was initially found on oxide supported GNPs [123], and triggered extensive studies and considerable debates [124–126]. It is generally believed that the low-coordinated Au sites [127], the metal-support interaction [128], and quantum size effect for the small site (less than ~2 nm) [129] are responsible for the enhanced activities. By establishing a "safe" electron dose range in the ETEM so that electron-induced morphological changes of Au are minimized [117], Uchiyama et al. studied the shape changes of Au supported by CeO_2 and TiC in CO/air, N_2, O_2, and vacuum, as shown in Figure 18(A) and (B) [118]. No significant morphological change was observed for TiC-supported GNPs in different gas environments. On the other hand, CeO_2-supported GNPs became rounded upon exposure to O_2. As CeO_2-supported GNPs usually show much higher CO oxidation activities than TiC-supported GNPs, the distinct, morphological behaviors of the GNPs on two supports may be correlated to the catalytic processes. By increasing the CO/O ratio, the GNPs on CeO_2 supports become more faceted. An experimental shape diagram showing the effects of CO and O partial pressures on GNP shape is depicted in Figure 18(C). The adsorption of CO molecules on the GNP was believed to stabilize the polyhedral shape of GNPs. In our opinion, it is also possible that additional O vacancies were created under the reaction conditions. The O vacancies also help to stabilize the Au particles [41]. The dissociative adsorption of O_2 on the perimeter of the Au–CeO_2 interface may cause formation of rounded or multifaceted GNP surfaces. The understanding of the reaction mechanism was further developed by Ta et al. [43]. Their mass spectroscopy study showed that CO_2 was immediately produced when CeO_2-supported GNPs were exposed to pure CO at room temperature, which implies the lattice O of CeO_2 can participate in the CO oxidation. Note that for pure CeO_2, the conversion of CO typically starts at ~500 K [130]. If the isotopic $^{18}O_2$ was injected into the system, the mass spectra revealed the presence of $C^{16}O^{18}O$. This indicates that O dissociation is facilitated by CeO_2. The CeO_2-promoted dissociative adsorption of O species improves CO oxidation compared to the use of only Au particles.

Recently the mobility of CeO_2-supported GNPs was studied at atomic resolution by Kuwauchi et al. [10]. Using ETEM, it was observed that the GNPs are displacing and rotating back and forth between equivalent sites of the Au–CeO_2 (111) interface under 1 vol% CO/air. The frequency of this reversible motion is ~6.4 s^{-1}. In contrast, the frequency of this stepwise displacement is ~14.4 s^{-1} in O_2 gas. The lower frequency was attributed to the oxygen vacancies created under these reaction conditions. Their DFT

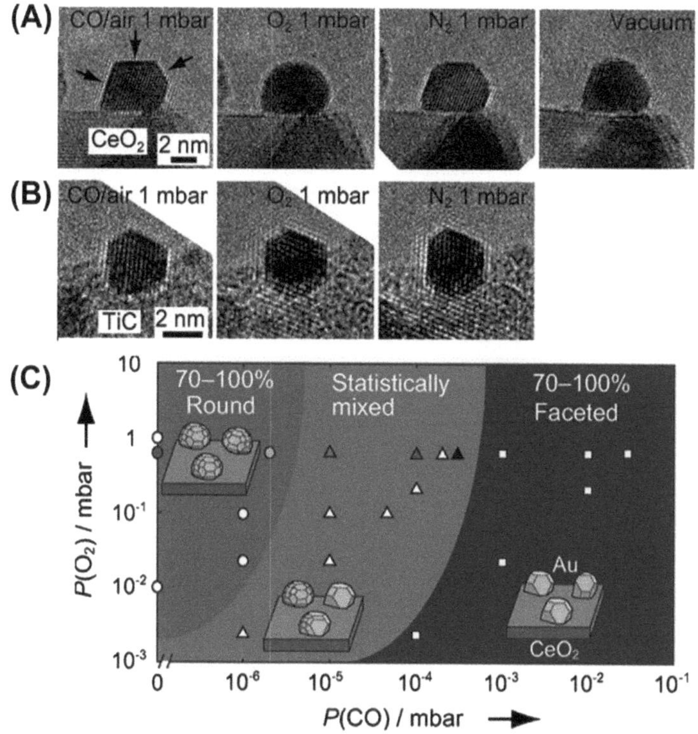

Figure 18 Shape of supported gold nanoparticles (GNPs) in vacuum and gaseous environments. (A) The shapes of GNPs on CeO_2 supports under different gases and vacuum conditions. (B) The shapes of GNPs on TiC supports in different gases. (C) A phase diagram showing the shapes of the CeO_2-supported GNPs versus the partial pressure of CO and O_2. *Reprinted by permission from Uchiyama et al. Angew. Chem. Int. Ed. 50 (2011) 10157–10160, copyright (2011) Wiley-VCH Verlag GmbH & Co. KGaA, Weinheim.*

study shows that the energy to displace GNPs on a Ce-terminated surface is much larger than for the O-terminated surface. Note that experimental studies show that the CeO_2 (111) surface is O terminated in both single crystals [131,132] and nanoparticles [13,55]. However studies show that linear O vacancies are present after annealing [61]. The result further confirms that the O vacancies on the CeO_2 (111) surface provide strong anchoring of the supported GNPs. The strong anchoring enhances the stability of the catalyst during reactions.

4.3.3 Other Thoughts: Manually Controlling the Shapes of Supported Metals

In the past decades, both metals and supports with well-defined shapes have been synthesized [133,134]. However, manual control of supported metal systems is relatively rare. Yamada et al. employed a Langmuir–Boldgett (LB) trough to physically pack the well-defined supports and metals to generate active interfaces for catalytic reactions [96].

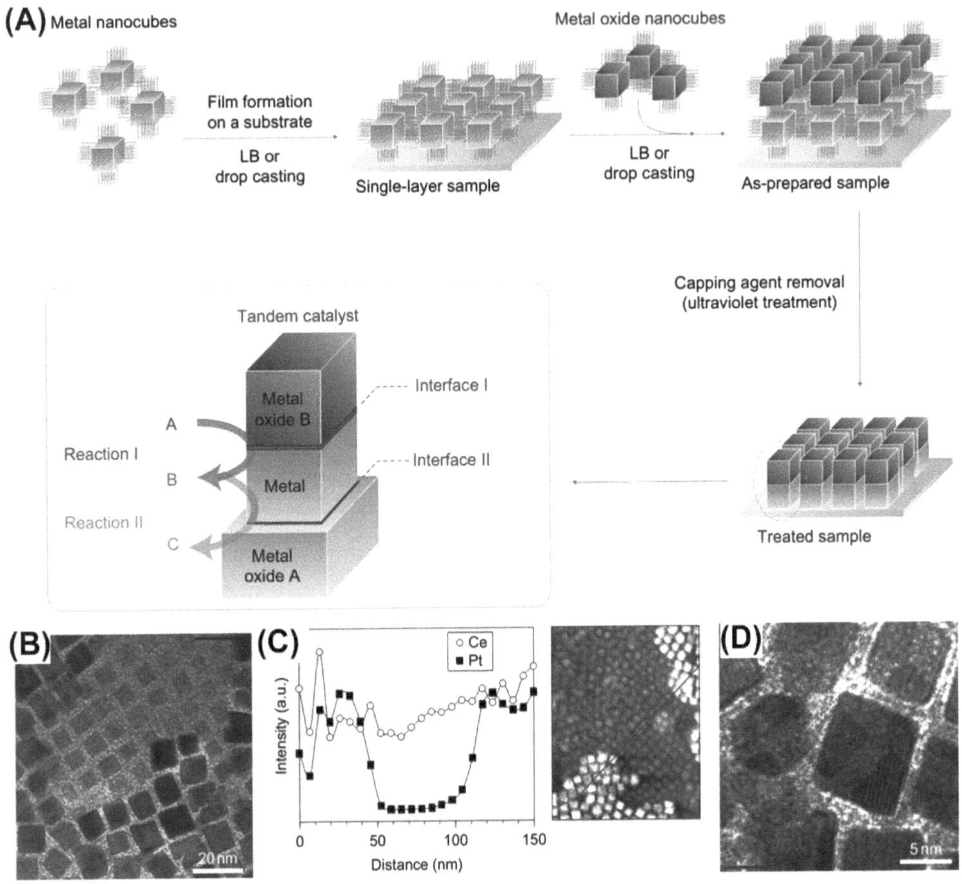

Figure 19 Design of a tandem catalyst system with two active interfaces for different reactions. (A) A schematic view of preparing a tandem catalyst. (B) A transmission electron microscopy image of a defect area of a bilayer film. (C) Energy dispersive X-ray spectroscopy line scan (left) along the defect area (right). (D) A high resolution electron microscopy image of a bilayer area with the Moiré pattern generated by the Pt and CeO_2 nanocubes. *Reprinted by permission from Macmillan Publishers Ltd: Yamada et al. Nat. Chem. 3 (2011) 372–376, copyright 2011.*

This may provide a possible method for developing well-defined supported metal catalysts. The LB trough may be utilized to deposit monolayers of amphiphiles of interest onto solid supports. These amphiphiles can include organic ligands often used in catalyst synthesis. Figure 19(A) shows the schematic of preparing the tandem catalyst systems using the LB approach. In their study, oleylamine-capped Pt nanocubes were dispersed in chloroform. The suspension was dropped into an LB trough with water as a subphase. After evaporating the chloroform, an LB thin film consisting of Pt nanocubes on the water surface was generated. The thin film was then transferred onto a Si wafer. A similar approach was used to generate a CeO_2 nanocube LB thin film. The thin film was transferred to the as-prepared Si-Pt substrate. The residual capping agents which helped to uniformly separate the nanocubes were removed by ultraviolet treatment. The formation of active metal oxide interfaces was further confirmed based on changes observed in sum frequency generation vibrational spectroscopy and the apparent activation energy of CO oxidation. Thus a tandem catalyst containing CeO_2, Pt, and SiO_2 (as the surface of Si wafer is oxidized) was prepared with two active interfaces including SiO_2–Pt and CeO_2–Pt. The tandem morphology was confirmed by plan-view HREM and EDS. At the defect areas where the tandem system was not generated, the contrast difference is obvious, as shown in Figure 19(B). The EDS line-scan (Figure 19(C)) shows the difference in amount of Ce and Pt between the tandem and defect areas. In Figure 19(D), the HREM lattice fringes confirm the presence of CeO_2 and Pt, as well as the interfacial area according to the Moiré pattern.

The tandem catalyst system has multiple benefits. The CeO_2–Pt interface is active for the decomposition of methanol to CO and H_2, while the Pt–SiO_2 interface is able to catalyze olefin hydroformylation using CO and H_2. The tandem catalyst can thus be used to form propanal (ethylene hydroformylation) by in situ methanol decomposition. In contrast, the use of only Pt–SiO_2 catalysts to form propanal from CO and H_2 would cause the formation of methanol in addition to the propanal formation. The overall selectivity of propanal formation was more than 94%. The enhanced properties are probably due to the high concentration of methanol which reduces the rate of methanol formation, as well as the high concentrations of CO and H_2 which accelerate propanal formation.

This study indicates that synthesis techniques which are often used in other areas may be employed to geometrically control supported catalytic systems. The LB trough in this tandem catalyst synthesis as well as some other techniques such as ALD and electron lithography [102] may be used for the future design of functional catalytic systems. For example, ALD, which is widely used for layer-by-layer thin film growth, has been used for supported catalyst design to selectively coat oxide surfaces while leaving the metal particles exposed [135]. Owing to the coating design, coking and sintering under reaction conditions are significantly suppressed. On the other hand, electron microscopy helps to verify if the expected configurations have been realized.

4.4 Bimetallic Catalysts

Bimetallic systems have attracted great interest in heterogeneous catalysis over the past few decades. The advantages of using bimetallic catalysts have been demonstrated over a variety of reactions such as catalytic reforming [136,137], oxygen reduction reaction (ORR, in cathode of fuel cell, O_2 to O^- and O^{2-}) [138,139], preferential CO oxidation in excess hydrogen (PROX) (by $CO + ½O_2 \rightarrow CO_2$ in excess H_2) [140–142], and hydrodesulphurization [143]. Incorporating another metal species into the parent metal can help to reduce the cost, improve stability, and enhance activity and selectivity. The

Pt-based bimetallic system is a classic example as Pt is widely used in catalysis. The natural abundance of Pt is extremely low: the concentration of Pt is only 0.005 ppm in earth's crust (~30 times less than Au), and the resulting production of useable Pt is only 133 tons per year (~13 times less than Au) [133]. Minimizing the use of Pt by introducing other cheaper metals is economically favorable for commercializing Pt-based catalysts. Pt was found to be a good catalyst for the hydrogen oxidation reaction (HOR, in anode of fuel cell, H_2 to H^+) and ORR in the proton exchange membrane (PEM) fuel cells [144]. However, the H_2 is usually generated by reforming hydrocarbons and is therefore mixed with a large amount of CO. The CO impurity can severely poison Pt catalysts in the anodes of the fuel cells [145–147]. To produce H_2 without CO, the PROX of CO becomes a key reaction for the better performance of fuel cells. Experimental and theoretical studies have demonstrated that the use of a bimetallic catalyst with a Pt shell on a Ru core has better performance in PROX relative to Pt alone and some other bimetallic systems [140]. Moreover, owing to the acidic environment and high potential of startup/shutdown cycles, the Ru particles can dissolve or agglomerate. The dissolution and agglomeration results in degradation of the Pt–Ru alloy catalyst [148]. The protection provided by the Pt shell increases the stability of the Pt–Ru bimetallic system. Usually the second metal introduced has a slightly different lattice parameter than the parent metal. The lattice strain may have a major impact on the electronic structure of the parent metal. In the Pt-Pd bimetallic system, Pt has a slightly larger lattice constant (3.92 Å for Pt compared to 3.89 Å for Pd). The compressive strain on the Pt surface due to Pd can result in a downshift of the d-band center, which weakens the binding strength of adsorbents [149–151]. The catalytic activities can increase or decrease depending on whether the adsorption or desorption of the intermediates determine the reaction rate [152,153]. For the ORR in acid fuel cells, multifold enhancement of the specific activity was reported for Pt-coated Pd cores and Pt-coated Pd_3Co cores compared to Pt nanoparticles alone [139]. The enhancement was attributed to the lattice mismatch-induced contraction on the Pt surface, which is in agreement with experimental measurements and DFT calculations.

Experimental and theoretical studies have shown that the catalytic performances of bimetallic systems significantly depend on the geometry of the nanoalloys [154,155]. For example, DFT studies suggest that optimum ORR activity occurs with a bilayer of Pt over the Ru core ($Pt_{2ML}Ru$) [156]. Hsieh et al. have shown that ordered $Pt_{2ML}Ru$ nanoparticles can be synthesized and may have applications in PEM fuel cell catalysts [141]. In their synthesis, the Ru nanoparticles were synthesized and successively overcoated with Pt. No organic ligands were used and the synthesis and coating were managed by controlling the reaction temperature and pH values in the solutions. However, based on their X-ray diffraction (XRD) studies, an alloyed interface with intermixing of Pt and Ru was found on the as-prepared core/shell nanoparticles. The authors attributed the intermixing to the surface defects on the Ru nanoparticles. To address this problem,

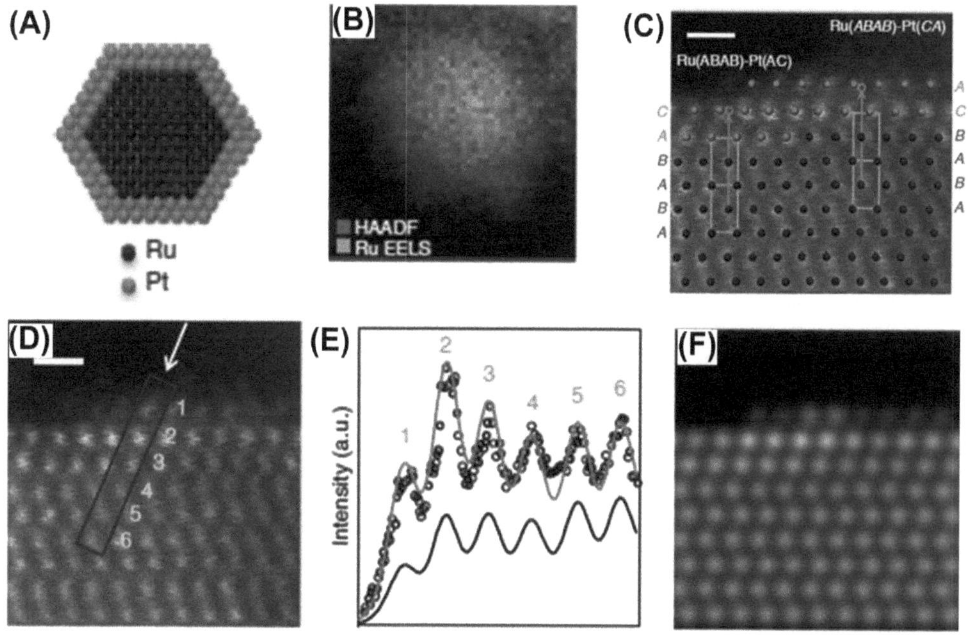

Figure 20 A Pt–Ru core–shell bimetallic catalyst for proton exchange membrane fuel cells. (A) Schematic view of a bilayer Pt overcoating on a Ru core. (B) High angle annular dark field (HAADF) and electron energy loss spectroscopy studies of a Pt–Ru core–shell bimetallic system. (C) An atomic resolution HAADF profile-view image of a bilayer Pt overcoating on Ru core. Scale bar 0.5 nm. (D), (E) Measurement of the intensity profile of surface spots in the HAADF images, as indicated by the red (black in print versions) box in (D). In (E), the black circles denote the intensities on experimental images, the red (black in print versions) line represents the intensities of simulated HAADF images of the Pt–Ru bimetallic system. The blue (gray in print versions) line denotes the intensity profile of a simulated image with only Ru. Scale bar 0.5 nm (F) The simulated ADF images using Pt–Ru core–shell system. *Reprinted by permission from Macmillan Publishers Ltd: Hsieh et al. Nat. Commun. 4 (2013)* http://dx.doi.org/10.1038/ncomms3466, *copyright 2013.*

the Ru nanoparticles were annealed at 450 °C in order to remove the surface lattice defects. The resulting Pt/Ru interface became much shaper based on their XRD, HAADF, and EDS studies. Figure 20(A) shows a schematic model of the $Pt_{2ML}Ru$ system. HAADF and EELS mapping (Figure 20(B)) show a rather uniform Pt shell with a thickness of approximately 2–3 MLs. In addition to the surface chemical composition, the atomic interfacial structure between Pt (111) and Ru (0001) was visualized. Ru has a hcp structure with AB stacking in the [0001] direction while Pt has a fcc structure with an ABC stacking in the [111] direction. By setting the terminating surface of Ru as the B layer, atomic resolution HAADF shows the Pt overlayer can either occupy the A site or C site, as shown in Figure 20(C). The energy difference between the two configurations is small according to their DFT calculations. The stacking shift should be caused by the Pt coating rather than the Ru, as the shifting of the Ru layer is energetically unfavorable based on their DFT studies as well as previous calculations on the segregations of single crystal surfaces [157]. As has been mentioned, the intensity of the atom spots in ADF images is thickness-dependent. The authors performed image simulation to take into account the decreasing thickness at the edge. The decrease in thickness results in a decrease in intensity. Therefore the increased intensity of the surface atom spots could be confidently assigned to the Pt atoms, as shown in Figure 20(E) and (F).

The $Pt_{2ML}Ru$ catalyst was added into PEM fuel cells. The $Pt_{2ML}Ru$ was demonstrated to be superior to $Pt_{1ML}Ru$ and Pt nanoparticles with regard to HOR activity. After 2500 startup/shutdown cycles, the catalytic performance showed little change. This was attributed to the stable ordered $Pt_{2ML}Ru$ structure. It might be interesting to further develop the bimetallic catalysts, as studies show the introduction of Pd can further prevent the dissolution of Pt to an extent.

For some simple reactions, imaging the surface of a bimetallic catalyst under reaction conditions in ETEMs provides fruitful information toward understanding the atomic origin of the enhanced catalytic activity. Au–Ag is one of the most studied bimetallic systems. The Au–Ag system was found to be more active for CO oxidation than using Au alone [128]. Moreover, the activity can be tuned by varying the ratio of Au and Ag [158–162]. Fujita et al. performed ETEM observations on nanoporous Au (NPG) with 20 at% Ag and NPG with 1.2 at% Ag [163]. Under 1 vol% CO in an air gas mixture at 30 pa, the (110) surface of NPG undergoes significant {111} faceting, as shown in Figure 21(A) and (B). However, NPG with 20 at% Ag maintains a flat surface under the same atmosphere, as shown in Figure 21(C) and (D). The authors attributed the stabilization of the Au surface to the surface Ag atoms, although the contrast difference of Au and Ag is too small to be identified in their experimental HREM images (although it may be detectable if a range of defoci was used). The activity enhancement of NPG in CO oxidation originates in part from the stabilized low-coordinate surface Au atoms at the kinks and step edges. With the help of single-crystal surface studies, the role of surface Ag is better understood. The sputtered Au (111) surface is still not as active as supported Au nanoparticles unless the O is pre-adsorbed [164–167], although the sputtered surface has a large amount of low-coordinate surface Au atoms. DFT studies show that the adsorption of CO may not be influenced significantly by the Ag impurities. However the dissociative adsorption of O_2 can be enhanced by surface Ag impurities [168]. It is worth noting that the surface structure models usually employed for DFT calculations are Au surfaces with Ag atom patches, which is a rather reasonable assumption, as surface segregation of Au–Ag bimetallic nanoparticles is verified by atomic simulations using semiempirical potentials [169]. As one caveat, there are some ambiguities in understanding alloy segregation in nanoparticles including possible substantial deviations from a simple Wulff shape [170]. This is currently a very active area of research with new data from aberration-corrected instruments as well as from other techniques and modeling methods changing how we understand these systems.

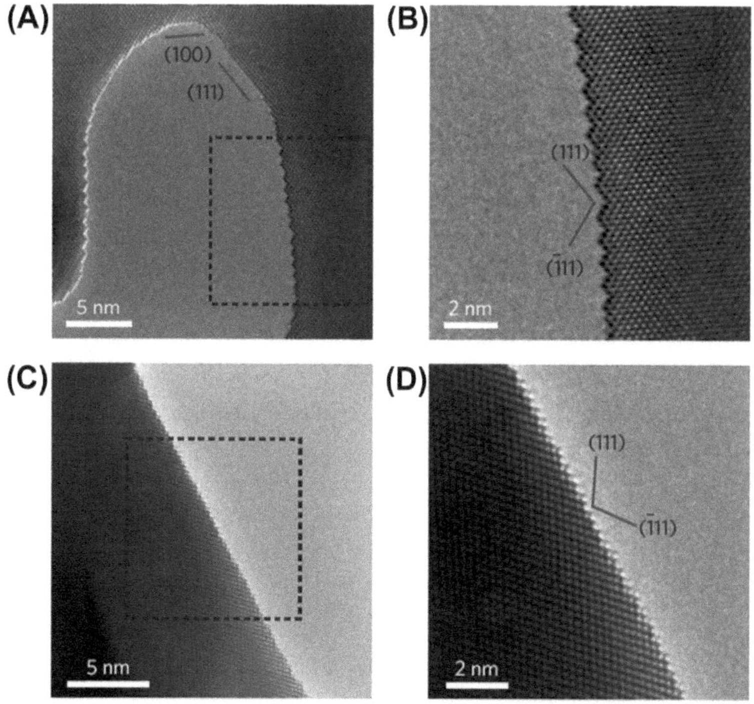

Figure 21 The surface structures of nanoporous gold (NPG) under CO oxidation. (A), (B) High resolution electron microscopy (HREM) images show the presence of (111) faceting on Low-Ag NPG containing 1.2 at% Ag. (C), (D) HREM images show the (111) faceting is not as obvious on the NPG with 20 at% Ag. *Reprinted by permission from Macmillan Publishers Ltd: Fujita et al. Nat. Mater. 11 (2012) 775–780, copyright 2012.*

4.5 Single Atom Catalysts

In heterogeneous catalysis, the active sites of the supported metal are often located on the low-coordinate positions such as the surface, corners, and edge, although there are some catalytic systems where this is not the case; for instance the activity of the catalyst may increase, but the selectivity may be worse. With decreasing size, the density of low-coordinate sites increases. Sometimes an efficient use of a catalytic metal is achieved when the size is reduced to single atoms. Stable single atom catalysts (SACs) have been obtained on oxide supports [7,171,172], metal supports [173,174], graphene sheets [175], zeolites, and metal organic frameworks [176–178]. Superior activities are observed for SACs compared to their multi-atom counterparts when normalized to the number of active sites. Interaction of the metal atom with the support should be considered to obtain a better understanding of SACs. Owing to the extremely small size of the SACs, the electronic effect induced by the support is more significant. For an oxide support, strong bonding between the metal catalyst and the O atoms of the support can result in a positive charge on the SACs [7,179–186]. The positive charge can facilitate gas adsorption and affect catalytic performance. It was demonstrated that the positively charged Au and Pt species on CeO_2 supports are responsible for high activity in the WGS [181]. For metal and graphene supports, the lattice strain can result in a change of d-band states [173,175], which can also influence catalytic performance. A common concern about the SACs is the stability issue, as low-coordinate SACs should result in high surface free energy. However some SACs appear to be rather stable: high activity is generally maintained after multiple reaction cycles at elevated temperature. The stability is usually attributed to strong anchoring of the single atom site by the support. The anchoring mechanisms have been investigated for Au–CeO_2 [41,43,61,181,187,188], Pt–CeO_2 [181,189], Pt–Al_2O_3 [190,191], Pd–Al_2O_3 [179], Ag–MnO_2 [192], and other supported metal systems [193]. The strong anchoring may not only help to stabilize the SACs, but also the support. For example, it was found that La single atoms on the surface of γ-Al_2O_3 can help to prevent the phase transformation of the γ-Al_2O_3 to the more stable α-Al_2O_3 [194]. The exact anchoring sites vary from system to system. Experimental verification and theoretical calculations should be conducted for each specific system to develop a complete understanding.

Aberration-corrected HAADF imaging at sub-Å resolution is capable of resolving single atoms, which provides a convenient method to check whether or not the catalysts are atomically well dispersed. Moreover, plan-view imaging is capable of visualizing the anchoring sites of catalyst atoms on a support. A good example is demonstrated by Chang et al. in studying the adsorption site of Pt on the (110) surface of a rutile TiO_2 (110) single crystal [195]. TiO_2 is widely used in chemical catalysis [117,123] and photocatalysis [196–198]. The (110) surface of TiO_2 is treated as a model surface and has been extensively studied [199]. The structural model is shown in Figure 22(A). The (1 × 1) TiO_2 (110) surface consists of two types of Ti and O atoms. The extrusion twofold coordinated O (O_{2c}) are bridging atoms on Ti atoms (Ti_{6c}) at the base layer. The rows of Ti_{6c} atoms running in the [001] direction separate the fivefold coordinated Ti atoms (Ti_{5c}), which are bonded with threefold coordinated O atoms (O_{3c}). The O_{2c} atoms can act as Brønsted bases which can adsorb H while the Ti_{5c} atoms act as Lewis acid sites for adsorbing organic molecules [200]. O_{3c} and Ti_{6c} are usually considered as bulk species because they have the same coordination number as in the bulk. On vacuum-prepared TiO_2 (110) surfaces, a small percentage of O_{2c} atoms are missing, which is often observed in STM studies [199,201]. The resulting vacancies (O_{br} vacancies) are often considered to be stable adsorption sites for Pt atoms in DFT studies [202,203]. Surprisingly, Chang et al. observed that Pt atoms are preferentially adsorbed onto the position of missing O_{3c} atoms (O_{ba} vacancies). If eight possible adsorption sites are considered on both stoichiometric and reduced surfaces, five of them can be clearly identified by their HAADF images, as shown in Figure 22(C). The unidentifiable adsorption sites are as such because plan-view HAADF images are not very sensitive to the difference in thickness between one or several layers (it is possible that the surface is not

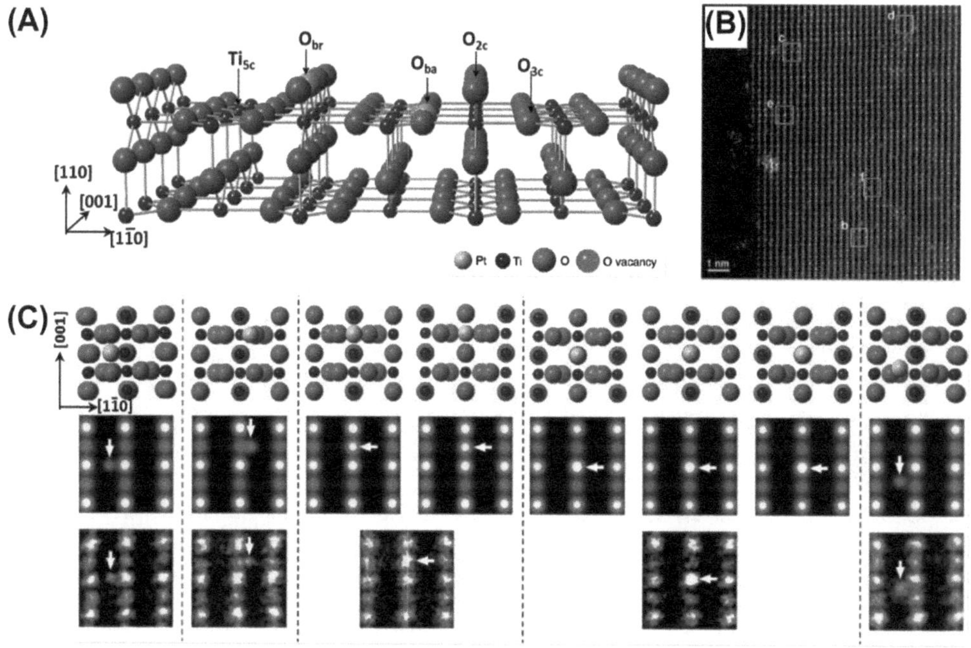

Figure 22 Adsorption sites of Pt atoms on a rutile TiO_2 (110) surface. (A) A structural model of atoms and defects on a TiO_2 (110) surfaces. (B) An experimental atomic resolution high angle annular dark field (HAADF) image of Pt on a TiO_2 (110) surfaces. (C) Structural models, experimental and simulated HAADF images of Pt adsorption sites on TiO_2 (110), as indicated in (B). *Panels (B) and (C) were adapted with permission from Chang et al. Nano Lett. 14 (2014) 134–138. Copyright (2014) American Chemical Society.*

atomically flat). However, the frequency of adsorption at O_{br} vacancies is statistically the highest. Their DFT calculation (GGA+U) shows that creating O_{br} vacancies requires less energy than the formation of O_{ba} vacancies (0.15 eV), which implies that the predominant O vacancies on the vacuum annealed surface are O_{br} vacancies. However, the highest binding energy of Pt on the surface was on the adsorption on O_{ba} vacancies in all of the eight sites considered, which is 0.18 eV higher than the adsorption on the O_{br} vacancies. Thus, it is possible that the O_{ba} vacancies are created due to the presence of Pt atoms and provide the anchoring sites for Pt atoms.

Tao Zhang's group has investigated SACs on FeO_x supports [7]. The Pt–FeO_x SACs exhibit significantly higher activity for CO oxidation compared to Pt–FeO_x with 3D Pt clusters as well as the standard Au–Fe_2O_3 catalyst. The Pt atoms were found to occupy the Fe sites on the (100) surface based on plan-view HAADF images. One type of FeO_x, α-Fe_2O_3, has a hexagonal corundum structure. Under the [0001] direction, it has a repeated stacking of Fe–Fe–O_3–⋯ layers. Hence the termination surface can be a single layer of Fe, a double layer of Fe, or an O_3 layer. Experimental and theoretical studies indicated that the surface can have mixed terminations [204–209]. It is possible that under O-rich conditions, O_3 terminated surface is preferred. The Fe terminated surface is favorable under O-deficient conditions [204]. Based on the HAADF results (see Figure 4(B)), the DFT calculations show that the most stable position for Pt is on the O_3-terminated (0001) surface. The Pt atoms occupy the threefold hollow sites on the O_3 layer, which can be understood as a replacement of the Fe atoms for the single Fe termination. The Pt atoms on the O_3 layer can also reduce the surface O vacancy formation energy and enhance the catalytic activity for CO oxidation. The Bader charge analysis suggests that the Pt is positively charged (+0.45|e|), which was confirmed experimentally by FT-IR.

The same group demonstrated that Ir–FeO_x SACs show even higher catalytic activity for the WGS reaction than Pt–FeO_x SACs [171]. The activity is one order of magnitude higher than that was observed for Ir nanoparticles or clusters on FeO_x supports. By increasing the loading of Ir, the size of the Ir catalysts can be tuned from single atoms to a mixture of single atoms and clusters, and decreasing the loading results in a higher percentage of single atoms sites. By comparing the TOFs of the catalysts with difference loadings, an interesting finding is that the single atom sites contribute to ~70% of the activity. This suggests that the single atom sites are the most important active sites for WGS. The high activities can be maintained over 20 h of reaction time. Additionally, HAADF-STEM images confirmed that after that reaction, the catalysts were primarily SACs. The long-term stability and enhanced activity were explained the similar way as for the Pt SACs on FeO_x. The Ir SACs were found to occupy the Fe positions and they were positively charged according to X-ray absorption near-edge structure (XANES). In addition, the H_2 TPR shows that the formation of O vacancies is facilitated by the Ir SACs.

HAADF imaging has also been used to study SACs on porous zeolites [210,211]. As zeolites are more sensitive to the electron beam compared to oxide supports, low dose imaging was usually employed for characterizing such materials. The signal-to-noise ratios in these images are relatively low due to the low electron dose. However, useful structural information can still be extracted based on image processing. For example, Ortalan et al. investigated the anchoring positions of Ir atoms on a dealuminated HY (DAY) zeolite [210]. They found that more Ir atoms were anchored to the T6 sites compared to the T5 sites of the zeolite crystal. After treating the Ir–zeolite sample in H_2 flow at 393 K for 30 min, Ir clusters start to form preferentially on the T6 sites. Therefore, the authors deduced that some Ir atoms anchored to the T5 sites migrate to the T6 sites during the treatment.

5. SUMMARY AND OUTLOOK

In this review, we discussed how the structures and chemical compositions of catalytic systems can be characterized using TEM. The development of TEM has enabled the characterization of catalytic nanoparticles with atomic resolution under reaction conditions, which has significantly helped with understanding catalyst structures and reaction mechanisms. Currently, direct characterization of catalytic reactions using TEM is limited to the catalyst itself. Adsorption can be observed by TEM by imaging the surface relaxation or other morphological changes on the catalyst. However, direct imaging of the adsorption of reactant molecules on the catalytic active sites is still a great difficulty. This is partly because the contrast of the small amount of light molecules is too low compared to that of the catalyst itself, both in the imaging and spectroscopic techniques of TEM. Recently Yoshida et al. suggested that the adsorbed CO molecules could be directly imaged by using aberration-corrected ETEM [119]. However, the signal from the CO is rather low and should be confirmed by other techniques. Increasing the electron dose during characterization may increase the contrast. However, high electron dose can cause desorption of molecules [212,213] and even damage the catalyst [55,214]. New techniques such as exit wave reconstruction, which numerically "adds" a set of low dose images to increase the contrast [215–218], may allow for imaging of the adsorbates. While it is relatively difficult using TEM, other techniques such as FT-IR and Raman are good at characterizing the adsorption and desorption phenomena. A comprehensive understanding of the properties of catalysts requires the employment of different characterization techniques and theoretical studies.

The main focus of this chapter has been methods of looking at catalysts using electron microscopy, mainly at the atomic scale. It is appropriate at the end to step back slightly with a few words of caution, the three R's: Radiation damage, Rigor, and Reality.

Radiation Damage: The environment of any electron microscope is harsh, not just the vacuum but also because of energetic energy loss processes. It is sometimes easy to damage materials, and on rare occasions to heat them using the electron beam. The simplest process is direct momentum transfer, called knock-on damage, which occurs whenever the electron energy is above some threshold which depends upon not just the atomic mass but also the bonding. In many cases this can be minimized by using lower voltages. However, there is a second damage process, radiolysis, which becomes more common if the voltage is reduced and is very significant for oxides as well as chemisorbed species. Some years ago the group of Yagi et al. built a UHV electron microscope which was capable of doing both lower resolution reflection electron microscopy (REM) of surfaces and photoelectron electron microscopy (PEEM), the aim being to image chemisorbed species on silicon using PEEM and correlate this with surface structure using REM. Unfortunately it turned out that the chemisorbed species were completely removed by the microscope beam [219]. The electron flux that they used was several orders of magnitude smaller than what one has in contemporary instruments. While it is possible that there will be an equilibrium chemisorbed surface concentration in ETEM experiments, the extent to which they are imaged is unclear.

Rigor: The first images of single atoms in a home built STEM instrument was reported by the group of Crewe in the 1970s (e.g., Refs [220–226]) and when commercial instruments became available these instruments became a staple method of examining catalysts (e.g., Refs [227,228]). Atomic imaging of small particles became relatively easy in the early 1980s, although there were papers such as the elegant work of Tsutomu [229] and surface studies at lower resolution [230–237] much earlier. Profile imaging was discovered in the early 1980s [11,238–240], and was rapidly used and extended by many groups around the world for both profile and plan-view imaging using instruments without aberration correctors (e.g., Refs [235,241–256]), see also [257] for other early references. Because those instruments had lanthanum hexaboride sources it was relatively easy to image the surface (as well as the bulk) [258]. When brighter and more coherent field emission sources became common, it in fact became much harder to image surfaces; it was only when aberration-correctors became common that surfaces could be imaged as easily as they were in the 1980s. For certain the newer corrected instruments are better, but at the same time they are sometimes used on problems which were extensively studied some years earlier, for instance the reconstruction on the gold (001) surface [259].

Reality: It is sometimes possible to generate a model to explain what images of nanoparticles used in catalysts mean in terms of the local structure. Unfortunately that does not prove that the model is unique. While coupling electron microscopy with DFT calculations is very powerful, at the same time this still does not make the interpretation unique. Methods such as DFT only lead to a local minimum, so if the initial atomic structure model used is not appropriate then the results they yield will only be better than the initial guess, not necessarily right. Sometimes DFT can be problematic by itself, for instance the well-known errors in chemisorption energies with simple functionals. One always has to worry whether an interpretation conforms to how the sample was made, basic thermodynamics as well as Occam's razor.

Are we at the limits of what can be achieved in imaging catalysts? The answer is no, and in the next few years there may well be much more information. We have already crossed the barrier of being able to image nanoparticles in three-dimensions at the atomic scale [260] with a three dimensional resolution of $2 \times 2 \times 2$ Å^3 voxels. We expect that within a few years we will see measurements of, for instance, the positions of all atoms in bimetallic nanoparticles with three dimensional precision of 0.5 Å, as well as three dimensional chemical mapping and perhaps even mapping in three dimensions of chemical shifts.

ACKNOWLEDGMENTS

Part of this work is supported by the funding from the Northwestern University Institute for Catalysis in Energy Processes (ICEP) on grant number DOE DE-FG02-03-ER15457. K.R.P. would like to thank the funding from the Institute for Atom-Efficient Chemical Transformations (IACT), an Energy Frontier Research Center funded by the U.S. Department of Energy, Office of Science, Office of Basic Energy Sciences. The authors also thank the contribution from Jianguo Wen, James A. Enterkin, Zili Wu, Linhua Hu, and Robert M. Kennedy. Part of HREM images are obtained using the facilities at the Electron Microscopy Center for Materials Research at Argonne National Laboratory.

REFERENCES

[1] Y. Zhu, H. Inada, K. Nakamura, J. Wall, Nat. Mater. 8 (2009) 808–812.
[2] S. Takeda, H. Yoshida, Microscopy 62 (2013) 193–203.
[3] M. De Graef, Introduction to Conventional Transmission Electron Microscopy, Cambridge University Press, Cambridge, 2003.
[4] D.B. Williams, C.B. Carter, Transmission Electron Microscopy: A Textbook for Materials Science, Springer, New York, 2009.
[5] S.J. Pennycook, P.D. Nellist, Scanning Transmission Electron Microscopy: Imaging and Analysis, Springer, New York, 2011.
[6] J.M. Cowley, Appl. Phys. Lett. 15 (1969) 58–59.
[7] B. Qiao, A. Wang, X. Yang, L.F. Allard, Z. Jiang, Y. Cui, J. Liu, J. Li, T. Zhang, Nat. Chem. 3 (2011) 634–641.
[8] G. Jayaram, R. Plass, L.D. Marks, Interface Sci. 2 (1995) 379–395.
[9] E. Bengu, R. Plass, L.D. Marks, T. Ichihashi, P.M. Ajayan, S. Iijima, Phys. Rev. Lett. 77 (1996) 4226.
[10] Y. Kuwauchi, S. Takeda, H. Yoshida, K. Sun, M. Haruta, H. Kohno, Nano Lett. 13 (2013) 3073–3077.
[11] L.D. Marks, D.J. Smith, Nature 303 (1983) 316–317.
[12] R. Yu, L.H. Hu, Z.Y. Cheng, Y.D. Li, H.Q. Ye, J. Zhu, Phys. Rev. Lett. 105 (2010) 226101.
[13] Y. Lin, Z. Wu, J. Wen, K.R. Poeppelmeier, L.D. Marks, Nano Lett. 14 (2014) 191–196.
[14] Y. Lin, J. Wen, L. Hu, R.M. Kennedy, P.C. Stair, K.R. Poeppelmeier, L.D. Marks, Phys. Rev. Lett. 111 (2013) 156101.
[15] L.P. Hansen, Q.M. Ramasse, C. Kisielowski, M. Brorson, E. Johnson, H. Topsøe, S. Helveg, Angew. Chem. Int. Ed. 50 (2011) 10153–10156.
[16] N. Alem, Q.M. Ramasse, C.R. Seabourne, O.V. Yazyev, K. Erickson, M.C. Sarahan, C. Kisielowski, A.J. Scott, S.G. Louie, A. Zettl, Phys. Rev. Lett. 109 (2012) 205502.
[17] N. Lu, Q. Wan, J. Zhu, J. Phys. Chem. Lett. 1 (2010) 1468–1471.
[18] M.-R. He, R. Yu, J. Zhu, Nano Lett. 12 (2012) 704–708.
[19] O. Scherzer, J. Appl. Phys. 20 (1949) 20–29.
[20] M. Haider, H. Rose, S. Uhlemann, B. Kabius, K. Urban, J. Electron Microsc. 47 (1998) 395–405.
[21] O.L. Krivanek, N. Dellby, A.R. Lupini, Ultramicroscopy 78 (1999) 1–11.
[22] C. Kisielowski, B. Freitag, M. Bischoff, et al., Microsc. Microanal. 14 (2008) 469–477.
[23] D. Van Dyck, M. Op de Beeck, Ultramicroscopy 64 (1996) 99–107.
[24] C.-L. Jia, Markus Lentzen, K. Urban, Microsc. Microanal. 10 (2004) 174–184.
[25] R. Kilaas, MacTempas, http://www.totalresolution.com/.
[26] P.A. Stadelmann, JEMS, http://cimewww.epfl.ch/people/stadelmann/jemsWebSite/jems.html.
[27] K. Ishizuka, xHREM, http://www.hremresearch.com/Eng/simulation.html.
[28] J.M. Zuo, J.C. Mabon, Microsc. Microanal. 10 (Suppl. 2) (2004).
[29] C.T. Koch, QSTEM, http://elim.physik.uni-ulm.de/?page_id=834.
[30] M.A. Okeefe, R. Kilaas, Scanning Microsc. (1988) 225–244.
[31] S.D. Findlay, N. Shibata, H. Sawada, E. Okunishi, Y. Kondo, Y. Ikuhara, Ultramicroscopy 110 (2010) 903–923.
[32] E. Okunishi, H. Sawada, Y. Kondo, Micron 43 (2012) 538–544.
[33] A. Trovarelli, Catal. Rev. 38 (1996) 439–520.
[34] C. Ratnasamy, J.P. Wagner, Catal. Rev. 51 (2009) 325–440.
[35] J. Beckers, G. Rothenberg, Green Chem. 12 (2010) 939–948.
[36] L. Vivier, D. Duprez, ChemSusChem 3 (2010) 654–678.
[37] M. Mogensen, N.M. Sammes, G.A. Tompsett, Solid State Ionics 129 (2000) 63–94.
[38] S. Park, J.M. Vohs, R.J. Gorte, Nature 404 (2000) 265–267.
[39] A. Atkinson, S. Barnett, R.J. Gorte, J.T.S. Irvine, A.J. McEvoy, M. Mogensen, S.C. Singhal, J. Vohs, Nat. Mater. 3 (2004) 17–27.
[40] J. Kašpar, P. Fornasiero, M. Graziani, Catal. Today 50 (1999) 285–298.
[41] J.A. Farmer, C.T. Campbell, Science 329 (2010) 933–936.
[42] C.T. Campbell, C.H.F. Peden, Science 309 (2005) 713–714.
[43] N. Ta, J. Liu, S. Chenna, P.A. Crozier, Y. Li, A. Chen, W. Shen, J. Am. Chem. Soc. 134 (2012) 20585–20588.

[44] H.-F. Wang, X.-Q. Gong, Y.-L. Guo, Y. Guo, G.Z. Lu, P. Hu, J. Phys. Chem. C 113 (2009) 10229–10232.
[45] H.-F. Wang, H.-Y. Li, X.-Q. Gong, Y.-L. Guo, G.-Z. Lu, P. Hu, PCCP 14 (2012) 16521–16535.
[46] Y. Tang, H. Zhang, L. Cui, C. Ouyang, S. Shi, W. Tang, H. Li, J.-S. Lee, L. Chen, Phys. Rev. B 82 (2010) 125104.
[47] S. Agarwal, L. Lefferts, B.L. Mojet, et al., ChemSusChem 6 (2013) 1898–1906.
[48] P.W. Tasker, J. Phys. C: Solid State 12 (1979) 4977.
[49] J.A. Enterkin, A.E. Becerra-Toledo, K.R. Poeppelmeier, L.D. Marks, Surf. Sci. 606 (2012) 344–355.
[50] C. Noguera, J. Phys. Condens. Matter 12 (2000) R367.
[51] J. Goniakowski, F. Finocchi, C. Noguera, Rep. Prog. Phys. 71 (2008) 016501.
[52] P.A. Crozier, R. Wang, R. Sharma, Ultramicroscopy 108 (2008) 1432–1440.
[53] N.V. Skorodumova, S.I. Simak, B.I. Lundqvist, I.A. Abrikosov, B. Johansson, Phys. Rev. Lett. 89 (2002) 166601.
[54] S. Turner, S. Lazar, B. Freitag, R. Egoavil, J. Verbeeck, S. Put, Y. Strauven, G. Van Tendeloo, Nanoscale 3 (2011) 3385–3390.
[55] S.J. Haigh, N.P. Young, H. Sawada, K. Takayanagi, A.I. Kirkland, ChemPhysChem 12 (2011) 2397–2399.
[56] Z. Wu, M. Li, D.R. Mullins, S.H. Overbury, ACS Catal. 2 (2012) 2224–2234.
[57] D.J.M. Bevan, J. Inorg. Nucl. Chem. 1 (1955) 49–59.
[58] C. Binet, M. Daturi, J.-C. Lavalley, Catal. Today 50 (1999) 207–225.
[59] H. Nörenberg, J.H. Harding, Surf. Sci. 477 (2001) 17–24.
[60] O. Dulub, U. Diebold, G. Kresse, Phys. Rev. Lett. 90 (2003) 016102.
[61] F. Esch, S. Fabris, L. Zhou, T. Montini, C. Africh, P. Fornasiero, G. Comelli, R. Rosei, Science 309 (2005) 752–755.
[62] S. Torbrügge, M. Reichling, A. Ishiyama, S. Morita, Ó. Custance, Phys. Rev. Lett. 99 (2007) 056101.
[63] D.C. Grinter, R. Ithnin, C.L. Pang, G. Thornton, J. Phys. Chem. C 114 (2010) 17036–17041.
[64] G.E. Murgida, M.V. Ganduglia-Pirovano, Phys. Rev. Lett. 110 (2013) 246101.
[65] A. Ohtomo, D.A. Muller, J.L. Grazul, H.Y. Hwang, Nature 419 (2002) 378–380.
[66] S. Thiel, G. Hammerl, A. Schmehl, C.W. Schneider, J. Mannhart, Science 313 (2006) 1942–1945.
[67] N. Reyren, S. Thiel, A.D. Caviglia, et al., Science 317 (2007) 1196–1199.
[68] J.G. Mavroides, J.A. Kafalas, D.F. Kolesar, Appl. Phys. Lett. 28 (1976) 241–243.
[69] B. Ura, J. Trawczyński, A. Kotarba, W. Bieniasz, M.J. Illán-Gómez, A. Bueno-López, F.E. López-Suárez, Appl. Catal. B: Environ. 101 (2011) 169–175.
[70] F.A. Rabuffetti, H.-S. Kim, J.A. Enterkin, Y. Wang, C.H. Lanier, L.D. Marks, K.R. Poeppelmeier, P.C. Stair, Chem. Mater. 20 (2008) 5628–5635.
[71] F. Dang, K.-i. Mimura, K. Kato, H. Imai, S. Wada, H. Haneda, M. Kuwabara, CrystEngComm 13 (2011) 3878–3883.
[72] L. Hu, C. Wang, S. Lee, R.E. Winans, L.D. Marks, K.R. Poeppelmeier, Chem. Mater. 25 (2013) 378–384.
[73] E. Ringe, R.P. Van Duyne, L.D. Marks, J. Phys. Chem. C 117 (2013) 15859–15870.
[74] J.A. Enterkin, W. Setthapun, J.W. Elam, S.T. Christensen, F.A. Rabuffetti, L.D. Marks, P.C. Stair, K.R. Poeppelmeier, C.L. Marshall, ACS Catal. 1 (2011) 629–635.
[75] N. Erdman, K.R. Poeppelmeier, M. Asta, O. Warschkow, D.E. Ellis, L.D. Marks, Nature 419 (2002) 55–58.
[76] K. Johnston, M.R. Castell, A.T. Paxton, M.W. Finnis, Phys. Rev. B 70 (2004) 085415.
[77] B. Cord, R. Courths, Surf. Sci. 162 (1985) 34–38.
[78] F. Silly, D.T. Newell, M.R. Castell, Surf. Sci. 600 (2006) 219–223.
[79] Q.D. Jiang, J. Zegenhagen, Surf. Sci. 338 (1995) L882–L888.
[80] C.H. Lanier, A. van de Walle, N. Erdman, E. Landree, O. Warschkow, A. Kazimirov, K.R. Poeppelmeier, J. Zegenhagen, M. Asta, L.D. Marks, Phys. Rev. B 76 (2007) 045421.
[81] N. Erdman, O. Warschkow, M. Asta, K.R. Poeppelmeier, D.E. Ellis, L.D. Marks, J. Am. Chem. Soc. 125 (2003) 10050–10056.
[82] G.-z. Zhu, G. Radtke, G.A. Botton, Nature 490 (2012) 384–387.
[83] M.R. Castell, Surf. Sci. 516 (2002) 33–42.

[84] R. Herger, P.R. Willmott, O. Bunk, C.M. Schleputz, B.D. Patterson, B. Delley, Phys. Rev. Lett. 98 (2007) 076102.
[85] D.M. Kienzle, A.E. Becerra-Toledo, L.D. Marks, Phys. Rev. Lett. 106 (2011) 176102.
[86] M. Kawasaki, K. Takahashi, T. Maeda, R. Tsuchiya, M. Shinohara, O. Ishiyama, T. Yonezawa, M. Yoshimoto, H. Koinuma, Science 266 (1994) 1540–1542.
[87] S.B. Lee, F. Phillipp, W. Sigle, M. Rühle, Ultramicroscopy 104 (2005) 30–38.
[88] J.D. Baniecki, M. Ishii, K. Kurihara, K. Yamanaka, T. Yano, K. Shinozaki, T. Imada, K. Nozaki, N. Kin, Phys. Rev. B 78 (2008) 195415.
[89] R. Sum, R. Lüthi, H.P. Lang, H.J. Güntherodt, Phys. C: Supercond. 235–240 (Part 1) (1994) 621–622.
[90] S.S. Sheiko, M. Möller, E.M.C.M. Reuvekamp, H.W. Zandbergen, Phys. Rev. B 48 (1993) 5675–5678.
[91] K. Szot, W. Speier, Phys. Rev. B 60 (1999) 5909–5926.
[92] K. Szot, W. Speier, U. Breuer, R. Meyer, J. Szade, R. Waser, Surf. Sci. 460 (2000) 112–128.
[93] K. Szot, W. Speier, J. Herion, C. Freiburg, Appl. Phys. A 64 (1996) 55–59.
[94] A. Gunhold, K. Gömann, L. Beuermann, M. Frerichs, G. Borchardt, V. Kempter, W. Maus-Friedrichs, Surf. Sci. 507–510 (2002) 447–452.
[95] J.A. Lopez-Sanchez, N. Dimitratos, C. Hammond, et al., Nat. Chem. 3 (2011) 551–556.
[96] Y. Yamada, C.-K. Tsung, W. Huang, Z. Huo, S.E. Habas, T. Soejima, C.E. Aliaga, G.A. Somorjai, P. Yang, Nat. Chem. 3 (2011) 372–376.
[97] D. Zanchet, H. Tolentino, M.C. Martins Alves, O.L. Alves, D. Ugarte, Chem. Phys. Lett. 323 (2000) 167–172.
[98] G. Corthey, A.A. Rubert, A.L. Picone, et al., J. Phys. Chem. C 116 (2012) 9830–9837.
[99] T. Bligaard, J.K. Nørskov, Electrochim. Acta 52 (2007) 5512–5516.
[100] F. Behafarid, J. Matos, S. Hong, L. Zhang, T.S. Rahman, B. Roldan Cuenya, ACS Nano 8 (2014) 6671–6681.
[101] A.D. Polli, T. Wagner, T. Gemming, M. Rühle, Surf. Sci. 448 (2000) 279–289.
[102] V. Komanicky, H. Iddir, K.-C. Chang, A. Menzel, G. Karapetrov, D. Hennessy, P. Zapol, H. You, J. Am. Chem. Soc. 131 (2009) 5732–5733.
[103] C.J. Kliewer, M. Bieri, G.A. Somorjai, J. Am. Chem. Soc. 131 (2009) 9958–9966.
[104] C.-L. Jia, S.-B. Mi, K. Urban, I. Vrejoiu, M. Alexe, D. Hesse, Nat. Mater. 7 (2008) 57–61.
[105] X. Xiao, X. Liu, H. Zhao, D. Chen, F. Liu, J. Xiang, Z. Hu, Y. Li, Adv. Mater. 24 (2012) 5762–5766.
[106] Y. Wang, H.J. Zhang, J. Wei, C.C. Wong, J. Lin, A. Borgna, Energy Environ. Sci. 4 (2011) 1845–1854.
[107] L. Hu, Q. Peng, Y. Li, J. Am. Chem. Soc. 130 (2008) 16136–16137.
[108] L. Hu, K. Sun, Q. Peng, B. Xu, Y. Li, Nano Res. 3 (2010) 363–368.
[109] X. Xie, Y. Li, Z.-Q. Liu, M. Haruta, W. Shen, Nature 458 (2009) 746–749.
[110] J. Xiao, Q. Kuang, S. Yang, F. Xiao, S. Wang, L. Guo, Sci. Rep. 3 (2013).
[111] L. Vitos, A.V. Ruban, H.L. Skriver, J. Kollár, Surf. Sci. 411 (1998) 186–202.
[112] R. Kaischew, Arbeitstag. Festköper Phys. (1952) 81.
[113] W.L. Winterbottom, Acta Metall. 15 (1967) 303–310.
[114] J.A. Enterkin, K.R. Poeppelmeier, L.D. Marks, Nano Lett. 11 (2011) 993–997.
[115] J. Enterkin, R. Kennedy, J. Lu, J. Elam, R. Cook, L. Marks, P. Stair, C. Marshall, K. Poeppelmeier, Top. Catal. 56 (2013) 1829–1834.
[116] P.L. Hansen, J.B. Wagner, S. Helveg, J.R. Rostrup-Nielsen, B.S. Clausen, H. Topsøe, Science 295 (2002) 2053–2055.
[117] Y. Kuwauchi, H. Yoshida, T. Akita, M. Haruta, S. Takeda, Angew. Chem. Int. Ed. 51 (2012) 7729–7733.
[118] T. Uchiyama, H. Yoshida, Y. Kuwauchi, S. Ichikawa, S. Shimada, M. Haruta, S. Takeda, Angew. Chem. Int. Ed. 50 (2011) 10157–10160.
[119] H. Yoshida, Y. Kuwauchi, J.R. Jinschek, K. Sun, S. Tanaka, M. Kohyama, S. Shimada, M. Haruta, S. Takeda, Science 335 (2012) 317–319.
[120] B. Hammer, J.K. Norskov, Nature 376 (1995) 238–240.
[121] G.C. Bond, Catal. Today 72 (2002) 5–9.
[122] A. Franceschetti, S.J. Pennycook, S.T. Pantelides, Chem. Phys. Lett. 374 (2003) 471–475.
[123] M. Haruta, S. Tsubota, T. Kobayashi, H. Kageyama, M.J. Genet, B. Delmon, J. Catal. 144 (1993) 175–192.

[124] M.S. Chen, D.W. Goodman, Science 306 (2004) 252–255.
[125] B. Yoon, H. Häkkinen, U. Landman, J. Phys. Chem. 107 (2003) 4066–4071.
[126] J. Guzman, B.C. Gates, J. Am. Chem. Soc. 126 (2004) 2672–2673.
[127] M. Mavrikakis, P. Stoltze, J.K. Nørskov, Catal. Lett. 64 (2000) 101–106.
[128] X.Y. Liu, A. Wang, T. Zhang, C.-Y. Mou, Nano Today 8 (2013) 403–416.
[129] O. Lopez-Acevedo, K.A. Kacprzak, J. Akola, H. Häkkinen, Nat. Chem. 2 (2010) 329–334.
[130] Z. Wu, M. Li, S.H. Overbury, J. Catal. 285 (2012) 61–73.
[131] K.-i. Fukui, Y. Namai, Y. Iwasawa, Appl. Surf. Sci. 188 (2002) 252–256.
[132] D.A. Siegel, W.C. Chueh, F. El Gabaly, K.F. McCarty, J. de la Figuera, M. Blanco-Rey, J. Chem. Phys. 139 (2013) 114703.
[133] H. Zhang, M. Jin, Y. Xia, Chem. Soc. Rev. 41 (2012) 8035–8049.
[134] C. Sun, H. Li, L. Chen, Energy Environ. Sci. 5 (2012) 8475–8505.
[135] J. Lu, B. Fu, M.C. Kung, G. Xiao, J.W. Elam, H.H. Kung, P.C. Stair, Science 335 (2012) 1205–1208.
[136] F. Besenbacher, I. Chorkendorff, B.S. Clausen, B. Hammer, A.M. Molenbroek, J.K. Nørskov, I. Stensgaard, Science 279 (1998) 1913–1915.
[137] G.W. Huber, J.W. Shabaker, J.A. Dumesic, Science 300 (2003) 2075–2077.
[138] L. Liu, G. Samjeske, S.-i. Nagamatsu, O. Sekizawa, K. Nagasawa, S. Takao, Y. Imaizumi, T. Yamamoto, T. Uruga, Y. Iwasawa, J. Phys. Chem. C 116 (2012) 23453–23464.
[139] J.X. Wang, H. Inada, L. Wu, Y. Zhu, Y. Choi, P. Liu, W.-P. Zhou, R.R. Adzic, J. Am. Chem. Soc. 131 (2009) 17298–17302.
[140] S. Alayoglu, A.U. Nilekar, M. Mavrikakis, B. Eichhorn, Nat. Mater. 7 (2008) 333–338.
[141] Y.-C. Hsieh, Y. Zhang, D. Su, et al., Nat. Commun. 4 (2013).
[142] H. Zhang, M. Jin, H. Liu, J. Wang, M.J. Kim, D. Yang, Z. Xie, J. Liu, Y. Xia, ACS Nano 5 (2011) 8212–8222.
[143] A.T. Bell, Science 299 (2003) 1688–1691.
[144] H.A. Gasteiger, S.S. Kocha, B. Sompalli, F.T. Wagner, Appl. Catal. B: Environ. 56 (2005) 9–35.
[145] L. Carrette, K.A. Friedrich, U. Stimming, ChemPhysChem 1 (2000) 162–193.
[146] S.J. Lee, S. Mukerjee, E.A. Ticianelli, J. McBreen, Electrochim. Acta 44 (1999) 3283–3293.
[147] S.J.C. Cleghorn, X. Ren, T.E. Springer, M.S. Wilson, C. Zawodzinski, T.A. Zawodzinski, S. Gottesfeld, Int. J. Hydrogen Energy 22 (1997) 1137–1144.
[148] R. Borup, J. Meyers, B. Pivovar, et al., Chem. Rev. 107 (2007) 3904–3951.
[149] L. Grabow, Y. Xu, M. Mavrikakis, PCCP 8 (2006) 3369–3374.
[150] M. Mavrikakis, B. Hammer, J.K. Nørskov, Phys. Rev. Lett. 81 (1998) 2819–2822.
[151] Y. Xu, A.V. Ruban, M. Mavrikakis, J. Am. Chem. Soc. 126 (2004) 4717–4725.
[152] J.X. Wang, J. Zhang, R.R. Adzic, J. Phys. Chem. A 111 (2007) 12702–12710.
[153] J.X. Wang, F.A. Uribe, T.E. Springer, J. Zhang, R.R. Adzic, Faraday Discuss. 140 (2009) 347–362.
[154] R. Ferrando, J. Jellinek, R.L. Johnston, Chem. Rev. 108 (2008) 845–910.
[155] W. Yu, M.D. Porosoff, J.G. Chen, Chem. Rev. 112 (2012) 5780–5817.
[156] L. Yang, M.B. Vukmirovic, D. Su, K. Sasaki, J.A. Herron, M. Mavrikakis, S. Liao, R.R. Adzic, J. Phys. Chem. C 117 (2012) 1748–1753.
[157] A.V. Ruban, H.L. Skriver, J.K. Nørskov, Phys. Rev. B 59 (1999) 15990–16000.
[158] J.-H. Liu, A.-Q. Wang, Y.-S. Chi, H.-P. Lin, C.-Y. Mou, J. Phys. Chem. B 109 (2004) 40–43.
[159] A.-Q. Wang, J.-H. Liu, S.D. Lin, T.-S. Lin, C.-Y. Mou, J. Catal. 233 (2005) 186–197.
[160] A.-Q. Wang, C.-M. Chang, C.-Y. Mou, J. Phys. Chem. B 109 (2005) 18860–18867.
[161] A. Wang, Y.-P. Hsieh, Y.-F. Chen, C.-Y. Mou, J. Catal. 237 (2006) 197–206.
[162] C.-W. Yen, M.-L. Lin, A. Wang, S.-A. Chen, J.-M. Chen, C.-Y. Mou, J. Phys. Chem. C 113 (2009) 17831–17839.
[163] T. Fujita, P. Guan, K. McKenna, et al., Nat. Mater. 11 (2012) 775–780.
[164] J.J. Pireaux, M. Chtaïb, J.P. Delrue, P.A. Thiry, M. Liehr, R. Caudano, Surf. Sci. 141 (1984) 211–220.
[165] N. Saliba, D.H. Parker, B.E. Koel, Surf. Sci. 410 (1998) 270–282.
[166] J. Kim, E. Samano, B.E. Koel, Surf. Sci. 600 (2006) 4622–4632.
[167] J. Gong, D.W. Flaherty, R.A. Ojifinni, J.M. White, C.B. Mullins, J. Phys. Chem. C 112 (2008) 5501–5509.
[168] L.V. Moskaleva, S. Rohe, A. Wittstock, V. Zielasek, T. Kluner, K.M. Neyman, M. Baumer, PCCP 13 (2011) 4529–4539.
[169] L. Deng, W. Hu, H. Deng, S. Xiao, J. Tang, J. Phys. Chem. C 115 (2011) 11355–11363.

[170] E. Ringe, R.P. Van Duyne, L.D. Marks, Nano Lett. 11 (2011) 3399–3403.
[171] J. Lin, A. Wang, B. Qiao, X. Liu, X. Yang, X. Wang, J. Liang, J. Li, J. Liu, T. Zhang, J. Am. Chem. Soc. 135 (2013) 15314–15317.
[172] A. Uzun, V. Ortalan, Y. Hao, N.D. Browning, B.C. Gates, J. Phys. Chem. C 113 (2009) 16847–16849.
[173] H. Zhang, T. Watanabe, M. Okumura, M. Haruta, N. Toshima, Nat. Mater. 11 (2012) 49–52.
[174] J.S. Jirkovský, I. Panas, E. Ahlberg, M. Halasa, S. Romani, D.J. Schiffrin, J. Am. Chem. Soc. 133 (2011) 19432–19441.
[175] S. Sun, G. Zhang, N. Gauquelin, et al., Sci. Rep. 3 (2013).
[176] J.M. Thomas, R. Raja, D.W. Lewis, Angew. Chem. Int. Ed. 44 (2005) 6456–6482.
[177] M. Ranocchiari, C. Lothschütz, D. Grolimund, J.A. van Bokhoven, Proc. Roy. Soc. A: Math. Phys. Eng. Sci. 468 (2012) 1985–1999.
[178] M. Flytzani-Stephanopoulos, B.C. Gates, Annu. Rev. Chem. Biomol. Eng. 3 (2012) 545–574.
[179] S.F.J. Hackett, R.M. Brydson, M.H. Gass, I. Harvey, A.D. Newman, K. Wilson, A.F. Lee, Angew. Chem. Int. Ed. 46 (2007) 8593–8596.
[180] Y.-T. Kim, K. Ohshima, K. Higashimine, T. Uruga, M. Takata, H. Suematsu, T. Mitani, Angew. Chem. Int. Ed. 45 (2006) 407–411.
[181] Q. Fu, H. Saltsburg, M. Flytzani-Stephanopoulos, Science 301 (2003) 935–938.
[182] K.T. Rim, D. Eom, L. Liu, E. Stolyarova, J.M. Raitano, S.-W. Chan, M. Flytzani-Stephanopoulos, G.W. Flynn, J. Phys. Chem. C 113 (2009) 10198–10205.
[183] T.K. Ghosh, N.N. Nair, ChemCatChem 5 (2013) 1811–1821.
[184] Z. Hu, H. Metiu, J. Phys. Chem. C 115 (2011) 17898–17909.
[185] V. Shapovalov, H. Metiu, J. Catal. 245 (2007) 205–214.
[186] W. Song, A.P.J. Jansen, E.J.M. Hensen, Faraday Discuss. 162 (2013) 281–292.
[187] X. Liu, K. Zhou, L. Wang, B. Wang, Y. Li, J. Am. Chem. Soc. 131 (2009) 3140–3141.
[188] C. Zhang, A. Michaelides, D.A. King, S.J. Jenkins, J. Phys. Chem. C 113 (2009) 6411–6417.
[189] D. Pierre, W. Deng, M. Flytzani-Stephanopoulos, Top. Catal. 46 (2007) 363–373.
[190] J.H. Kwak, J. Hu, D. Mei, C.-W. Yi, D.H. Kim, C.H.F. Peden, L.F. Allard, J. Szanyi, Science 325 (2009) 1670–1673.
[191] M. Moses-DeBusk, M. Yoon, L.F. Allard, D.R. Mullins, Z. Wu, X. Yang, G. Veith, G.M. Stocks, C.K. Narula, J. Am. Chem. Soc. 135 (2013) 12634–12645.
[192] Z. Huang, X. Gu, Q. Cao, P. Hu, J. Hao, J. Li, X. Tang, Angew. Chem. Int. Ed. 51 (2012) 4198–4203.
[193] C. Zhang, F. Liu, Y. Zhai, H. Ariga, N. Yi, Y. Liu, K. Asakura, M. Flytzani-Stephanopoulos, H. He, Angew. Chem. Int. Ed. 51 (2012) 9628–9632.
[194] S. Wang, A.Y. Borisevich, S.N. Rashkeev, M.V. Glazoff, K. Sohlberg, S.J. Pennycook, S.T. Pantelides, Nat. Mater. 3 (2004) 143–146.
[195] T.-Y. Chang, Y. Tanaka, R. Ishikawa, K. Toyoura, K. Matsunaga, Y. Ikuhara, N. Shibata, Nano Lett. 14 (2013) 134–138.
[196] A.L. Linsebigler, G. Lu, J.T. Yates, Chem. Rev. 95 (1995) 735–758.
[197] A. Fujishima, X. Zhang, D. Tryk, Surf. Sci. Rep. 63 (2008) 515–582.
[198] K. Nakata, A. Fujishima, J. Photoch. Photobio. C 13 (2012) 169–189.
[199] U. Diebold, Surf. Sci. Rep. 48 (2003) 53–229.
[200] S.-C. Li, L.-N. Chu, X.-Q. Gong, U. Diebold, Science 328 (2010) 882–884.
[201] U. Diebold, S.-C. Li, M. Schmid, Annu. Rev. Phys. Chem. 61 (2010) 129–148.
[202] H. Iddir, V. Skavysh, S. Öğüt, N.D. Browning, M.M. Disko, Phys. Rev. B 73 (2006) 041403.
[203] V. Çelik, H. Ünal, E. Mete, Ş. Ellialtıoğlu, Phys. Rev. B 82 (2010) 205113.
[204] X.G. Wang, W. Weiss, S.K. Shaikhutdinov, M. Ritter, M. Petersen, F. Wagner, R. Schlögl, M. Scheffler, Phys. Rev. Lett. 81 (1998) 1038–1041.
[205] S. Yamamoto, T. Kendelewicz, J.T. Newberg, et al., J. Phys. Chem. C 114 (2010) 2256–2266.
[206] J. Jin, X. Ma, C.Y. Kim, D.E. Ellis, M.J. Bedzyk, Surf. Sci. 601 (2007) 3082–3098.
[207] L. Maike, M. Wolfgang, J. Phys. Condens. Matter 21 (2009) 134010.
[208] E. Wasserman, J.R. Rustad, A.R. Felmy, B.P. Hay, J.W. Halley, Surf. Sci. 385 (1997) 217–239.
[209] S. Thevuthasan, Y.J. Kim, S.I. Yi, S.A. Chambers, J. Morais, R. Denecke, C.S. Fadley, P. Liu, T. Kendelewicz, G.E. Brown Jr., Surf. Sci. 425 (1999) 276–286.
[210] V. Ortalan, A. Uzun, B.C. Gates, N.D. Browning, Nat. Nanotechnol. 5 (2010) 506–510.
[211] J. Lu, C. Aydin, N.D. Browning, B.C. Gates, Angew. Chem. Int. Ed. 51 (2012) 5842–5846.

[212] Y. Kondo, K. Yagi, K. Kobayashi, H. Kobayashi, Y. Yanaka, K. Kise, T. Ohkawa, Ultramicroscopy 36 (1991) 142–147.
[213] K. Yagi, Surf. Sci. Rep. 17 (1993) 307–362.
[214] A.K. Petford, L.D. Marks, M. O'Keeffe, Surf. Sci. 172 (1986) 496–508.
[215] W.M.J. Coene, A. Thust, M. Op de Beeck, D. Van Dyck, Ultramicroscopy 64 (1996) 109–135.
[216] C.L. Jia, A. Thust, Phys. Rev. Lett. 82 (1999) 5052–5055.
[217] A. Thust, Phys. Rev. Lett. 102 (2009) 220801.
[218] A. Thust, W.M.J. Coene, M. Op de Beeck, D. Van Dyck, Ultramicroscopy 64 (1996) 211–230.
[219] Y. Shidahara, K. Aoki, Y. Tanishiro, H. Minoda, K. Yagi, Surf. Sci. 357–358 (1996) 820–824.
[220] A.V. Crewe, J. Wall, J. Langmore, Science 168 (1970) 1338–1340.
[221] E. Zeitler, M.G.R. Thomson, Optik 31 (1970) 258.
[222] A.V. Crewe, J. Microsc.-Oxf 100 (1974) 247–259.
[223] J. Wall, J. Langmore, M. Isaacson, A.V. Crewe, Proc. Natl. Acad. Sci. 71 (1974) 1–5.
[224] V. Beck, A.V. Crewe, Ultramicroscopy 1 (1975) 137–144.
[225] M. Isaacson, D. Kopf, M. Utlaut, N.W. Parker, A.V. Crewe, Proc. Natl. Acad. Sci. 74 (1977) 1802–1806.
[226] J.S. Wall, Chem. Scr. 14 (1979) 271–278.
[227] L.A. Freeman, A. Howie, M.M.J. Treacy, J. Microsc.-Oxf 111 (1977) 165–178.
[228] M.M.J. Treacy, A. Howie, C.J. Wilson, Philos. Mag. A 38 (1978) 569–585.
[229] K. Tsutomu, Jpn. J. Appl. Phys. 7 (1968) 27.
[230] D. Cherns, Philos. Mag. 30 (1974) 549–556.
[231] G. Lehmpfuhl, Y. Uchida, Ultramicroscopy 4 (1979) 275–282.
[232] K. Yagi, K. Takayanagi, K. Kobayashi, N. Osakabe, Y. Tanishiro, G. Honjo, Surf. Sci. 86 (1979) 174–181.
[233] S. Iijima, Ultramicroscopy 6 (1981) 41–52.
[234] Y. Tanishiro, H. Kanamori, K. Takayanagi, K. Yagi, G. Honjo, Surf. Sci. 111 (1981) 395–413.
[235] K. Takayanagi, Ultramicroscopy 8 (1982) 145–161.
[236] K. Yagi, Scanning Electron Microsc. (1982) 1421–1428.
[237] A.F. Moodie, C.E. Warble, Philos. Mag. 16 (1967) 891.
[238] L.D. Marks, Phys. Rev. Lett. 51 (1983) 1000–1002.
[239] L.D. Marks, Acta Crystallogr. Sect. A 40 (1984) C390.
[240] L.D. Marks, Surf. Sci. 139 (1984) 281–298.
[241] K. Abdelmoula, J.J. Metois, G. Nihoul, Phys. Status Solidi A-Appl. Res. 77 (1983) K5.
[242] K. Takayanagi, K. Yagi, Trans. Jpn. Inst. Metals 24 (1983) 337–348.
[243] G. Nihoul, K. Abdelmoula, J.J. Metois, Ultramicroscopy 12 (1984) 353–366.
[244] J.M. Gibson, M.L. McDonald, F.C. Unterwald, Phys. Rev. Lett. 55 (1985) 1765–1767.
[245] S.A. Nepijko, T.N. Bondarenko, H. Hofmeister, J. Cryst. Growth 71 (1985) 735–738.
[246] P. Lu, D.J. Smith, Phys. Rev. Lett. 59 (1987) 2177–2179.
[247] M. Shiojiri, T. Isshiki, K. Okashita, Y. Hirota, T. Maeda, S. Sekimoto, Ultramicroscopy 23 (1987) 355–363.
[248] M. Shiojiri, T. Isshiki, Y. Hirota, Ultramicroscopy 30 (1989) 329–336.
[249] D.J. Smith, R.W. Glaisher, P. Lu, M.R. McCartney, Ultramicroscopy 29 (1989) 123–134.
[250] K. Takayanagi, J. Electron Microsc. 38 (1989) S58–S65.
[251] M. Mitome, K. Takayanagi, Y. Tanishiro, Ultramicroscopy 33 (1990) 255–260.
[252] X.G. Ning, H.Q. Ye, Philos. Mag. A 62 (1990) 431–446.
[253] P. Lu, D.J. Smith, Surf. Sci. 254 (1991) 119–124.
[254] T. Isshiki, M. Toyoshima, M. Tsujikura, H. Saijo, M. Shiojiri, Ultramicroscopy 41 (1992) 201–209.
[255] K. Yagi, J. Electron Microsc. 44 (1995) 269–280.
[256] T. Mishima, T. Osaka, Surf. Sci. 395 (1998) L256–L260.
[257] L.D. Marks, Rep. Prog. Phys. 57 (1994) 603.
[258] L.D. Marks, Surf. Sci. 139 (1984) 281–298.
[259] M. Mitome, K. Takayanagi, Y. Tanishiro, Phys. Rev. B 42 (1990) 7238–7241.
[260] C.-C. Chen, C. Zhu, E.R. White, C.-Y. Chiu, M.C. Scott, B.C. Regan, L.D. Marks, Y. Huang, J. Miao, Nature 496 (2013) 74–77.

CHAPTER 8

Catalysis by Atomically Precise Gold Nanoclusters

Yuxiang Chen, Gao Li, Huifeng Qian, Rongchao Jin*
Department of Chemistry, Carnegie Mellon University, Pittsburgh, PA, USA
*Corresponding author: E-mail: rongchao@andrew.cmu.edu

1. INTRODUCTION

Metal nanoparticle-based catalysis is of major importance in chemical industry. Among the precious metals, platinum (Pt) has long been widely used for many decades, while gold (Au) did not receive much attention until the 1980–1990s [1,2]. Recent decade has witnessed the "golden rush" in catalytic research [3–7]. In fundamental studies, the polydispersity of conventional nanoparticle-based catalysts (or nanocatalysts) poses major challenges in obtaining the deep insight into the catalytic mechanism [7]. First of all, polydisperse nanoparticles only give an average catalytic reactivity, and the precise size-dependent reactivity cannot be obtained. Second, the surface of the nanocatalyst is hard to be mapped out and thus remains poorly defined, unlike bulk single-crystal surfaces which expose well-defined [111] and [100] faces, and can be readily mapped by scanning probe techniques. Third, the conventionally synthesized nanocatalysts via solution-phase methods bear some capping agents on the particle surfaces but the exact information is hard to obtain. Overall, the polydispersity of nanoparticles and their poorly defined surface structures as well as unknown surface compositions preclude atomic/molecular-level insight into the catalytic mechanism, and sometimes it is even difficult to relate the observed catalytic properties to the metal nanoparticles or the support. To overcome these major obstacles, well-defined nanocatalysts are highly desirable, especially when one pursues the fundamental studies of catalysis.

In recent years, *atomically precise* gold nanoparticles (often called nanoclusters) have emerged as a new class of nanomaterial, notably the thiolate-protected gold nanoclusters [8,9]. Specific-size Au nanoclusters are composed of precise numbers of gold atoms (n) and of thiolate (-SR) ligands (m), denoted as $Au_n(SR)_m$, with n ranging from ten to a few hundred atoms (equivalent size from subnanometer to 2–3 nm). Therefore, these unique nanoparticles very much resemble molecular compounds, since both the formula and atomic structure are well defined, while conventional nanoparticles only possess an average size and no exact formula can be determined. More importantly, the attainment of molecularly pure nanoclusters permits the growth of single crystals of nanoclusters for X-ray crystallographic determination of

Catalysis by Materials with Well-Defined Structures
http://dx.doi.org/10.1016/B978-0-12-801217-8.00008-6

the total structures of nanoclusters, that is, every atom of the nanocluster can be precisely pinned down. Based upon the total structures of nanoclusters, significant progress of atomic/molecular level understanding of nanocatalysis is expected to be achieved in future work [7].

Compared to conventional metallic Au nanoparticle catalysts, $Au_n(SR)_m$ nanoclusters possess several distinct features that are of particular interest to catalysis. First, metallic gold nanoparticles (2–100 nm) adopt a face-centered cubic (FCC) structure, but $Au_n(SR)_m$ nanoclusters often adopt different atom-packing structures [9]; for example, the icosahedral $Au_{25}(SR)_{18}$ and biicosahedral $Au_{38}(SR)_{24}$ nanoclusters (see detailed discussion in Section 2) [9]. Second, the ultrasmall size induces strong electron-energy quantization in nanoclusters, as opposed to the continuous conduction band in metallic gold nanoparticles [8]. Thus, $Au_n(SR)_m$ nanocluters become semiconductors and possess a sizable band gap (e.g., ~1.3 eV for $Au_{25}(SR)_{18}$ and ~0.9 eV for $Au_{38}(SR)_{24}$) [9]. Third, unique structures have been discovered in the crystal structures of nanoclusters, such as the staple-like motifs and volcano-like atomic sites—which can act as catalytically active centers in chemical reactions [7].

In this Chapter, we first briefly discuss the synthesis of $Au_n(SR)_m$ nanoclusters and then focus on the catalytic reactions that can be catalyzed by such nanoclusters. The catalysis research of $Au_n(SR)_m$ nanoclusters is still in its infancy, but the reported work has shown great promise, and future work is expected to advance fundamental understanding of catalytic mechanism at the atomic/molecular level.

2. SYNTHESIS OF ATOMICALLY PRECISE GOLD NANOCLUSTERS

We briefly introduce the "size focusing" methodology for synthesizing atomically precise $Au_n(SR)_m$ nanoclusters [10,11]. In this method, an appropriate distribution of size-mixed nanoclusters is first prepared by kinetic control over the reduction reaction of gold precursor (typically Au(I)-SR complexes or polymers) with $NaBH_4$ or other reducing agents, which is the key step for the attainment of single-sized nanoclusters as the final product. Then, these size-mixed nanoclusters are subjected to size-focusing under harsh conditions (e.g., at 80–90 °C and in the presence of excess thiol); under such conditions the unstable nanoclusters are decomposed or converted to the more stable ones, and eventually only the most stable size in the initial size distribution survives the "focusing" process. The initial size distribution should be properly controlled (e.g., not too broad—otherwise, two or multiple stable sizes would result after size focusing).

A series of size-discrete, robust $Au_n(SR)_m$ nanoclusters have been available through intense research in the past years, such as $Au_{25}(SR)_{18}$ [11], $Au_{28}(SR)_{20}$ [12], $Au_{36}(SR)_{24}$ [13], $Au_{38}(SR)_{24}$ [14], $Au_{144}(SR)_{60}$ [15], and $Au_{333}(SR)_{79}$ [16]. Herein, we choose a few examples for more detailed discussions.

2.1 The $Au_{25}(SR)_{18}$ Nanocluster

Size-focusing was first observed in the synthesis of $Au_{25}(SR)_{18}$ nanoclusters in our early work [17,18]. In the synthesis, the first step is the spontaneous reduction of Au(III) to Au(I) at 0 °C after adding thiol (typically 3–5 equivalents per mole of gold), forming Au(I)-SR complexes or polymers (i.e., the intermediates). The subsequent $NaBH_4$ reduction of Au(I)-SR to Au(0) produced polydisperse nanoclusters protected by thiolate. Over the prolonged aging process of the reaction product, a spontaneous "size-focusing" process was observed, evidenced by the evolution of the optical absorption spectra. Initially, the polydisperse product gave a decay-like spectrum (note: the absorption peaks of individual sizes were washed out), but with time, we observed several prominent peaks in the UV–vis spectrum of the crude product, indicating a gradual growth of monodisperse nanoclusters, Figure 1. Three absorption peaks were observed at 670, 450, and 400 nm, and the final product consists of molecularly pure $Au_{25}(SR)_{18}$ nanoclusters (Figure 2).

The case of the $Au_{25}(SR)_{18}$ synthesis has illustrated the basic principle of the size-focusing growth process, which is based upon the stability property of different-sized nanoclusters. In order to attain atomic monodispersity, synthetic control is needed so that only one specific size of nanoclusters survives the size-focusing process. We found that it is critical to control the size distribution of the starting $Au_n(SR)_m$ mixture. Below, we further elaborate on the importance of the initial size distribution for the final molecular purity of nanoclusters with the case of $Au_{38}(SR)_{24}$ synthesis.

2.2 The $Au_{38}(SR)_{24}$ Nanocluster

In the size-focusing synthesis of $Au_{38}(SR)_{24}$, similar to the case of $Au_{25}(SR)_{18}$, the first step is to obtain a crude mixture (in this case, glutathionate (SG)-protected

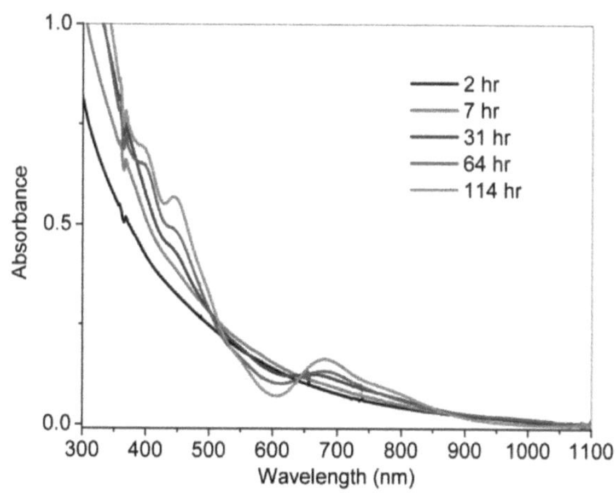

Figure 1 UV–vis spectroscopic monitoring of the "size focusing" process for the growth of $Au_{25}(SR)_{18}$ nanoclusters. *Adapted with permission from Ref. [18].*

$Au_n(SG)_m$ nanoclusters) [20]. The size-mixed nanoclusters were then subjected to a thermal thiol etching process [21,22] (e.g., 80 °C) in a two-phase (water/toluene) system, during which the ligand exchange (from -SG to $-SCH_2CH_2Ph$) on the nanoclusters was completed, and subsequently the etching process by thiol caused size-focusing of polydisperse nanoclusters into the molecularly pure $Au_{38}(SC_2H_4Ph)_{24}$ nanoclusters (Figure 3) [20,23].

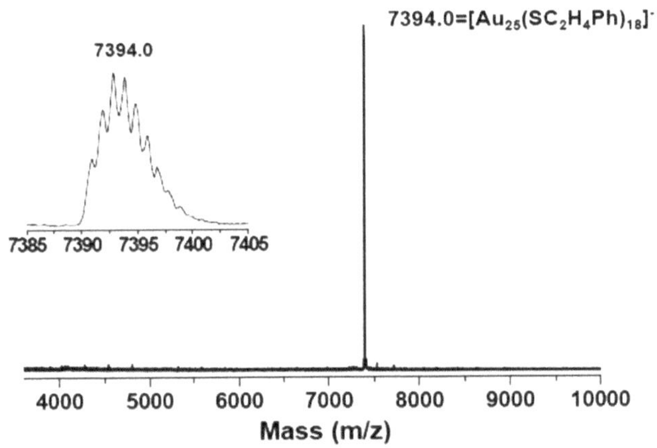

Figure 2 Electrospray ionization (ESI) mass spectrum of $Au_{25}(SC_2H_4Ph)_{18}^-$ nanoclusters. The experimental formula weight is 7394 Da (most abundant ion in the isotope pattern; theoretical value: 7394). *Adapted with permission from Ref. [19].*

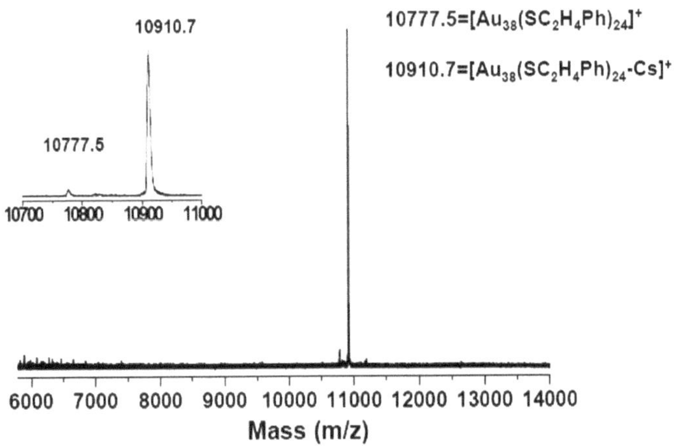

Figure 3 ESI-MS spectrum of charge-neutral $Au_{38}(SC_2H_4Ph)_{24}$ nanoclusters. The experimental formula weight of $Au_{38}(SC_2H_4Ph)_{24}$ is 10,777.5 Da (theoretical value: 10,778.05). In mass spectrometric analysis, CsOAc was added to form positively charged [cluster-Cs]$^+$ adducts. The observed small peak of $[Au_{38}(SCH_2CH_2Ph)_{24}]^+$ was due to induced ionization by Cs^+ salt. *Adapted with permission from Ref. [20].*

A detailed mechanistic investigation on the conversion process clearly showed a size-focusing process [20], evidenced both by optical spectroscopy and MALDI-MS analyses (Figure 4). The optical spectrum of the starting $Au_n(SC_2H_4Ph)_m$ showed a decaying curve, indicating a mixture. With reaction going on, several distinct peaks started to emerge in the spectrum, and the final product showed a distinct optical spectrum characteristic of $Au_{38}(SC_2H_4Ph)_{24}$ nanoclusters (Figure 4 (A)) [21,22]. MALDI-MS analysis reveals that, with the increase of reaction time, the relatively large Au nanoclusters ($n > 38$) were decomposed and converted to $Au_{38}(SC_2H_4Ph)_{24}$ (Figure 4(B)) [20]. After ~40 h, molecularly pure $Au_{38}(SC_2H_4Ph)_{24}$ nanoclusters were obtained (Figure 4(B)); note that the fragment at 9342 Da was caused by the MALDI method.

A key condition to obtain single-sized $Au_{38}(SC_2H_4Ph)_{24}$ nanoclusters is to control the size distribution of the $Au_n(SG)_m$ mixture prior to the size-focusing step. In this case, the solvent was found to play an important role in controlling the size range of the $Au_n(SG)_m$ starting mixture [20]. Specifically, acetone produced an appropriate size distribution of $Au_n(SG)_m$ (Figure 5 (B)) that permitted high-yielding synthesis of $Au_{38}(SR)_{24}$. Comparison between acetone and methanol (solvents) showed that the acetone-mediated synthesis of $Au_n(SG)_m$ (the starting mixture) had a dominant size range from 8 to 18 kDa, while the methanol system produced a dominant size range below 8 kDa, that is, smaller than Au_{38} (Figure 5(A)). The acetone-mediated high yield synthesis of $Au_{38}(SC_2H_4Ph)_{24}$ is attributed to the down-conversion of those higher-mass $Au_n(SG)_m$ nanoclusters ($38 < n < 100$), while such components were missing in the

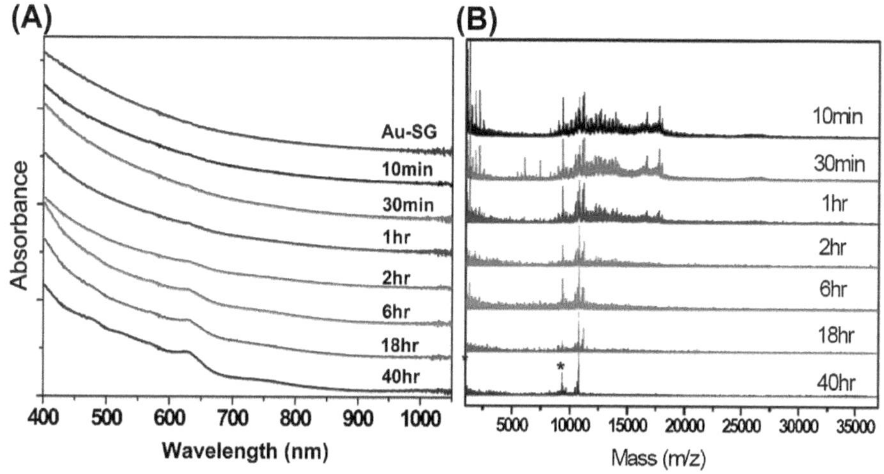

Figure 4 UV–vis spectra (A) and MALDI mass spectra (B) of Au nanoclusters during the size-focusing process. Au–SG stands for the starting $Au_n(SG)_m$ mixture in aqueous solution; 10 min to 40 h indicates the time that $Au_n(SG)_m$ reacts with excess PhC_2H_4SH. The asterisk (*) shows a fragment of $Au_{38}(SC_2H_4Ph)_{24}$ which was caused by the MALDI method. *Adapted with permission from Ref. [20].*

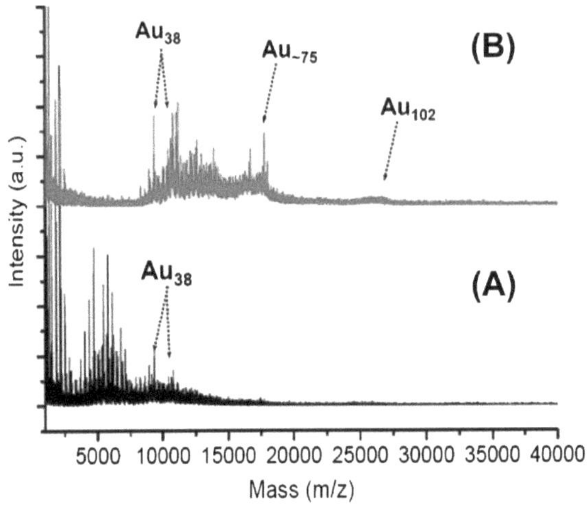

Figure 5 MALDI mass spectra of Au nanoclusters obtained from ligand exchange of $Au_n(SG)_m$ with PhC_2H_4SH (reaction for 10 min). The starting $Au_n(SG)_m$ nanoclusters were made (A) in methanol and (B) in acetone. *Adapted with permission from Ref. [20].*

methanol-mediated synthesis (Figure 5), hence, low yield of Au_{38}. Of note, the initial size range of $Au_n(SG)_m$ should not be too large; otherwise, $Au_{144}(SR)_{60}$ nanoclusters could be produced, leading to a mixture with $Au_{38}(SR)_{24}$ nanoclusters.

The above examples explicitly demonstrate the importance of the size distribution of the starting nanoclusters (i.e., prior to the size-focusing step) for the eventual formation of monodisperse nanoclusters. The first step is essentially kinetically controlled, while the second step (i.e., size focusing) is essentially controlled by thermodynamics [24]; Thus, the discovery of the size-focusing mechanism has established a universal methodology for size-controlled synthesis of nanoclusters [10]. Nevertheless, the reaction parameters (e.g., reaction temperature, kinetics, ligand bulkiness, etching time, and thiol/gold ratio) should be carefully balanced in the size-focusing synthesis [25,26].

2.3 The $Au_{144}(SR)_{60}$ Nanocluster

To obtain the proper size distribution of polydisperse $Au_n(SR)_m$ for size-focusing into monodisperse $Au_{144}(SR)_{60}$ nanoclusters [27], we adjusted the gold salt to thiol ratio (Au/SR) in the first step. By adjusting the $HAuCl_4$ to PhC_2H_4SH ratio to 1:3, the size range was controlled around 24–36 kDa. In the subsequent size-focusing step, the initial polydispersed nanoclusters were incubated in large amounts of thiol at 80 °C, and after 24 h of size-focusing we obtained pure $Au_{144}(SC_2H_4Ph)_{60}$ nanoclusters (Figure 6). The initial polydispersed nanoclusters showed a decay-like UV–vis spectrum; after size focusing, step-like bands were observed at 510 and 700 nm [27].

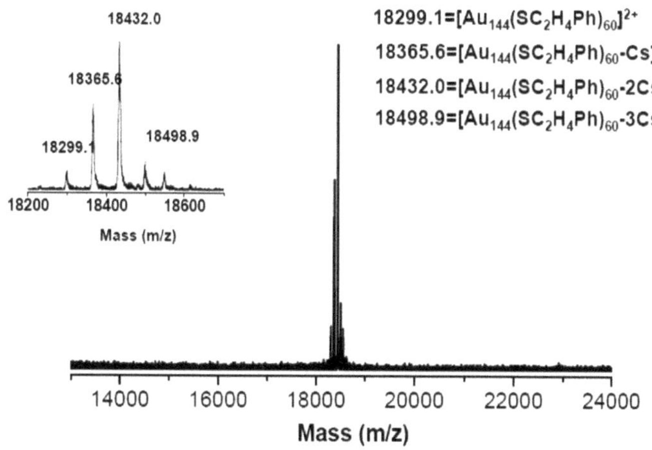

Figure 6 ESI-MS spectrum of $Au_{144}(SC_2H_4Ph)_{60}$ nanoclusters. CsOAc was added to form positively charged $[\text{cluster-Cs}]^+$ adducts for ESI analysis. The experimental molecular weight is $18{,}299.1 \times 2 = 36{,}598.2$ (theoretical: 36,596.5). *Adapted with permission from Ref. [27].*

3. ATOMIC STRUCTURES OF GOLD NANOCLUSTERS

The molecularly well-defined nanoclusters permit the growth of single crystals for total structure (i.e., every atom) determination of the nanoclusters. X-ray crystallography reveals that the $Au_{25}(SR)_{18}$ nanocluster comprises a Au_{13} icosahedral core (Figure 7(A), magenta) and a $Au_{12}(SR)_{18}$ shell (Figure 7(A), cyan (white in print version) and yellow (light gray in print version)) [11]. Some of the surface atoms are indeed exposed, such as the triangular Au_3 site as shown in the space-filling model (Figure 4(B)). Such sites are readily accessible by reactant molecules in catalytic reactions, thus these cavities (much resembling "pockets") may serve as the catalytically active sites.

The electronic structure of $Au_{25}(SR)_{18}$ exhibits discrete energy levels caused by the quantum-size effect [11]. The molecular orbitals can be roughly divided into the Au-core orbitals (i.e., primarily contributed by Au_{13}) and the shell orbitals (i.e., primarily by $Au_{12}(SR)_{18}$), as reflected in the geometric core–shell structure. It has been proved that, regardless of the thiolate type, $Au_{25}(SR)_{18}$ nanoclusters with different R groups exhibit the same geometric structure [25]. The distinct oxidation states of surface Au atoms (electron-deficient, formal Au(I) sites) and the core Au atoms (electron-rich) have been found to be important in catalytic processes (vide infra).

The $Au_{38}(SCH_2CH_2Ph)_{24}$ nanocluster comprises a biicosahedral Au_{23} core and six dimeric staples –SR–Au–SR–Au–SR– ($Au_2(SR)_3$ for short) and three monomeric staples –SR–Au–SR– ($Au(SR)_2$ for short) protecting the Au_{23} core (Figure 8) [14]. Again, the gold atoms are distributed in two kinds of chemical environments (surface and core).

The atomic structure of the $Au_{144}(SC_2H_4Ph)_{60}$ nanocluster has not been solved, as the larger nanoclusters pose major challenges in crystallization. Future effort is to overcome the difficulties via searching for good solvent/antisolvent pairs and crystallization conditions.

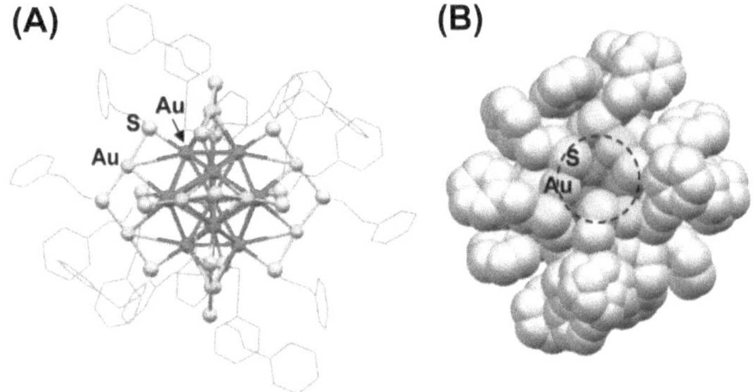

Figure 7 X-ray crystal structure of $Au_{25}(SCH_2CH_2Ph)_{18}$ nanoclusters (magenta (dark gray in print version): gold atoms of the core, cyan (white in print version): gold atoms of the surface): (A) ball-stick model and (B) space-filling model. *Adapted with permission from Refs [9] and [11].*

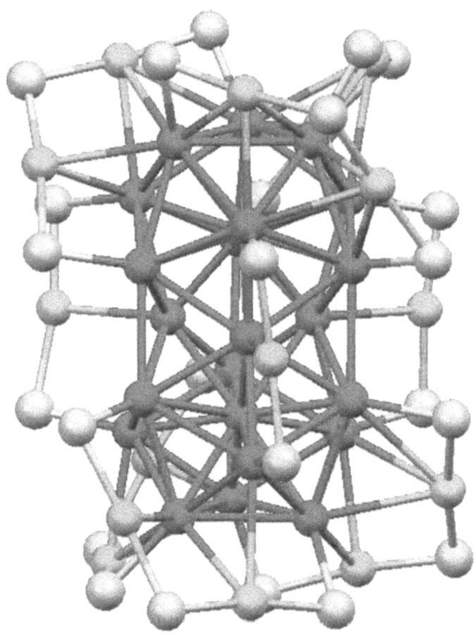

Figure 8 Crystal structure of $Au_{38}(SC_2H_4Ph)_{24}$ (magenta (dark gray in print version): gold atoms of the Au_{23} core, cyan (light gray in print version): gold atoms of the shell, yellow (white in print version): sulfur). All $-C_2H_4Ph$ tails are omitted for clarity. *Adapted with permission from Ref. [14].*

4. THERMAL STABILITY OF $Au_n(SR)_m$ NANOCLUSTERS

The thiolate-protected gold nanoclusters are thermally robust as they are synthesized by size-focusing under harsh conditions. The thermal stability of $Au_n(SR)_m$ nanoclusters is important for catalytic applications as many reactions are run under thermal conditions.

An important issue is at which temperature the surface ligands will be desorbed. Thermogravimetric analysis (TGA) shows that $Au_n(SR)_m$ (e.g., $Au_{25}(SR)_{24}$, $Au_{38}(SR)_{24}$, and $Au_{144}(SR)_{60}$) starts to lose ligands at ~200 °C and the ligand loss completes at ~250 °C (Figure 9(A)) [28]. The ligand-loss temperature was found to be not affected by the atmosphere (e.g., N_2, air, O_2, and H_2) [29].

We further tested the isothermal stability of $Au_{25}(SR)_{18}$ nanoclusters (i.e., maintained at 150 °C for 60 min in air atmosphere) [30]. No weight loss was found, which indicated that the gold nanoclusters remain stable after the annealing process, Figure 9(B); note that the structure of $Au_{25}(SR)_{18}$ is thermodynamically stable, thus, structural isomerization without ligand loss should not occur. Therefore, the thermal tests confirm that the nanoclusters are robust and thermally stable below the ligand-loss temperature [26].

5. CATALYTIC PROPERTIES OF $Au_n(SR)_m$ NANOCLUSTERS

Our interest in exploring $Au_n(SR)_m$ for catalysis was inspired by the observation of reversible conversion between $[Au_{25}(SR)_{18}]^0$ and $[Au_{25}(SR)_{18}]^-$ [31,32]. The redox properties of nanoclusters are particularly important for those catalytic reactions involving electron transfer for activating reactants.

The native $Au_{25}(SR)_{18}$ nanoclusters from the synthesis are anionic (i.e., $[Au_{25}(SR)_{18}]^-$, counterion = tetraoctylammonium (TOA^+)) [11]. Interestingly, we found that when a solution of $[Au_{25}(SR)_{18}]^-$ was exposed to air, the nanoclusters were gradually converted to charge-neutral $[Au_{25}(SR)_{18}]^0$ clusters [31], which was first discovered by the color change of the solution (see also the UV–vis spectral changes in Figure 10). Single crystal X-ray crystallography [31] revealed that the product is $[Au_{25}(SCH_2CH_2Ph)_{18}]^0$ and its

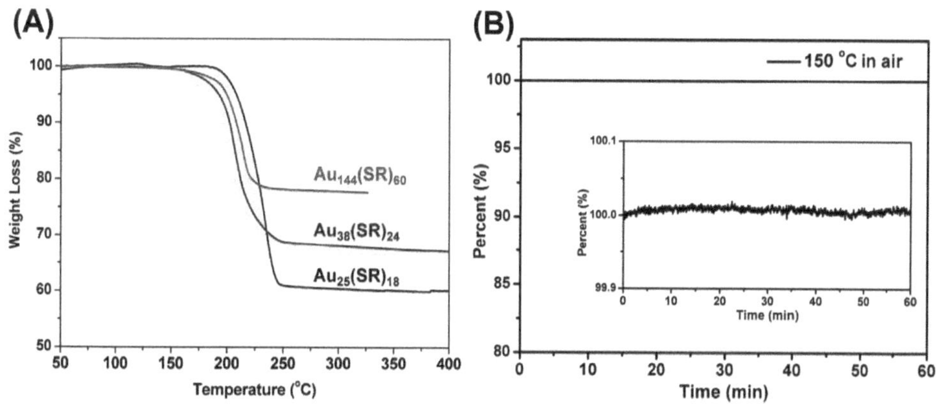

Figure 9 (A) Thermogravimetric analysis of $Au_n(SR)_m$ nanoclusters (under N_2 atmosphere). (B) Isothermal stability analysis of pure $Au_{25}(SR)_{18}$ nanoclusters (maintained at 150 °C, in air atmosphere for 60 min). No discernible loss of ligands (starting: 100%) was observed. The inset shows a zoom-in of the TGA curve at constant 150 °C. *Adapted with permission from Refs [28] and [30].*

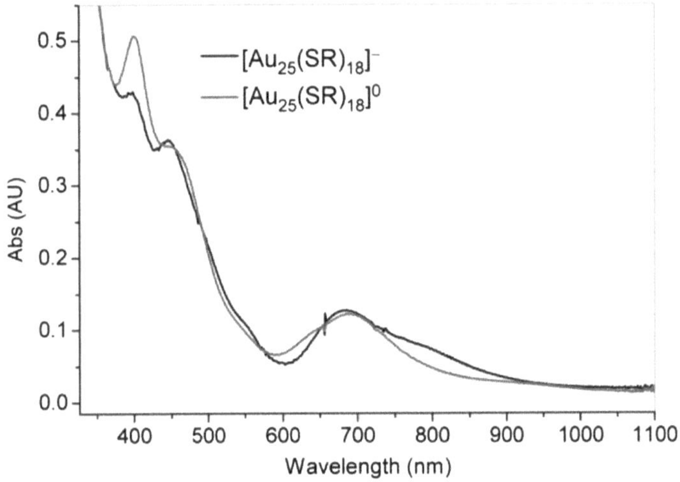

Figure 10 Reversible conversion between $[Au_{25}(SR)_{18}]^0$ and $[Au_{25}(SR)_{18}]^-$. *Adapted with permission from Ref. [32].*

structure shows the same framework as that of $[Au_{25}(SCH_2CH_2Ph)_{18}]^-$. Further work [32] revealed that the negative charge of $[Au_{25}(SCH_2CH_2Ph)_{18}]^-$ resides on the Au_{13} core of the cluster, rather than on the ligands. This explains why the gold core, instead of the thiolate ligands, was first oxidized (i.e., one-electron loss to yield neutral $[Au_{25}(SCH_2CH_2Ph)_{18}]^0$); normally, one would expect that O_2 would first oxidize the thiolate ligands on the Au surface. The redox process between $[Au_{25}(SCH_2CH_2Ph)_{18}]^0$ and $[Au_{25}(SCH_2CH_2Ph)_{18}]^-$ is completely reversible [32]. If using peroxide as the oxidant, the anion-to-neutral conversion is much faster than that using O_2.

5.1 Catalytic Oxidation

As discussed above, $Au_n(SR)_m$ nanoclusters can interact with oxidants (e.g., O_2 and H_2O_2) even at room temperature. The oxidation–reduction reactivity of $Au_{25}(SR)_{18}$ nanoclusters inspired us to pursue the catalytic application of $Au_{25}(SR)_{18}$ and other $Au_n(SR)_m$ nanoclusters in catalytic oxidation. The activation of oxygen (O_2) plays an important role in the gold nanocluster-catalyzed selective oxidation processes. Below we discuss the catalytic properties of nanoclusters in styrene oxidation [28,33,34] and sulfide oxidation [35]. Recent work has also evaluated the CO oxidation reaction catalyzed by $Au_n(SR)_m$ nanoclusters [29,36–38].

5.1.1 Selective Oxidation of Styrene

We first investigated solution-phase styrene oxidation catalyzed by $Au_n(SR)_m$ nanocluster catalysts (free and SiO_2-supported gold nanoclusters) using O_2 as the oxidant, Scheme 1 [33]. The catalytic reaction was run at 80–100 °C for 12–24 h, which gave rise to benzaldehyde as the major product (up to ~70% selectivity) and styrene epoxide (~25% selectivity) and acetophenone (<5% selectivity) using $Au_{25}(SR)_{18}$ nanoclusters.

Using free (i.e., unsupported) Au_{25}, Au_{38}, and Au_{144} nanocluster catalysts, we observed a strong size-dependence in the catalytic oxidation reaction of styrene, and the smaller $Au_n(SR)_m$ nanocluster catalyst exhibited much higher catalytic activity [28]. The SiO_2-supported $Au_{25}(SR)_{18}$ catalyst gave rise to comparable performance as that of the free nanocluster catalyst, but the advantage of $Au_{25}(SR)_{18}/SiO_2$ catalyst is its recyclability and reuse in catalytic reaction; no apparent deterioration in activity and selectivity of the reused supported catalyst was observed after multiple cycles [28].

Furthermore, Zhu et al. [28] compared three oxidant systems using $Au_{25}(SR)_{18}/SiO_2$ catalyst: (A) *tert*-butyl hydroperoxide (TBHP) as the oxidant; (B) TBHP as the initiator and O_2 as the main oxidant; (C) O_2 as the oxidant (without initiator). As shown in Table 1, system **A** gave a high conversion of styrene (e.g., 86%) with 100% selectivity for benzaldehyde, system **B** gave rise to a much lower activity (25% conversion with 100% selectivity for benzaldehyde), and system **C** was even lower (18% conversion with 80% selectivity for benzaldehyde). The catalytic results were consistent with the fact that TBHP is much more reactive than O_2. A similar phenomenon was also found in the

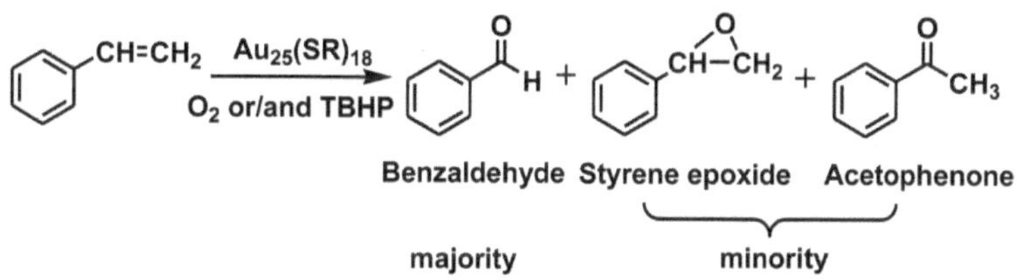

Scheme 1 Selective oxidation of styrene catalyzed by $Au_{25}(SR)_{18}$ nanocluster catalysts.

Table 1 Comparison of the three oxidant systems in catalytic oxidation of styrene

Catalyst	Ingredients of the oxidant	Conversion of styrene	Selectivity for benzaldehyde
Au_{25}	TBHP only	86%	100%
	TBHP as initiator + O_2 as the primary oxidant	25%	100%
	O_2 only	18%	80%
Au_{38}	TBHP only	85	100%
	TBHP as initiator + O_2 as the primary oxidant	16	100%
	O_2 only	5	79%
Au_{144}	TBHP only	91	100%
	TBHP as initiator + O_2 as the primary oxidant	9	94%
	O_2 only	5	85%

three oxidant systems using $Au_{38}(SR)_{24}/SiO_2$ and $Au_{144}(SR)_{60}/SiO_2$ catalysts (Table 1) [28]. The results indicate that the ingredients of the oxidant play a critical role in the oxidation reaction, and that the oxidant (denoted as [O]) activation is a key step for achieving high conversion of styrene. The results also imply that the active oxygen species in the catalytic cycles may be peroxide-like or hydroperoxide-like species [28].

Tsukuda and coworkers [34] investigated $Au_{25}(SG)_{18}/HAP$ catalyst for selective oxidation of styrene, where HAP refers to hydroxyapatite ($Ca_{10}(PO_4)_6(OH)_2$). The ligand-off Au_{25}/HAP catalyst was obtained by removing the ligands via 300 °C thermal treatment. The catalyst was explored in styrene oxidation (with anhydrous *tert*-butyl hydroperoxide (TBHP) as the oxidant). The reaction yielded styrene oxide with a selectivity of ~92%, in contrast with the results obtained by Zhu et al. in which the benzaldehyde (as opposed to the epoxide) constitutes the major product.

5.1.2 Selective Oxidation of Sulfides

Selective oxidation of sulfides to sulfoxides is an important organic transformation since the resulting sulfoxides are versatile intermediates for the preparation of biologically and medicinally important products. The main side-reaction of sulfoxidation is the formation of sulfone. Therefore, synthetic methods for mild and selective oxidation of sulfides to sulfoxides that minimize production of sulfones and side-reactions are highly desirable.

Sulfides have much less binding affinity toward gold than thiolates in terms of bonding, but sulfides can be chemisorbed onto the surface of gold nanoclusters via the interaction between the sulfur atom (–S–) and the gold surface, while sulfoxides (–S(=O)–) only weakly bind to the gold surface; hence the conversion from sulfide to sulfoxide on the clusters surface should be feasible. Li et al. [35] investigated the oxide-supported ~1 wt% $Au_{25}(SR)_{18}$ nanocluster catalyst for selective oxidation of sulfides to sulfoxides using iodosylbenzene (PhIO) as oxidant (Scheme 2).

The TiO_2-supported $Au_{25}(SR)_{18}$ nanocluster catalyst gave rise to a high catalytic activity (e.g., ~97% conversion of sulfide with ~92% selectivity for the sulfoxide product). Meanwhile, different types of substrates (i.e., sulfides) have been investigated to test the effects of electronic and steric factors in the selective oxidation. Those sulfides with electron-rich groups were oxidized with higher conversion than those bearing electron deficient groups, and sulfides with greater steric groups showed lower conversion than those with less steric groups. The support effect was also investigated, and the catalyst activity of $Au_{25}(SR)_{18}$/oxide followed the order of

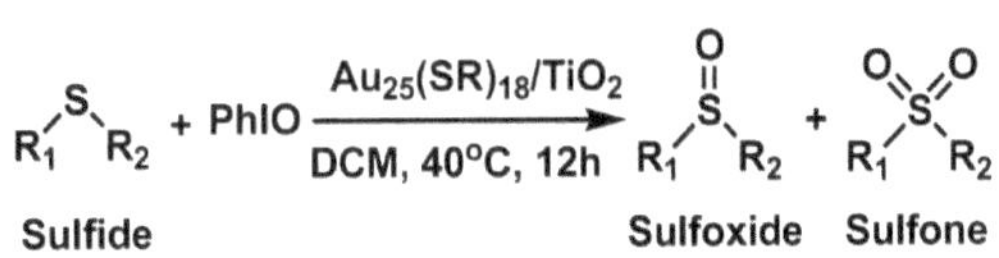

Scheme 2 Oxidation of sulfides by iodosylbenzene using $Au_{25}(SR)_{18}/TiO_2$ catalyst.

$TiO_2 > Fe_2O_3 > CeO_2 > MgO$. The oxide-supported $Au_{25}(SR)_{18}$ catalysts showed excellent recyclability in the sulfoxidation process [35].

Recent work by Kawasaki et al. [39] reported the nanocluster-induced generation of highly reactive singlet oxygen ($^1O_2$) which can be utilized in catalytic oxidation. The $^1O_2$ was efficiently produced through the direct photosensitization by $Au_{25}(SR)_{18}^-$ nanoclusters (SR = phenylethanethiol, SC_2H_4Ph, or captopril) via visible or near-IR irradiation (e.g., 532, 650, and 808 nm). $^1O_2$ was successfully detected using direct observation of characteristic $^1O_2$ emission around 1276 nm as well as three different $^1O_2$ selective probes. Catalytic selective oxidation of sulfide to sulfoxide by $^1O_2$ was demonstrated on the photoexcited $Au_{25}(SC_2H_4Ph)_{18}^-$ nanoclusters. When the reactions were carried out in open vessels in *dark* with O_2 bubbling, almost no sulfoxidation of sulfides was observed at 25 or 55 °C. With the assistance of photoexcited $Au_{25}(SR)_{18}^-$ at 532 nm irradiation, $Au_{25}(SR)_{18}^-$ exhibited ~100% selective catalytic oxidation of sulfides to sulfoxides. The conversion increased with higher reaction temperatures or longer reaction times. The ~100% selectivity achieved in the sulfoxidation of sulfides was in contrast to the use of other oxidants such as PhIO (87% only).

We also examined the catalytic oxidation of sulfides to sulfoxides by $^1O_2$ in the presence of $Au_{25}(SR)_{18}^-$ using different wavelengths of photoexcitation (532, 650, and 808 nm). Nearly 100% selectivity was achieved in the sulfoxidation of sulfide, irrespective of the excitation wavelength, and the conversion is the largest for photoexcited $Au_{25}(SR)_{18}^-$ upon 650 nm photoexcitation, which is very close to the absorbance at the HOMO-LUMO peak of $Au_{25}(SR)_{18}^-$ (Scheme 3) [39].

The above catalytic results of the selective oxidation by $Au_n(SR)_m$ nanoclusters allow us to gain some valuable insights at the atomic level. In the $[Au_{25}(SR)_{18}]^q$ nanocluster, the Au_{13} core possesses 8 (when q = −1) or 7 (when q = 0) delocalized valence electrons—which originate from Au(6s) and are primarily distributed within the Au_{13} core [32], while the Au_{12} shell bears positive charges due to charge transfer from gold to sulfur atoms of the thiolate ligands. The electron-rich Au_{13} core should facilitate the activation of the oxidant [O] (e.g., O_2, TBHP, and PhIO) by electron transfer to [O], accompanied by $[Au_{25}(SR)_{18}]^-$ conversion to neutral $[Au_{25}(SR)_{18}]^0$; of note, the oxidation readily occurs at elevated temperatures (e.g., under catalytic reaction conditions). For catalytic oxidation on $Au_{25}(SR)_{18}$, activation of reactants (e.g., styrene, sulfide, and alcohol) should be facilitated by the partial positive charge ($Au^{\delta+}$, δ = ~0.3) [40] on the triangular Au_3 motif of gold atoms.

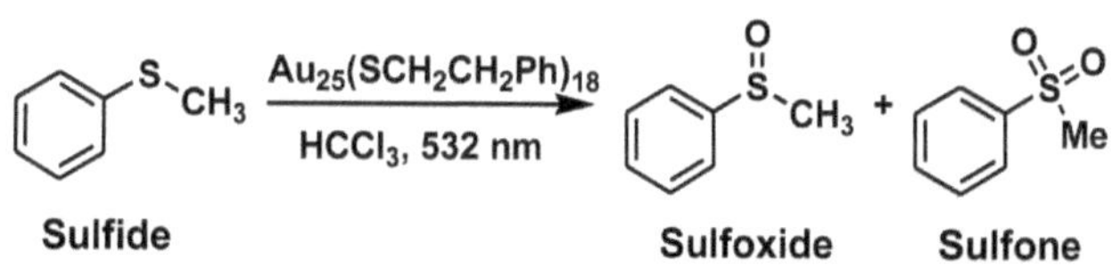

Scheme 3 Oxidation of sulfides by O_2 using photoexited $Au_{25}(PET)_{18}^-$ catalyst.

5.2 Catalytic Hydrogenation

5.2.1 Hydrogenation of Nitrophenol

Yamamoto et al. [41] compared the catalytic properties of $Au_{25}(SG)_{18}$ and dimethylformamide (DMF)-stabilized gold nanoclusters (not of atomic precision) in the reduction reaction of 4-nitrophenol to 4-aminophenol by $NaBH_4$ in water at room temperature, Scheme 4. The reduction process was monitored by UV–vis spectroscopy. The $Au_{25}(SG)_{18}$ nanoclusters exhibited higher catalytic activity than DMF-capped clusters in the nitrophenol reduction; the pseudo-first-order rate constant was $8\times10^{-3}\,s^{-1}$ for $Au_{25}(SG)_{18}$ versus $3\times10^{-3}\,s^{-1}$ for DMF-capped Au clusters. High catalytic activity of $Au_{25}(SG)_{18}$ catalyst can be achieved even at a very low catalyst concentration (e.g., 1.0 μM). No induction time was observed in the case of $Au_{25}(SG)_{18}$ despite –SG being a strongly binding ligand due to less steric hindrance, while the DMF-stabilized Au clusters exhibited an induction time (~4000 s), which was attributed to the impedance by the surface DMF layer on the clusters when reactants access the cluster surface [41]. For $Au_{25}(SG)_{18}$ nanoclusters, the unique core–shell structure has less steric hindrance and thus renders the catalytic active sites more accessible to reactants.

Shivhare et al. [42] demonstrated intact $Au_{25}(SR)_{18}$ (where, $R=C_6H_{13}$, C_8H_{17}, and $C_{12}H_{25}$) nanoclusters for the reduction of nitrophenol by $NaBH_4$ in mixed THF/water and under N_2 atmosphere. The $Au_{25}(SR)_{18}$ nanoclusters were stable and retained their structural integrity toward $NaBH_4$ treatment, while the larger $Au_{\sim180}(SC_6H_{13})_{\sim100}$ nanoclusters were not.

5.2.2 Selective Hydrogenation of Aldehydes and Ketones

The $Au_n(SR)_m$ nanocluster catalysts were also demonstrated to be capable of catalyzing chemoselective hydrogenation reactions, such as the selective hydrogenation of α,β-unsaturated ketones to α,β-unsaturated alcohols under mild conditions (60 °C, in mixed solvents (1:1 toluene/acetonitrile)) [33]. The $Au_{25}(SR)_{18}$ nanoclusters were found to preferentially hydrogenate the C=O bond of benzalacetone against the C=C bond, and the product (i.e., unsaturated alcohol) was obtained with 76% selectivity, whereas 14% selectivity was observed for unsaturated ketone, and 10% for saturated alcohol [33]. Further exploration of the catalytic reaction led to a nearly complete selectivity for α,β-unsaturated alcohol in a mixed toluene/ethanol (1:1) solvent at room temperature [43]. The UV–vis spectra and MS analysis of the free $Au_{25}(SR)_{18}$ nanocluster catalyst confirmed that the $Au_{25}(SR)_{18}$ nanoclusters remain intact during the hydrogenation reaction. The

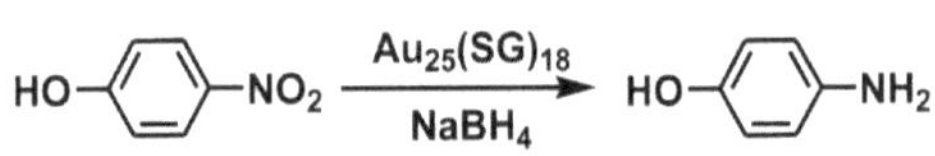

PNP

Scheme 4 Hydrogenation reaction of 4-nitrophenol to 4-aminophenol catalyzed by $Au_{25}(SR)_{18}$ nanocluster catalysts.

oxide-supported catalysts (including $Au_{25}(SR)_{18}/Fe_2O_3$, $Au_{25}(SR)_{18}/TiO_2$, and $Au_{25}(SR)_{18}/SiO_2$) were also investigated (Table 2) [43]. It was found that the catalytic activity was improved (from 8% to 16%) except the case of the $Au_{25}(SR)_{18}/SiO_2$ catalyst. The catalytic activities of the $Au_{25}(SR)_{18}/Fe_2O_3$ and $Au_{25}(SR)_{18}/TiO_2$ catalysts were almost equivalent (Table 2). In all cases, the chemoselectivity was 100%. In addition, the chemoselective hydrogenation was tested with a range of substituted α, β-unsaturated ketones and aldehydes; moderate yields and high selectivity were obtained (Table 2).

Zhu et al. [44] further expanded the $Au_{25}(SR)_{18}$ nanocluster catalyst for stereoselective hydrogenation of a bicyclic ketone (i.e., 7-(phenylmethyl)-3-oxa-7- azabicyclo [3.3.1] nonan-9-one) to exo-alcohol (Scheme 5).The stereoselective hydrogenation was run at room temperature in an ethanol/toluene (5:1, V/V) mixed solvent using H_2 (1 atm) as hydrogen source; ~18% conversion of the bicyclic ketone with 100% selectivity for the exo-alcohol was obtained; note that the stereoselective product was determined by nuclear magnetic resonance (NMR) spectroscopy [44].The high stereoselectivity (100%) was rationalized to be effected by the atomic spatial restraints and the unique core–shell configuration of the $Au_{25}(SR)_{18}$ nanocluster.

Table 2 Chemoselective hydrogenation reaction of α,β-unsaturated ketones to α,β-unsaturated alcohols catalyzed by $Au_{25}(SR)_{18}$ nanocluster catalysts

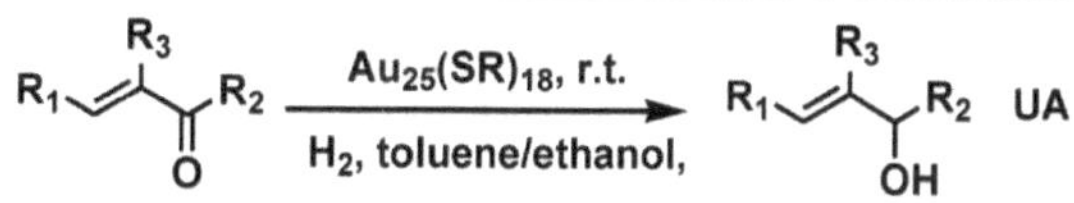

Entry	Substrates	Selectivity for UA	Coversion (%)		
			Au_{25}	Au_{25}/Fe_2O_3	Au_{25}/TiO_2
1	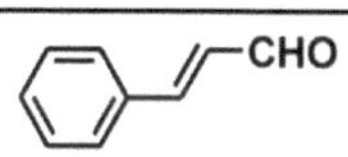	100	38	49	46
2	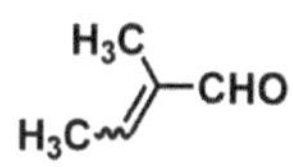	100	43	53	51
3	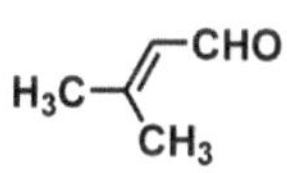	100	44	54	53
4	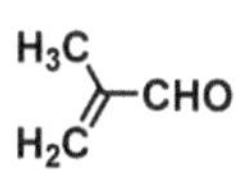	100	39	52	52
5	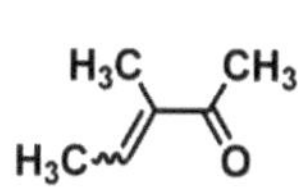	100	29	45	44
6	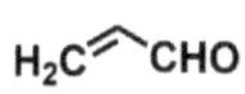	91[a]	46	47	47

Source: Adapted with permission from Ref. [43].

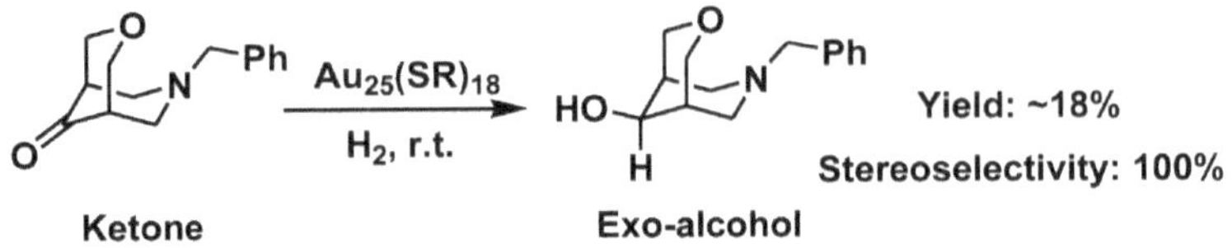

Scheme 5 Stereoselective hydrogenation reaction of bicyclic ketone to exo-alcohol catalyzed by $Au_{25}(SR)_{18}$ nanocluster catalysts.

5.2.3 Selective Hydrogenation of Nitrobenzaldehyde Derivatives

In recent work, Li et al. synthesized thermally robust $Au_{99}(SPh)_{42}$ nanoclusters and investigated catalytic application in chemoselective hydrogenation of nitrobenzaldehyde derivatives [45]. The $Au_{99}(SPh)_{42}$ nanoclusters show absorption peaks at ~920 nm (1.35 eV), 730 nm (1.70 eV), 600 nm (2.07 eV), and 490 nm (2.53 eV), which is in contrast with the conventional gold nanoparticles that exhibit a plasmon resonance band at 520 nm (for spherical particles), thus $Au_{99}(SPh)_{42}$ nanoclusters are semiconducting (band gap ~0.7 eV) [45].

The ceria-supported $Au_{99}(SPh)_{42}$ nanoclusters were utilized as a catalyst for chemoselective hydrogenation of nitrobenzaldehyde to nitrobenzyl alcohol in water using H_2 gas as the hydrogen source [45]. The catalytic hydrogenation reaction was carried out under mild conditions (80 °C for 4 h). Using free (i.e., unsupported) $Au_{99}(SPh)_{42}$ nanoclusters as the catalyst, we obtained an 8.7% conversion of 4-nitrobenzaldehyde with ~100% selectivity for 4-nitrobenzyl alcohol (Table 3, entry 1). The ~100% selective hydrogenation of the aldehyde group came as a surprise, because conventional nanogold catalysts instead give rise to the reduction product of the nitro group.

With supported catalysts on three different oxides (i.e., SiO_2, TiO_2, and CeO_2), $Au_{99}(SPh)_{42}$(1 wt%)/oxide catalysts were evaluated under identical reaction conditions as in the above case of unsupported nanocluster catalyst, except that H_2O was used as the solvent and the reaction was prolonged to 12 h. The catalysts gave much higher activity owing to the improved stability of nanoclusters and the promoting effect of supports (Table 3). Note that the chemoselectivity for product 4-nitrobenzyl alcohol is maintained at ~100% in all the cases. The results suggest that the cluster catalysts are influenced by the acidic properties of the oxide supports. Among the supports, SiO_2 is acidic, while TiO_2 is amphoteric and CeO_2 is basic. Among them, CeO_2 gave rise to the best performance. A range of nitrobenzaldehyde derivatives were tested on the $Au_{99}(SPh)_{42}$/CeO_2 catalyst and high catalytic activity was observed (Table 4). From the substrates, the catalytic activity of the $Au_{99}(SPh)_{42}$/CeO_2 catalyst is considerably influenced by the electrophilic aromatic substitution, instead of the electron-rich/deficient side groups. Finally, the supported catalysts show excellent recyclability.

We further tested the size-dependent catalytic performance of Au_{25}, Au_{36}, Au_{38}, and Au_{99} nanoclusters [45,46], and comparably high activity/selectivity were obtained. Based on the structures of $Au_{25}(SPh)_{18}$ and $Au_{36}(SPh)_{24}$, DFT modeling of the 4-nitrobenzaldehyde

Table 3 Hydrogenation of 4-nitrobenzaldehyde to 4-nitrobenzyl alcohol using the free and supported $Au_{99}(SPh)_{42}$ nanoclusters as catalysts[a]

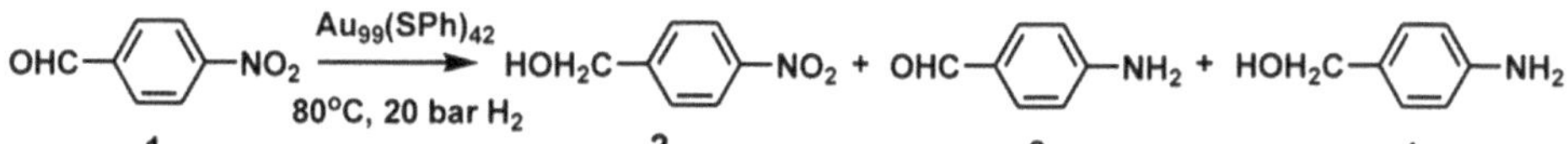

1 2 3 4

Entry	Support	Conv. (%)[b]	Selectivity (%)[b]		
			2	3	4
1[c]	none	8.7	100	n.d.	n.d.
2	SiO_2	27.9	100	n.d.	n.d.
3	TiO_2	69.8	100	n.d.	n.d.
4	CeO_2	93.1	100	n.d.	n.d.
5[d]	CeO_2	92.9	100	n.d.	n.d.
6[e]	CeO_2	93.4	100	n.d.	n.d.

[a]Reaction conditions: $Au_{99}(SPh)_{42}$/oxide (100 mg, one wt% loading) catalysts in 1 mL H_2O, 0.05 mmol 4-nitrobenzaldehyde, 0.1 mmol pyridine, 20 bar H_2, 80 °C, 12 h.

[b]The conversion (Conv.) of 4-nitrobenzaldehyde and selectivity for 4-nitrobenzyl alcohol were determined by NMR analysis. n.d. = not detected.

[c]Reaction conditions: 1 mg free $Au_{99}(SPh)_{42}$ nanoclusters in 1 mL toluene, 0.1 mL H_2O, 0.05 mmol 4-nitrobenzaldehyde, 0.1 mmol pyridine, 20 bar H_2, 80 °C, 4 h.

[d]2nd reuse.

[e]3rd reuse. The conversion and selectivity values are averaged over three measurements (error bar: ~0.5%).

adsorption [46] shows that both the –CHO and –NO_2 groups are adsorbed on the S–Au–S staples of the gold nanocluster (Figure 11). The adsorption of the 4-nitrobenzaldehyde molecule on the different-sized $Au_n(SR)_m$ nanoclusters are moderately strong and similar in strength. The DFT results suggest that the catalytic activity of the $Au_n(SR)_m$ nanoclusters is primarily determined by the surface areas of the Au nanoclusters, consistent with the observed trend of the conversion of 4-nitrobenzaldehyde versus the cluster size. Overall, experiment combined with theory offers the molecular insight into the hydrogenation of 4-nitrobenzaldehyde and the active site structure on gold nanocluster catalysts.

5.3 Catalytic Carbon–Carbon Coupling Reaction

Gold nanoclusters also hold great promise for the carbon–carbon coupling reactions, such as Ullmann-type homo-coupling and Sonogashira cross-coupling reactions. The catalytic application in C–C coupling reaction was due to the good capability of gold nanocluster catalysts in activating the C–I bond of iodobenzene and the terminal C–H bond of phenylacetylene. Below, we illustrate the C–C coupling reactions.

5.3.1 Ullmann-Type Homo-Coupling Reaction

Gold nanoparticles are good catalysts for activating the C–I bond of the iodobenzene in the carbon–carbon homo-coupling process. Li et al. [47] investigated the catalytic activity of the oxide-supported $Au_{25}(SR)_{18}$ catalysts for the Ullmann-type homo-coupling reactions of aryl

Table 4 Chemoselective hydrogenation of a range of substrates with nitro group and aldehyde group using $Au_{99}(SPh)_{42}/CeO_2$ as the catalyst. Reaction conditions: same as noted in Table 3

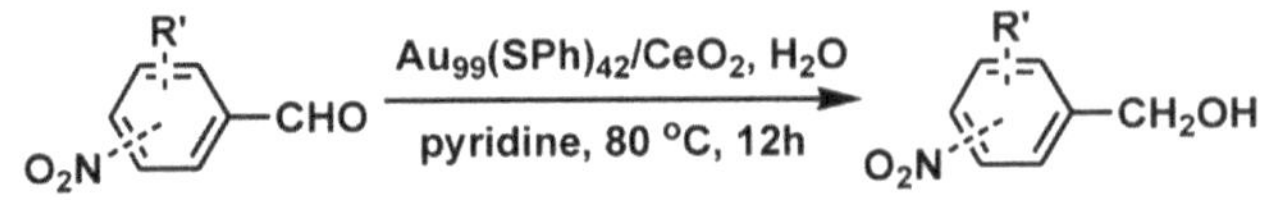

Entry	Substrate	Product	Conv. (%)	Select. (%)
1	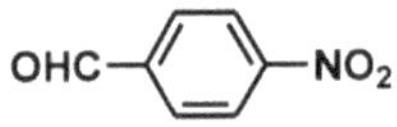	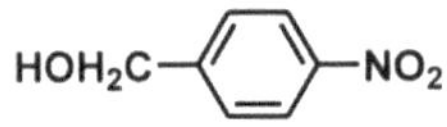	93.1	100
2	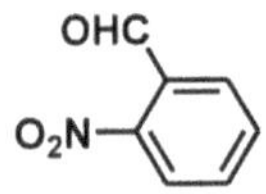	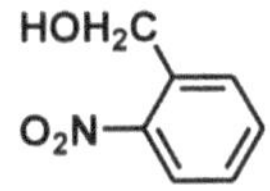	73.2	100
3	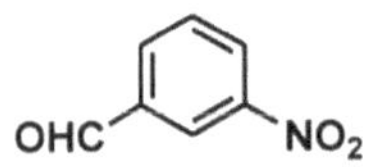	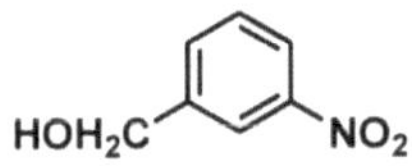	98.5	100
4	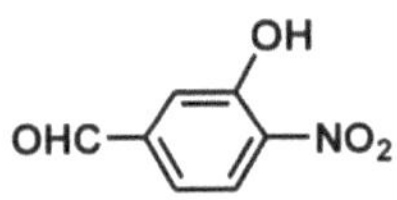	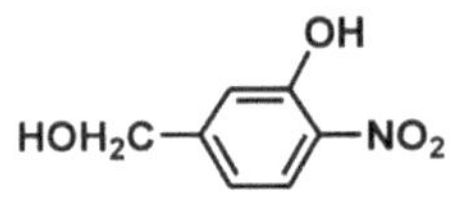	84.1	100
5	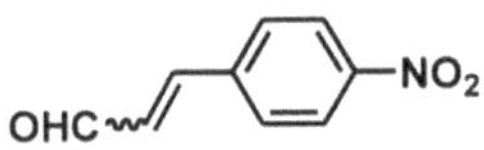	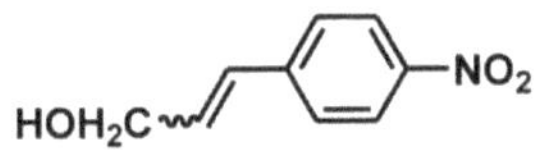	90.0	100
6	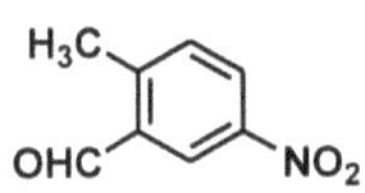	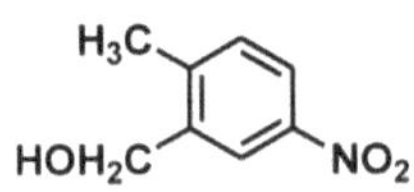	98.9	100
7	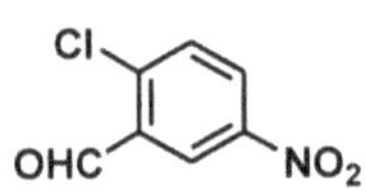	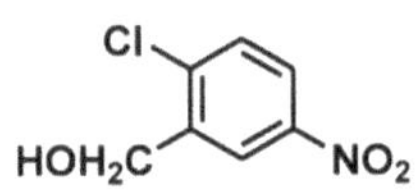	99.5	100

iodides. Optimization showed the suitable conditions for the homocoupling reaction as follows: DMF as solvent and K_2CO_3 as base, at 130°C under N_2 atmosphere. Under such conditions, the $Au_{25}(SR)_{18}/CeO_2$ catalyst gave rise to 99.8% conversion of iodobenzene (Table 5). The application of the $Au_{25}(SR)_{18}$ catalyzed homo-coupling reaction was tested with substrates bearing a range of substituents. The electron-rich substrates performed better than electron-deficient substrates in the reaction. The support effect was also investigated, but unlike the case of hydrogenation, no distinct effect of the oxide supports was observed on the investigated supports (including CeO_2, SiO_2, TiO_2, and Al_2O_3).

5.3.2 Sonogashira Cross-Coupling Reaction

Li et al. [30] expanded the catalytic application of the oxide-supported $Au_{25}(SR)_{18}$ nanocluster catalysts to the Sonogashira cross-coupling reaction between iodobenzene

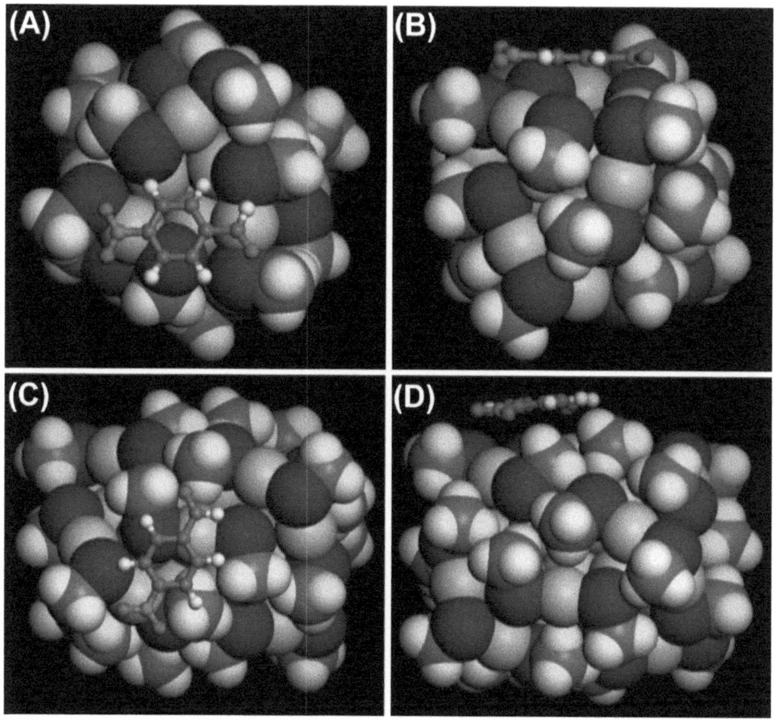

Figure 11 The 4-nitrobenzaldehyde molecule adsorbed on $Au_{25}(SCH_3)_{18}^-$ nanocluster in top view (A) and side view (B), and on $Au_{38}(SCH_3)_{24}$ nanocluster in top view (C) and side view (D). Color code: Au, yellow (light gray in print version); S, blue (black in print version); C, gray (dark gray in print version); H, white (white in print version); O, red (black in print version); N, pink (dark gray in print version).

Table 5 Homo-coupling of aryl iodides using Au_{25}/CeO_2 catalyst

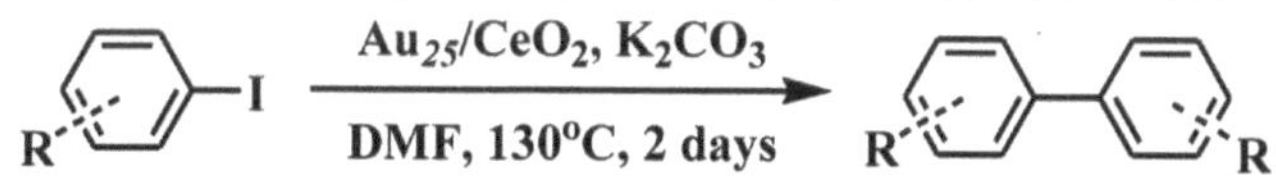

Entry	Substrates	Conversion[%]
1	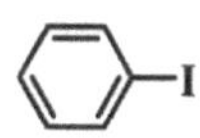	99.8
2	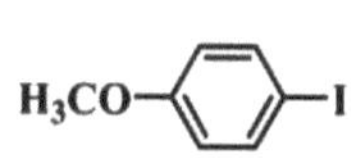	99.5
3	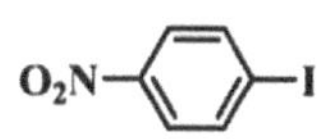	67.5
4	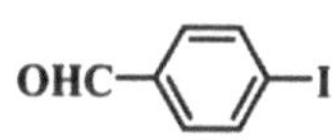	78.2
5	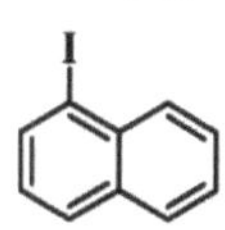	99.7

and phenylacetylene. The supported catalyst was made by impregnation of oxide powders (including CeO_2, TiO_2, MgO, and SiO_2) in a CH_2Cl_2 solution of $Au_{25}(SR)_{18}$ (~1 wt% loading), and then annealed at 150 °C (1 h) in a vacuum oven. The supported gold particle size after the annealing was confirmed by scanning transmission electron microscopy (STEM) to be identical with that of Au_{25}, and thermogravimetric analysis (TGA) confirmed that the protecting thiolate ligand remained on the surface of gold nanoclusters after the annealing. The $Au_{25}(SR)_{18}$/oxide catalysts after thermal treatment was applied in the Sonogashira cross-coupling reaction of iodobenzene and phenylacetylene.

After optimization of the conditions, the reaction was carried out under N_2 atmosphere at 160 °C using DMF as solvent and K_2CO_3 as base. A high conversion of *p*-iodoanisole (up to 96.1%) with excellent selectivity for cross-coupling product (up to 88.1%) was obtained when it was catalyzed by $Au_{25}(SR)_{18}/CeO_2$ catalyst (Table 6, entry 1). The $Au_{25}(SR)_{18}$ nanocluster catalyst performed much better than the larger sized 2–3 nm Au nanoclusters (Table 6, entry 1 vs 5) and CeO_2-supported Au nanoparticles (bare, ~20 nm). Support effects were also investigated, and no distinct effect of the oxide supports was observed (i.e., CeO_2, SiO_2, TiO_2, and MgO). In the recyclability test, the conversion of *p*-iodoanisole exhibited no significant loss, while the selectivity was decreased from 88.1% to 64.5% after five cycles. The drop in selectivity should be caused by the gradual

Table 6 The catalytic performance of $Au_{25}(SR)_{18}$ ($R=CH_2CH_2Ph$) nanoclusters supported on various oxides as catalysts for Sonogashira cross-coupling reaction of p-iodoanisole and phenylacetylene

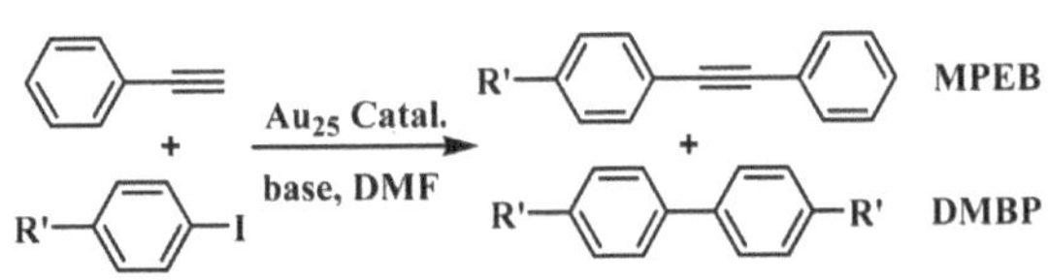

entry	catalyst	conv. (%)	selectivity (%) MPEB	selectivity (%) DMBP
1	$Au_{25}(SR)_{18}/CeO_2$	96.1	88.1	11.9
2	$Au_{25}(SR)_{18}/TiO_2$	92.8	82.9	17.1
3	$Au_{25}(SR)_{18}/SiO_2$	90.8	79.3	20.7
4	$Au_{25}(SR)_{18}/MgO$	93.3	80.6	19.4
5	AuNC 2-3 nm $(SC_6H_{13})/CeO_2$	65.5	57.2	42.8

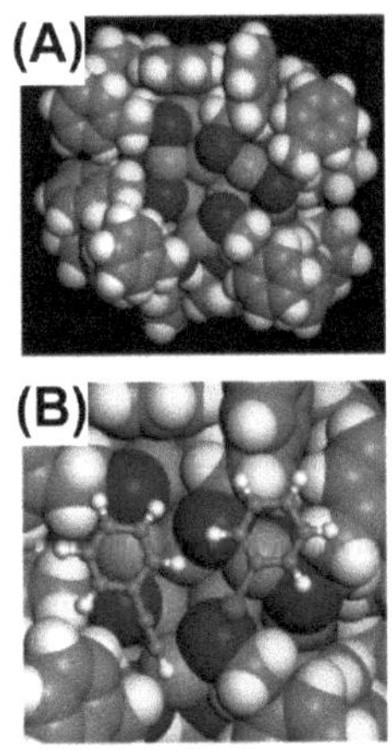

For the DFT-simulated adsorption of reactants: (A) top view of one of the two open facets of $Au_{25}(SR)_{18}$ nanoclusters where three external gold atoms are exposed; the other facet is on the back side. (B) top view of the co-adsorption of phenylacetylene and iodobenzene on the surface of the $Au_{25}(SR)_{18}$ nanoclusters. $R'=OCH_3$. Au, yellow (light gray in print version); S, blue (black in print version); C, gray (dark gray in print version); H, white (white in print version); I, green (light gray in print version).

Source: Adapted with permission from ref. [30]

degradation of nanoclusters in the multiple recycling tests, as larger Au clusters give a much lower selectivity.

Density functional theory (DFT) calculation of the reactant adsorption shows that both reactants (i.e., iodobenzene and phenylacetylene) prefer to adsorb on the open facet with the phenyl ring facing a surface gold atom. A total adsorption energy reaches −0.90 eV when the two reactants co-adsorb on the $Au_{25}(SR)_{18}$ catalyst. The DFT results suggested that the catalytic active sites are associated with the $Au_{25}(SR)_{18}$, which is consistent with the experimental results.

5.4 Catalytic Effect of Single-Atom-Doped Nanoclusters

Doping of the atomically precise gold nanoclusters with a second metal (e.g., Pd and Pt) for catalysis has been reported [48,49]. Tsukuda and coworkers compared Au_{25} and Pd_1Au_{24} nanoclusters supported on multi-walled carbon nanotubes for the aerobic oxidation of benzyl alcohol [48]. They showed that single Pd atom doping significantly improved the catalytic performance of the 25-atom nanocluster catalyst. Herein, we illustrate the catalytic activity of the monoplatinum-doped $Pt_1Au_{24}(SR)_{18}$ nanoclusters for the selective oxidation of styrene using $PhI(OAc)_2$ as the oxidant [49]. Various analyses confirmed that the single platinum atom is located in the center of the core (Figure 12), which implies that the shell (i.e., $Au_{12}(SR)_{18}$) of the $Pt_1Au_{24}(SR)_{18}$ nanocluster is the same as that of $Au_{25}(SR)_{18}$. The catalytic reaction was carried out at 70 °C for 10 h in acetonitrile under N_2 atmosphere. Significantly, $Pt_1Au_{24}(SR)_{18}/TiO_2$ catalyst gave much higher catalytic activity (i.e., 90.8% conversion of styrene) and higher selectivity (89.9%) for benzaldehyde than that of $Au_{25}(SR)_{18}/TiO_2$ catalyst (conversion 58.9% and selectivity 54%) under the identical reaction conditions. These results show that the $PtAu_{24}(SR)_{18}/TiO_2$ catalyst has a remarkable enhancement in the catalytic activity/selectivity for styrene oxidation than the undoped $Au_{25}(SR)_{18}/TiO_2$ catalyst.

In future work, the tunable characteristic of bimetallic nanoclusters [50] will enable improvements in the catalytic activity, selectivity, and stability for specific reaction processes.

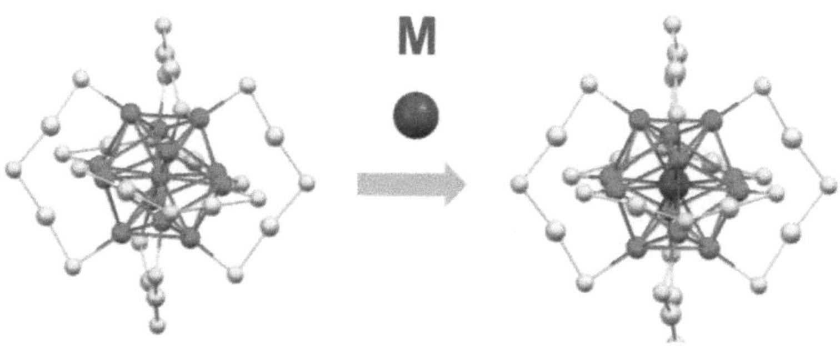

Figure 12 Single atom doping of gold nanocluster (e.g., $Pt_1Au_{24}(SR)_{18}$).

6. SUMMARY

In this chapter, we have discussed the size-focusing methodology for synthesizing atomically precise gold nanoclusters (such as $Au_{25}(SR)_{18}$ and $Au_{38}(SR)_{24}$) and their catalytic application. One of the important properties of metal nanoclusters is the quantum size effect. Gold nanoclusters with size of a few dozen atoms exhibit discrete electronic energy levels, as opposed to the continuous band of conduction electrons in the metallic state. The new electronic properties of nanoclusters, as well as their unique atom-packing structures may generate significant effects in nanocatalysis. The attainment of X-ray structures of nanoclusters permits precise reactivity–structure correlation.

The $Au_n(SR)_m$ nanoclusters have been demonstrated in several major reactions (Scheme 6), such as (1) catalytic oxidation (such as selective oxidation of styrene to benzaldehyde or styrene epoxide, selective oxidation of sulfide to sulfoxide, and CO oxidation to CO_2), (2) catalytic hydrogenation (such as nitrophenol to aminophenol and chemoselective hydrogenation), (3) carbon–carbon coupling reactions (such as Ullmann homo-coupling of iodobenzene and Sonogashira cross-coupling of iodobenzene and phenylacetylene). Other reactions not discussed above in detail include electrochemical catalysis [51], photocatalysis [52–55], and aerobic oxidation of cyclohexane [56].

The demonstrated catalytic reactions strongly indicate that the catalytic activity and selectivity are associated with the unique structure and electronic properties of $Au_n(SR)_m$ nanoclusters. The $Au_n(SR)_m$ nanoclusters are expected to become a promising class of model catalysts, although much work remains to be carried out in future. As a new class of catalysts, $Au_n(SR)_m$ nanoclusters will bridge the gap between organometallic complex-based homogeneous catalysis and nanoparticle-based heterogeneous catalysis, and also bridge the gap between the bulk crystal structure model catalysts and real-world catalysts. These atomically defined gold nanoclusters provide exciting opportunities for achieving fundamental understanding of nanocatalysis at the atomic level, including the insight into the size-dependence and deep understanding of the molecular activation,

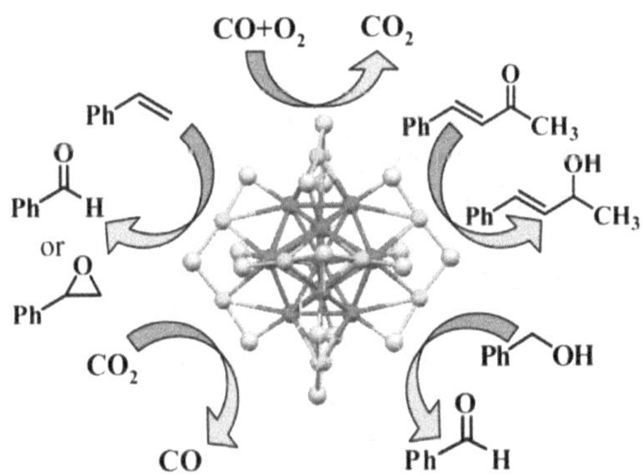

Scheme 6 Nanocluster-catalyzed chemical reactions.

catalytic active sites, and catalytic mechanism by correlation with the structures of gold nanoclusters. Future research on atomically precise nanocluster catalysts will contribute to the fundamental catalysis and the new design of highly selective catalysts for specific chemical processes.

ACKNOWLEDGMENTS

The work was supported by U.S. Department of Energy-Office of Basic Energy Sciences (Grant DE-FG02-12ER16354).

REFERENCES

[1] G.C. Bond, C. Louis, D.T. Thompson, Catalysis by Gold, Imperial College Press, London, 2006.
[2] M. Haruta, S. Tsubota, T. Kobayashi, H. Kageyama, M.J. Genet, B. Delmon, J. Catal. 144 (1993) 175.
[3] R. Jin, Nanotechnol. Rev. 1 (2012) 31.
[4] S. Yamazoe, K. Koyasu, T. Tsukuda, Acc. Chem. Res. 47 (2014) 816.
[5] G.J. Hutchings, C.J. Kiely, Acc. Chem. Res. 46 (2013) 1759.
[6] D. Astruc, F. Lu, J.R. Aranzaes, Angew. Chem. Int. Ed. 44 (2005) 7852.
[7] G. Li, R. Jin, Acc. Chem. Res. 46 (2013) 1749.
[8] R. Jin, Nanoscale 2 (2010) 343.
[9] H. Qian, M. Zhu, Z. Wu, R. Jin, Acc. Chem. Res. 45 (2012) 1470.
[10] R. Jin, H. Qian, Z. Wu, Y. Zhu, M. Zhu, A. Mohanty, N. Garg, J. Phys. Chem. Lett. 1 (2010) 2903.
[11] M. Zhu, C.M. Aikens, F.J. Hollander, G.C. Schatz, R. Jin, J. Am. Chem. Soc. 130 (2008) 5883.
[12] C. Zeng, T. Li, A. Das, N.L. Rosi, R. Jin, J. Am. Chem. Soc. 135 (2013) 10011.
[13] C. Zeng, H. Qian, T. Li, G. Li, N.L. Rosi, B. Yoon, R.N. Barnett, R.L. Whetten, U. Landman, R. Jin, Angew. Chem. Int. Ed. 51 (2012) 13114.
[14] H. Qian, W.T. Eckenhoff, Y. Zhu, T. Pintauer, R. Jin, J. Am. Chem. Soc. 132 (2010) 8280.
[15] H. Qian, R. Jin, Nano Lett. 9 (2009) 4083.
[16] H. Qian, Y. Zhu, R. Jin, Proc. Natl. Acad. Sci. U.S.A. 109 (2012) 696.
[17] M. Zhu, E. Lanni, N. Garg, M.E. Bier, R. Jin, J. Am. Chem. Soc. 130 (2008) 1138.
[18] Z. Wu, J. Suhan, R. Jin, J. Mater. Chem. 19 (2009) 622.
[19] M.A. MacDonald, P. Zhang, H. Qian, R. Jin, J. Phys. Chem. Lett. 1 (2010) 1821.
[20] H. Qian, Y. Zhu, R. Jin, ACS Nano 3 (2009) 3795.
[21] N.K. Chaki, Y. Negishi, H. Tsunoyama, Y. Shichibu, T. Tsukuda, J. Am. Chem. Soc. 130 (2008) 8608.
[22] H. Qian, M. Zhu, U.N. Andersen, R. Jin, J. Phys. Chem. A 113 (2009) 4281.
[23] H. Qian, C. Liu, R. Jin, Sci. China: Chem. 55 (2012) 2359.
[24] Z. Wu, M. MacDonald, J. Chen, P. Zhang, R. Jin, J. Am. Chem. Soc. 133 (2011) 9670.
[25] Z. Wu, C. Gayathri, R. Gil, R. Jin, J. Am. Chem. Soc. 131 (2009) 6535.
[26] C. Zeng, Y. Chen, G. Li, R. Jin, Chem. Mater. 26 (2014) 2635.
[27] H. Qian, R. Jin, Nano Lett. 9 (2009) 4083.
[28] Y. Zhu, H. Qian, R. Jin, Chem. Eur. J. 16 (2010) 11455.
[29] X. Nie, H. Qian, Q. Ge, H. Xu, R. Jin, ACS Nano 6 (2012) 6014.
[30] G. Li, D-e. Jiang, C. Liu, C. Yu, R. Jin, J. Catal. 306 (2013) 177.
[31] M. Zhu, W.T. Eckenhoff, T. Pintauer, R. Jin, J. Phys. Chem. C 112 (2008) 14221.
[32] M. Zhu, C.M. Aikens, M.P. Hendrich, R. Gupta, H. Qian, G.C. Schatz, R. Jin, J. Am. Chem. Soc. 131 (2009) 2490.
[33] Y. Zhu, H. Qian, M. Zhu, R. Jin, Adv. Mater. 22 (2010) 1915.
[34] Y. Liu, H. Tsunoyama, T. Akita, T. Tsukuda, Chem. Commun. 46 (2010) 550.
[35] G. Li, H. Qian, R. Jin, Nanoscale 4 (2012) 6714.
[36] X. Nie, C. Zeng, X. Ma, H. Qian, Q. Ge, H. Xu, R. Jin, Nanoscale 5 (2013) 5912.

[37] G. Ma, A. Binder, M. Chi, C. Liu, R. Jin, D-e. Jiang, J. Fan, S. Dai, Chem. Commun. 48 (2012) 11413.
[38] Z. Wu, D-e. Jiang, A.K.P. Mann, D.R. Mullins, Z.-A. Qiao, L.F. Allard, C. Zeng, R. Jin, S.H. Overbury, J. Am. Chem. Soc. 136 (2014) 6111.
[39] H. Kawasaki, S. Kumar, G. Li, C. Zeng, D. Kauffman, J. Yoshimoto, Y. Iwasaki, R. Jin, Chem. Mater. 26 (2014) 2777.
[40] D-e. Jiang, M.L. Tiago, W. Luo, S. Dai, J. Am. Chem. Soc. 130 (2008) 2777.
[41] H. Yamamoto, H. Yano, H. Kouchi, Y. Obora, R. Arakawa, H. Kawasaki, Nanoscale 4 (2012) 4148.
[42] A. Shivhare, S.J. Ambrose, H. Zhang, R.W. Purves, R.W.J. Scott, Chem. Commun. 49 (2013) 276.
[43] Y. Zhu, H. Qian, B.A. Drake, R. Jin, Angew. Chem. Int. Ed. 49 (2010) 1295.
[44] Y. Zhu, Z. Wu, G.C. Gayathri, H. Qian, R.R. Gil, R. Jin, J. Catal. 271 (2010) 155.
[45] G. Li, C. Zeng, R. Jin, J. Am. Chem. Soc. 136 (2014) 3673.
[46] G. Li, D-e. Jiang, S. Kumar, Y. Chen, R. Jin, ACS Catal. 4 (2014) 2463.
[47] G. Li, C. Liu, Y. Lei, R. Jin, Chem. Commun. 48 (2012) 12005.
[48] S. Xie, H. Tsunoyama, W. Kurashige, Y. Negishi, T. Tsukuda, ACS Catal. 2 (2012) 1519.
[49] H. Qian, D-e. Jiang, G. Li, C. Gayathri, A. Das, R.R. Gil, R. Jin, J. Am. Chem. Soc. 134 (2012) 16159.
[50] R. Jin, K. Nobusada, Nano Res. 7 (2014) 285.
[51] D.R. Kauffman, D. Alfonso, C. Matranga, H. Qian, R. Jin, J. Am. Chem. Soc. 134 (2012) 10237.
[52] M. Lee, P. Amaratunga, J. Kim, D. Lee, J. Phys. Chem. C 114 (2010) 18366.
[53] A. Kogo, N. Sakai, T. Tatsuma, Electrochem. Commun. 12 (2010) 996.
[54] J. Lee, H. Seop Shim, M. Lee, J.K. Song, D. Lee, J. Phys. Chem. Lett. 2 (2011) 2840.
[55] C. Yu, G. Li, S. Kumar, H. Kawasaki, R. Jin, J. Phys. Chem. Lett. 4 (2013) 2847.
[56] Y. Liu, H. Tsunoyama, T. Akita, S. Xie, T. Tsukuda, ACS Catal. 1 (2011) 2.

CHAPTER 9

Catalysis on Single Supported Atoms

Chaitanya K. Narula[1,*], Melanie Moses-DeBusk[2]
[1]Materials Science & Technology Division, Oak Ridge National Laboratory, Oak Ridge, TN, USA; [2]Energy & Transportation Sciences Division, Oak Ridge National Laboratory, Oak Ridge, TN, USA
*Corresponding author: E-mail: narulack@ornl.gov

1. INTRODUCTION

Emission treatment catalysts are some of the most successful examples of the deployment of heterogeneous catalysts [1]. Heterogeneous catalysts are generally metals supported on inexpensive supports. For example, platinum and rhodium supported on alumina and ceria-zirconia are core components of emission treatment catalyst [1]. Fresh heterogeneous catalysts contain a mix of single particles, nanoparticles, and well-defined particles [2]. Extensive work over the years has led to the development of durable and stable supports that retain their surface area and pore morphology over the operating life of the catalyst. The dependence of catalytic performance on metal particle size has also been well known. Advances in electron microscopy have enabled researchers to extend size correlation to nano- and sub-nanometer metal particles supported on metal oxide.

The decrease in metal particle size and concurrent increase in the number of active catalyst sites has enabled reduction in the precious metal component of emission treatment catalysts. The incentive to reduce metal particle size came from the observation that reduction in size lead to improved performance [3–5]. In fact, theoretical and experimental studies have shown that sub-nanometer particles tend to have higher activity than their nanometer size counterparts [6–12]. The smallest sub-nanometer particle is a single atom supported on an oxide substrate. There are only a few well-studied examples in literature, but a comparison of turnover frequency (TOF) data clearly shows that supported single atoms are significantly more reactive than even nanoparticles for CO oxidation (Figure 1, data from Ref. [13]). This is an important discovery because this shows that single atoms and rafts, present in fresh emission treatment catalysts, are not mere spectators but also participate in catalytic process. Furthermore, single supported atoms are isoelectronic with organometallic compounds and one can envision a rich chemistry of single supported atoms at par with organometallic chemistry developing over time.

The increase in activity for single atoms, as compared with nanoparticles, is as significant as the increase in activity for nanoparticles when compared with large particles for Au supported on ZrO_2 [13]. At present, a generalization to all single atom catalyst cannot be made due to the limited number of TOF data available on single supported atom

Catalysis by Materials with Well-Defined Structures
http://dx.doi.org/10.1016/B978-0-12-801217-8.00009-8

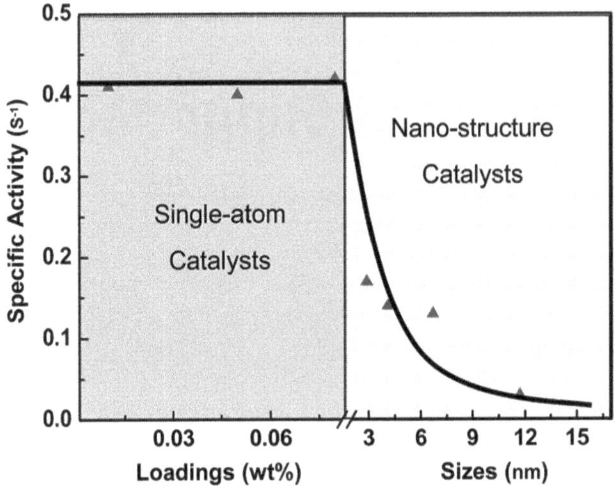

Figure 1 The activity of Au/ZrO_2 as a function of loading (left) and particle size (right). *Reprinted with permission from Acc. Chem. Res. 46 (2013) 1740. Copyright 2013 American Chemical Society.*

systems. But, the reported TOF for single supported atoms are in general higher than supported nanoparticles for comparable systems and reactions.

Prior to advances in electron microscopy, the presence of single atoms in supported catalysts was indirectly inferred by extended X-ray absorption fine structure (EXAFS) data. While availability of aberration corrected electron microscopes have made it possible to directly observe single supported atoms, the challenge has been in the synthesis of mono-disperse single supported atoms. Advances in synthetic methods and characterization tools have made it possible to prepare mono-disperse supported single atoms on various supports [14–37]. Recent reviews provide a summary of catalytic activities on single atom catalysis [38,39]. In the next sections, synthetic routes, characterization, and some example of single supported atoms will be summarized.

2. SYNTHESIS

Synthesis of mono-disperse single supported atoms on oxide supports is quite challenging. For example, synthesis of size selected atomic Pd cations on MgO thin films could be deposited by low-kinetic energy (0.2 eV) deposition, resulting in 0.45×10^{-2} of a monolayer of Pd at 90 K [38]. This method falls in the category of soft landing. Wet chemistry methods have also been moderately successful in the synthesis of single supported atoms. Early approaches involved employing mononuclear organometallic compounds as the source of metal atoms [13,40,41].

A few examples of mono-disperse single atoms supported on metal oxides substrates, prepared by wet chemical methods, have appeared in recent literature.

In general, metal loading is kept low to keep single atoms isolated as they form during decomposition of precursors. For example, single atoms of gold on ZrO_2 substrates are synthesized from $HAuCl_4$ by the deposition precipitation method with Au loading at <0.1% [13]. The co-precipitation method has been employed to prepare Pt_1/FeO_x materials [14] and Ir/FeO_x materials [7,42]. Platinum on an α-alumina single crystal surface was prepared by a drop-wise impregnation method using a CH_2Cl_2 solution of [Pt_4(μ-$CH_3COO)_8$] in an in situ changer under a flow of high purity Ar. The solvent was removed under vacuum at room temperature to obtain mono-disperse single atoms on alumina surface. The sample formed 1.5 nm rods upon hydrogen reduction at 373 K [43]. Stable mono-disperse samples of single atoms on θ-alumina were prepared from impregnation of θ-alumina with aqueous chloroplatinic acid and subsequent pyrolysis at 450 °C (rate 1 °C min^{-1}) after drying [44]. Finally, mesoporous alumina supported palladium catalyst (Pd/meso-Al_2O_3) was prepared by a modified surfactant-templated route through hydrolysis of aluminum sec-butoxide and subsequent aging in the presence of lauric acid. The palladium was loaded by wet impregnation using a tetraammine palladium(II) nitrate solution and subsequent calcination [17].

An important feature of all mono-dispersed single supported atoms, regardless of the wet synthetic route employed, is that the metal loading is quite low and varies from 0.01 to 0.17–0.18 wt%. The high end loading of single Pt atoms at about 0.17–0.18% can be obtained for Pt on FeO_x and θ-alumina. On the other hand, the loading of Ir on FeO_x is only 0.01%. Higher loadings than these limits, in each case, show metal clusters. It is possible that higher loadings in future will be achieved by wet methods by modifying substrates and carefully controlling the deposition process.

3. CHARACTERIZATION

The characterization of single supported atoms requires a combination of techniques. First, samples are examined by advanced electron microscopy techniques that show only single atoms on the support in several areas of the TEM sample. In general, no rafts or particles are observed. Figure 2 shows the atomic resolution high-angle annular-dark field images (HAADF-STEM) taken on a sub-ångström resolution aberration corrected scanning transmission electron microscope [42]. The Ir sample loaded on FeO_x at 0.01 wt% exhibits only individual Ir atoms uniformly dispersed on the surface of FeO_x substrate. Higher loadings revealed single Ir atoms as well as larger rafts.

Since electron microscopy can only examine a limited amount of sample, it is necessary to substantiate data with bulk techniques such as EXAFS and X-ray absorption near edge structure (XANES). The absence of metal–metal interaction is a clear indication of the absence of rafts or particles. In general, the single supported atoms are oxidized, and if metal–metal interaction is not seen even after exposing the sample to reducing

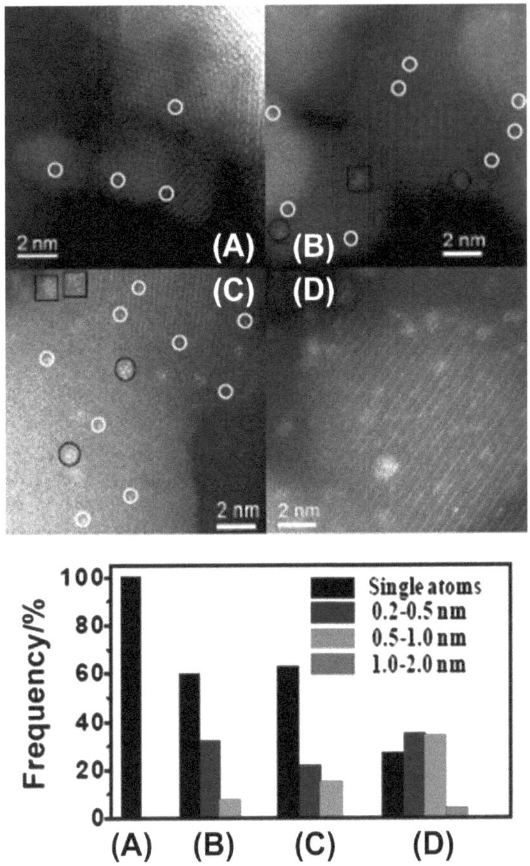

Figure 2 HAADF-STEM images (top) and frequencies of the observed size (bottom) of Ir/FeO_x with Ir wt% loadings (A) 0.01, (B) 0.22, (C) 0.32, and (D) 2.4. *Reprinted with permission from J. Am. Chem. Soc. 135 (2013) 15314. Copyright 2013 American Chemical Society.*

atmosphere at high temperatures, the presence of rafts can be ruled out. Figure 3 shows EXAFS and XANES spectra of Pt/FeO_x samples with 0.17 and 2.5 wt% Pt loading [14]. The low loading samples contain only Pt single atoms while the higher loading sample primarily consists of Pt rafts on iron oxide.

The EXAFS of both samples showed a prominent peak at ~1.7 Å due to Pt–O contribution and a weak peak at ~2.5 Å for Pt–Pt or Pt–Fe contribution. The analysis of the data shows Pt–O at 2.02 Å for 0.17 wt% Pt/FeO_x and 1.98 Å for 2.5 wt% Pt/FeO_x samples. Since the samples were pre-reduced, the Pt–O contribution is primarily from Pt interaction with oxygen of support. In addition, the 2.5 wt% Pt/FeO_x sample shows a Pt–Pt contribution at 2.53 Å. This corresponds to Pt–Pt distances in rafts observed in HAADF-STEM images of this sample. The second shell for 0.17 wt% Pt/FeO_x does not fit with Pt–Pt bonding and has been assigned to Pt–Fe at a distance of 2.88 Å with an average coordination

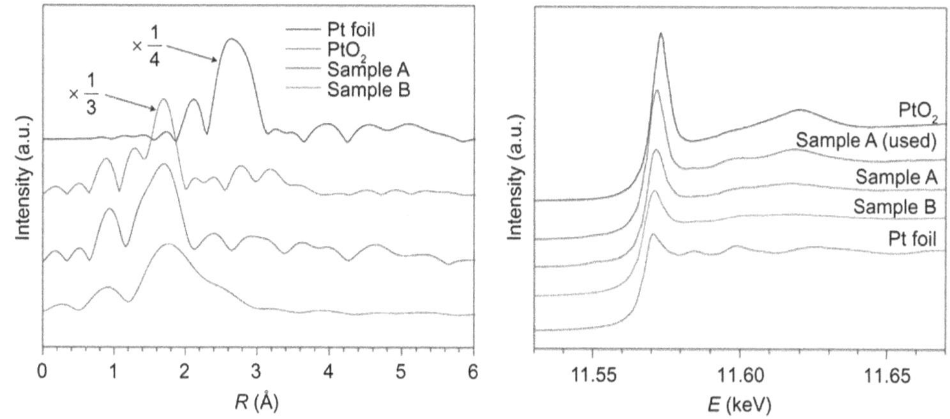

Figure 3 EXAFS (left) and XANES (right) of Pt/FeO_x. Samples A and B contain 0.17 and 2.5 wt% Pt, respectively. *Reprinted with permission from MacMillan Publishers Ltd: Nat. Chem. 3 (2011) 634. Copyright 2011.*

number of 0.9. This interaction originates from Pt bonding with nearest Fe atoms via oxygen bridges.

XANES of 0.17 wt% Pt/FeO_x shows white line intensities between that of Pt foil and PtO_2 suggesting single Pt atoms carry a positive charge. In contrast, the white line intensity of 2.5 wt% Pt/FeO_x is similar to that of Pt foil indicative of Pt–Pt bonds in the sample.

4. DENSITY FUNCTIONAL THEORETICAL MODELING OF SINGLE SUPPORTED ATOMS

Insights into the structure of single supported atoms can also be obtained from density functional theoretical modeling (DFT). Typically, Vienna ab initio software package (VASP) is employed for such calculation. DFT calculations confirm that the lowest energy locations for single Pt atoms on an FeO_x surface is the threefold hollow sites of the O3-terminated surface (Figure 4, top). Here, each Pt atom is coordinated by three oxygen atoms on the surface. A good explanation of the structure of Pt/FeO_x is that the Pt atoms replace Fe on the FeO_x surface [14]. In contrast, the single Pt atoms supported on θ-alumina are described as surface adsorbed atoms. For Pt single atoms on θ-alumina, the calculations were performed in the slab geometry that describes (010) surface of θ-alumina. In the (010) direction, bulk θ-alumina can be described as alternative layers of O and Al atoms in a 2:3 ratio. The slab geometry is defined by a charge neutral supercell of 180 atoms which is two cells wide and four cells deep. The slabs are separated by a 15 Å vacuum to minimize spurious interactions by periodic images (Figure 4, bottom). Lowest energy location for a Pt single atom is between two oxygen atoms in adjacent rows [45]. The Pt–O bond distances are 2.18 and 2.13 Å with an O2-Pt-O1 angle of 140.55°. The absorption energy for Pt atoms is $-75.25\,\mathrm{kcal\,mol^{-1}}$ and there is no magnetic moment associated with the system.

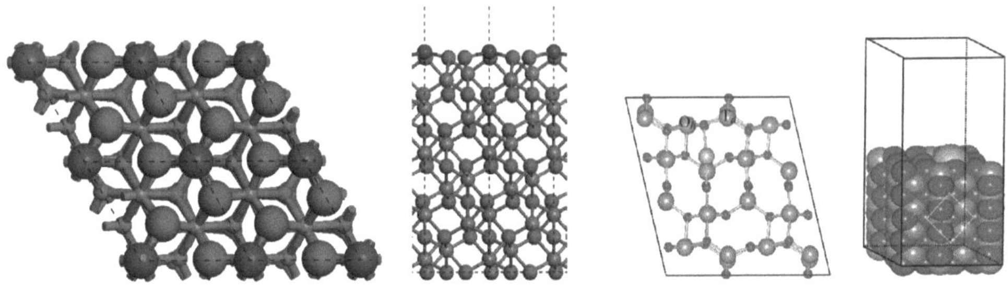

Figure 4 DFT models of Pt/FeO_x (left) and Pt/θ-alumina (right). *Left: Reprinted with permission from MacMillan Publishers Ltd: Nat. Chem. 3 (2011) 634. Copyright 2011. Right: Reprinted with permission from J. Phys. Chem. C 116 (2012) 5628. Copyright 2012 American Chemical Society.*

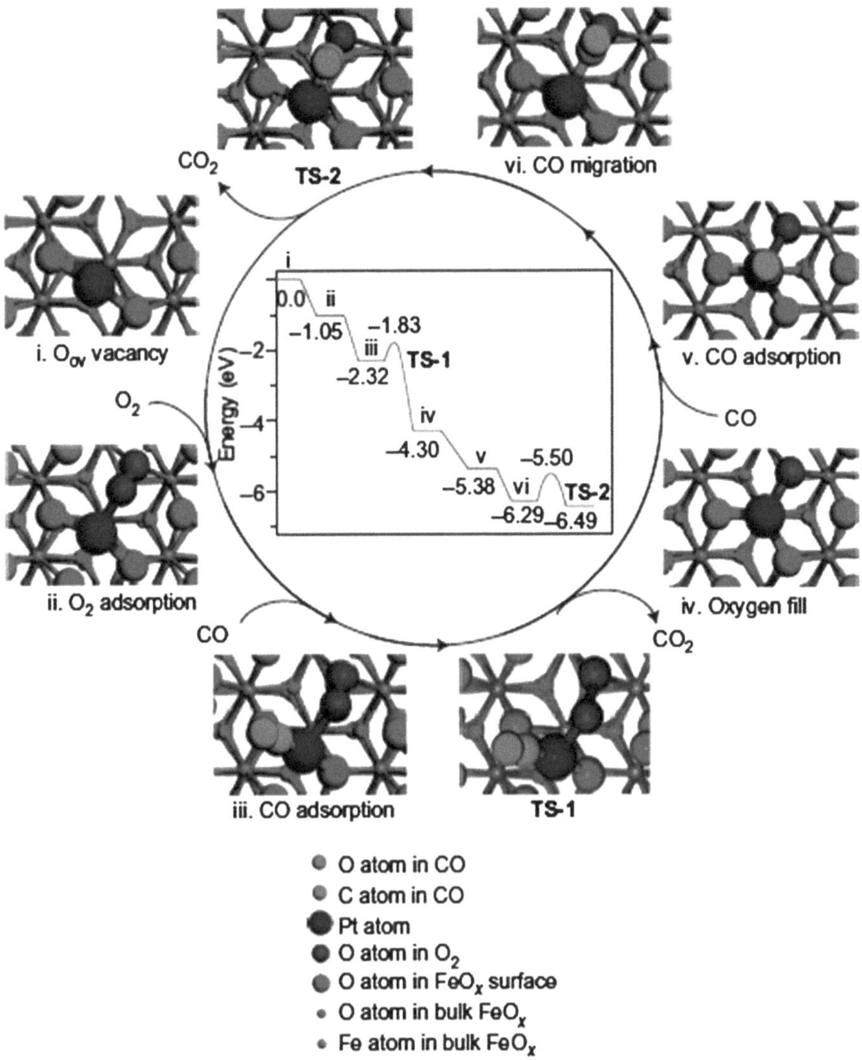

Figure 5 Proposed pathway for CO oxidation on Pt/FeO_x. *Reprinted with permission from MacMillan Publishers Ltd: Nat. Chem. 3 (2011) 634. Copyright 2011.*

The Density of State analysis shows a loss of degeneracy of d orbitals and a shift to higher energy for adsorbed Pt atoms as compared to free Pt atoms. All d orbitals are filled and there is overlap of Pt d orbitals with O 2p orbitals. The evolution of O 2p orbitals from the alumina surface to Pt shows that O 2pz loses electrons to platinum. The strong bonding of Pt with oxygen of alumina can be ascribed to a pronounced tendency of Pt to form s-d hybrids.

5. CATALYTIC REACTION CHEMISTRY

CO oxidation is the most commonly studied example of catalysis for new heterogeneous catalyst systems; therefore, it is not surprising that single supported atoms have also been investigated for CO oxidation. The conventional mechanism of CO oxidation is the Langmuir–Hinshelwood (L–H) mechanism [46–49]. In the L–H mechanism, CO is first adsorbed on a Pt particle surface and then oxygen dissociatively adsorbs on the particles' CO-free Pt atoms. Thus, at least two Pt atoms are necessary for the L–H pathway to enable CO and O to bond before reacting to form CO_2. Oxygen atoms then migrate to CO, and CO_2 is released. CO oxidation indeed occurs on Pt/FeO_x and Pt/θ-Al_2O_3, and the measured TOFs show the single supported atoms to contain significantly more reactive catalyst sites than supported Pt nanoparticles. The 0.17 wt% Pt/FeO_x sample exhibits a TOF (reported as TOF $\times 10^2 s^{-1}$) of 13.6 while TOF of 2.5 wt% Pt/FeO_x is 8.01 [14]. The TOF of 0.18 wt% Pt/θ-Al_2O_3 sample is 18.7 s^{-1} [44].

Since the conventional L–H scheme cannot operate on single supported atoms, modified L–H schemes have been proposed for CO oxidation on Pt/FeO_x and Pt/θ-Al_2O_3 (Figure 5). For CO oxidation on Pt/FeO_x, CO is first adsorbed on a Pt atom. Then, the stoichiometric hematite surfaces near the Pt atoms are reduced enabling the adsorption of an oxygen molecule. The first CO_2 molecule is released through an activation barrier of 0.49 eV leaving a single oxygen on the hematite surface. The second CO molecule is adsorbed on Pt which then migrates to this oxygen atom to form a transition state with a barrier of 0.79 eV. After the release of the second CO_2 molecule, the Pt(0) atom is regenerated. Thus, in this modified L–H mechanism, the substrate participates and takes over the role of location that can adsorb oxygen after CO coverage of supported platinum particles. The experimental support for this mechanism comes from agreement between observed and calculated CO stretching vibration frequency of 2062 cm^{-1}.

For inert substrate, such as alumina, the conventional L–H mechanism or the modified version described in the previous paragraph for Pt/FeO_x cannot work since alumina is not able to act as an oxygen storage material for CO oxidation. The proposed mechanism for CO oxidation on Pt/θ-Al_2O_3 was inspired by organometallic chemistry. Considering that the Pt atom adsorbed on an alumina surface is isoelectronic with $[(Ph_3P)_2Pt]$, it was proposed to be catalytically active [45]. Two intermediates isolated in the reactions of CO and oxygen over $[(Ph_3P)_2Pt]$ have been isolated and characterized by single

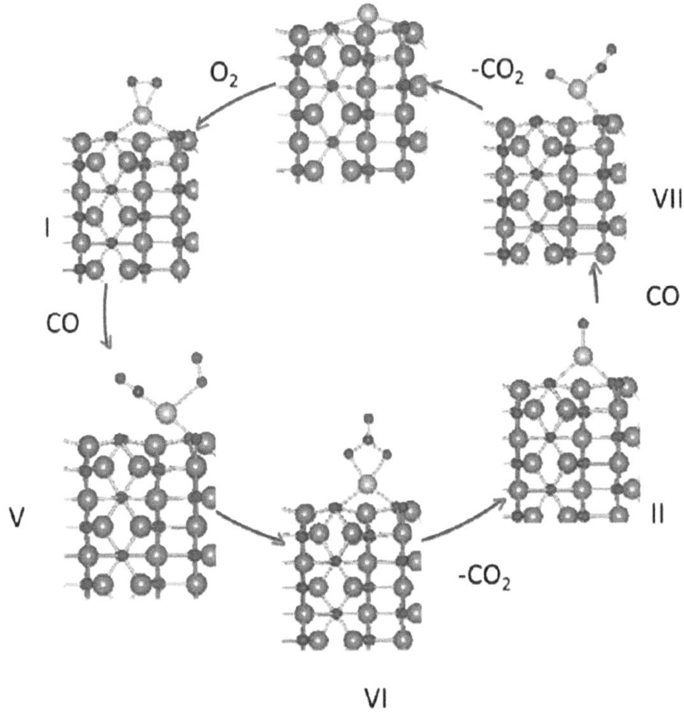

Figure 6 Proposed pathway for CO oxidation of Pt/θ-Al_2O_3. *Reprinted with permission from J. Am. Chem. Soc. 135 (2013) 12634. Copyright 2013 American Chemical Society.*

crystal X-ray diffraction techniques. The structures of these intermediates are $[(Ph_3P)_2Pt(O_2)]$ and $[(Ph_3P)_2Pt(CO_3)]$. Thus, analogues of these structures could be intermediates for CO oxidation over single supported Pt atoms in Pt/θ-Al_2O_3. The proposed mechanism, based on DFT calculation, indeed finds these intermediates to be energetically stable structures [44].

Unlike Pt/FeO_x, the experimental data clearly showed that Pt atoms in Pt/θ-Al_2O_3 do not get reduced to Pt atoms which are only bonded to surface oxygen atoms. Instead, it is an oxidized species even after a reduction step in hydrogen at 150 °C. This led the researchers to first explore various oxygen bonding modes, and the lowest energy mode involved side on bonding of an oxygen molecule. Attempts to optimize terminal oxygen of the oxygen molecule bonded to Pt via a monodentate bond were not successful because the oxygen molecule continued to bend, moving gradually to a bidentate mode. The energetics of various reactions in the proposed mechanism show that all steps in this pathway are exothermic except decomposition of the carbonate species (VI in Figure 6). This suggests that CO oxidation will proceed to carbonate species and will stop after all of the Pt sites have been covered with carbonate. External energy will need to be applied to decompose carbonate and release CO_2. In the case of Pt/θ-Al_2O_3, the IR studies

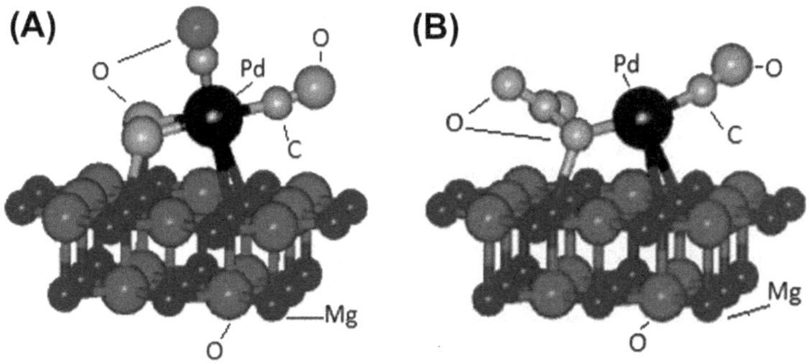

Figure 7 Intermediates in CO oxidation over Pd/MgO. The Pd atom is black, Mg and O ions of the substrate are purple and red, respectively, adsorbed O_2 is yellow, carbons are gray. Oxygen atoms which originate from the two CO molecules are purple and green, respectively. *Reprinted with permission from S. Abett et al. Phys. Rev. Lett. 86 (2001) 5950. Copyright 2013 American Physical Society.*

show that platinum atoms indeed get covered with carbonate at room temperature. If the reaction is carried out at 170 °C, no carbonate intermediate is observed.

$$*\mathrm{Pt} + \mathrm{O_2} \rightarrow *\mathrm{Pt(O_2)}\ (\mathbf{I})\quad -1.8736\mathrm{eV}$$

$$*\mathrm{Pt(O_2)}\ (\mathbf{I}) - \mathrm{O} \rightarrow *\mathrm{Pt(O)}\ (\mathbf{II})\quad 5.6875\mathrm{eV}$$

$$*\mathrm{Pt(O_2)}\ (\mathbf{I}) + \mathrm{CO} \rightarrow *\mathrm{Pt(O_2)(CO)}\ (\mathbf{V})\quad -1.9798\mathrm{eV}$$

$$*\mathrm{Pt(O_2)(CO)}\ (\mathbf{V}) \rightarrow *\mathrm{Pt(CO_3)}\ (\mathbf{VI})\quad -2.6396\mathrm{eV}$$

$$*\mathrm{Pt(CO_3)}\ (\mathbf{VI}) - \mathrm{CO_2} \rightarrow *\mathrm{Pt(O)}\ (\mathbf{II})\quad 2.2204\mathrm{eV}$$

$$*\mathrm{Pt(O)}\ (\mathbf{II}) + \mathrm{CO} \rightarrow *\mathrm{Pt(O)(CO)}\ (\mathbf{VII})\quad -1.8102\mathrm{eV}$$

$$*\mathrm{Pt(O)(CO)}\ (\mathbf{VII}) - \mathrm{CO_2} \rightarrow *\mathrm{Pt}\quad -1.4884\mathrm{eV}$$

CO oxidation on MgO supported single Pd atom has also been proposed to proceed via a modified L–H scheme [40]. In this modification, two CO molecules bond to a Pd atom while an O_2 molecule bonds to both the Pd atom and Mg atoms of the substrate (A in Figure 7). This intermediate then forms a carbonate species which bonds to both the Pd atom and one Mg atom through one oxygen (monodentate carbonate) (B in Figure 7). The elimination of CO_2 leaves a CO molecule on the Pd atom and the remaining oxygen atom from the carbonate after loss of CO_2 fills an oxygen vacancy in the MgO substrate.

The preferential oxidation of CO in hydrogen (PROX) has also been explored over Pt/FeO_x [14]. In general, the TOF shows that Pt/FeO_x is more active than even Pt nanoparticles. The 0.17 wt% Pt/FeO_x sample exhibits a TOF (reported as TOF $\times 10^2\,s^{-1}$) of 21.2 and 31.1 at 27 °C and 80 °C, respectively. In comparison, the corresponding TOF of 2.5 wt% Pt/FeO_x is 9.15 and 16.2 at 27 °C and 80 °C, respectively [14].

The TOF has also been used to compare the activity of zirconia supported Au atoms at 120 °C for 1,3-butadiene hydrogenation reactions. While single Au atoms exhibit TOF of 0.4–0.42 s^{-1}, the nanoparticles (particle size, 2.9–6.7 nm) show only a TOF of 0.13–0.17 s^{-1} [13]. The larger particles of 11.7 nm show a TOF of only 0.03 s^{-1}. The authors conclude that single Au atoms are active sites for selective hydrogenation of 1,3-butadiene.

The turnover number (TON) for single supported Pd atoms on mesoporous alumina for crotyl alcohol oxidation is 1305 while that of 0.8 nm Pd particles over mesoporous alumina is 1908. The increase in particle size to 1.3 nm leads to a dramatic decrease to TON of 163 only [17].

Finally, the TOF of 0.01 wt% Ir/FeO_x for the water gas shift reaction at 300 °C for single supported Ir atoms is 2.31 s^{-1} which drops to 0.89 s^{-1} when Ir loading increases to 0.22 wt% resulting in a dramatic decrease in the population of single Ir atoms [42].

These examples clearly show that catalyst sites populated by single atoms on metal oxide supports are highly active. Conventional mechanisms cannot easily explain the reaction pathways and required investigations into new reaction mechanisms which provided new insights into catalytic pathways.

6. CONCLUSIONS

The results presented in this chapter show that it is possible to produce mono-disperse single supported atoms by wet chemical methods suggesting it would be economical to scale up the process. Although there are only few examples known, it is obvious that single atoms are significantly more reactive per catalyst site than even nanoparticles. A general statement to the high reactivity will become possible only after several more examples of the reactivity of single supported atoms appear in literature. The mechanistic pathways for most of the reactions studied so far (except CO oxidation) have not been explored, and it would not be surprising if the mechanisms follow pathways that are seen homogeneous catalysis. Even for CO oxidation, it became necessary to modify conventional L–H pathway to explain the results.

ACKNOWLEDGMENTS

This research is sponsored by the U.S. Department of Energy, Assistant Secretary for Energy Efficiency and Renewable Energy, Office of Vehicle Technologies, as part of the Propulsion Materials Program, U.S. Department of Energy, under contract DE-AC05-00OR22725 with UT-Battelle, LLC.

REFERENCES

[1] R.F. Heck, R.J. Farrauto, Catalytic Air Pollution Control, Van Nostrand Reinhold, New York, 2002.
[2] C.K. Narula, L.F. Allards, D.A. Blom, M.J. Moses, W. Shelton, W.F. Schneider, Y. Xu, SAE 2007-01-1018.
[3] M. Haruta, T. Kobayashi, H. Sano, N. Yamada, Chem. Lett. (1987) 405–408.
[4] G.A. Somorjai, J.Y. Park, Chem. Soc. Rev. 37 (2008) 2155–2162.

[5] M. Crespo-Quesada, A. Yarulin, M. Jin, Y. Xia, L. Kiwi-Minsker, J. Am. Chem. Soc. 133 (2011) 12787–12794.
[6] A.A. Herzing, C.J. Kiely, A.F. Carley, P. Landon, G.J. Hutchings, Science 321 (2008) 1331–1335.
[7] J. Lin, B. Qiao, J. Liu, Y. Huang, A. Wang, L. Li, W. Zhang, L.F. Allard, X. Wang, T. Zhang, Angew. Chem. Int. Ed. 51 (2012) 2920–2924.
[8] U. Heiz, A. Sanchez, S. Abbet, W.D. Schneider, J. Am. Chem. Soc. 121 (1999) 3214–3217.
[9] Y. Lei, F. Mehmood, S. Lee, J. Greeley, B. Lee, S. Seifert, R.E. Winans, J.W. Elam, R.J. Meyer, P.C. Redfern, D. Teschner, R. Schlogl, M.J. Pellin, L.A. Curtiss, S. Vajda, Science 328 (2010) 224–228.
[10] B. Qiao, L. Liu, J. Zhang, Y. Deng, J. Catal. 261 (2009) 241–244.
[11] M. Turner, V.B. Golovko, O.P.H. Vaughan, P. Abdulkin, A. Berenguer-Murcia, M.S. Tikhov, B.F.G. Johnson, R.M. Lambert, Nature 454 (2008) 981–983.
[12] W.E. Kaden, T. Wu, W.A. Kunkel, S.L. Anderson, Science 326 (2009) 826–829.
[13] X. Zhang, H. Shi, B.Q. Xu, Angew. Chem. Int. Ed. 44 (2005) 7132–7135.
[14] B. Qiao, A. Wang, X. Yang, L.F. Allard, Z. Jiang, Y. Cui, J. Liu, J. Li, T. Zhang, Nat. Chem. 3 (2011) 634–641.
[15] S. Abbet, A. Snaches, U. Heiz, W.D. Schneider, A.M. Ferrari, G. Pacchioni, N. Rosch, J. Am. Chem. Soc. 122 (2000) 3453–3457.
[16] J. Lu, C. Aydin, N.D. Browning, B.C. Gates, Angew. Chem. Int. Ed. 51 (2012) 5842–5846.
[17] S.F. Hackett, R.M. Brydson, M.H. Gass, I. Harvey, A.D. Newman, K. Wilson, A.F. Lee, Angew. Chem. Int. Ed. 46 (2007) 8593–8596.
[18] Y.-T. Kim, K. Ohshima, K. Higashimine, T. Uruga, M. Takata, H. Suematsu, T. Mitani, Angew. Chem. Int. Ed. 118 (2006) 421–425.
[19] S. Wang, A.Y. Borisevich, S.N. Rashkeev, M.V. Glazoff, K. Sohlberg, S.J. Pennycook, S.T. Pantelides, Nat. Mater. 3 (2004) 143–146.
[20] P.E. Batson, Microsc. Microanal. 14 (2008) 89–97.
[21] J.H. Kwak, J. Hu, D. Mei, C.W. Yi, H. Kim do, C.H. Peden, L.F. Allard, J. Szanyi, Science 325 (2009) 1670–1673.
[22] Q. Fu, H. Saltsburg, M. Flytzani-Stephanopoulos, M. Science 301 (2003) 935–938.
[23] M. Yang, L.F. Allard, M. Flytzani-Stephanopoulos, J. Am. Chem. Soc. M. 135 (2013) 3768–3771.
[24] H. Zhang, T. Watanabe, M. Okumura, M. Haruta, N. Toshima, Nat. Mater. 11 (2012) 49–52.
[25] Z. Novotny, G. Argentero, Z. Wang, M. Schmid, U. Diebold, G. Parkinson, Phys. Rev. Lett. 108 (2012) 216103.
[26] A. Hashimoto, M. Takeguchi, J. Electron Microsc. 61 (2012) 409–413.
[27] E. Bayram, J. Lu, C. Aydin, A. Uzun, N.D. Browning, B.C. Gates, R.G. Finke, ACS Catal. 2 (2012) 1947–1957.
[28] S. Sun, G. Zhang, N. Gauquelin, N. Chen, J. Zhou, S. Yang, W. Chen, X. Meng, D. Geng, M.N. Banis, R. Li, S. Ye, S. Knights, G.A. Botton, T.K. Sham, X. Sun, Sci. Rep. 3 (2013) 1775.
[29] L. Wang, S. Zhang, Y. Zhu, A. Patlolla, J. Shan, H. Yoshida, S. Takeda, A.I. Frenkel, F. Tao, ACS Catal. 3 (2013) 1011–1019.
[30] Y. Zhai, D. Pierre, R. Si, W. Deng, P. Ferrin, A.U. Nilekar, G. Peng, J.A. Herron, D.C. Bell, H. Saltsburg, M. Mavrikakis, M. Flytzani-Stephanopoulos, Science 329 (2010) 1633–1636.
[31] C. Zhang, F. Liu, Y. Zhai, H. Ariga, N. Yi, Y. Liu, K. Asakura, M. Flytzani-Stephanopoulos, H. He, Angew. Chem. Int. Ed. 51 (2012) 9628–9632.
[32] G. Wang, J. Su, Y. Gong, M. Zhou, J. Li, Angew. Chem. Int. Ed. 49 (2010) 1302–1305.
[33] Z. Huang, X. Gu, Q. Cao, P. Hu, J. Hao, J. Li, X. Tang, Angew. Chem. Int. Ed. 51 (2012) 4198–4203.
[34] C.-T. Wu, K.M.K. Yu, F. Liao, N. Young, P. Nellist, A. Dent, A. Kroner, S.C.E. Tsang, Nat. Commun. 3 (2012) 1050.
[35] W. Tang, Z. Hu, M. Wang, G.D. Stucky, H. Metiu, E.W. McFarland, J. Catal. 273 (2010) 125–137.
[36] G. Kyriakou, M.B. Boucher, A.D. Jewell, E.A. Lewis, T.J. Lawton, A.E. Baber, H.L. Tierney, M. Flytzani-Stephanopoulos, E.C. Sykes, Science 335 (2012) 1209–1212.
[37] J.S. Jirkovsky, I. Panas, E. Ahlberg, M. Halasa, S. Romani, D.J. Schiffrin, J. Am. Chem. Soc. 133 (2011) 19432–19441.
[38] X.-F. Yang, A. Wang, B. Qiao, J. Li, J. Liu, T. Zhang, T. Acc, Chem. Res. 46 (2013) 1740–1748.
[39] M. Flytzani-Stephanopolos, B.C. Gates, Annu. Rev. Chem. Biomol. Eng. 3 (2012) 545–574.

[40] S. Abbet, U. Heiz, H. Hakkinen, U. Landman, Phys. Rev. Lett. 86 (2001) 5950.
[41] J.M. Thomas, R. Raja, D.W. Lewis, Angew. Chem. Int. Ed. 44 (2005) 6456.
[42] J. Lin, A. Wang, B. Qiao, X. Yang, X. Wang, J. Liang, J. Li, J. Liu, T. Zhang, J. Am. Chem. Soc. 135 (2013) 15314.
[43] K. Asakura, W.J. Chun, M. Shirai, K. Tomishige, Y. Iwasawa, J. Phys. Chem. B. 101 (1997) 5549.
[44] M.M. DeBusk, M. Yoon, L.F. Allard, D.R. Mullins, Z. Wu, Z.X. Yang, G. Veith, G.M. Stocks, C.K. Narula, J. Am. Chem. Soc. 135 (2013) 12634.
[45] C.K. Narula, G.M. Stocks, J. Phys. Chem. C 116 (2012) 5628–5636.
[46] I. Langmuir, Trans. Faraday Soc. 17 (1922) 621–654.
[47] A. Alavi, P. Hu, T. Deutsch, P.L. Silvestrelli, J. Hutter, Phys. Rev. Lett. 80 (1998) 3650–3653.
[48] A. Eichler, J. Hafner, Phys. Rev. B 59 (1999) 5960–5967.
[49] P. Sadeghi, K. Dunphy, C. Punckt, H.H. Rotermund, J. Phys. Chem. C 116 (2012) 4686–4691.

CHAPTER 10

Tailoring Nanoparticle Electrocatalysts for Proton Exchange Membrane Fuel Cells

Huiyuan Zhu, Yu-Xi Huang, Sen Zhang, Shouheng Sun*
Department of Chemistry, Brown University, Providence, Rhode Island, USA
*Corresponding author: E-mail: ssun@brown.edu

1. INTRODUCTION

A fuel cell is an energy conversion device that is designed to convert chemical energy into electric energy. It contains two electrodes, anode and cathode, and an electrolyte separator that is sandwiched by these two electrodes. Fuel is oxidized at anode, releasing electrons that travel along an external circuit to power an electronic device. These electrons are "pulled" and "captured" by an electronegative component, often oxygen, O_2, at cathode. In this process, the anode and the cathode must be properly separated to prevent direct electron transfer from anode to cathode across the separator, i.e., the separator must be an electron insulator, but it should be an excellent ion conductor. Depending on the fuel used and ions transported, various forms of fuel cells have been designed and tested, among which a proton exchange membrane fuel cell (PEMFC) attracts particular attention due to its low operation temperature that is suitable for operation in ambient conditions, high energy density stored in fuel, high conversion efficiency with much less thermal energy loss, and low environmental impact [1–3]. Figure 1 illustrates the schematics of the PEMFC with hydrogen (H_2) as the fuel and O_2 as an electron-sink [4]. When H_2 is oxidized, electrons and protons, H^+ are produced. While electrons are drawn to the external circuit by the electronegative O_2 at the cathode in a process known as oxygen reduction reaction (ORR), the H^+ migrates from anode to cathode from inside through the separator to neutralize the reduced O_2 to form water, H_2O. To facilitate H^+ conduction, the separator is designed to contain an acidic electrolyte and an acidic polymer-membrane, Nafion, a copolymer of tetrafluoroethylene and perfluoro-2-(fluorosulfonylethoxy) propylvinylether [4–6]. Therefore, the H_2-based PEMFC is the device to convert the energy stored in H_2 into electric energy via chemical reaction, $2H_2 + O_2 \rightarrow 2H_2O$, with H_2O as the only product. To avoid the problematic issues associated with H_2 production, transportation, and storage, other organic fuels, such as formic acid, methanol, or even ethanol, have been explored as alternative fuels to be used in PEMFCs. Compared with H_2, these organic fuels are heavier and less "cleaner" (they generate not only H_2O but also CO_2 upon their complete oxidation), but they are much easier to produce, transport, and store [7–8].

Catalysis by Materials with Well-Defined Structures
http://dx.doi.org/10.1016/B978-0-12-801217-8.00010-4

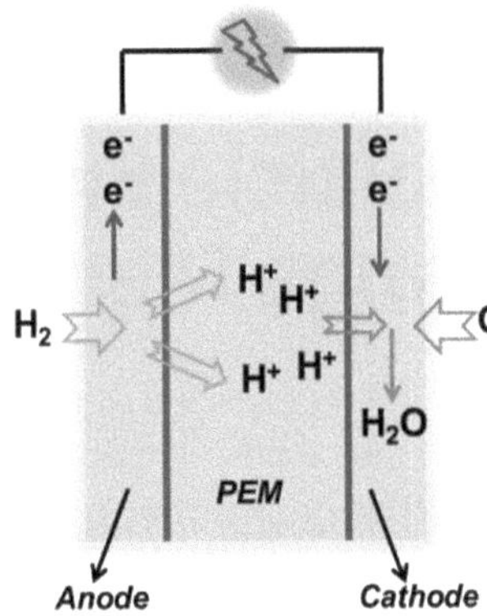

Figure 1 Schematic illustration of a typical PEM fuel cell with H_2 as fuel.

Despite the promising perspectives shown as an alternative energy source and the great efforts devoted to the device developments, the PEMFC technology is still considered too expensive for large-scale commercial applications. A major limitation is in its dependence on platinum (Pt), an expensive, yet not easily available, noble metal, for normal fuel cell operations [5,9]. For fuel to release electrons (be oxidized) at anode and for these electrons to be captured easily by O_2 at cathode in ambient conditions, Pt catalysts must be present at both electrodes to lower the reaction overpotentials and to maximize voltage output. However, Pt is neither cost-efficient nor highly durable (prone to be oxidized, dissolved, agglomerated, or poisoned) to use in the fuel cell reaction conditions [3,10,11]. Consequently, recent research and development efforts have been dedicated heavily to minimize the use of Pt by controlling Pt NP size and shape, by alloying Pt with other early transition metals, and/or by creating core/shell morphologies to obtain the optimal catalysis. Ideally, Pt catalysts can be replaced totally by those based on earth-abundant non-noble elements. To achieve this goal requires both fundamental understanding of fuel cell reactions and easy production of NP catalysts that are small in size and uniform in composition and shape.

Recent advances in synthetic methodology have made the preparation of monodisperse NPs and NP catalysts possible. Specifically, solution phase reactions have been well-developed and demonstrated to be successful in producing sub-10 nm monodisperse NPs [12–15]. The key to the preparation of monodisperse NPs is in controlling two processes during the reaction: atom nucleation and nuclei growth into NPs. The nucleation process contains two thermodynamically competing routes in packing discrete atoms into ordered clusters, which is thermodynamically unfavored due to the entropy decrease, and the formation of chemical bonds to stabilize the atoms, which is thermodynamically favored. By controlling reaction conditions, nuclei can be formed. The growth process is then spontaneous due to the reduction of free energy. However, this spontaneous growth will not stop till NP grows into bulk-like aggregates and precipitate out from the reaction solution. To prevent this uncontrolled NP growth, surfactant is usually added in the reaction system. Due to the surface passivation by the surfactant, the NPs can be stabilized and separated as a product. By tuning the NP's nucleation/growth condition and the interaction between the NP and the surfactant, the precise control over NP's physical parameters such as size, shape, composition, and structure has been achieved. With high surface area and tunable surface profile, these NPs can be used as ideal catalysts to study catalytic reactions and to obtain structure-catalysis correlation and optimal catalytic performance [12–15]. In this chapter, we demonstrate how these well-defined NPs are designed and chemically synthesized for both the fundamental understanding of catalytic processes and the fabrication of advanced electrocatalyst for PEMFCs.

2. NP CATALYSTS FOR ORR

2.1 ORR Overview

When electrons released from fuel reach the cathode, they are captured by O_2. In this process, O_2 is reduced and the reduced species further react with H^+, generated from fuel oxidation and "swimming" through the electrolyte separator to the cathode, to form H_2O. The process is referred to as ORR. It is a common chemical reaction at cathode to accept electrons in a PEMFC and many other types of fuel cell devices. For O_2 to be reduced completely, four electrons are required to break the O=O double bond with additional energy needed to reverse the triplet spin in O_2 to singlet in H_2O [16]. This complicates the reduction pathways and makes the ORR process kinetically more sluggish than the H_2 oxidation one at the anode [17]. As a result, ORR contributes in large part to electrochemical overpotentials observed in the PEMFC, especially at high current density. Figure 2 illustrates a typical voltage loss in the state-of-the-art H_2 PEMFC under automotive operation conditions, where the cell potential is decreased by the overpotentials of ORR (η_{ORR}), hydrogen oxidation (η_{HOR}), electronic resistance ($i{\cdot}R_{\Omega}$), proton conduction resistance ($i{\cdot}R_{H+}$), and gas diffusion ($\eta_{tx,\,gas}$) [17]. At a current density of 1.5 A/cm^2, which is required for many practical applications, the overpotential arisen from ORR is responsible for over 70% voltage loss of this H_2 PEMFC [17].

ORR in an acid electrolyte can undergo via either a direct 4-electron pathway to form H_2O (1) or two 2-electron pathways with H_2O_2 produced as an intermediate (2a and 2b) [18]. The two-electron pathways should be avoided because in the presence of a trace amount of metal ions, H_2O_2 can be decomposed, generating hydroxyl and hydroperoxyl radicals (OH• & OOH•) that is strongly oxidative and can cause severe degradation of proton exchange membrane [19]. ORR on the Pt catalyst surface is mainly a 4-electron process; however, its exact mechanism is still elusive with both associative (proton attacks O_2 first) and dissociative (O_2 dissociates first) pathways been proposed [18].

$$O_2 + 4H^+ + 4e^- \rightarrow 2H_2O \tag{1}$$

$$O_2 + 2H^+ + 2e^- \rightarrow H_2O_2 \tag{2a}$$

$$H_2O_2 + 2H^+ + 2e^- \rightarrow 2H_2O \tag{2b}$$

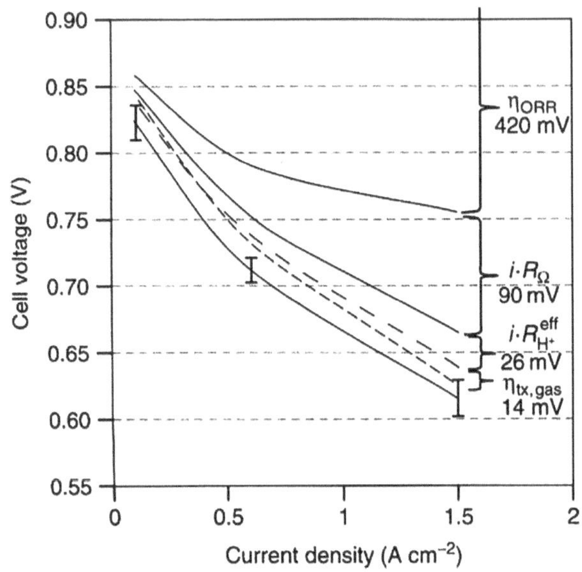

Figure 2 Voltage losses in the state-of-the-art H_2-air PEMFC full active area shortstack (~20 cells) operated under representative automotive conditions. *Reprinted from Ref. [17] with permission by John Wiley & Sons, Ltd.*

2.2 Pt NP Shape Effect

The shape control achieved in Pt NP synthesis allows the detailed study of ORR on different crystal facets. The face-centered cubic (*fcc*) Pt NPs prepared from the solution phase reaction often adopt polyhedral shape that is enclosed by a mixture of low-index crystallographic planes of {111} and {100} [20,21]. By controlling molecule binding to the crystal facets, the Pt NP growth direction can be tailored, leading to NPs with controlled shapes. This shape-controlled synthesis has been demonstrated in the preparation of tetrahedral, truncated octahedral, and cubic Pt NPs via the reduction of Pt acetylacetonate, $Pt(acac)_2$, in the presence of polyvinylpyrrolidone (PVP) or sodium polyacrylate [22,23]. The shape control can also be achieved by controlling the nucleation step at different reaction temperatures. For example, when trace amount of $Fe(CO)_5$ is present in the solution phase reduction of $Pt(acac)_2$, polyhedral Pt nuclei are formed easily. CO produced by the thermal decomposition of $Fe(CO)_5$ binds strongly on the Pt {100} plane and allows the faster growth along with <111> direction. When $Fe(CO)_5$ is added at 120 °C, the CO-mediated growth yields 7 nm cubic Pt NPs. However, if $Fe(CO)_5$ is added at a higher temperature at 160 °C or at 180 °C, 5 nm truncated cubes or 3 nm polyhedral Pt NPs are obtained. Here, the high temperature addition of $Fe(CO)_5$ causes a fast nucleation/growth of NPs and generate a thermodynamically more favored polyhedral morphology [24,25]. The cubic Pt NPs are more active ORR catalyst than the polyhedral ones in 0.5 M H_2SO_4, as shown in Figure 3. At the half-wave potential of ORR polarization curve from the polyhedral Pt-catalyzed ORR, the

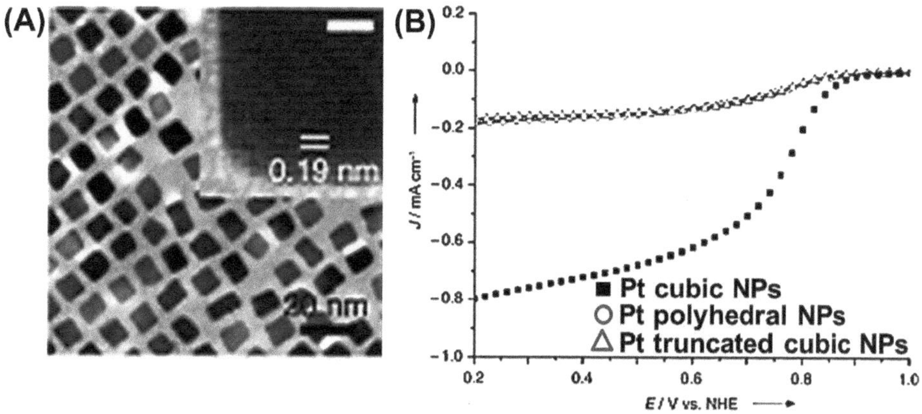

Figure 3 (A) TEM image of the 7 nm Pt nanocubes. (B) ORR polarization curves of the polyhedral, truncated cubic, and cubic Pt NPs in 0.5 M H_2SO_4. *Reprinted from Ref. [25] with permission by Wiley.*

specific activity of the cubic Pt NPs is 4 times higher than the polyhedral ones. The results support the conclusion obtained from the thin film studies that bisulfate anion SO_4^{2-} binds to {111} plane slightly more strongly than to {100}, making the {111} plane less active for O_2 adsorption and activation [26]. This trend is reversed in the $HClO_4$ electrolyte, where ClO_4^- anions are only weakly adsorbed on {111} compared to {100}. As a result, the polyhedral Pt NPs are more active for ORR catalysis in the $HClO_4$ solution [25].

NPs with exposure of high-index facets are also studied. Comparing with low-index facets, high-index facets contain more unsaturated atomic steps, terraces, kinks, and edges, which can potentially enhance the catalytic activity. Pt concave nanocubes are reported with exposure of high-index planes ({730}, {210}, and {520} (Figure 4(A)) [27]. In the synthesis, mixing of K_2PtCl_4 and $Na_2H_2P_2O_7$ yields Pt pynophosphate complex, which is critical to form Pt concave nanocubes. Slow addition of the precursor ensures the kinetic favorable growth instead of thermodynamic favorable growth of Pt nanocubes. Br^- ions also play an important role to block the growth along <100> axis. The as-synthesized concave Pt nanocubes are enclosed by {510}, {720}, and {830} facets and exhibit a much enhanced ORR activity compared with NPs enclosed by low-index facets (Figure 4(B) and 4(C)). It is worth noting that although these NPs with high-index facets are catalytically active, they are plagued with morphology evolution and activity degradation during fuel cell operations. The unsaturated atomic steps, terraces, edge, and corner sites are active for crystal growth and dissolution.

2.3 PtM Alloy Effect

To enhance ORR activity it is required to improve adsorptive activation of O_2 and desorption of the oxygenated intermediates, such as −OH and −OOH, from the Pt surface [28]. An exquisite balance between these two processes can give the "volcano"

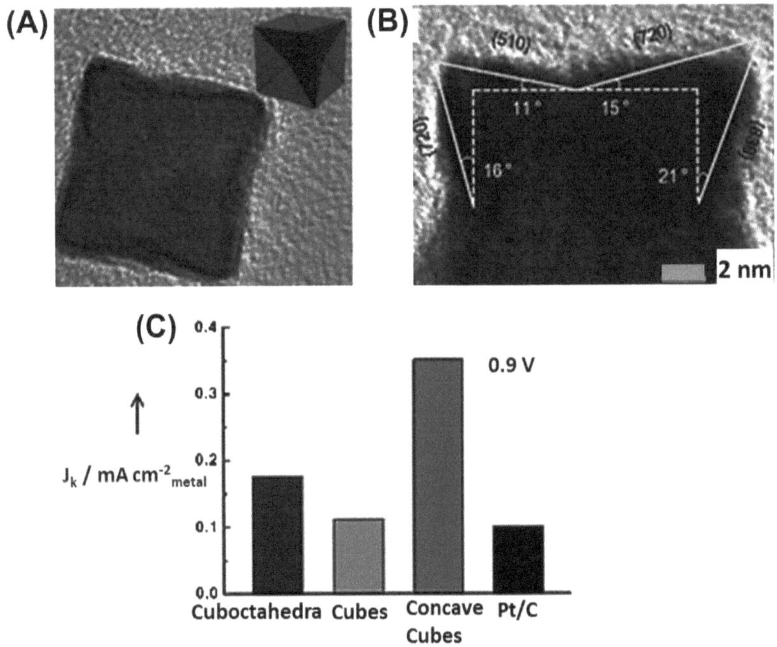

Figure 4 (A) TEM and (B) HRTEM images of Pt concave nanocubes. (C) Specific activities of these catalysts at 0.9 V versus reversible hydrogen electrode (RHE). The metal loading on the glassy carbon electrode was 15.3 μg cm^{-2}. *Reprinted from Ref. [27] with permission by Wiley.*

type activity profile and reach an optimal level on ORR catalysis. These two processes are strongly affected by Pt surface energetics. Recent density function theory (DFT) calculation indicates Pt (111) surface has a Pt-O binding energy 0.2 eV stronger than the energy level required for optimal ORR catalysis, making the desorption of oxygenated intermediates from the Pt surface more difficult during the reaction [28–30]. Alloying Pt with 3d transition metals can downshift the Pt d-band center and decreases the Pt-O binding energy [30]. This alloy-induced electronic effect is strongly dependent on the alloy composition. In the study of single crystalline thin film of Pt_3M alloys (M = Ti, V, Fe, Co, Ni), the proper downshift of the d-band center leads to the less coverage of oxygenated species on the surface of the alloy and the higher active surface accessibility. As a result, Pt_3Co reaches the best energy level and the highest ORR activity (Figure 5(A)) [31]. This alloy composition effect is also observed in the studies of PtM (M = Fe, Ni, Co) electrodes and 5 nm Pt_3M NPs (Figure 5(B) and (C)) [31–35].

Similar to Pt NPs, the alloy MPt NPs also have a strong shape effect on ORR. Studies on single-crystalline thin film of Pt_3Ni has demonstrated that the alloy (111) has a downshift d-band center and thus a much lower OH_{ad} surface coverage than Pt (111), leading to a 10-fold increase in activity than Pt (111) [36]. Furthermore, Pt_3Ni (111) is more active

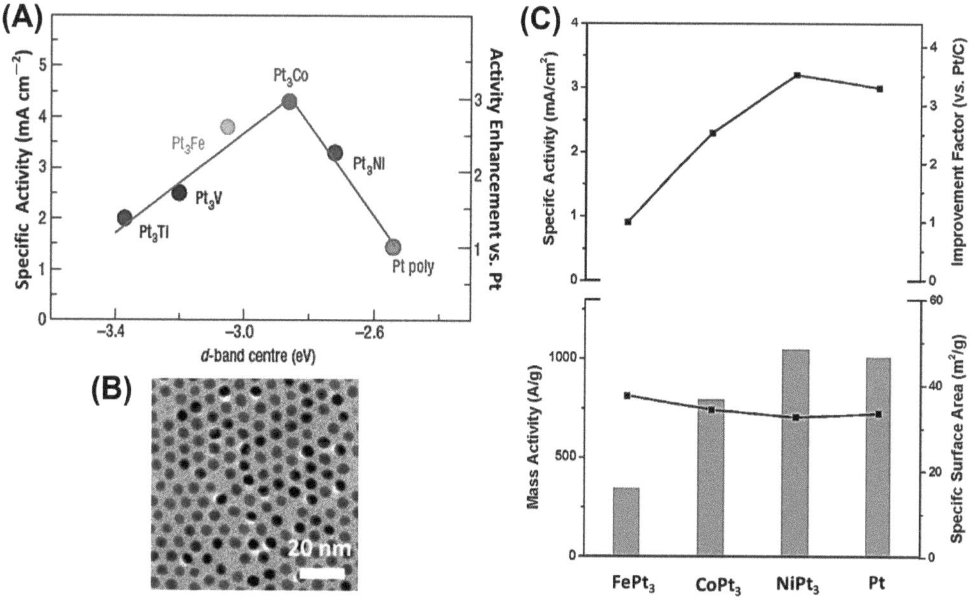

Figure 5 (A) Relationships between experimentally measured specific activity for the ORR on Pt_3M single crystalline surfaces in 0.1 M $HClO_4$ versus the d-band center position. *Adapted by permission from Macmillan Publishers Ltd: (Nature Materials) (Ref. [31], copyright (2007)).* (B) TEM images of the as-synthesized 5 nm Pt_3Fe, (C) M-dependent specific activity, mass activity, and specific active area of MPt_3 NPs. *Reprinted with permission from Ref. [34]. Copyright (2011) American Chemical Society.*

than the Pt_3Ni (100) and (110) in $HClO_4$ due to its proper d-band energy position and surface geometry [36]. Such a high activity on sing-crystalline Pt_3Ni (111) has motivated the synthesis of Pt_3Ni NPs with tailored surface facet. Figure 6 shows the TEM images of octahedral and cubic Pt_3Ni NPs. The Pt_3Ni octahedral NPs are synthesized by the co-reduction of $Pt(acac)_2$ and $Ni(acac)_2$ by oleylamine in the presence of trace amount of $W(CO)_5$. Here $W(CO)_5$ is used to promote growth along the <100> direction. The octahedral Pt_3Ni NPs have a specific activity 5.1 times higher than the cubic Pt_3Ni NPs and 6.5 times higher than the cubic Pt NPs. This trend is also held true in the mass activity [37].

Compared with the zero-dimensional NPs, one-dimensional (1D) Pt-based alloy nanowires (NWs) or nanorods (NRs) have been found to be active catalyst for ORR with much enhanced stability, which is believed to arise from strong NW-support interactions. This is demonstrated in the enhanced ORR catalysis of 2.5-nm wide MPt (M = Fe, or Co) alloy NWs [38]. These NWs are synthesized through the thermal decomposition of $Fe(CO)_5$ or $Co_2(CO)_8$ and reduction of $Pt(acac)_2$ in the presence of oleylamine and sodium oleate. Figure 7 shows their typical TEM images and ORR performances in $HClO_4$. At 0.9V versus RHE, FePt NWs have a specific activity of 1.53 mA cm^{-2} and a mass activity of 844 mA mg^{-1}Pt, much higher than those of the commercial Pt NPs. More importantly, due to the strong physical adsorption

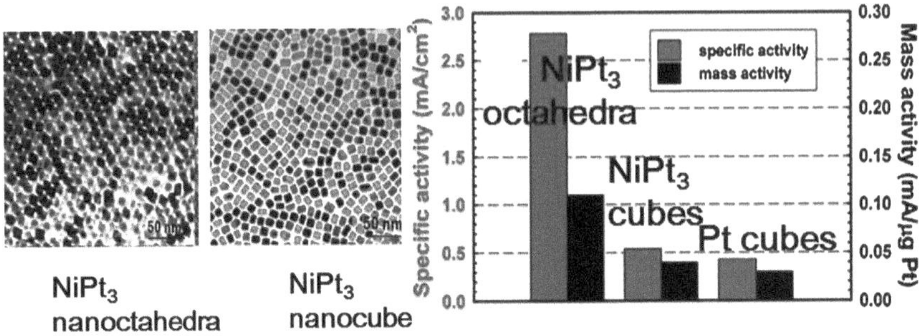

Figure 6 TEM images and ORR-specific and mass activities of octahedral and cubic $NiPt_3$ NPs. *Reprinted with permission from Ref. [37]. Copyright (2010) American Chemical Society.*

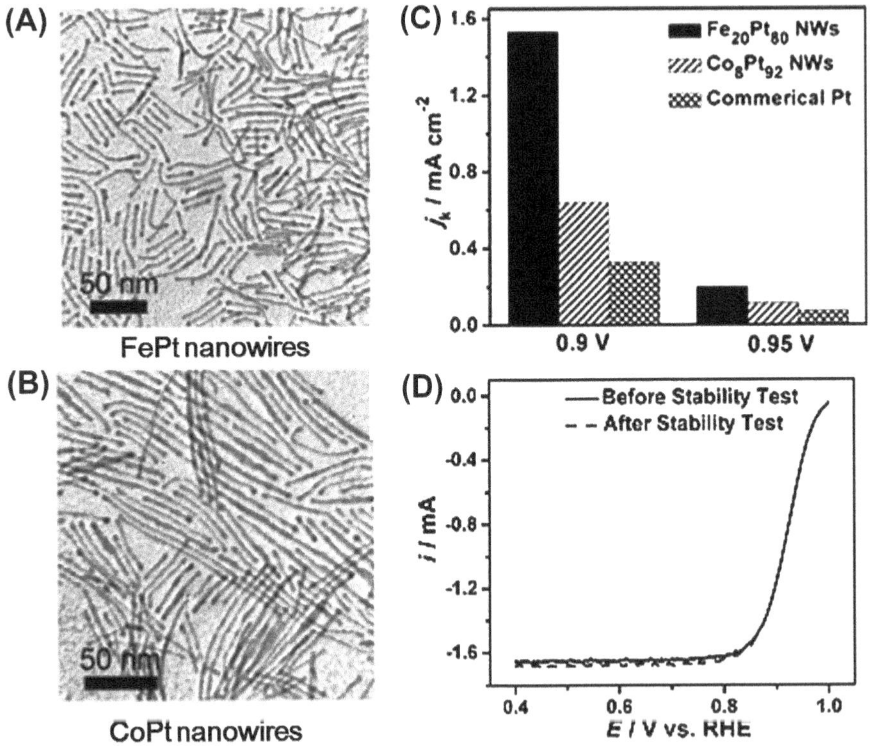

Figure 7 TEM images of (A) FePt NWs; (B) CoPt NWs; and (C) specific activities of FePt, CoPt NWs, and commercial Pt catalysts; and (D) ORR polarization curves of the FePt and CoPt NWs before and after stability test. *Reprinted from Ref. [38] with permission by Wiley.*

between NWs and carbon support, the NWs show an impressive increase in stability – they show no activity drop after 4000 electrochemical cycles between 0.6 and 1.0 V versus RHE (Figure 7(D)) [38].

2.4 Core/Shell Effect on ORR

Core/shell structure with Pt as a shell and other components as a core is an appealing motif in the design of highly efficient catalyst for ORR. It has a maximal Pt effective utilization ratio (percentage of Pt atoms as catalytic sites out of the overall Pt atoms) compared to conventional Pt NPs. Furthermore, the activity and durability of the Pt shell can be strongly influenced by the core component due to the electronic and strain effects. This core/shell effect can be easily realized by dealloying MPt NPs to form MPt/Pt core/shell NPs [31]. For example, in the electro-anodization, the redox potential difference of M and Pt results in a preferential oxidation and dissolution of M at the surface, generating a Pt-rich surface around the NPs. This strategy has been demonstrated in the CuPt/Pt NPs [39]. CuPt NPs with tunable compositions are first made through a liquid-metal salt impregnation and thermal annealing of commercial Pt with $Cu(NO_3)_2$ on carbon. The CuPt NPs are then electrochemically scanned in the range of 0.05–1.2 V in $HClO_4$ to remove Cu at the surface. After 200 scan cycles, the CuPt/Pt NPs are formed. These core/shell NPs exhibit the higher ORR activity than the Pt NPs, and this high activity is correlated with the strain effect in the CuPt/Pt structure, which arises from lattice mismatch between the CuPt core (with a smaller crystalline lattice) and the Pt shell, causing a compression strain on Pt shell and the downshift of Pt d-band center. This strain is tunable by CuPt compositions. This electrochemical dealloying method has also been applied to other Pt alloy nanostructures, such as CuFePt alloy NRs [40]. CuFePt NRs are synthesized via the reduction of $Cu(acac)_2$ by oleylamine in the presence of FePt NRs. The reduction leads to Cu atom deposition along the rod structure, followed by Cu diffusion into the FePt structure. After loading on carbon and treated with acetic acid at 70 °C for overnight to neutralize and remove oleylamine, the Fe in the surface layers of the NRs is readily etched, but the Cu and Pt are stable in the same solvent, leading to the formation of CuPt-rich shell. Further electrochemical cyclic voltammetry (CV) cycling promotes Cu oxidation/etching and enables the formation of pure Pt shell. Figure 8(A) and (B) shows the typical TEM image of the as-synthesized CuFePt NRs and the elemental mapping of the CuFePt/Pt core/shell NRs after the acidic and electrochemical treatment. With the desired core/shell structure and 1D morphology, the CuFePt/Pt NRs demonstrated 7- and 5- fold higher mass and specific actives than commercial Pt NPs (Figure 8(C)) and a superior durability after 5000 potential cycles (Figure 8(D)) [40].

Pt shell catalysis in the core/shell structure can also be tuned by controlling Pt shell strain via core crystal structure. This is demonstrated in the core/shell FePt/Pt catalyst system. FePt as an alloy can adopt *fcc*-type solid solution structure or face-centered tetragonal (*fct*)-type intermetallic structure (Figure 9(A) and (B)) [41,42]. These two structures have different crystal lattices. When they are formed as a core, Pt strain in the FePt/Pt structure can be fine-tuned by simply controlling the *fcc*-*fct* structure transition. To prepare FePt/Pt NPs, *fcc*-FePt NPs are first synthesized by decomposition of $Fe(CO)_5$ and

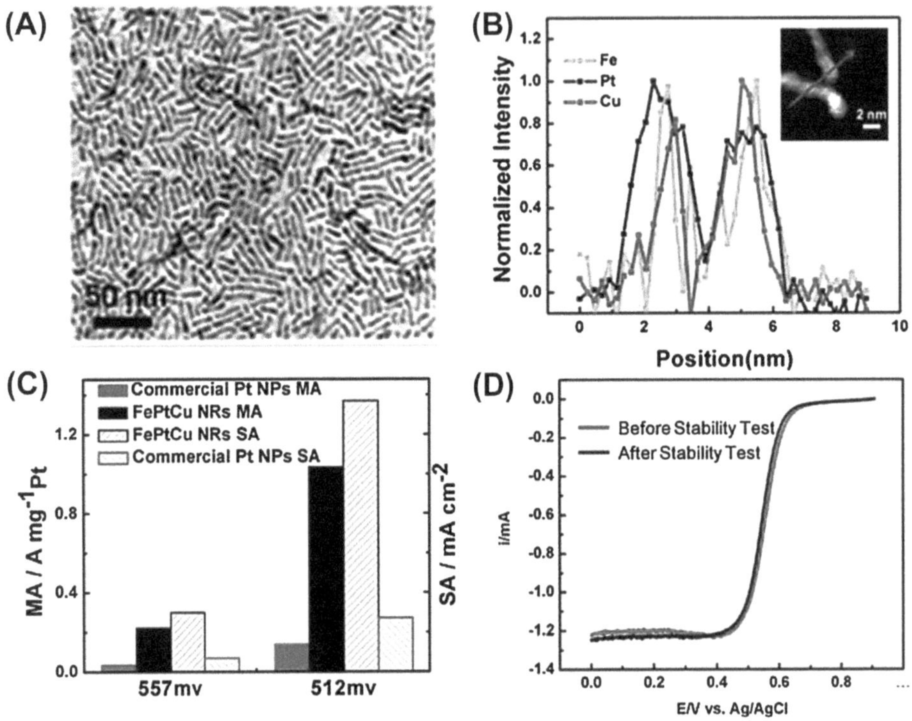

Figure 8 (A) TEM image of as-synthesized FePtCu NRs, (B) STEM–EELS line scan of the representative FePtCu/Pt NRs obtained after acid washing and CV cycling, (C) mass and specific ORR activities, and (D) ORR polarization curves of FePtCu before and after stability test. *Reprinted with permission from Ref. [40]. Copyright (2014) American Chemical Society.*

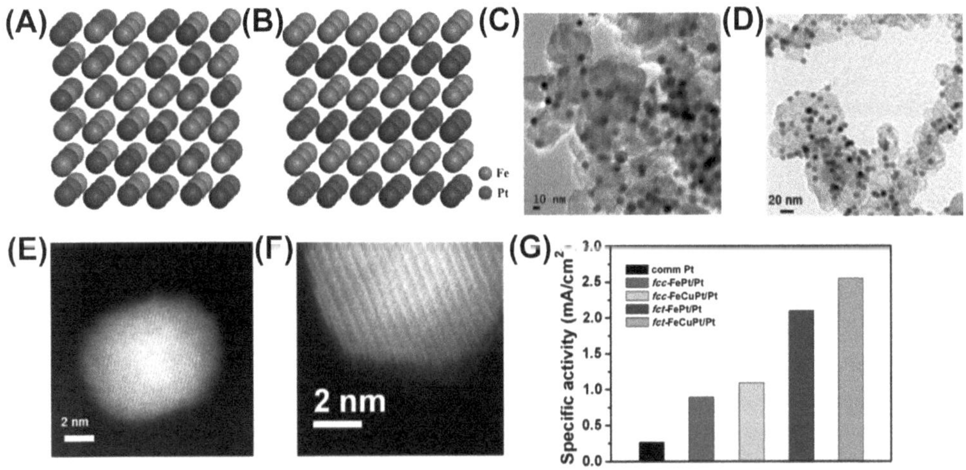

Figure 9 Schematic illustrations of the periodic supercells of *fcc*-$Fe_{50}Pt_{50}$ (A) and *fct*-$Fe_{50}Pt_{50}$ (B). TEM images of the C-$Fe_{51}Pt_{49}$ NPs annealed at 400 °C (C) and 650 °C (D). HR-STEM-HAADF images of the C-$Fe_{51}Pt_{49}$ NPs annealed at 400 °C (E) and 650 °C (F). ORR specific activities of the NP catalysts at 0.531V ($E_{1/2}$ of the commercial Pt catalyst) (G). *Reprinted with permission from Ref. [42]. Copyright (2014) American Chemical Society.*

reduction of $Pt(acac)_2$ [42]. Then these FePt NPs are deposited on a carbon support and are subject to high temperature annealing at 650 °C to convert the *fcc*-FePt to *fct*-FePt. Then the annealed FePt NPs are subject to electro-anodization in 0.1 M $HClO_4$ from −0.2 to 1.0V versus Ag/AgCl [42]. Once CV behavior becomes stabilized, FePt/Pt core/shell NPs are formed. Figure 8(C)–8(F) is the TEM images and HR-STEM-HAADF images of these FePt/Pt NPs. The images show clearly the Fe/Pt intermetallic structure (dark (Fe) and bright (Pt) strips due to the Z-contrast difference between Fe and Pt) in the *fct*-FePt and this *fct*-FePt is surrounded by ~3 atomic layers of Pt. ORR test in 0.1 $HClO_4$ indicates the *fct*-FePt/Pt NPs have over two times higher specific activity than the *fcc*-FePt/Pt NPs, even though their size, shape, and composition are identical. Theoretical simulations suggest that the *fcc*-FePt core results in too much compression on Pt strain and this strain over-compression is released by the *fct*-FePt core. This surface strain and therefore the Pt-O binding energy can be further optimized in *fct*-FeCuPt/Pt NPs. In this structure, the presence of Cu further expands the *fct*-lattice to release the over-compression on Pt shell. As a result, the *fct*-FeCuPt/Pt NPs become the most active ORR catalyst among the FePt/Pt NP systems with the specific activity early 10 times of that from the commercial Pt NP catalyst (Figure 9(G)).

Core/shell structure is also ideal for designing new ORR catalyst with minimum Pt usage. This is achieved by coating a thin layer of Pt over a non-Pt metal core. This core/shell structure ensures the maximum exposure of Pt for ORR, and core-shell interactions can be used to further tune Pt electronic structure, strain, Pt-O binding energy, and Pt catalysis. One approach to deposit Pt over a non-Pt core is via galvanic replacement [43]. Figure 10 illustrates the principle of this approach. The core NPs are first synthesized and modified by a mono-layer of Cu via the underpotential deposition. In the presence of H_2PtCl_4, Cu is oxidized while H_2PtCl_4 is reduced and Pt deposited. The net result is that Cu serves as a sacrifice template and is replaced by Pt, forming the core/shell NPs with a monolayer shell of Pt (Pt_{ML}). Using this method, a variety of non-Pt core/shell NPs have been synthesized and studied for ORR. They are all more active than the pure Pt catalyst. For example, Pd/Pt_{ML} NPs exhibit a 3-fold mass activity at 0.9V versus RHE and a higher durability due to the slight compressive strain on Pt_{ML} produced by Pd core [44]. Adding 10% Au in the core (Pd_9Au_1/Pt_{ML}) can further enhance the core-shell NPs durability because Au helps to make the Pt_{ML} more resistant to oxidation under the ORR condition [45].

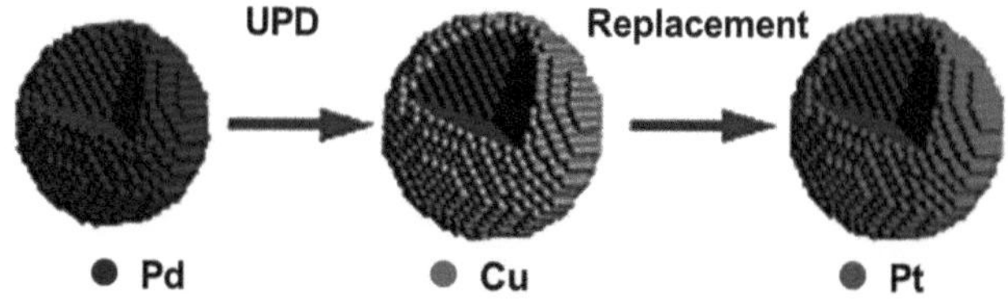

Figure 10 The galvanic replacement method for the synthesis of core/shell multimetallic NPs. *Reprinted from Ref. [43] with permission by Wiley.*

Alternatively, the core/shell NPs can be prepared by seed-mediated synthesis. In this approach, the seeding NPs are first prepared and then mixed with the shell materials in the reaction solution for shell deposition around the seeding NPs. The key to this synthesis is that the surfactant used to stabilize the seeding NPs should be weakly bound to the NP surface to allow shell deposition at a high reaction temperature. This is demonstrated in the synthesis of Au/$FePt_3$ NPs. In the approach, the monodisperse 7 nm Au NPs are synthesized by borane tert-butylamine reduction of $HAuCl_4$ in the mixture of oleylamine and tetraline. Then these Au NPs are mixed with $Fe(CO)_5$ and $Pt(acac)_2$ in octadecene with oleylamine and oleic acid as surfactant and the reaction solution is heated at 200 °C. This reaction gives Au/$FePt_3$ NPs with the $FePt_3$ shell being ~1.5 nm (Figure 11(A) and (B)) [46]. These 10 nm Au/$FePt_3$ NPs show higher ORR activity than the 10 nm $FePt_3$ and 10 nm Pt (Figure 11(C)), suggesting the enhancement is from the Au core. More importantly, Au/$FePt_3$ NPs are more durable with an enhancement factor of 7 compared to Pt after 60,000 potential cycles between 0.6 and 1.1 V versus RHE (Figure 11(D)). The results show that the existence of Au core can effectively hinder the formation of subsurface Pt oxide and prevent Pt from dissolution [45]. This synthesis has been extended to Pd/FePt [47], Pd/Au/FePt [48], Au/CuPt [49], AuCu/Pt [50], and AgPd/Pt [51] NPs, allowing deep studies on the core-shell interactions and ORR catalysis. For example, the activity of the stable Au/Pt NP catalyst can be further increased by alloying Cu with Au, forming AuCu/Pt NPs [50]. The same core/shell concept on enhanced ORR catalysis has been extended to 1D NWs. In a recent demonstration, 2.5-nm wide FePtM (M = Pd, Au) NWs are made and then coated with a layer of FePt using the seed-mediated synthesis with shell thickness controlled from 0.3 to 1.3 nm [52]. The core/shell NWs show shell thickness and M-dependent ORR catalysis and the FePtPd/FePt NWs with a shell thickness of 0.8 nm become the most active catalyst with a specific activity of 3.47 mA/cm^2 and mass activity of 1.68 A/mg Pt at 0.5 V (vs Ag/AgCl) and show no activity decrease after 5000 potential sweeps between 0.4 and 0.8 V (vs Ag/AgCl).

3. NP CATALYSTS FOR FUEL OXIDATION REACTIONS

3.1 Methanol Oxidation Reaction and Formic Acid Oxidation Reaction

Using H_2 as a fuel in PEMFCs has its advantages in the fast kinetics of H_2 oxidation (less than 5 mV overpotential on the state-of-the-art Pt catalyst), the highest mass energy density (120 MJ/kg at room temperature), and the clean final product (H_2O). But it does have the issues in production, purification, storage, and transportation. Small liquid organic molecules, such as formic acid and methanol, are promising fuel alternatives [53]. Besides their safety and ease in use, they are also renewable (can be obtained by biomass reforming).

When methanol (CH_3OH) is oxidized completely at anode, it produces proton, electrons, and CO_2 (3). This process is often referred to as methanol oxidation reaction (MOR).

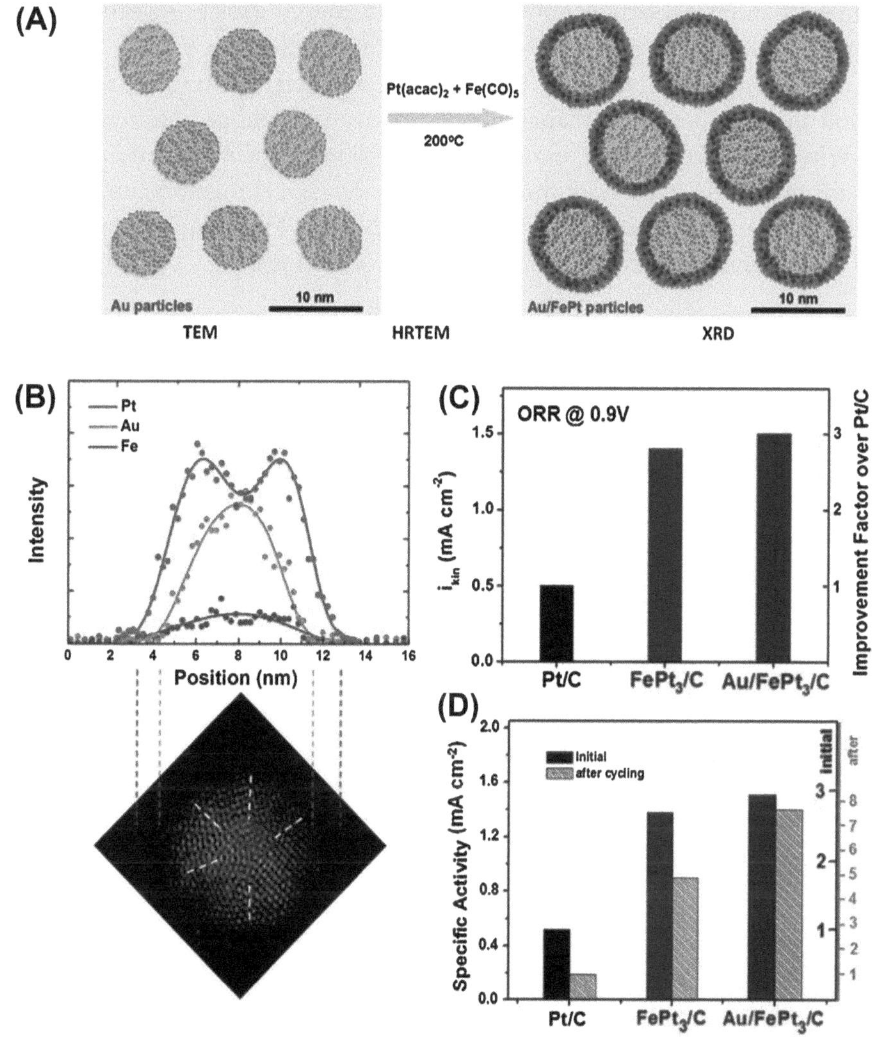

Figure 11 (A) Schematic illustration of core/shell Au/$FePt_3$ NPs synthesis. (B) Line profile-elemental distribution along a single Au/$FePt_3$ NP. (C) Specific activities for ORR at 0.9 V versus RHE. (D) Specific activities for ORR before and after 60,000 potential cycling. *Reprinted with permission from Ref. [46]. Copyright (2011) American Chemical Society.*

$$CH_3OH + H_2O \rightarrow CO_2 + 6H^+ + 6e^- \quad (3)$$

Mechanism study indicates that MOR on Pt involves several sequential steps: (1) dehydrogenation of CH_3OH, forming Pt-CO_{ads}; (2) dehydrogenation of water, forming Pt-OH_{ads}; and (3) oxidative recombination of CO_{ads} and OH_{ads} to form CO_2 and H^+ followed by CO_2 desorption. In this MOR process, Pt is fast to catalyze (1) but slow in (2), leading to detrimental CO accumulation on the Pt surface and low MOR efficiency.

Formic acid, HCOOH, has lower volumetric energy density (2086 kW/L) than methanol (4690 kW/L), but in PEMFC operation condition, it has less a problem in "cross-over" than methanol [53,54]. The "cross-over" is referred to as the fuel (CH_3OH) migration through proton exchange membrane to the cathode, interfering with the ORR. When HCOOH is used, its cross-over problem can be minimized as the energy barrier for HCOOH to migrate through the Nafion-SO_3H membrane is high due to electrostatic repulsion of $HCOO^-$ and -SO_3^- in Nafion. The complete anode reaction for the formic acid oxidation reaction (FAOR) is as follows (4):

$$HCOOH \rightarrow CO_2 + 2H^+ + 2e^- \tag{4}$$

This is commonly called dehydrogenation pathway. But on the Pt catalyst, HCOOH tends to undergo nonfaradaic dehydration first, generating CO (5) that poisons the Pt surface by forming Pt-CO, and needs more energy input to oxidize CO (6) and to regenerate active Pt surface [50,51].

$$HCOOH \rightarrow CO_{ads} + H_2O \tag{5}$$

$$CO_{ads} + H_2O \rightarrow CO_2 + 2H^+ + 2e^- \tag{6}$$

An ideal catalyst for FAOR should be able to efficiently catalyze the FAOR in dehydrogenation pathway without generating CO.

3.2 Pt-Based NP Catalysts for MOR and FAOR

Considering the slow kinetics and CO-poisoning problems of single component Pt NPs in their catalysis for MOR and FAOR, Pt NPs are extensively studied in the form of alloys to tune Pt catalytic performance. One example is the study on PtPd NPs [55]. Monodisperse PtPd NPs are synthesized via a co-reduction of $Pt(acac)_2$ and $Pd(acac)_2$, with reducing agent morpholineborane (MB) in oleyamine. In the synthesis, precursors of $Pd(acac)_2$ and $Pt(acac)_2$ in different ratios are dissolved in oleylamine and then heated to 60 °C at which MB dissolved in oleylamine is quickly injected to initiate NP nucleation. The solution is further incubated at 90 °C to allow the formation of $Pd_{88}Pt_{12}$ NPs and at 180 °C for $Pd_{67}Pt_{33}$, $Pd_{52}Pt_{48}$, and $Pd_{34}Pt_{66}$ NPs. Figure 12(A) and (B) shows the typical TEM image of the 5 nm $Pd_{67}Pt_{33}$ NPs and their STEM-HAADF 2D elemental mapping image confirming their homogeneous solid solution structures. The MOR studies in 0.1 M $HClO_4$ and 0.1 M methanol of these PtPd NPs are shown in Figure 12(C). All of the PtPd NPs are more active than the pure Pt or Pd components, and their peak potentials and current densities are strongly related to Pt/Pd ratio. This correlation shown in Figure 12(D) reveals a "maximum activity plateau" for NPs with 40–60% of atomic Pt. This composition-dependent MOR catalysis is explained by the bifunctional synergy of Pt and Pd: the water dehydrogenation is catalyzed by Pd to form Pd-OH, while the methanol

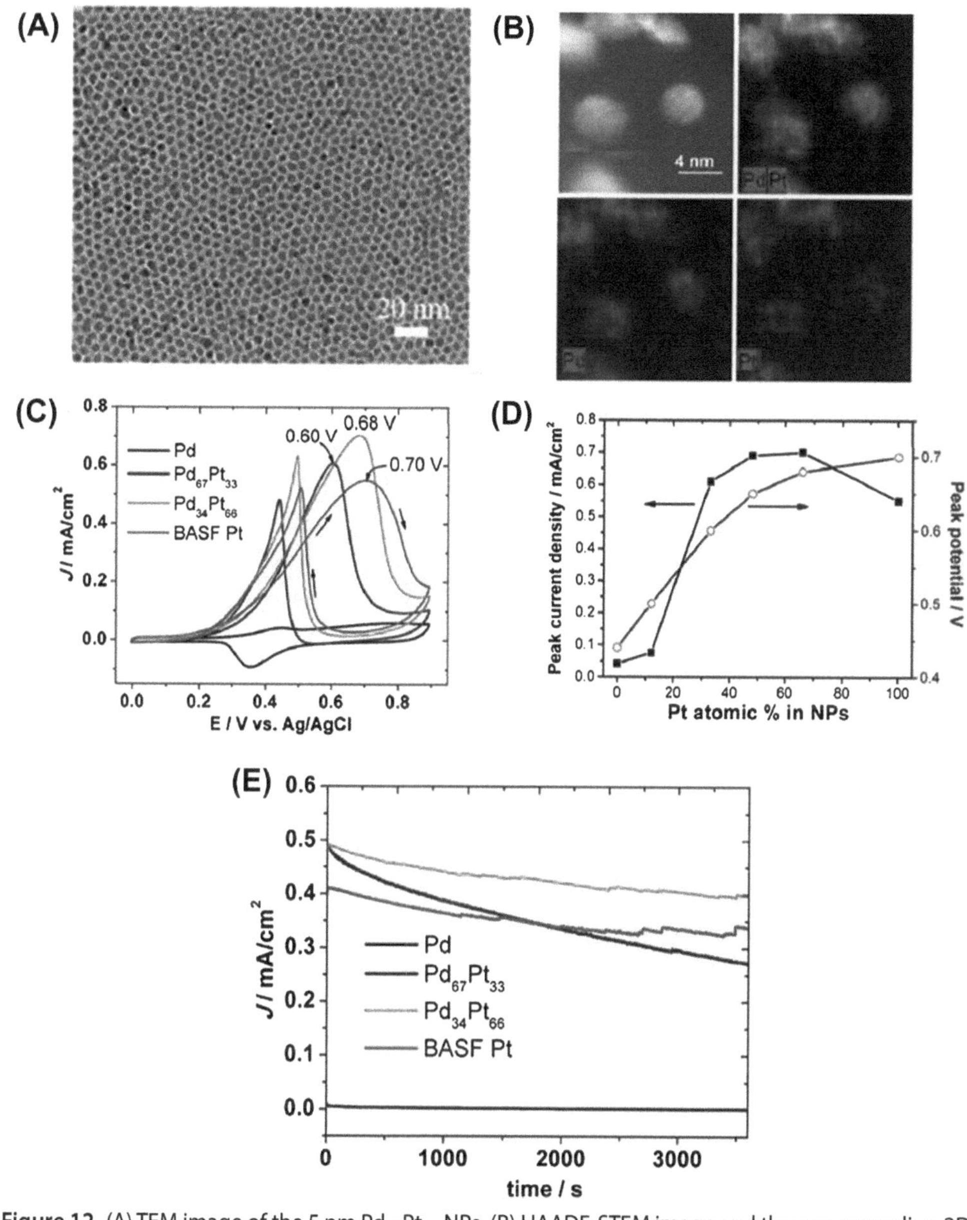

Figure 12 (A) TEM image of the 5 nm $Pd_{67}Pt_{33}$ NPs. (B) HAADF-STEM image and the corresponding 2D elemental mapping of several representative 5 nm $Pd_{67}Pt_{33}$ NPs. (C) *J-V* curves reflecting MOR catalysis of NP catalysts in 0.1 M $HClO_4$ and 0.1 M methanol. (D) Methanol oxidation peak current density and peak potential versus the Pt atomic percentage in PdPt NPs. (E) Chronoamperometric curves reflecting the stability of the NP catalysts in 0.1 M $HClO_4$ and 0.1 M methanol at constant voltage 0.5 V (vs Ag/AgCl) for 1 h. *Reprinted with permission from Ref. [55]. Copyright (2011) American Chemical Society.*

dehydrogenation is catalyzed by Pt to form Pt-CO. The adjacent Pd-OH and Pt-CO expedite the CO_2 formation and desorption to complete MOR. If no Pd, pure Pt can only catalyze the water dehydrogenation at a higher potential, leading to the higher overpotential in pure Pt than in PtPd. The stability of the PtPd NPs is tested by the 1-h chronoamperometric test at a constant potential of 0.5 V. As shown in Figure 11(E), $Pd_{52}Pt_{48}$ and $Pd_{34}Pt_{66}$ NPs have higher current densities than the commercial Pt, at both the start and the end of the stability test. Similarly, this alloy-induced MOR catalysis enhancement is also found in other Pt NPs alloyed with transition metals such as Ru [56], Rh [57], and some non-noble transition metals such as Mn [58], Zn [59], and Cu [60].

This alloy composition control has also been extended to the Pt NPs alloyed with main group metals. For example, Pt_3Sn NPs are synthesized through the solution-based hot-injection and co-reduction of $Pt(acac)_2$ and $Sn(acac)_2$ with the size tuned from 4 to 7 nm by controlling injection temperature [61]. Figure 13(A) shows a TEM image of the as-synthesized 5.4-nm Pt_3Sn. The Pt_3Sn NPs exhibit a negative shift of the onset potential for CO oxidation compared to Pt, indicating the weakened CO binding onto Pt due to the incorporation of Sn (Figure 13(B)–(D)). Besides, the Sn can accelerate the water dehydrogenation and therefore facilitate the CO oxidation to complete the MOR. In MOR test, the Pt_3Sn NPs show much enhanced catalytic activity and stability compared to pure Pt. Moreover, annealing the Pt_3Sn NPs at 400 °C generates a more uniform alloy structure and further enhances the Pt_3Sn MOR activity, as shown in Figure 13(E). This Pt_3Sn represents a new class of alloy NP catalysts with strong CO tolerance and high efficiency for MOR.

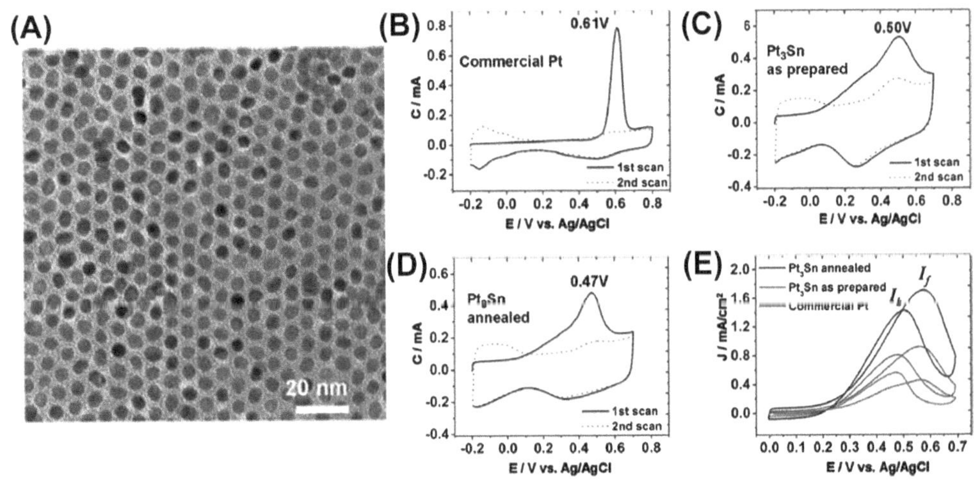

Figure 13 (A) TEM image of the as-synthesized 5.2 ± 0.4 nm Pt_3Sn NPs. CO strippings of the commercial Pt (B), the Pt_3Sn as prepared (C), and the annealed Pt_3Sn (D). (E) *J-V* curves of the Pt_3Sn as prepared, the annealed Pt_3Sn, and the commercial Pt in 0.1 M $HClO_4$ and 0.1 M methanol with the potential sweeping rate at 50 mV/s. *Reprinted with permission from Ref. [61]. Copyright (2011) American Chemical Society.*

MOR catalysis can also be enhanced by tailoring Pt alloy NP shape. The PtPd alloy NPs have been made in tetrahedral or cubic shapes [62]. In the synthesis, the alloy growth is controlled by small ions as efficient facet-selective agents and PVP as the capping ligand. When sodium oxalate and formaldehyde are applied to make PtPd, the stabilization effect of $C_2O_4^{2-}$ on the {111} leads to the formation of tetrahedral NPs. When a mixture of Br^- and I^- anions are present, {100} growth is inhibited and cubic PtPd NPs are formed (Figure 14(A)–(C)). The shape-dependent catalytic performance of the PtPd NPs toward MOR is shown in Figure 14(D) with cubic NPs being more active and stable than the

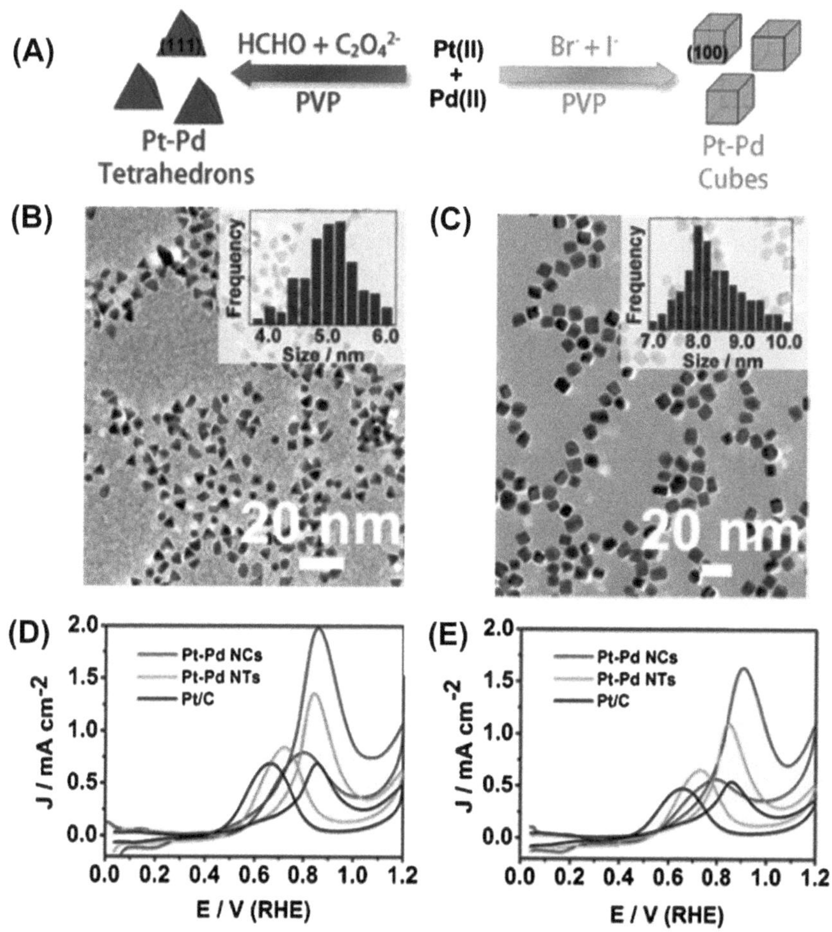

Figure 14 (A) Schematic illustration of shape-selective synthesis of PtPd nanotetrahedrons (NTs) and nanocubes. TEM images of the Pt-Pd NTs (B) and nanocubes (C). (D) *J-V* curves obtained for the Pt-Pd nanocubes, NTs, and Pt/C in the electrolyte of 0.1 M $HClO_4$ and 1 M CH_3OH at the sweep rate of 50 mV/s (E) *J-V* curves obtained after 4000 additional cycles. *Reprinted with permission from Ref. [62]. Copyright (2011) American Chemical Society.*

tetrahedral NPs and much more efficient than the commercial Pt NPs (Figure 14(E)). This study suggests that the {111} facet is more efficient for the MOR catalysis.

Pt's activity for FAOR is strongly suppressed by the CO-poisoning associated with the undesired nonfaradic dehydration pathway. Alloying Pt with other metals can change the electronic structure of Pt and make the direct FAOR pathway more dominant. Studies show that Au [63], Ag [64], Pb, Bi [65], especially Pb and Bi, are effective in alloying with Pt to improve FAOR efficiency by eliminating the CO-poisoning on Pt. These monodisperse Pt_3Pb NPs are synthesized by reducing $Pt(acac)_2$ and $Pb(acac)_2$ in benzyl ether in the presence of oleylamine, oleic acid, and hexadecanethiol [66]. These 3.7 nm Pt_3Pb can be used as the seeds and generate 4 nm Pt_3Pb/Pt core/shell NPs (type A) and 4.9 nm Pt_3Pb/Pt core/shell NPs with Pt islands (type B) (Figure 15(A)). In catalyzing FAOR in 0.1 M H_2SO_4 and 0.5 M HCOOH, both Pt_3Pb and Pt_3Pb/Pt NPs (type A) show a complete CO-poisoning elimination and high activity toward FAOR, whereas the Pt_3Pb/Pt (type B) NPs are partially poisoned during the catalysis process. The DFT calculations suggest that the HCOOH dehydrogenation is more energy favorable on the surface of Pt_3Pb alloy or a thin layer of Pt supported by Pt_3Pb alloy (type A). However, this CO-resistant effect disappears when the Pt surface atoms are far from the Pt_3Pb alloy surface (type B).

Tailoring alloy NP's crystalline structure can also enhance the NP's CO tolerance and catalytic efficiency for FAOR and MOR. The *fct*-FePtAu NPs are recently reported as an example to demonstrate this strategy [67]. The *fcc*-FePtAu alloy NPs are first synthesized by co-reduction of $Pt(acac)_2$ and $HAuCl_4$ and thermal decomposition of $Fe(CO)_5$ with the NPs compositions precisely controlled by varying metal precursor ratios. Figure 16(A) shows the representative TEM image of the monodisperse 4 nm *fcc*-FePtAu. The *fcc*-FePtAu NPs are then deposited on carbon support and transformed to *fct*-FePtAu

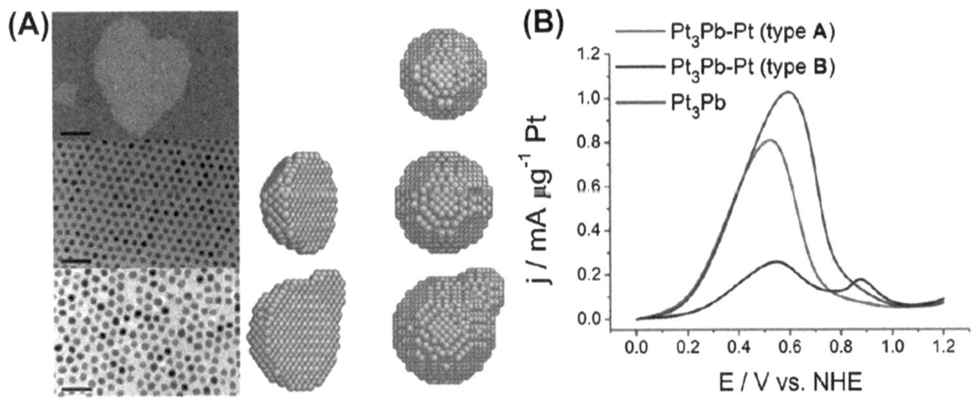

Figure 15 (A) TEM images Pt_3Pb NPs (top, scale bar: 100 nm), Pt_3Pb-Pt (type A) nanocubes (middle, scale bar: 20 nm), and Pt_3Pb-Pt (type B) nanocubes (bottom, scale bar: 20 nm). (B) FAOR on PtPb and PtPb-Pt NPs. *Reprinted with permission from Ref. [66]. Copyright (2012) American Chemical Society.*

NPs by thermal annealing at 600 °C. The STEM-EELS and STEM-EDS analyses reveal that after the phase transformation the Au diffuses out and aggregated into cluster islands around the shell to form a *fct*-FePtAu/PtAu core/shell structure (Figure 16(B) and (C)). It turns that such a unique structure is essential for FAOR catalytic enhancement. As shown in Figure 16(D) and (E), the FAOR study shows that the Au segregation on the NP's surface helps to eliminate the CO poisoning on Pt. Correspondingly, the CO-oxidation peak in FAOR disappears on *fct*-FePtAu/PtAu NPs, and the oxidation current density at low potential is increased to 2809.9 mA/mg_{Pt}, much higher than that of the Pt (less than 200 mA/mg_{Pt}, 20% mass loading on Vulcan carbon, 2.5–3.5 nm in diameter obtained from Fuel Store) NP catalysts. These *fct*-FePt/PtAu NPs are also highly stable (Figure 16(F)). After 13-h chronoamperometric test at 0.2 V (vs Ag/AgCl), the mass activity from the *fct*-FePt/PtAu NPs only drops about 7% to 2600 mA/mg_{Pt}, the best one among the NP catalysts reported for FAOR.

3.3 Pd-Based NP Catalyst for FAOR

Like Pt, Pd can also catalyze FAOR. Unlike Pt, FAOR on Pd is dominated by direct hydrogenation pathway, which makes Pd a promising alternative as non-Pt catalyst for FAOR. To maximize the catalytic potential of Pd, fine tuning of the size and shape of the Pd NPs is necessary. Previous studies on Pd electrode have shown that for the FAOR, Pd

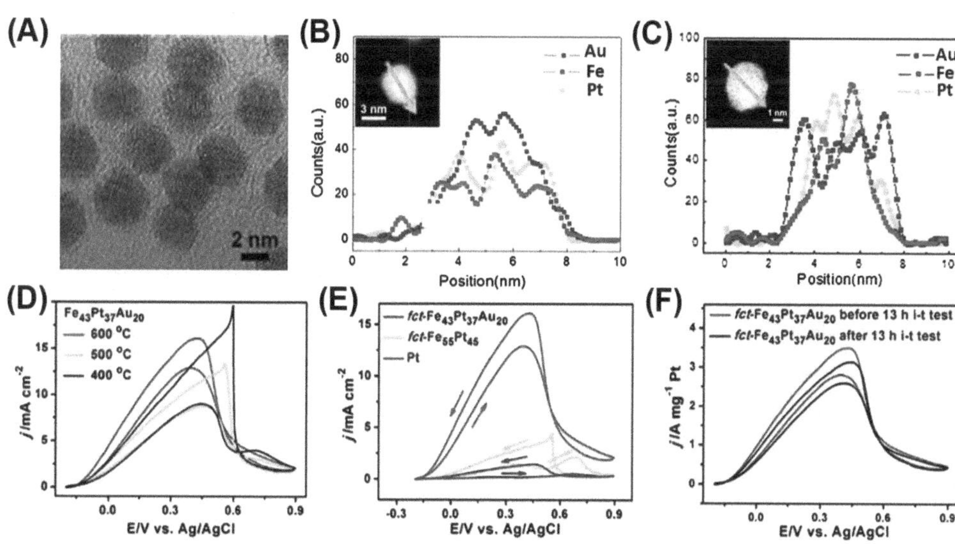

Figure 16 (A) TEM images of the as-synthesized 4 nm *fcc*-$Fe_{43}Pt_{37}Au_{20}$ NPs. STEM-EDS line scans crossing *fcc*-$Fe_{43}Pt_{37}Au_{20}$ (B) and *fct*-$Fe_{43}Pt_{37}Au_{20}$ (C) NPs. (D) *J-V* curves of the $Fe_{43}Pt_{37}Au_{20}$ NPs annealed at different temperatures. (E) *J-V* curves of the specific activity of the *fct*-$Fe_{43}Pt_{37}Au_{20}$, *fct*-$Fe_{55}Pt_{45}$, and commercial Pt catalysts. (F) *J-V* curves of the *fct*-$Fe_{43}Pt_{37}Au_{20}$ NPs before and after a 13-h *I-t* stability test. The studies were performed in N_2-saturated 0.5 M H_2SO_4 + 0.5 M HCOOH solutions. *Reprinted with permission from Ref. [67]. Copyright (2012) American Chemical Society.*

(100) plane shows higher current density, while Pd (111) exhibits lower overpotentials [68]. Thus, synthesizing a Pd NP catalyst that can simultaneously maintain both high current density and low overpotential. Pd polyhedra including truncated cubes, cuboctahedra, truncated octahedra, and octahedra have been prepared by controlled growth of Pd onto the pre-made Pd nanocubes in the presence of PVP and formaldehyde [69]. Such a shape control from a cube to an octahedron is mainly resulted from selective deposition of Pd atoms on the {100} facets (Figure 17(A)–(E)). Further experiments indicate that the aldehyde group plays an important role in blocking the <111> direction and determining the formation of final octahedral NPs. FAOR over these Pd polyhedra shows the decrease in both maximum specific current density and peak potential when the surface ratio of {100} to {111} facets decrease (Figure 17(F)). These studies suggest that Pd nanocubes with the slight truncations at the corners should be the best one for FAOR. However, these truncated cubic Pd NPs have a relatively low mass activity due to their large sizes. To make small Pd NPs, $Pd(acac)_2$ dissolved in oleylamine is reduced by borane tert-butylamine, in which oleylamine acts both as a solvent and as a surfactant [70].

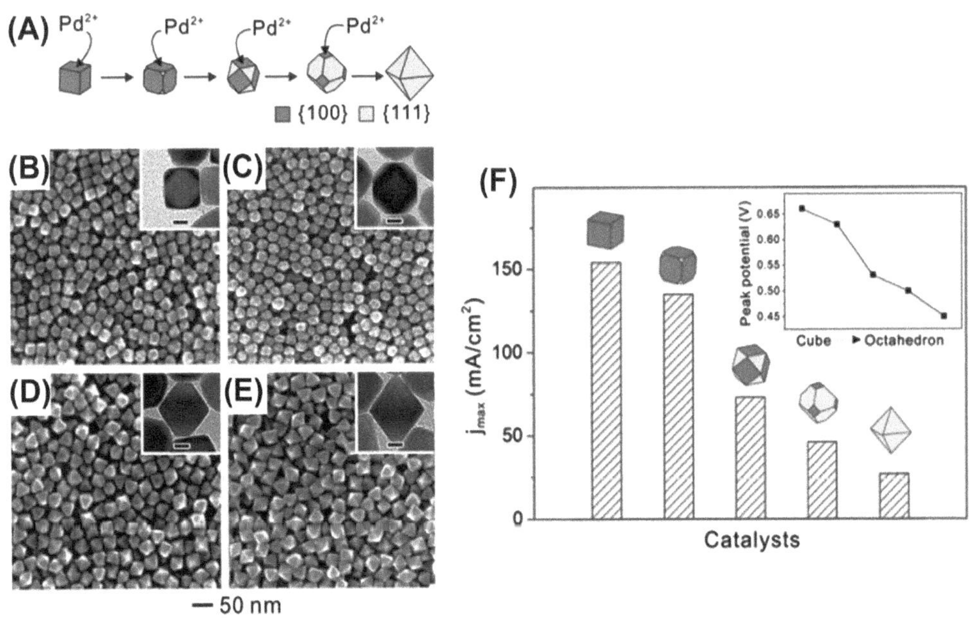

Figure 17 (A) Schematic illustration of shape evolution from Pd nanocube bound by {100} facets to Pt nanooctahedron enclosed by {111} facets. SEM images of the Pd polyhedrons obtained by controlling the ratio of Na_2PdCl_4/Pt nanocube seed: 5.8 mg (B), 8.7 mg (C), 17.4 mg (D), and 29.0 mg (E). TEM images of these Pd polyhedrons are given in the insets (scale bars: 10 nm). (F) Maximum current densities of formic acid oxidation over Pd polyhedrons enclosed by {111} and {100} facets in different proportions. The inset shows their corresponding peak potentials. *Reproduced from Ref. [69] with permission of The Royal Society of Chemistry.*

When the reduction is held at 60 °C and the reaction solution is aged at 90 °C for 1 h, monodisperse 4.5-nm Pd NPs are obtained. In addition to their smaller size, the oleylamine-coated Pd NPs are also readily cleaned by acetic acid washing, in which oleylamine is neutralized and washed off. The resulted Pd NPs are electrochemically active and show higher activity for FAOR [70].

Alloying Pd with other metals has been explored to enhance NP's catalytic performance. Monodisperse MPd (M = Co, Cu) NPs are successfully made in the solution synthesis by reducing $PdBr_2$ and $M(acac)_2$ in the presence of oleylamine and trioctylphosphine (TOP) [71]. TOP is a strong surfactant for metal, especially for Pd, and is used to facilitate conucleation and growth of Pd and Co (or Cu). In the synthesis, the reaction solution is aged at 260 °C and the MPd alloy NPs are obtained (Figure 18(A)). Moreover, the NP compositions (for instance, from $Co_{10}Pd_{90}$ to $Co_{60}Pd_{40}$) are controlled by varying the precursor ratio of $PdBr_2/M(acac)_2$, while the NP sizes (5–12 nm) are tailored by the heating rate (Figure 18(B)–(C)). The FAOR study in $HClO_4$ show that CoPd NPs have better catalytic activities than Pd and CuPd and their FAOR mass activities change with the trend of $Co_{50}Pd_{50} > Co_{60}Pd_{40} > Co_{10}Pd_{90} >$ Pd (Figure 18(D)). The $Co_{50}Pd_{50}$ NPs (8 nm) have an FAOR oxidation peak at 0.4V (vs Ag/AgCl) with a peak current density of 774 A/g_{Pd}.

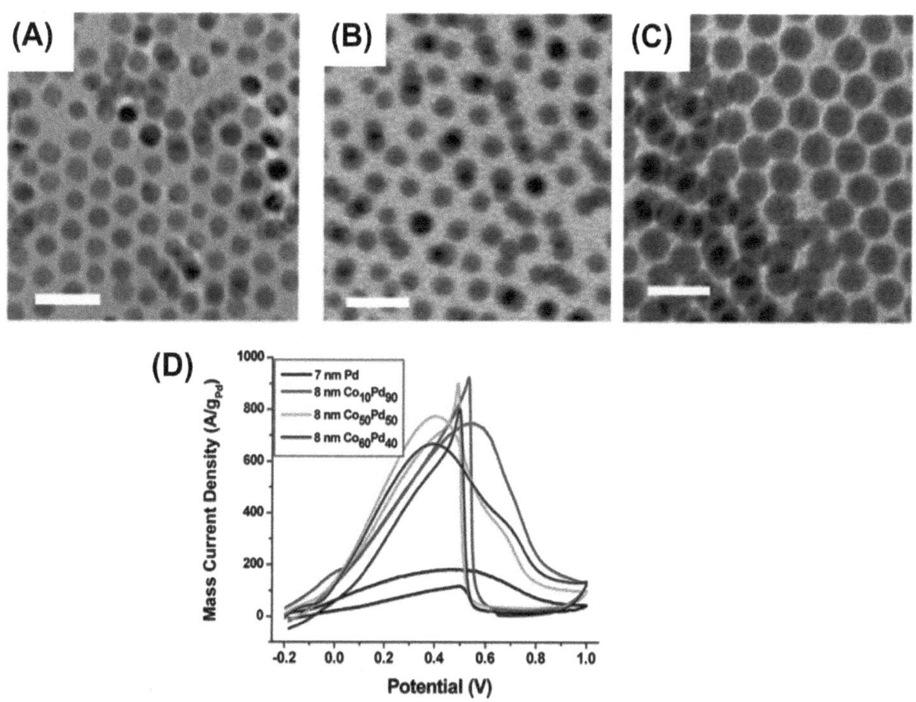

Figure 18 TEM images of 8 nm (A), 5 nm (B), and 12 nm (C) $Co_{60}Pd_{40}$ NPs (scale bar = 25 nm). (D) NP mass current densities versus applied potential in 0.1 M $HClO_4$ and 2 M HCOOH at 25 °C (scan rate: 50 mV/s). *Reprinted with permission from Ref. [71]. Copyright (2012) American Chemical Society.*

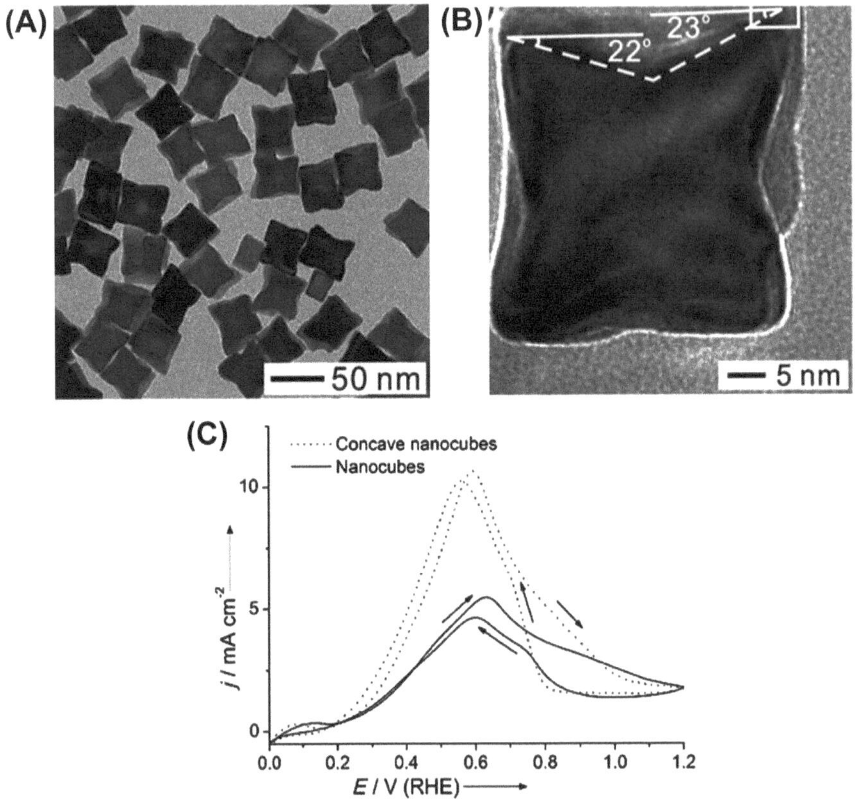

Figure 19 TEM image (A) and HRTEM image (B) of Pd concave nanocubes. (C) *J-Vs* of FAOR catalyzed by Pd concave nanocubes in N_2-saturated 0.1 M $HClO_4$ and 2 M HCOOH solution at a scan rate of 50 mV s^{-1}. *Reprinted from Ref. [72] with permission by Wiley.*

As a comparison, commercial Pd catalysts showed an oxidation peak at 0.75 V (vs Ag/AgCl) with peak current density of only 254 A/g_{Pd} under the same condition.

Pd NPs with high-index facet exposure was studied for FAOR. Due to the presence of the high density of low-coordinated atoms in the forms of atomic steps and kinks, high-index facets are generally more active than low-index planes that are composed of closely packed surface atoms [72]. As an example, in the study of concave palladium NPs for FAOR, the Pd nanocrystals with high-index {730} facets were prepared by reduction of Na_2PdCl_4 with Pd nanocubes as seeds in the presence of PVP, KBr, and ascorbic acid. Because of the exposure of the high-index {730}facets, the as-prepared concave Pd NPs exhibit higher specific activity than Pd nanocubes in the electro-oxidations of formic acid (Figure 19) [72].

The catalytic efficiency of Pd NPs for FAOR can be further increased by controlling NP-support interactions. In a recent demonstration, 4.5 nm Pd NPs loaded on graphene (G) show two times higher FAOR specific activity than the same Pd NPs deposited on C [73]. More importantly, after 250 potential cycles between −0.3 and 0.9 V

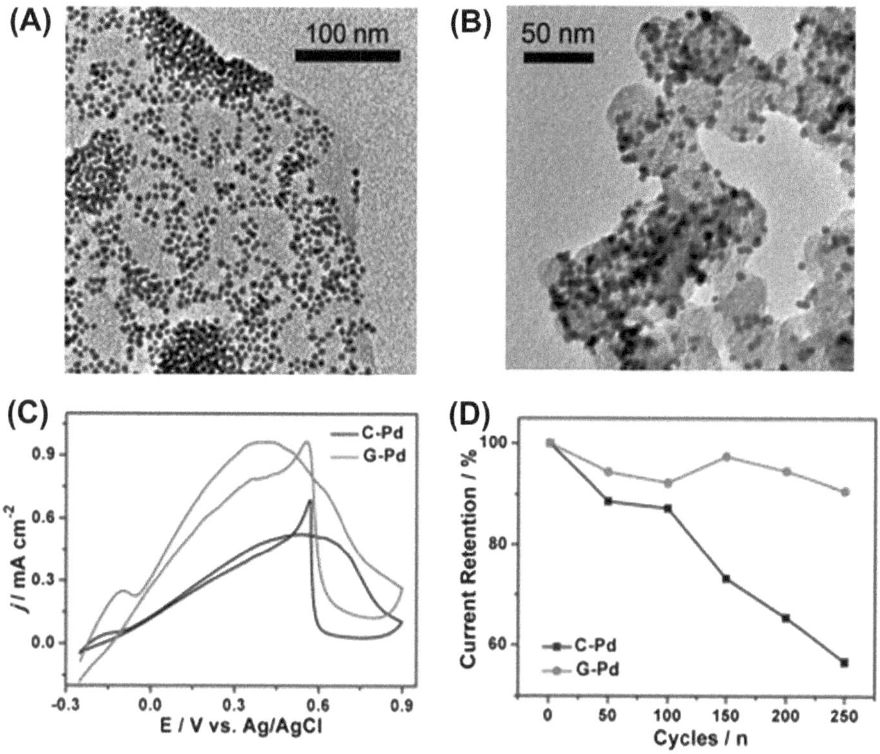

Figure 20 TEM images of G-Pd (A) and C-Pd (B). (C) *J-Vs* of FAOR catalyzed by G-Pd and C-Pd in N_2-saturated 0.1 M $HClO_4$ and 0.1 M HCOOH solution at a scan rate of 50 mV s^{-1}. (D) Electrocatalytic cycling stability of G-Pd and C-Pd. Current retention values in (D) were obtained through normalizing the currents of different cycles to the first one. *Reproduced from Ref. [73] with permission of The Royal Society of Chemistry.*

(vs Ag/AgCl), the G-Pd catalyst retains over 90% of its initial activity while the C-Pd has only about 55% of its activity left (Figure 20). Meanwhile, the electrochemically active surface area of the G-Pd is reduced by about 4.9%, much less than the 45.8% loss in the C-Pd. Despite what have been observed, it is still elusive on why G-Pd becomes more efficient in catalyzing FAOR. The interaction between G and Pd may help to stabilize Pd from easy oxidation and dissolution in acid.

4. SUMMARY

Recent advance in synthetic methodology in solution chemistry makes it possible to prepare various NP catalysts with controlled sizes, shapes, compositions, and structures. With the size control, surface atom ratios, and surface atom interactions with reactants can now be better studied and optimal NP size be selected for catalytic reactions. With the shape control, chemical reactions on different crystal facets can be compared and the results have been used to direct the synthesis of NP catalyst with the desired shapes. More importantly, the chemical syntheses allow the incorporation of different metals into one nanostructure, forming alloy and core/shell NPs. When alloy NPs are used as the catalyst, their catalysis can be correlated with their Pt d-band center change, making it possible to rationalize NP catalysis with their electronic structure. With core/shell structure, Pt atom spacing on the NP surface and therefore the Pt strain can be tuned to control surface energetic on Pt's interaction with oxygenated species to enhance the electrocatalysis. The well-tuned NP structures based on Pt have been studied extensively to catalyze ORR, MOR, and FAOR and have reached unprecedented activity and durability compared to the single component Pt NP catalyst. Despite these exciting progresses made in NP design, fabrication, and catalysis, Pt is still the essential element needed for electrocatalysis and is a huge limiting factor in pursuing large-scale commercial applications of these catalysts in PEMFCs. The studies and more understanding on Pt catalysis have motivated the search for new NP catalysts consist of earth-abundant elements. For example, recent studies have indicated that carbon-based catalysts, including carbon nanotubes, metal-carbon composites, and graphene, are a promising new class of nanocatalysts for fuel cell reactions. Once their structure is better defined and their activity/stability is further optimized, they will replace Pt and make it possible to commercialize fuel cell technology for renewable energy applications.

REFERENCES

[1] G. Lota, K. Fic, E. Frackowiak, Energy Environ. Sci. 4 (2011) 1592–1605.
[2] B. Seger, P.V.J. Kamat, J. Phys. Chem. C 113 (2009) 7990–7995.
[3] Y.-J. Wang, D.P. Wilkinson, J. Zhang, J. Chem. Rev. 111 (2011) 7625–7651.
[4] S. Guo, S. Zhang, S. Sun, Angew. Chem. Int. Ed. 52 (2013) 8526–8544.
[5] Y. Bing, H. Liu, L. Zhang, D. Ghosh, J. Zhang, J. Chem. Soc. Rev. 39 (2010) 2184–2202.
[6] Z. Chen, D. Higgins, A. Yu, L. Zhang, J. Zhang, Energy Environ. Sci. 4 (2011) 3167–3192.
[7] X. Zhao, M. Yin, L. Ma, L. Liang, C. Liu, J. Liao, T. Lu, W. Xing, Energy Environ. Sci. 4 (2011) 2734–2753.
[8] N.V. Rees, R.G. Compton, J. Solid State Electrochem. 15 (2011) 2095–2100.
[9] R. Bashyam, P. Zelenay, Nature 443 (2006) 63–66.
[10] Z. Chen, M. Waje, W. Li, Y. Yan, Angew. Chem. Int. Ed. 46 (2007) 4060–4063.
[11] C. Koenigsmann, E. Sutter, T.A. Chiesa, R.R. Adzic, Nano Lett. 12 (2012) 2013–2020.
[12] C.B. Murray, C.R. Kagan, M.G. Bawndi, Annu. Rev. Mater. Sci. 30 (2000) 545–610.
[13] J. Park, J. Joo, S.G. Kwon, Y. Jang, T. Hyeon, Angew. Chem. Int. Ed. 46 (2007) 4630–4660.
[14] S. Zhang, J. Lee, S. Sun, Open Sur. Sci. J. 4 (2012) 26–34.
[15] J. Lee, S. Zhang, S. Sun, Chem. Mater. 25 (2013) 1293–1304.
[16] C.-J. Zhong, J. Luo, B. Fang, B.N. Wanjala, P.N. Njoki, R. Loukrakpam, J. Yin, Nanotechnology 21 (2010) 062001–062010.
[17] W. Gu, D.R. Baker, Y. Liu, H.A. Gasteiger, in: W. Vielstich, H.A. Gasteiger, H. Yokokawa (Eds.), Handbook of Fuel Cells – Fundamentals, Technology and Applications, vol. 6, John Wiley & Sons, Ltd, Chichester, 2009, p. 631.
[18] N.M. Markovic, T. Schmidt, V.R. Stamenkovic, P.N. Ross, Fuel Cells 1 (2) (2001) 105–116.
[19] V.A. Sethuraman, J.W. Weidner, A.T. Haung, S. Motupally, L.V. Protsailo, J. Electrochem. Soc. 155 (2008) B50–B57.

[20] K.J. Kinoshita, J. Electrochem. Soc. 137 (1990) 845–848.
[21] V. Mazumder, Y. Lee, S. Sun, Adv. Funct. Mater. 20 (2010) 1224–1231.
[22] T.S. Ahmadi, Z.L. Wang, T.C. Green, A. Henglein, M.A. El-Sayed, Science 272 (1996) 1924–1926.
[23] J.M. Peoski, Z.L. Wang, T.C. Green, M.A. El-Sayed, J. Phys. Chem. B 102 (1998) 3316–3320.
[24] C. Wang, H. Daimon, Y. Lee, J. Kim, S. Sun, J. Am. Chem. Soc. 129 (2007) 6974–6975.
[25] C. Wang, H. Daimon, T. Onodera, T. Koda, S. Sun, Angew. Chem. Int. Ed. 47 (2008) 3588–3591.
[26] N.M. Markovic, H.A. Gasteiger, P.N. Ross, J. Phys. Chem. 99 (1995) 3411–3415.
[27] T. Yu, D.Y. Kim, Y. Xia, Angew. Chem. Int. Ed. 50 (2011) 2773–2777.
[28] J.K. Norskov, J. Rossmeisl, A. Logadottir, L. Lindqvist, J. Phys. Chem. B 108 (2004) 17886–17892.
[29] H. Wang, Y. Guo, G. Lu, P. Hu, J. Chem. Phys. 130 (2009) 224701–224706.
[30] V.R. Stamenkovic, B.S. Mun, K.J.J. Mayrhofer, P.N. Ross, N.M. Markovic, J. Rossmeisl, J. Greeley, J.K. Norskov, Angew. Chem. Int. Ed. 45 (2006) 2897–2901.
[31] V.R. Stamenkovic, B.S. Mun, M. Arenz, K.J.J. Mayrhofer, C.A. Lucas, G. Wang, P.N. Ross, N.M. Markovic, Nat. Mater. 6 (2007) 241–247.
[32] T. Toda, H. Igarashi, M. Watanabe, J. Electrochem. Soc. 145 (1998) 4185–4188.
[33] T. Toda, H. Igarashi, H. Uchida, M. Watanabe, J. Electrochem. Soc. 146 (1999) 3750–3756.
[34] C. Wang, M. Chi, D. Li, D. van der Vliet, G. Wang, Q. Lin, J.F. Mitchell, K.L. More, N.M. Markovic, V.R. Stamenkovic, ACS Catal. 1 (2011) 1355–1359.
[35] C. Wang, D. van der Vliet, K.-C. Chang, H. You, D. Strmcnik, J.A. Schlueter, N.M. Markovic, V.R. Stamenkovic, J. Phys. Chem. C 113 (2009) 19365–19368.
[36] V.R. Stamenkovic, B. Fowler, B.S. Mun, G. Wang, P.N. Ross, C.A. Lucas, N.M. Markovic, Science 315 (2007) 493–497.
[37] J. Zhang, H. Yang, J. Fang, S. Zou, Nano Lett. 10 (2010) 638–644.
[38] S. Guo, D. Li, H. Zhu, S. Zhang, N.M. Markovic, V.R. Stamenkovic, S. Sun, Angew. Chem. Int. Ed. 52 (2013) 3465–3468.
[39] P. Strasser, S. Koh, T. Anniyev, J. Greeley, K.L. More, C. Yu, Z. Liu, S. Kaya, D. Nordlund, H. Ogasawara, M.F. Toney, A. Nilsson, Nat. Chem. 2 (2010) 454–460.
[40] H. Zhu, S. Zhang, S. Guo, D. Su, S. Sun, J. Am. Chem. Soc. 135 (2013) 7130–7133.
[41] S. Sun, C.B. Murray, D. Weller, L. Folks, A. Moser, Science 287 (2000) 1989–1992.
[42] S. Zhang, X. Zhang, G. Jiang, H. Zhu, S. Guo, D. Su, G. Lu, S. Sun, J. Am. Chem. Soc. 136 (2014) 7734–7739.
[43] H. Yang, Angew. Chem. Int. Ed. 50 (2011) 2674–2676.
[44] J. Zhang, Y. Mo, M.B. Vukmirovic, R. Klie, K. Sasaki, R.R. Adzic, J. Phys. Chem. B 108 (2004) 10955–10964.
[45] K. Sasaki, H. Naohara, Y. Cai, Y.M. Choi, P. Liu, M.B. Vukmirovic, J.X. Wang, R.R. Adzic, Angew. Chem. Int. Ed. 49 (2010) 8602–8607.
[46] C. Wang, D. van der Vliet, K.L. More, N.J. Zaluzec, S. Peng, S. Sun, H. Daimon, G. Wang, J. Greeley, J. Pearson, A.P. Paulikas, G. Karapetrov, D. Strmcnik, N.M. Markovic, V.R. Stamenkovic, Nano Lett. 11 (2011) 919–926.
[47] V. Mazumder, M. Chi, K.L. More, S. Sun, J. Am. Chem. Soc. 132 (2010) 7848–7849.
[48] V. Mazumder, M. Chi, K.L. More, S. Sun, Angew. Chem. Int. Ed. 49 (2010) 9368–9372.
[49] X. Sun, D. Li, Y. Ding, W. Zhu, S. Guo, Z. Wang, S. Sun, J. Am. Chem. Soc. 136 (2014) 5745–5759.
[50] J. Yang, X. Chen, X. Yang, J.Y. Ying, Energy Environ. Sci. 5 (2012) 8976–8981.
[51] S. Khanal, N. Bhattarai, J.J. Velazquez-Salazar, D. Bahena, G. Soldano, A. Ponce, M.M. Mariscal, S. Mejie-Rosales, M. Jose-Yacaman, Nanoscale 5 (2013) 12456–12463.
[52] S. Guo, S. Zhang, D. Su, S. Sun, J. Am. Chem. Soc. 135 (2013) 13879.
[53] A. Chen, P. Holt-Hindle, Chem. Rev. 110 (2010) 3767–3804.
[54] X. Yu, P.G. Pickup, J. Power Sources 182 (2008) 124–132.
[55] Y. Liu, M. Chi, V. Mazumder, K.L. More, S. Soled, J.D. Henao, S. Sun, Chem. Mater. 23 (2011) 4199–4203.
[56] Y.H. Lee, G. Lee, J.H. Shim, S. Hwang, J. Kwak, K. Lee, H. Song, J.T. Park, Chem. Mater. 18 (2006) 4209–4211.
[57] Y. Zhang, M. Janyasupab, C.-W. Liu, X. Li, J. Xu, C.-C. Liu, Adv. Funct. Mater. 22 (2012) 3570–3575.
[58] Y. Kang, C.B. Murray, J. Am. Chem. Soc. 132 (2010) 7568–7569.

[59] Y. Kang, J.B. Pyo, X. Ye, T.R. Gordon, C.B. Murray, ACS Nano 6 (2012) 5642–5647.
[60] D. Xu, Z. Liu, H. Yang, Q. Liu, J. Zhang, J. Fang, S. Zou, K. Sun, Angew. Chem. Int. Ed. 48 (2009) 4217–4221.
[61] Y. Liu, D. Li, V.R. Stamenkovic, S. Soled, J.D. Henao, S. Sun, ACS Catal. 1 (2011) 1719–1723.
[62] A.-X. Yin, X.-Q. Min, Y.-W. Zhang, C.-H. Yan, J. Am. Chem. Soc. 133 (2011) 3816–3819.
[63] N. Kristian, Y. Yanb, X. Wang, Chem. Commun. (2008) 353–355.
[64] J.B. Xu, T.S. Zhao, Z.X. Liang, J. Phys. Chem. C 112 (2008) 17363–17367.
[65] E. Casado-Rivera, D.J. Volpe, L. Alden, C. Lind, C. Downie, T. Vázquez-Alvarez, A.C.D. Angelo, F.J. DiSalvo, H.D. Abrunã, J. Am. Chem. Soc. 126 (2004) 4043–4049.
[66] Y. Kang, L. Qi, M. Li, R.E. Diaz, D. Su, R.R. Adzic, E. Stach, J. Li, C.B. Murray, ACS Nano 6 (2010) 2818–2825.
[67] S. Zhang, S. Guo, H. Zhu, D. Su, S. Sun, J. Am. Chem. Soc. 134 (2012) 5060–5063.
[68] M. Baldauf, D.M. Kolb, J. Phys. Chem. 100 (1996) 11375–11381.
[69] M. Jin, H. Zhang, Z. Xie, Y. Xia, Energy Environ. Sci. 5 (2012) 6352–6357.
[70] V. Mazumder, S. Sun, J. Am. Chem. Soc. 131 (2009) 4588–4589.
[71] V. Mazumder, M. Chi, M.N. Mankin, Y. Liu, Ö. Metin, D. Sun, K.L. More, S. Sun, Nano Lett. 12 (2012) 1102–1106.
[72] M. Jin, H. Zhang, Z. Xie, Y. Xia, Angew. Chem. Int. Ed. 50 (2011) 7850–7854.
[73] T. Jin, S. Guo, J.-L. Zuo, S. Sun, Nanoscale 5 (2013) 160–163.

CHAPTER 11

Nanofaceted Metal Surfaces: Structure, Reactivity, and Applications

Wenhua Chen*, Hao Wang, Robert A. Bartynski
Department of Physics and Astronomy, and Laboratory for Surface Modification, Rutgers, The State University of New Jersey, Piscataway, NJ, USA
*Corresponding author: E-mail: wchen@physics.rutgers.edu

1. INTRODUCTION

Shape control of nanoparticles (nanoclusters) has received increasing attention due to the challenge for achieving high selectivity to desired reaction products in catalysis [1–7]. Recent studies have demonstrated that selectivity in catalysis can be tuned by controlling the nanoparticle shape [8,9] or the morphology (shape) of a single-crystal surface [10,11]. Bratlie and coworkers have investigated benzene hydrogenation on Pt nanoparticles of different shapes and discovered that both cyclohexane and cyclohexene were formed on cuboctahedral nanoparticles, whereas only cyclohexane was produced on cubic nanoparticles [8]. Studies on metal single-crystal surfaces have demonstrated that selectivity to N_2 is higher on faceted Ir(210) exposing three-sided pyramids than on planar Ir(210) in decomposition of NO at high coverage [10] and in reduction of NO by CO at low CO coverage [11].

Highly dispersed supported catalysts that are used in industry are usually composed of oxide-supported nanoparticles that have a wide distribution of sizes and shapes. It has been shown that the shape, size, and composition of the nanoparticles, as well as the support material, all influence the reactivity and selectivity of the catalysts. To elucidate the reaction mechanisms, which is in turn beneficial to the design and development of new catalysts, macroscale model catalysts (planar surfaces of metal single crystals) or nanoscale model catalysts (metallic nanoclusters grown on planar single-crystal surfaces) are usually used to investigate the reactions. In the last decade, various innovative methods have been developed for synthesizing nanoparticles and great progress has been made in controlling their shape and size [4,5,7,12–14]. The shape-controlled nanoparticles have well-defined surfaces as they are the result of a synthetic process that enables well-regulated nucleation and growth at the atomic level. As an example, Figure 1 shows Ag, Au, and Pt nanoparticles synthesized via a colloidal process, which have well-defined shapes and sizes [4].

Faceted metal single-crystal surfaces represent a unique type of nanoscale model catalysts [10,11,15–22]. Faceting occurs when an initially planar surface undergoes a

Catalysis by Materials with Well-Defined Structures
http://dx.doi.org/10.1016/B978-0-12-801217-8.00011-6

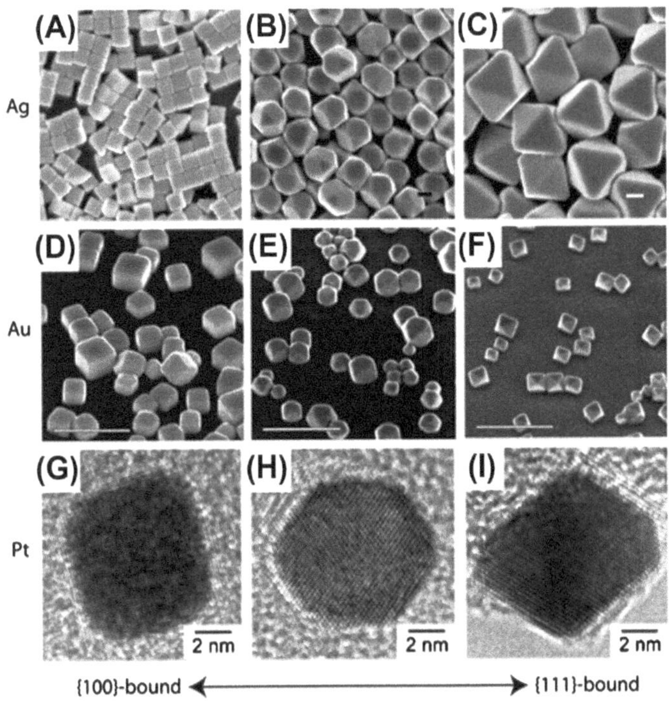

Figure 1 Nanoparticles of Ag (A-C, scale bar = 100nm), Au (D-F, scale bar = 1um), and Pt (G-I, scale bar = 2nm) with controlled shapes and sizes synthesized with colloidal process. *Reproduced from Ref. [4] with permission. Copyright (2008) Wiley-VCH Verlag GmbH & Co. KGaA, Weinheim.*

morphological rearrangement exposing new crystal faces (facets) of nanoscale dimensions. The phenomenon of faceting has been widely observed not only on many planar metal surfaces [23–25] but also on curved metal surfaces [26–33], when these surfaces interact with strongly adsorbed species at elevated temperatures. Moreover, the formation of facets has also been observed on supported catalyst particles under reaction conditions [34,35]. Prolonged heating at elevated temperatures can cause a small particle to achieve its equilibrium crystal shape, which is dictated by minimization of the total surface free energy. These metal particles can undergo reversible shape changes when reaction conditions vary [34,36,37]. Studies under ultrahigh vacuum (UHV) conditions have revealed that the faceted $O/Re(11\bar{2}1)$ surface undergoes a reversible morphological transition as a function of surface oxygen coverage and the structural connections between the various facets observed on different surface morphologies shed light on the mass transport pathways in this morphological evolution [38]. These studies enable investigation of the mass transport pathways in reversible morphological transitions, which in turn will aid in the understanding of reversible shape changes of supported nanoparticles under reaction conditions. Faceted surfaces of metal single crystals have been used as nanoscale model catalysts for a series of chemical reactions [10,11,15–22], as nanotemplates for growth of metallic nanoclusters [38–40] and for synthesis of nanostructured metallic monolayer (ML) electrocatalysts [41]. The well-defined facets allow for detailed experimental and theoretical characterization. A good example is a recent investigation of adsorption and desorption of hydrogen on clean planar Ir(210) and clean faceted Ir(210), in which calculations successfully explained, in semiquantitative fashion, why planar Ir(210) exhibits a single peak in its H_2 thermal desorption spectrum, while faceted Ir(210) exhibits two [22]. Studies of chemical reactions on planar and faceted metal surfaces can provide insights into structure–reactivity/selectivity relationships, which is crucial for improving reactivity and selectivity by controlling the nanoscale morphology and size distribution of the heterogeneous catalysts.

Faceting is a general phenomenon on many surfaces including planar surfaces of semiconductors, oxides, and metals as well as curved surfaces. In this review, we focus on faceted and planar metal single-crystal surfaces and their relevance to catalysis. First, an overview of adsorbate-induced faceting of metal surfaces is provided and methods for the identification of the facet orientations are illustrated. This is followed by chemical reactions on faceted and planar metal single-crystal surfaces with focus on structure sensitivity and size effects, two important issues in catalysis. Finally, the utilization of faceted metal surfaces as nanotemplates for growth of metallic nanoclusters with narrow size distribution and regular spacing as well as for synthesis of nanostructured metallic ML electrocatalysts is discussed.

2. ADSORBATE-INDUCED FACETING OF METAL SURFACES

2.1 Overview of Faceting

In general, *morphology* refers to the shape or topography of a surface (e.g., planar or faceted surface) while *structure* refers to the arrangement of atoms on the surface (e.g., atomic structures of Ir(210), Ir(110), and Ir(311)). The formation of facets on an initially planar surface leads to a substantial change in both surface morphology and surface structure since the new crystal planes exposed on the facets are usually different from that on the initial planar surface.

The shape of a crystal with minimum free energy, i.e., the so-called equilibrium crystal shape (ECS), can be determined by the Wulff construction [42], and the formation of facets on solid surfaces is often discussed in terms of the ECS. The surfaces in the ECS all have small γ (specific surface free energy). When facets form on an initially planar surface with specific surface free energy γ_0 and surface area A_0, the total surface free energy of the faceted surface, $\Sigma_i \gamma_i A_i$, is smaller than that of the original planar surface, i.e.,

$$\Sigma_i \gamma_i A_i < \gamma_0 A_0 \tag{1}$$

where γ_i and A_i are the specific surface free energy and surface area of the ith facet, respectively. If we denote the tilt angle between the ith facet and the planar surface as

θ_i and the projected area of the ith facet on the planar surface as A_i^p, Eqn (1) can be rewritten as

$$\Sigma_i \gamma_i \lambda_i / cos\theta_i < \gamma_0 \quad (2)$$

where $\lambda_i = A_i^p / A_0$ is the partial contribution of the ith facet to the total projected surface area so that $\Sigma_i \lambda_i = 1$. As an example, Figure 2 shows the ECS of Mo constructed with surface energies calculated from first principals [43].

Faceting of a clean metal surface is not commonly observed because the anisotropy in surface free energy is generally too small for spontaneous faceting to occur. However, the presence of an adsorbate can enhance the anisotropy in surface free energy considerably, making faceting more energetically favorable. Notably, in the presence of sufficient thermal energy, surfaces can spontaneously rearrange to minimize their total surface free energies (e.g., by developing facets) even if this may involve an increase in the total surface area. There are a series of reports of metal-induced faceting of single crystal metal surfaces, including metal (Pt, Pd, Au, Ir, Rh) on W(111) and Mo(111) [44–50], metal (K) on Cu(115) [51] and metal (Pd and Au) on Ta(111) [52,53], as well as Pd and Pt overlayer-induced faceting of curved metal surfaces (e.g., a W tip [28,33]). These studies reveal that, when the overlayer coverage of a certain metal on the metal substrates reaches or exceeds one physical monolayer (PML), the substrate metals become fully faceted and the metal overlayer remains on top of (i.e., floats on) the substrates and does not intermix with them [46,47,54]. Here, one PML is defined to be the critical (minimum) coverage

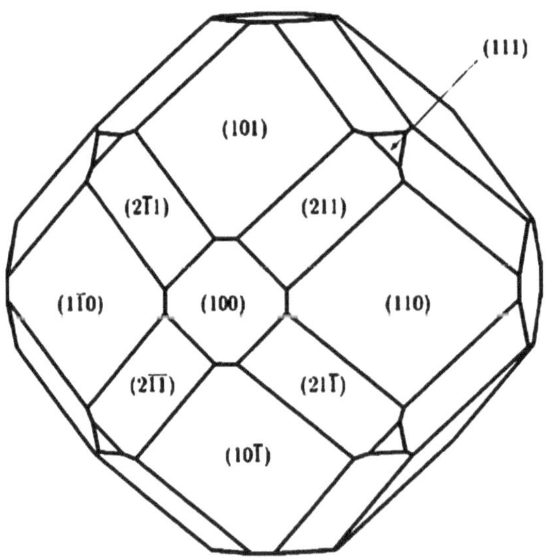

Figure 2 Equilibrium crystal shape of Mo constructed with surface energies calculated from first principles. *Reprinted with permission from Ref. [43]. Copyright (1998) by the American Physical Society.*

of the overlayer metal that is needed so that no substrate atoms are "visible" through the overlayer in the temperature range of facet formation, which was verified using low energy electron diffraction (LEED), low energy ion scattering (LEIS), and soft X-ray photoelectron spectroscopy (SXPS) [54,55]. As an example for metal-induced faceting of metal single-crystal surface, Figure 3 shows typical large and small (with atomic resolution) scale scanning tunneling microscopy (STM) images from Pd-covered faceted W(111) with three-sided pyramidal facets, illustrating well-defined morphology and structure of the faceted surface with narrow facet size distribution [50].

Extensive studies have been carried out on faceting of metal single-crystal surfaces induced by adsorption of nonmetallic species, e.g., (O and Cl) on W(111) [28,56,57], (O, S, N and CO) on Mo(111) [58–60], (N and S) on Fe(111) [61–63], (O and N) on Cu(11n) ($n=5$ and 9) [64,65], N on Cu(410) [66], O/Ir(210) [17,67], O on Rh(210) and Rh(553) [68–70], (O and C) on Rh(755) [71], (O and N) on Ni(210) [72,73], N on Cu(210) [73], C on stepped Ni [74], O/Pt(210) [75], Cl/Ag(111) [76], O on Re$(12\bar{3}1)$ [77], (O, N and C) on Re$(11\bar{2}1)$ [38,41,78], and O on Ru$(11\bar{2}0)$ and Ru$(11\bar{2}1)$ [79,80]. Among these studies, oxygen-induced faceting of metal surfaces with face-centered cubic (fcc) and body-centered cubic (bcc) structures have been investigated the most. In contrast to the one PML coverage required for metal-induced faceting to occur, a fractional ML of a nonmetallic adsorbate can induce the metal substrate to become faceted [28,67,77]. Moreover, unlike metal-induced faceted metal surfaces that often have the same morphology for different adsorbed species [28,44,46,50], the morphology of

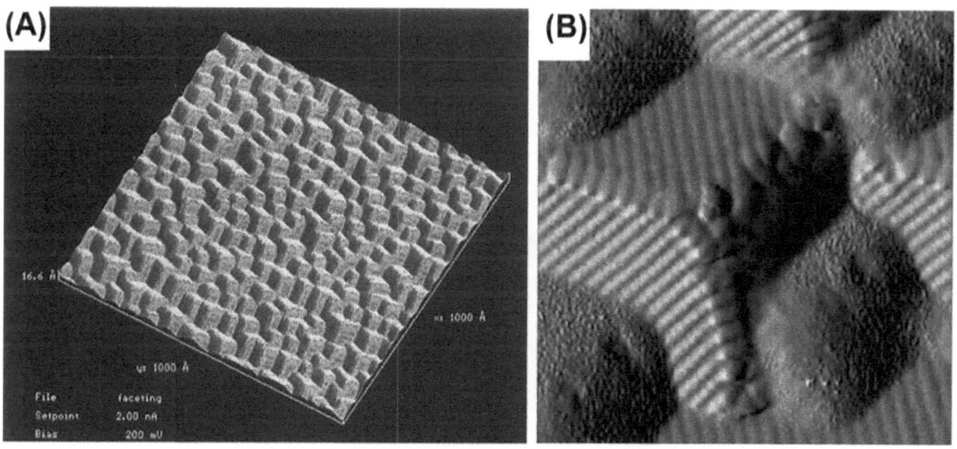

Figure 3 (A) Scanning tunneling microscopy (STM) image (100×100 nm) of Pd-covered faceted W(111), which is prepared by deposition of ~1.2 PML of Pd on W(111) followed by annealing in UHV at 1075 K for 3 min. *(Reprinted from Ref. [50] with permission from Elsevier Science B.V.)* (B) STM image (10×10 nm) of Pd-covered faceted W(111), which is prepared by deposition of ~1.5 PML of Pd on W(111) followed by annealing in UHV at 1075 K for 3 min. *Reprinted from Ref. [106] with permission from author.*

nonmetallic species-induced faceted metal surfaces can change dramatically with adsorbate species, adsorbate coverage and annealing condition (temperature and time). Recent studies have shown that the surface morphology of a metal surface can be tailored by kinetic control of nonmetallic adsorbates [38,41,78]. As schematically illustrated in Figure 4, two-sided ridges, three-sided pyramids, and four-sided pyramids can be formed on the same Re$(11\bar{2}1)$ substrate by kinetic control of the adsorbates (N, C, and O). Significant changes in morphology as a function of oxygen coverage have also been observed for the faceted Re$(12\bar{3}1)$ and Re$(11\bar{2}1)$ surfaces [38,77,81].

Faceting is a thermodynamically driven but kinetically limited process. Therefore, for adsorbate-induced faceting to occur, the system must meet or exceed threshold (lowest) values of both adsorbate coverage and substrate temperature. The adsorbed species enhances the anisotropy of the surface free energy, which is the driving force for faceting, while a sufficiently high substrate temperature is required for surface atoms to overcome the kinetic barriers to diffuse and facilitate facet nucleation. For example, annealing Re$(11\bar{2}1)$ in NH_3 at elevated temperatures leads to N-induced faceting of Re$(11\bar{2}1)$ [78] whereas no faceting of Ru$(11\bar{2}1)$ occurs upon annealing in NH_3 [80]. The lack of N-induced faceting of Ru$(11\bar{2}1)$ is attributed to insufficient nitrogen coverage on the Ru surface due to much lower desorption temperature of N_2 on Ru than that on the Re surface.

In general, the average facet size grows with increasing annealing temperature and time to minimize the contribution from edge and corner energies to the total energy [15,17,65,67,77,80]. Facet size can be characterized by LEED and STM after heating the substrate to elevated temperature to induce faceting followed by cooling to 300 K prior to the measurement. However, knowledge of kinetics of faceting in real space and real time at elevated temperature is extremely useful for getting insights into kinetic

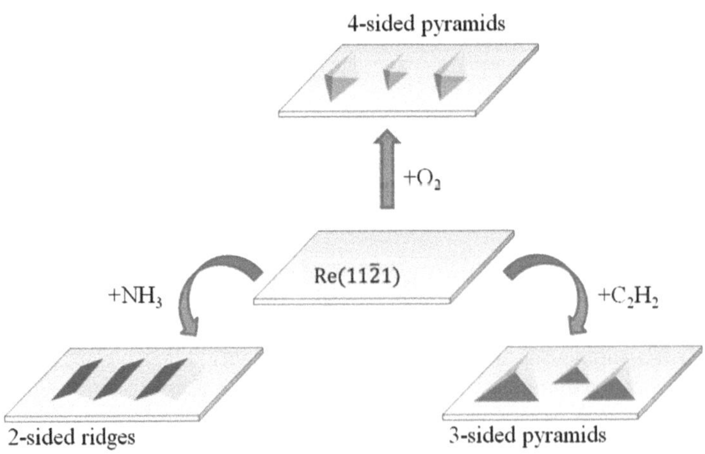

Figure 4 Schematic of O-, N-, and C-induced faceting of Re$(11\bar{2}1)$ by annealing in O_2, NH_3, and C_2H_2.

processes of facet growth. Low energy electron microscopy (LEEM) has proven to be a very powerful surface-sensitive technique for following the nucleation and growth of facets at elevated temperatures in situ, with lateral resolution of <10 nm, which has been successfully used to study faceting of Pt/W(111) [82] and O/Ir(210) [83]. For faceting of O/Ir(210), LEEM data demonstrate that facets nucleate and grow uniformly over the entire surface without formation of "patches" of facets coexisting with planar regions.

Investigation of adsorbate-induced faceting often requires multiple experimental techniques. Auger electron spectroscopy (AES), X-ray photoelectron spectroscopy (XPS), and LEIS are usually used to check the surface cleanliness and characterize the adsorbate coverage while LEED and STM are employed to monitor the surface morphology and structure. As an example, Figure 5 shows X-slope STM images (100×100 nm) from an oxygen-covered faceted $Ru(11\bar{2}1)$ surface [80], which are obtained from the same original STM topological image by differentiating the height along X (scanning direction) (Figure 5(A)) or −X direction (Figure 5(B)), respectively.

While experimental studies provide a wealth of information regarding the structure and morphology of surface faceting, theoretical calculations aid in understanding of the atomic-level aspects of the process. Chen studied adsorption of Pd, Pt, and Au on W(111) using the embedded atom method and the results support that the surface energy anisotropy is the driving force for faceting of the surface [84]. Oleky performed Monte Carlo simulations of metal-induced faceting of different bcc(111) surfaces based on a solid-on-solid model and the results predict a hysteresis effect on the faceting–defaceting phase transition temperature with respect to an alternating heat treatment [85]. First principles calculations based on density functional theory (DFT) have also been used to gain insights into adsorbate-induced faceting. Che et al.'s calculations on metal-induced faceting of

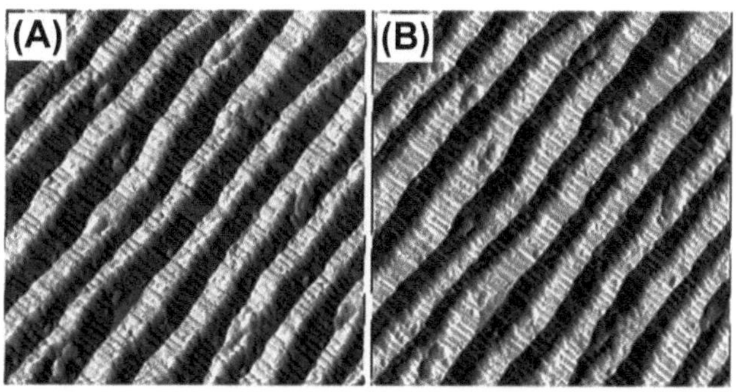

Figure 5 Scanning tunneling microscopy (STM) images (100×100 nm) of oxygen-covered faceted $Ru(11\bar{2}1)$, which is prepared by dosing with 60 L NO_2 at 883 K. (A) X-slope image of the faceted surface obtained from the original STM image by differentiating the height along X direction (scanning direction), which leads to the left side of the ridges brighter. (B) X-slope image obtained from the same original image as (A) by differentiating the height along −X direction, which leads to the right side of the ridges brighter. *Reprinted with permission from Ref. [80]. Copyright (2013) AIP Publishing LLC.*

Mo(111) confirm that faceting of the substrate is driven by the enhancement of the surface energy anisotropy [86]. Che and coworkers's calculations on faceting of W(111) and Mo(111) induced by an ultrathin metal overlayer (Pt, Pd, Rh) indicate that both surfaces undergo faceting to expose {211} facets when the metal (Pt, Pd, Rh) coverage reaches one PML [87,88], in agreement with earlier experimental findings [44–48,54]. Recently, Kaghazchi and Jacob have employed DFT calculations to study faceting of metal surfaces induced by nonmetallic adsorbates [78,89,90]. They investigated the stability of clean and nonmetallic adsorbates (O and N)-covered surfaces of the metal substrates (Ir(210), Re$(11\bar{2}1)$) and the corresponding facets that had been observed experimentally. The DFT calculations provide an atomistic understanding of facet formation in terms of binding sites and energies of adsorbate on the metal surfaces of the substrate and facets as well as the corresponding surface phase diagrams. As an example, Figure 6 shows the surface phase diagram for N-induced faceting of Re$(11\bar{2}1)$ which illustrates the surface free energy of planar and faceted surfaces as functions of the nitrogen chemical potential [78].

2.2 Identification of Facets

There are two geometric constraints for surface faceting to occur: (1) the faceted surface must retain the macroscopic orientation of the original planar surface; (2) the overall symmetry of the original planar substrate must be preserved by the faceted surface [24].

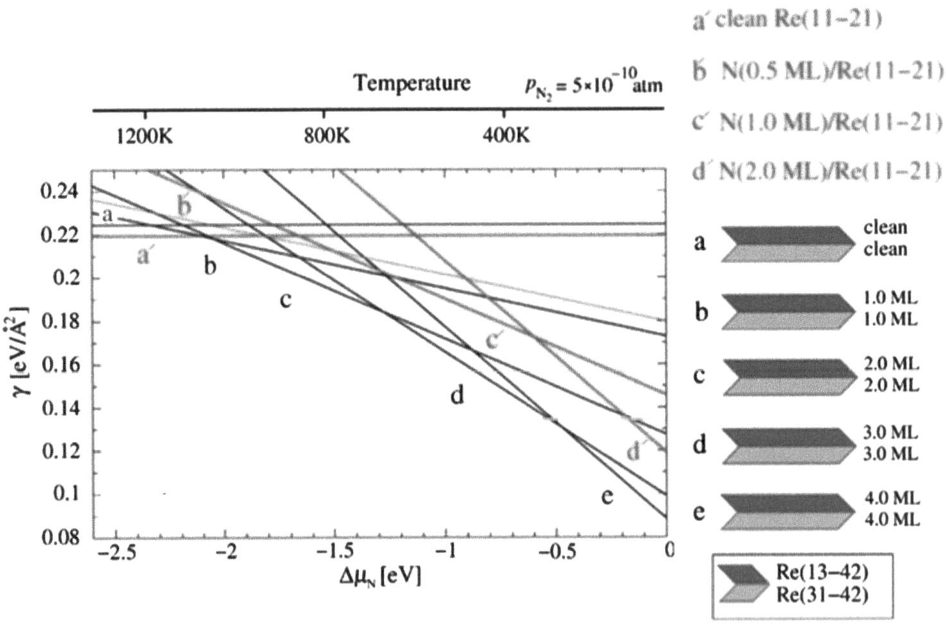

Figure 6 Surface phase diagram for N-induced faceting of Re$(11\bar{2}1)$ which illustrates the surface free energy of planar and faceted surfaces as function of the nitrogen chemical potential. *Reprinted with permission from Ref. [78]. Copyright (2014) AIP Publishing LLC.*

For example, oxygen-induced faceting of both W(111) and Ir(210) produce three-sided pyramids on the surfaces. However, since W(111) and Ir(210) have C_{3v} and C_{1h} symmetry, respectively, the W pyramids expose three {211} facets on each pyramid [28] while the Ir pyramids are elongated with one (110) and two {311} facets [17,67]. Here, the notation {hkl} denotes the set of all planes that are equivalent to (hkl) by the symmetry of the crystal [91]. These two constrains can simplify identification of facet orientations when the surface morphology is complicated.

Coverage dependence of adsorbate-induced morphological changes is observed in the faceting of $\text{Re}(12\bar{3}1)$ induced by oxygen. Depending on the oxygen coverage, a series of facet structures are formed ranging from two-sided ridges to complicated structures exposing five facets [38,77,81]. This complexity is in sharp contrast to the simple structures reported in faceting of many other metals, e.g., W(111) [28] and Ir(210) [17,67]. Below, as an example, we will illustrate how to identify facet orientations of two-sided ridges formed on $\text{Re}(12\bar{3}1)$ induced by oxygen at low coverage [38,77].

$\text{Re}(12\bar{3}1)$ is an atomically rough surface with six layers of atoms exposed and the top-layer atoms show a quasihexagonal arrangement with only C_1 symmetry [77]. Figure 7(A) shows a typical LEED pattern from a clean planar $\text{Re}(12\bar{3}1)$ surface taken at an incident electron energy (E_i) of 70 eV; all the diffraction spots converge to the center of the LEED screen as E_i increases. When the clean $\text{Re}(12\bar{3}1)$ surface is dosed with between 1 and 3 L O_2 (L refers to Langmuir) at 300 K followed by annealing at 900 K, all LEED spots from planar $\text{Re}(12\bar{3}1)$ totally disappear and new LEED spots emerge instead as shown in Figure 7(B). With increasing E_i, these new diffraction spots do not converge to the screen center but rather move toward either of two fixed spots marked by the two circles A and B, which indicates that the surface is no longer planar but completely faceted with a two-sided ridge like morphology. The two fixed spots are specular beams of the facets.

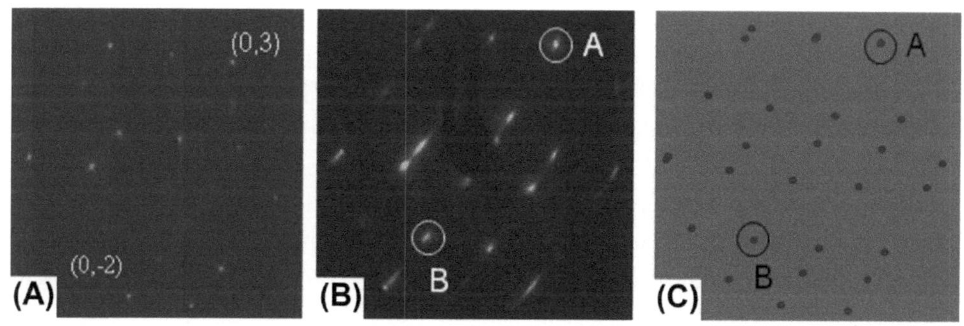

Figure 7 (A) Low energy electron diffraction (LEED) pattern from a clean planar $\text{Re}(12\bar{3}1)$ surface with the (0,3) and (0,−2) diffraction spots labeled; E_i=70 eV. (B) LEED pattern from a faceted O/$\text{Re}(12\bar{3}1)$ surface formed by dosing 3 L O_2 at 300 K followed by annealing at 900 K; E_i=70 eV. (C) The kinematical simulation of (B). *Reprinted with permission from Ref. [77]. Copyright (2006) by the American Physical Society.*

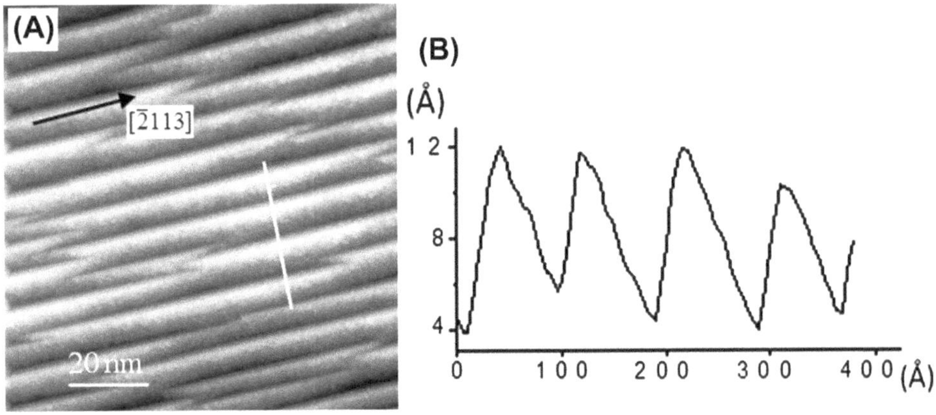

Figure 8 (A) Scanning tunneling microscopy image (100 × 100 nm) of a faceted O/Re$(12\bar{3}1)$ surface prepared by dosing with 1 L O_2 (0.7 ML O) at 300 K followed by annealing at 1000 K. (B) A cross-section profile of the line marked in (A). *Reprinted with permission from Ref. [77]. Copyright (2006) by the American Physical Society.*

The orientations of the two facets are determined by comparing the specular beam positions with the LEED spots from a planar Re$(12\bar{3}1)$ surface [38,77]. The (0,3) diffraction spot at $E_i = 75$ eV from the planar Re$(12\bar{3}1)$ surface is located almost in the same position on the LEED screen as specular beam A in Figure 7(B). The spatial orientations of the (0,3) beam can be determined simply from the structure of the reciprocal rods of Re$(12\bar{3}1)$ and the Ewald sphere construction, from which the tilt angle between the (0,3) beam and the normal direction of $(12\bar{3}1)$ is calculated to be 44°; therefore the tilt angle of the facet relative to $(12\bar{3}1)$ is 22°. Similarly, a good match is found between the (0,−2) spot from the planar Re$(12\bar{3}1)$ for $E_i = 85$ eV and specular beam B, and the tilt angle between this facet and $(12\bar{3}1)$ is then calculated to be 12.5°. Based on these measurements, the Miller indices of the facets are identified as $(01\bar{1}0)$ and $(11\bar{2}1)$, and the ridge direction is identified as $[\bar{2}113]$. Figure 7(C) shows a kinematical simulation of the LEED pattern in Figure 7(B) for $E_i = 70$ eV, based on 2D lattices of bulk-truncated $(01\bar{1}0)$ and $(11\bar{2}1)$. The similarity between the simulation and the experimental pattern not only confirms the orientations of the facets but also suggests that there is no reconstruction on either of the two facets (i.e., they retain their (1 × 1) structures). Figure 8(A) shows a typical STM image from an oxygen-covered faceted Re$(12\bar{3}1)$ surface having two-sided ridges. The tilt angles of the facets relative to $(12\bar{3}1)$ determined by measuring the slopes of the line segments in the cross-section profile of the STM image (Figure 8(B)) agree well with those identified in LEED experiments.

2.3 Preparation of Clean Faceted Metal Surfaces

An important feature of nonmetallic adsorbate-induced faceting of a metal single-crystal surface is that often the adsorbate can be removed so that the resulting clean faceted

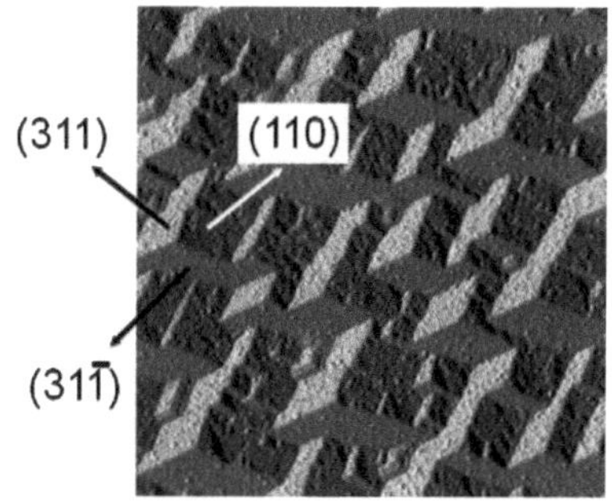

Figure 9 Scanning tunneling microscopy derivative image (100 × 100 nm) from an oxygen-covered faceted Ir(210) prepared by flash annealing Ir(210) in O_2 (5×10^{-8} Torr) to T > 1700 K and subsequent cooling in O_2 to 300 K. Following a flash annealing in O_2, facets form as the crystal cools below ~1150 K. *Reprinted with permission from Ref. [17]. Copyright (2005) American Chemical Society.*

metal surface can serve as a nanoscale unsupported model catalyst [10,11,16–22]. Clean faceted metal surfaces can be prepared by removing the nonmetallic species from the faceted surfaces through reaction with other gas at low temperatures such that the kinetic barrier of facet destruction is not reached. Below, as an example, we will demonstrate how to prepare clean faceted Ir(210).

As shown in Figure 9, upon annealing in O_2 at $T \geq 600$ K, three-sided pyramidal facets exposing (311), $(31\overline{1})$, and (110) faces develop on the initially planar Ir(210) surface and the average facet size increases with annealing temperature [17,67]. Clean faceted Ir(210) with tunable average facet sizes can routinely be prepared from Ir(210) crystal in situ via a three-step process [16,17,20], as illustrated in Figure 10. First, clean planar Ir(210) is prepared by cycles of flash annealing to 1700 K in O_2 (5×10^{-8} Torr) followed by flash annealing to 1700 K in UHV. Second, oxygen-covered faceted Ir(210) with average facet size of 5, 11, and 14 nm is obtained by annealing clean planar Ir(210) in O_2 (5×10^{-8} Torr) at 600 K (2 min), 950 K (2 min), and 1700 K (flash), respectively, and subsequent cooling in O_2 to 300 K. Third, clean faceted Ir(210) is obtained by removing surface oxygen via reaction with H_2 (5×10^{-9} Torr) at 400 K while facets retain their original structure and size. The fact that the size of facets on Ir(210) can be tuned on the nanometer scale without change in morphology and structures of the facets enables investigation of nanoscale size effects in catalytic reactions on unsupported Ir, which excludes not only the influence of the support material but also the effects from changes in catalyst morphology induced by the variation in particle sizes: two effects that are common to supported catalysts.

It should be noted that the above method to prepare clean faceted metal surfaces works only for certain metals with low O binding energies, on which the reaction between oxygen and hydrogen to form water as well as recombinative desorption of hydrogen occur at low temperatures so that the kinetic barrier of facet destruction is not reached. For example, Re has strong binding with O [92], so it is impossible to prepare a clean faceted Re surface.

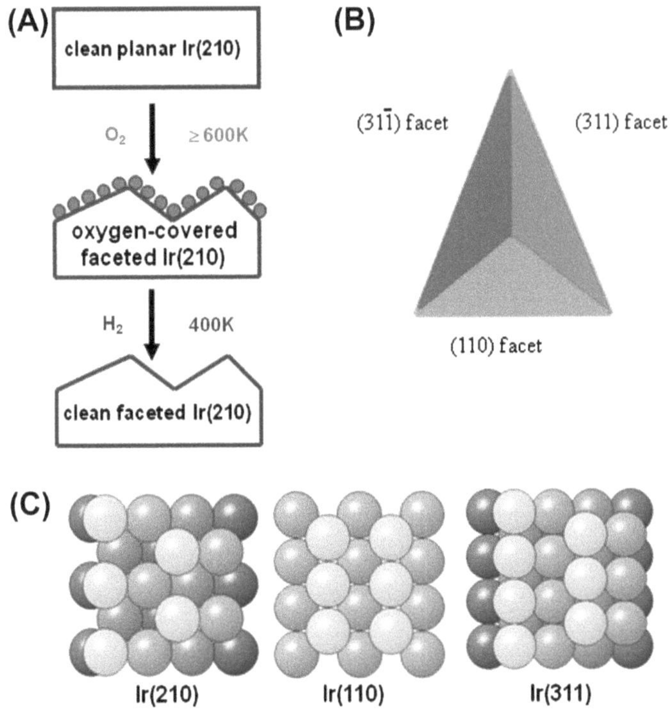

Figure 10 (A) Schematic of the process to prepare clean faceted Ir(210). (B) Schematic of a single pyramid on clean faceted Ir(210). (C) Top view of hard-sphere bulk truncation models of Ir(210), Ir(110), and Ir(311) surfaces.

3. REACTIONS ON FACETED AND PLANAR METAL SURFACES

In this section, we will review reactions performed on faceted and planar metal single-crystal surfaces with focus on ML metal-covered metal substrates as well as clean metal substrates. The variation of the facet sizes mainly leads to a change in density of edge and corner atoms between the facets, but leaves the morphology and facet structure unaffected. The reaction products are detected using temperature programmed desorption (TPD) and the adsorbed species on the surfaces are characterized by high-resolution electron energy loss spectroscopy (HREELS).

3.1 Reactions of Acetylene

Catalytic reactions of acetylene (C_2H_2) have attracted considerable interest because of their relevance to many important reactions. The acetylene reactions proceed through three competing pathways: cyclization to benzene ($3C_2H_2 \rightarrow C_6H_6$) [15,93–99], self-hydrogenation to ethylene ($2C_2H_2 \rightarrow C_2H_4 + 2C$) [15,99] and decomposition to hydrogen and atomic carbon ($C_2H_2 \rightarrow H_2 + 2C$) [16,100,101]. Of particular interest is the cyclization of C_2H_2 to C_6H_6, in which no C–C and C–H bond breaking occurs.

A summary of Pd-catalyzed cyclization of C_2H_2 from UHV to high pressure [102] and also a review of cyclization of C_2H_2 to C_6H_6 on transition metal surfaces including single crystals, supported particles, and bimetallic surfaces have been published elsewhere [98].

Earlier studies have revealed that acetylene cyclization occurs on metal single-crystal surfaces such as Pd(111) [93,103,104] and Cu(110) [96]. This reaction is extremely sensitive to both metal species and its surface structure, e.g., Pd(111) and Cu(110) are highly active toward acetylene cyclization [93,96,103,104] whereas Pd(110) is inactive for acetylene cyclization [100,104]. Reactions of C_2H_2 on Pd(111) studied by Tysoe et al. indicate that at low temperatures (<220 K) a flat-lying acetylenic surface species predominates, which behaves as the precursor for an unusual low temperature cyclization of acetylene to benzene [99]. Using HREELS for acetylene on Pd(111), Marchon has observed feature of the flat-lying benzene molecule at high coverage [105]. Pacchioni and Lambert have performed a theoretical study of the reaction mechanisms and surface intermediates for cyclization of acetylene over Pd(111) [94]. They have identified the structure of a stable surface intermediate C_4H_4 resulting from the combination of two C_2H_2 molecules and have obtained C_6H_6 by the addition of one C_2H_2 molecule to C_4H_4.

Barnes et al. have studied reactions of C_2H_2 on planar Pd/W(111) and faceted Pd/W(111) with different facet sizes as well as planar Pd/W(211) [15]. Annealing planar Pd/W(111) (with one PML Pd) at ≥700 K leads to formation of faceted Pd/W(111) consisting of three-sided pyramids exposing {211} facets as shown in Figure 3. These facets grow in average size with raising annealing temperature [50,106]. Substantial differences were found in the TPD spectra of C_2H_4 and C_6H_6 on faceted Pd/W(111) (≥700 K) as compared to planar Pd/W(111) (<700 K), indicating that both self-hydrogenation and cyclization of C_2H_2 are structure sensitive, as shown in Figure 11. In addition, the TPD spectra of C_2H_4 and C_6H_6 from C_2H_2 adsorbed on faceted Pd/W(111) change as a function of facet size, implying that self-hydrogenation and cyclization of C_2H_2 exhibit size effects on the nanometer scale. In particular, finite size effects are clearly seen for self-hydrogenation of C_2H_2 as evidenced by the fact that TPD spectra of C_2H_4 from faceted Pd/W(111) approach, but are different from planar Pd/W(211), when the size of the {211} facets increases. Using Monte Carlo simulations, Persson et al. have investigated the interplay between the reactions on adjacent facets of nanometer-sized supported catalyst particles [107]. The results indicate that the reaction kinetics on faceted nanocrystals can be remarkably different from those on single-crystal surfaces due to nontrivial coupling of the kinetics of the individual facets.

Abdelrehim et al. have investigated reaction pathways of C_2H_2 on planar Pd/W(211) using TPD and HREELS [108]. They have found that planar Pd/W(211) (with one PML Pd) is highly active for self-hydrogenation of C_2H_2 to C_2H_4 and cyclization of C_2H_2 to C_6H_6. The cyclization of C_2H_2 produces three TPD peaks of benzene at ~340, ~390 and ~430 K (spectra not shown), and the observation of a TPD peak

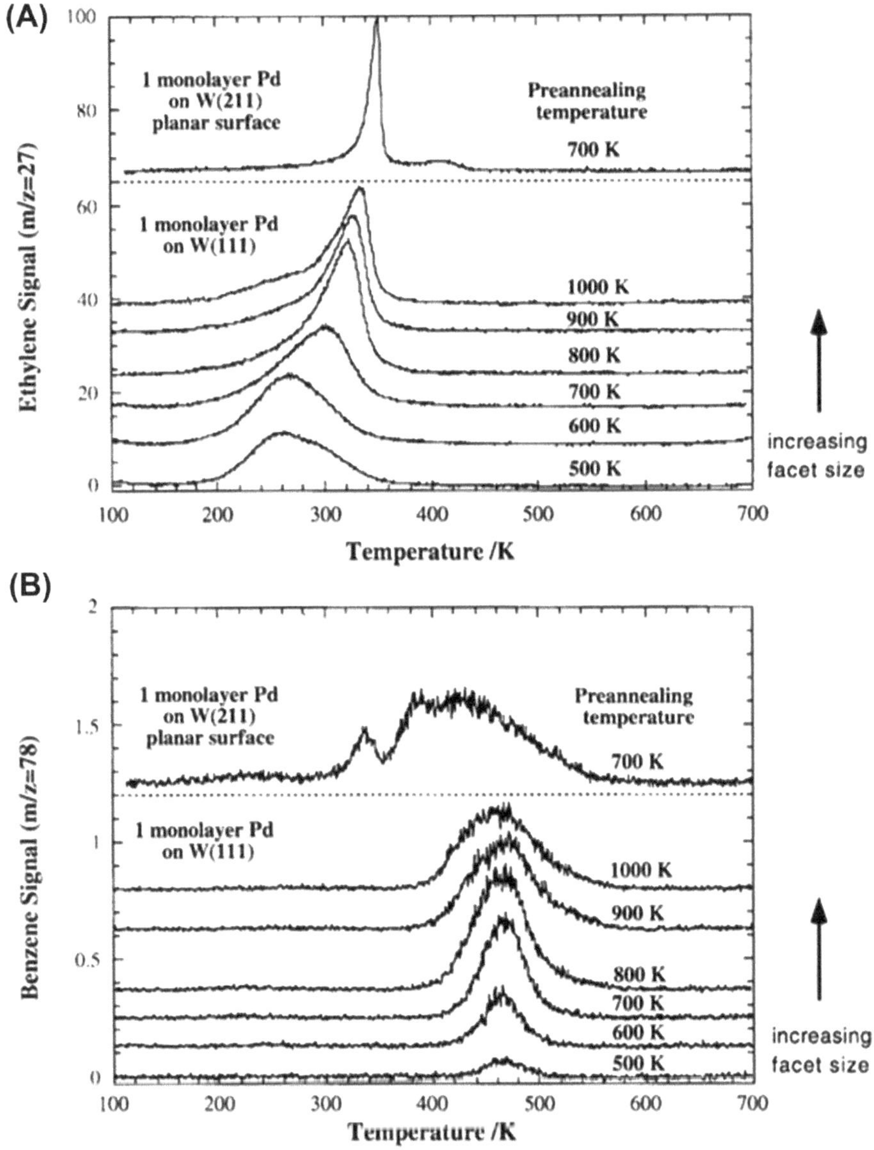

Figure 11 Temperature programmed desorption (TPD) spectra of (A) C_2H_4 and (B) C_6H_6 from adsorption of 3 L C_2H_2 on Pd-covered W(111). In each case, the W(111) substrate is precovered by 1 PML of Pd and then annealed to the indicated temperature prior to exposing C_2H_2. The sizes of {211} facets increase with annealing temperature >700 K. *Reprinted with permission from Ref. [15]. Copyright (2000) Plenum Publishing Corporation.*

corresponding to C_4H_6 strongly suggests that a C_4H_4 surface intermediate is involved in the formation of benzene whose desorption is reaction rate-limited.

Decomposition of C_2H_2 is often the main reaction pathway on metal single-crystal surfaces, which has been observed on Ir(111) [109,110], Ir(755) [109], planar and faceted

Ir(210) [16], Fe(100) [111–113], Fe(111) [112], Rh(111) [114,115], Rh(100) [114,116], Pt(111) [117], Ni(110) [118], Ru(001) and Ru(1110) [119], Pd(110) [100], Pd(100) [120], and W(211) [121]. The adsorption and reaction of C_2H_2 on clean planar Ir(210) and clean faceted Ir(210) with average facet size of ~14 nm have been investigated by TPD and HREELS [16]. Although an Ir organometallic complex is reported to catalyze cyclization of C_2H_2 [122], no evidence for cyclization of C_2H_2 is found on planar Ir(210) or faceted Ir(210). Instead, these two surfaces favor decomposition of C_2H_2. This was the first report of a surface reaction performed on clean faceted monometallic single-crystal surface with well-defined facet structures and controlled facet size on the nanometer scale. The studies also show that decomposition of C_2H_2 is structure sensitive as evidenced by the striking difference in TPD spectra of H_2 from C_2H_2 on the two Ir surfaces (spectra not shown). Upon adsorption of C_2H_2, the dominant surface hydrocarbon species formed on the planar and faceted Ir(210) surfaces at high coverage are mainly acetylide and ethylidyne while acetylide dominates at low coverage.

3.2 Decomposition of Ammonia

The catalytic decomposition of ammonia (NH_3) has been investigated extensively due to its potential for fuel cell applications [123–125]. There are potential benefits and technical advantages of using ammonia as a sustainable fuel for power generation on vehicles [126]. Ammonia decomposition has been reported on many metal single-crystal surfaces, e.g., Fe [112], Rh [127], Ni [128], Pt [129,130], Ru [131–134], and Ir [17,135–138]. In general, ammonia decomposes through stepwise dehydrogenation, and recombinative desorption of N_2 is the rate-limiting step.

In studies of supported metal catalysts, Yin et al. have concluded that supported Ru catalysts exhibit much higher NH_3 conversion than other supported metals (supported Rh, Pt, Pd, Ni, Fe) [139]; substantial effects of support and metal particle sizes have also been observed. Karim et al. have shown that with suitable pretreatment the particle shape of supported Ru/γ-Al_2O_3 catalysts evolves from round particles to flat, elongated ones when the particle size increases, and found that the activity of the Ru particles for NH_3 decomposition depends on the particle shape [140]. Egawa et al. have performed NH_3 decomposition on flat Ru(001) and stepped Ru(1110) surfaces under high pressure (2.7×10^{-5} Pa) at 520 K [131]. They have revealed that in the transient state the rate of N_2 formation on Ru(1110) is an order of magnitude faster than that on Ru(001), implying that stepped sites on the Ru surface play a key role in breaking the N–H bond. Using HREELS combined with TPD, Jacob et al. have identified molecular NH_3 on Ru$(11\bar{2}1)$ upon adsorption at 180 K, and NH_2, NH and N species on the Ru surface after successively annealing to elevated temperatures [132,133], which confirms that the decomposition of NH_3 on the Ru surface proceeds via a stepwise dehydrogenation pathway.

Based on reactor studies (0.01–1 Torr, 500–1900 K) over metal wires and foils, Papapolymerou and Bontozoglou have shown that the ammonia decomposition rate is faster on Ir by several orders of magnitude than that on other metals (Pd, Pt, and Rh) [141].

Choudhary et al. performed ammonia decomposition on Ir(100) over the pressure range from UHV to 1.5 Torr and at temperatures ranging from 200 to 800K [136]. They have found that the activation energy for ammonia decomposition on Ir(100) is in excellent agreement with that on supported Ir catalysts obtained from their earlier studies [135]. Carabineiro et al. have performed TPD measurements following adsorption of NH_3 on Ir(110) and Ir(510), respectively, and found that decomposition of NH_3 on these two surfaces occurs at or near-room temperature and N_2 desorbs around 640K [137,138]. Ammonia decomposition has also been investigated on clean planar Ir(210) and clean faceted Ir(210) with different facet sizes ranging from 5 to 14nm under UHV condition [17], where clean faceted Ir(210) with tunable facet sizes serves an an unsupported nanoscale model catalyst. As shown in Figure 12, the TPD spectra of H_2 and N_2 show structure sensitivity for thermal decomposition of NH_3 on faceted Ir(210) versus planar Ir(210) as well as size effects on faceted Ir(210). On planar Ir(210), there are two TPD peaks of H_2 labeled as α and β with the intensity higher for α than for β. On faceted Ir(210), there are also two TPD peaks of H_2 labeled as α' and β'. As the facet size increases, the relative α' and β' intensities changes in the way that the ratio of α' to β' decreases. In the case of N_2, both planar and faceted Ir(210) display broad peaks, but the peak temperature for faceted Ir(210) is higher than that for planar Ir(210). Furthermore, the N_2 peak temperature increases with facet size. The facet-size dependent changes in H_2 and N_2 spectra indicate that the decomposition of NH_3 on faceted Ir(210) exhibits size effects on the nanometer scale. This was the first observation of size effects in surface chemistry on an unsupported monometallic catalyst with well-defined (facet) structure and controlled (facet) size on the nanometer scale without switching from sample to sample.

3.3 Adsorption and Decomposition of Nitric Oxide

Catalytic conversion of NO to N_2 has been a topic of significant interest because it is a key step in reducing air pollution from automobile exhaust gases [142]. Extensive studies of NO decomposition have been carried out on various supported and unsupported

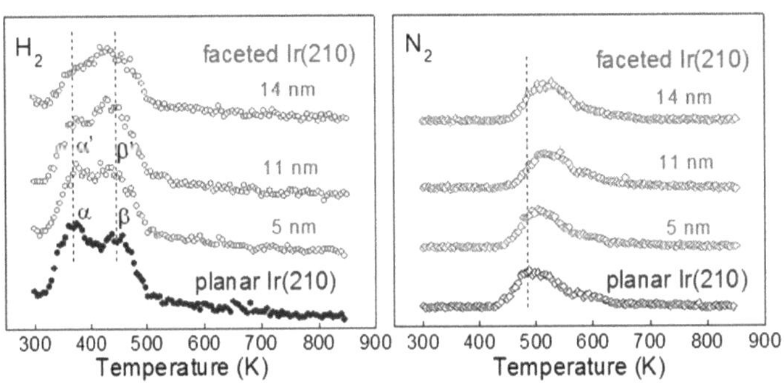

Figure 12 Temperature programmed desorption (TPD) spectra of H_2 and N_2 from clean planar Ir(210) and clean faceted Ir(210) with different facet sizes (5–14nm) following adsorption of 5.0L of NH_3 at 300K. *Reprinted with permission from Ref. [17]. Copyright (2005) American Chemical Society.*

metal catalysts, and it has been shown that both the support and metallic particle size influence the catalytic activity toward NO decomposition [142,143].

NO decomposes on many metal single-crystal surfaces such as Pt [144–148], Ir [10,149,150], Rh [71,114,151,152], Pd [153,154], Ni [155], Au [156,157], Cu [158,159], Ag [160,161], Mo [162], Ru [163,164], and Re [165]. Both experiments and calculations have demonstrated that the low-coordinated surface metal atoms of atomic steps are the most active sites for NO dissociation on hcp (hexagonal close-packed) Ru surfaces [164,166]. Similarly, studies of NO on fcc metal surfaces show that open surfaces containing low coordination sites of C_6 and C_7, which denote surface sites with six or seven nearest-neighbor metal atoms, are much more active compared to the more close-packed surfaces without such low coordination sites [148,154,167]. DFT calculations performed by Inderwildi et al. [168,169] illustrate that the activation barrier for NO dissociation is lower on Rh(311) than on Rh(111) for both clean and O-covered Rh surfaces, although the activation barrier increases with preadsorbed oxygen coverage on each Rh surface. The local geometric structures around the low-coordinated atoms are also important for N–O bond breaking. Using DFT calculations, Ge and Neurock have identified an ensemble of square-arranged Pt atoms as an important feature in activating the N–O bond on Pt surfaces [170].

On planar and faceted Ir(210) surfaces, adsorption of NO at 300 K is predominantly molecular [10]. On planar Ir(210), NO adsorbs at atop sites for the entire NO coverage. On faceted Ir(210) that contains (110) and {311} faces, NO adsorbs at bridge and atop sites at low NO coverage while only at atop sites at high NO coverage. As shown in TPD spectra (Figure 13), at low NO exposure (0.5 L), NO undergoes complete thermal decomposition yielding only N_2 as N-containing products from both surfaces. At high NO exposure (≥3 L), NO desorption from the faceted surface is either absent (3 L) or present (>3 L) in much smaller amounts than what is observed for planar Ir(210), although the amount of N_2 desorption is similar on both surfaces. In addition, no formation of N_2O is observed from faceted Ir(210) while traces of N_2O are formed on planar Ir(210), indicative of 100% selectivity to N_2 from faceted Ir(210). All together, the results illustrate that the thermal decomposition of NO is structure sensitive with faceted Ir(210) being more reactive and selective to N_2 than planar Ir(210) at high NO coverage. No evidence is found for size effects in thermal decomposition of NO over faceted Ir(210) for average facet size ranging from 5 to 14 nm, similar to oxidation of CO by preadsorbed oxygen on the same faceted Ir(210) [18]. This study [10] also concludes that Ir is more active and selective to N_2 formation than Pt, Rh, and Pd in NO decomposition based on the comparison of TPD spectra from adsorption of saturated NO on planar and faceted Ir(210) at 300 K with those from single crystal Pt [171,172], Rh [71], and Pd [173] surfaces presaturated by NO at room temperature. Notably, Pt, Rh, and Pd are constituents of commercial three-way catalysts. Using DFT calculations, Liu et al. have compared energy barriers for dissociation of NO and formation of N_2, N_2O, and NO_2

on Ir(211) and Pt(211), and confirmed that Ir indeed has higher reactivity and selectivity to N_2 in decomposition of NO than Pt having the same surface structure [174].

Catalytic conversion of NO to N_2 under oxygen-rich conditions has received increasing attention due to the challenge for future new exhaust catalysts needed for current diesel and lean-burn engines, which operate under high oxygen concentrations that

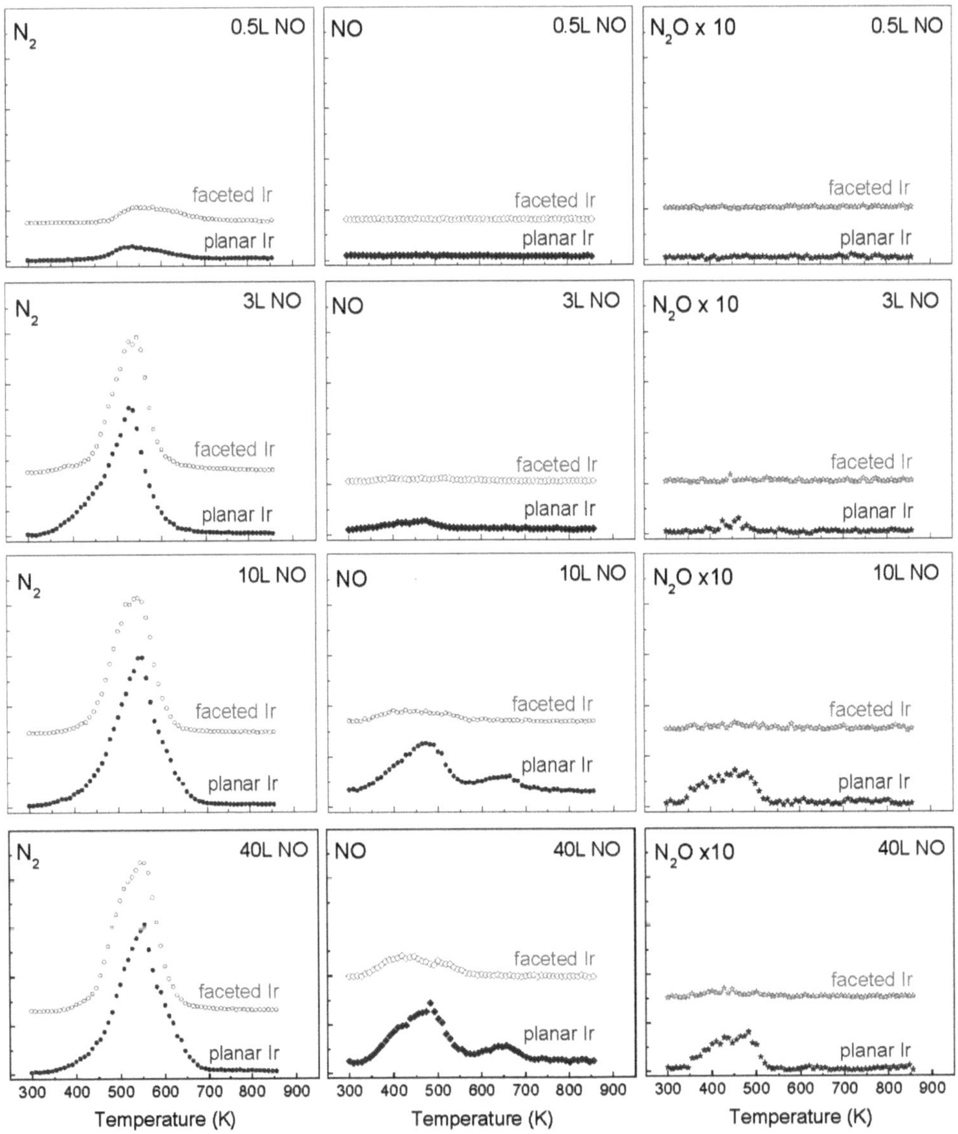

Figure 13 Temperature programmed desorption (TPD) spectra of N_2, NO and N_2O from adsorption of different exposures of NO at 300 K on planar Ir(210) and faceted Ir(210) with average facet size of 14 nm, respectively. Desorption of NO_2 is not observed. *Reprinted with permission from Ref. [10]. Copyright (2008) American Chemical Society.*

poison most platinum group metal catalysts [19]. The adsorption and decomposition of NO in the presence of preadsorbed oxygen have been studied both experimentally and theoretically on many metal single-crystal surfaces such as Pd [175], Pt [171,176–178], Rh [168,169,179–181], Ir [19,182,183], Ni [184,185], Cu [186], Ru [187,188], and Mo [189]. In general, in the presence of preadsorbed oxygen the surface becomes less active for NO dissociation. The reduced reactivity of the surface is most likely governed by the reduced degree of backbonding between the local d-band of the metal and the $2\pi^\star$ antibonding orbital of NO due to electrons withdrawn from the metal by preadsorbed oxygen [169,185]. Nevertheless, when comparing the activation barrier for NO dissociation on Rh single crystals with different surface structures precovered with oxygen, it is found that steric effects exceed electronic effects [169], i.e., the local geometrical structure of the surface plays a key role in determining the activation barrier for NO dissociation.

Figure 14 compares TPD spectra of N_2 and NO from 2L NO on O-covered faceted Ir(210) (with average facet size ~14 nm) with those on O-covered planar Ir(210) [19]. Desorption of N_2O and NO_2 are not observed. When precovered with up to 0.5 ML O, both planar and faceted Ir(210) surfaces exhibit unexpectedly high reactivity for NO decomposition. Upon increasing the oxygen coverage to 0.7 ML O, planar Ir(210) has little activity while faceted Ir(210) remains active toward NO decomposition, although NO decomposition is completely inhibited when both surfaces are preadsorbed by 1 ML O. In the presence of preadsorbed oxygen, a new N_2 peak (P2, F2) develops above 600 K accompanied by NO desorption from both surfaces, in addition to the N_2 peak (P1, F1) observed also on the clean Ir surfaces. The observation that the N_2 peak F2 appears at higher fractional oxygen coverage on faceted Ir(210) than P2 on planar Ir(210) correlates with higher reactivity of faceted Ir(210) for decomposition of NO at higher fractional oxygen coverage. No evidence for size effects in the decomposition of NO on O-covered faceted Ir(210) is observed for average facet size in the range of 5–14 nm.

3.4 Oxidation of CO by Oxygen and NO

3.4.1 Oxidation of CO by Oxygen

Catalytic oxidation of carbon monoxide (CO) by oxygen to carbon dioxide (CO_2) has been extensively studied in relation to air pollution control [190], which are catalyzed by both supported metal nanoparticles [191–193] and metal single-crystal surfaces [194–205]. It has been demonstrated that highly dispersed Ir/TiO_2 catalysts exhibit extremely high catalytic activity for CO oxidation below room temperature [206]. The activity of supported Ir depends on the support material: Ir/TiO_2 is more active than Ir/SnO_2, Ir/Al_2O_3, and Ir/Fe_2O_3 for CO oxidation. Size effects are seen in CO oxidation over supported Au [207], Pt [208], Pd [209], and Ru [210] catalysts and structure sensitivity has been reported in CO oxidation over Rh [211] and Pd [212,213] single-crystal surfaces with different crystal orientations. Moreover, CO oxidation has shown the continuity between UHV studies and high-pressure reactions (i.e., there is no pressure gap) [199] as

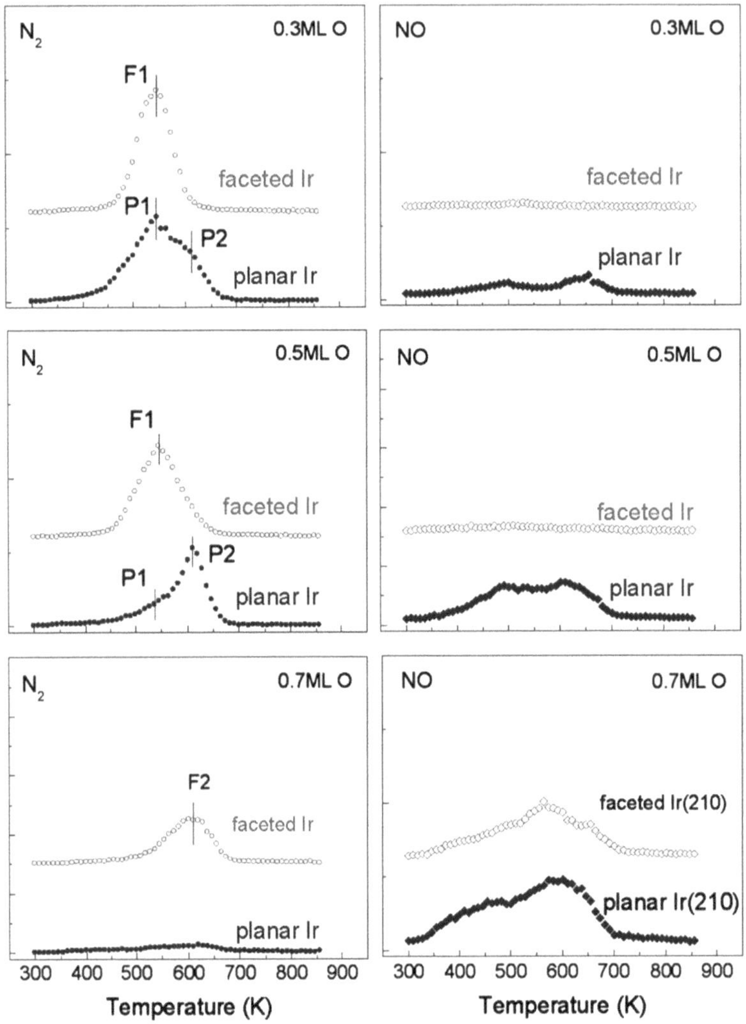

Figure 14 Temperature programmed desorption (TPD) spectra of N_2 and NO from adsorption of 2 L NO on planar Ir(210) and faceted Ir(210) (with average facet size of 14 nm) precovered by 0.3 ML O, 0.5 ML O and 0.7 ML O, respectively. *Reprinted from Ref. [19] with permission from Elsevier B.V.*

well as excellent agreement between results obtained for single-crystal surfaces and supported catalysts [214].

Yoshinobu and Kawai have shown that the desorption temperature of CO_2 formed from CO oxidation on Pt(111) strongly depends on the structure of adsorbed oxygen on Pt(111) [215]. They correlate the CO_2 desorption temperature from low to high with adsorbed O_2, disordered oxygen atoms and p(2 × 2)O, respectively. In contrast, Yasui et al. report that CO oxidation on Pd(110) occurs with various activation energies according to the different adsorbed sites of CO [216]. For CO oxidation on Ir(210), coadsorption

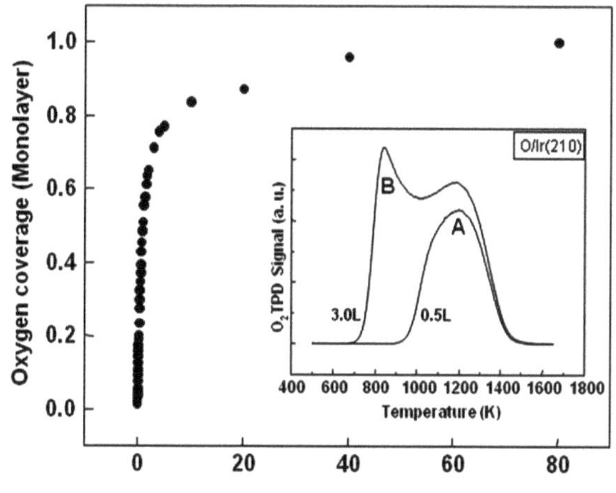

Figure 15 Oxygen uptake curve obtained on the basis of the integrated area under the Temperature programmed desorption (TPD) spectra of O_2 from Ir(210) following adsorption at 300 K. The inset contains the TPD spectra of O_2 following the adsorption of 0.5 L O_2 (0.3 ML O) and 3 L O_2 (0.7 ML O) on Ir(210) at 300 K. *Reprinted with permission from Ref. [18]. Copyright (2006) American Chemical Society.*

systems of O and CO at 300 K with disordered oxygen in different binding-energy states have been prepared [18]. Figure 15 shows the oxygen uptake curve on Ir(210), based on the integrated area under the TPD spectra of O_2 from Ir(210), with two selected spectra shown in the inset; here 1 ML is defined to be the saturation coverage upon exposure of ~80 L O_2 to the Ir(210) surface at 300 K [18,67]. There are two binding states for oxygen on Ir(210). For exposures ≤0.5 L (≤0.3 ML O), O_2 desorbs in a single peak from the high-binding-energy state (A), and the peak position shifts to lower temperature as the exposure increases. With exposures higher than 0.5 L (>0.3 ML O), an additional O_2 desorption peak associated with the low-binding-energy state (B) appears at lower temperatures. TPD studies from CO on O-covered planar Ir(210) [18] show that when O atoms occupy A states, the temperature T_i for the onset of CO_2 desorption decreases with increasing CO coverage. When O atoms occupy both A and B states, T_i is <330 K and independent of CO coverage. This observation indicates that the rate-limiting step for oxidation of CO by preadsorbed oxygen on Ir(210) is the O–Ir bond breaking. This conclusion is consistent with the results from DFT calculations which reveal that the predominant energetic barrier for CO oxidation by oxygen is the strength of the bond between the O atoms and the metal surface [217]. In addition, CO oxidation by oxygen is structure sensitive with planar Ir(210) being more active than faceted Ir(210) at low O precoverage (Figure 16(A)), but no evidence is found for size effects in oxidation of CO by preadsorbed oxygen on faceted Ir(210) for average facet sizes from 5 to 14 nm, as shown in Figure 16(B). For oxidation of CO by oxygen, the structure sensitivity on faceted Ir(210) versus planar Ir(210) as well as the absence of size effects on faceted

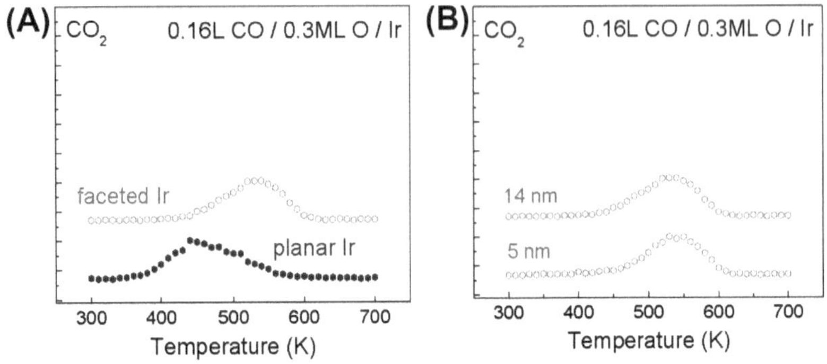

Figure 16 Temperature programmed desorption (TPD) spectra of CO_2 from adsorption of 0.16 L CO on 0.3 ML O-covered (A) planar Ir(210) and faceted Ir(210) with an average facet size of 14 nm (B) faceted Ir(210) with different facet sizes (5 and 14 nm). *Reprinted with permission from Ref. [18]. Copyright (2006) American Chemical Society.*

Ir(210) implies that if there is any size effects seen on supported Ir catalysts one needs to account for effects of the support and change in catalyst structure induced by changing Ir particle size.

3.4.2 Oxidation of CO by NO

Catalytic oxidation of CO by NO is of practical importance because toxic CO and NO can be removed simultaneously. In this reaction, oxidation of CO to CO_2 is accompanied by reduction of NO to N_2. So far, structure sensitivity in oxidation of CO by NO has been observed on single-crystal Pd [218] and Rh [219] surfaces with different surface orientations, while size effects in oxidation of CO by NO have been reported on Pd/MgO(100) [220–222] and faceted Ir(210) [20]. Using Pd clusters grown on MgO(100) as model catalysts for the CO + NO reaction, Picolo and Henry have found that medium sized particles (6.9 nm) exhibiting mainly (111) facets are more active while the largest particles (15.6 nm) exhibiting principally (100) facets are less active [220].

Both planar and faceted Ir(210) favor oxidation of CO to CO_2 by preadsorbed NO, which is accompanied by simultaneous reduction of NO with high selectivity to N_2 [20]. At low NO precoverage, CO_2 desorption temperature (T_i) as well as CO_2 desorption peak temperature (T_p) decreases with increasing CO exposure, and NO dissociation is affected by coadsorbed CO. At high NO precoverage, T_i and T_p are independent of CO exposure, and coadsorbed CO has no influence on dissociation of NO. Moreover, at low NO precoverage, planar Ir(210) is more active than faceted Ir(210) for oxidation of CO to CO_2: T_i and T_p are much lower on planar Ir(210) than those on faceted Ir(210), as shown in Figure 17(A). In addition, faceted Ir(210) with an average facet size of 5 nm is more active for oxidation of CO to CO_2 than faceted Ir(210) with an average facet size of 14 nm as shown in Figure 17(B), i.e., oxidation of CO by preadsorbed NO

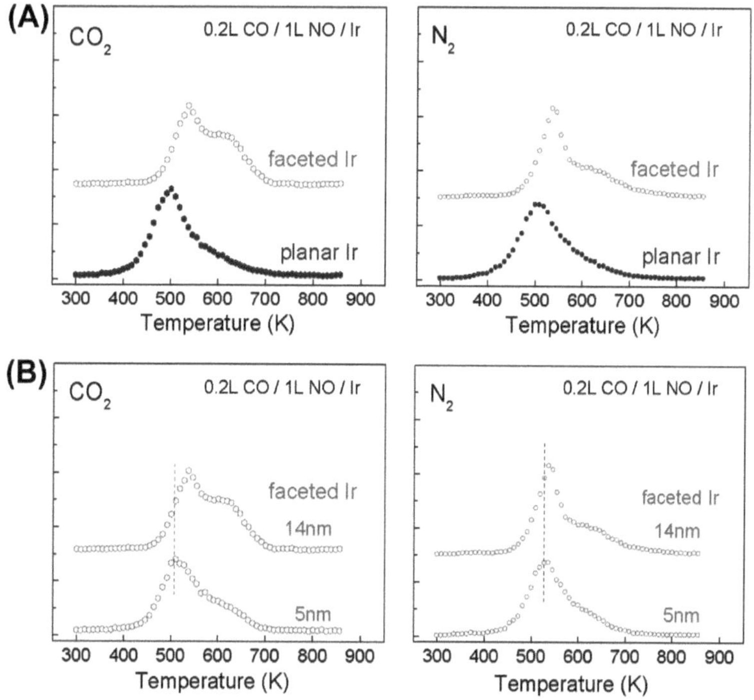

Figure 17 (A) Temperature programmed desorption (TPD) spectra of CO_2 and N_2 from adsorption of 0.2 L CO on 1 L NO predosed planar Ir(210) and faceted Ir(210) with average facet size of 14 nm at 300 K. (B) TPD spectra of CO_2 and N_2 from adsorption of 0.2 L CO on 1 L NO predosed faceted Ir(210) with different facet sizes (5 and 14 nm). *Reprinted with permission from Ref. [20]. Copyright (2012) American Institute of Physics.*

on faceted Ir(210) exhibits size effects on the nanometer scale. The TPD data indicate the same reaction pathway for CO_2 formation from CO + NO and CO + O reactions on planar Ir(210). In other words, formation of CO_2 from oxidation of CO by NO on Ir(210) proceeds via reaction of CO with O that is generated by decomposition of NO.

3.5 Reduction of NO by CO and C_2H_2

3.5.1 Reduction of NO by CO

Various supported and unsupported catalysts have been used to study the reduction of NO by CO both experimentally and theoretically, which is well documented in several review articles [142,223–225]. Among these studies, Pt-, Pd-, and Rh-based catalysts have been investigated the most. Supported Ir particles have shown unusually high reactivity for the reduction of NO by CO in the presence of oxygen [226–228]. In particular, Wang et al. have found that Ir/ZSM-5 is much more active than Pt/ZSM-5 for the reduction of NO by CO in the presence of excess of oxygen [227], signifying a potential application of Ir as a new exhaust catalyst.

Surface isocyanate (NCO), which often forms in the NO + CO reaction on supported catalysts, was found to be located on the supports and not on the metals [229]. NCO also forms on metal single-crystal surfaces under high reaction pressures, e.g., on Pd(111) under a total NO + CO pressure of at least 0.6 mbar [230] and on Ir(211) at an NO + CO pressure of 10 Torr [231]. However, such species were not found on Ir(211) [232] and Ir(111) [233] under UHV conditions. A so-called surface explosion [234] feature, manifested in the sharpness of desorption spectra, has been reported for NO + CO on Pt [234–237] and Pd [173,237–239] single-crystal surfaces under UHV conditions.

Both planar and faceted Ir(210) are highly active for reduction of NO by preadsorbed CO with high selectivity to N_2, which is accompanied by simultaneous oxidation of CO [11]. At low CO precoverage, faceted Ir(210) is more active for reduction of NO by CO with higher selectivity to N_2 than planar Ir(210), as evidenced by less desorption of NO and N_2O from faceted Ir(210) (see Figure 18(A)), and preadsorbed CO does not affect NO dissociation. Thus, the rate-limiting step for reduction of NO by CO on both surfaces is N–O bond breaking, whereby formation of N_2 and CO_2 proceeds through NO dissociation ($NO \rightarrow N + O$) followed by recombination and desorption of N yielding N_2 and reaction of O with CO producing CO_2. At high NO exposure and 1 ML CO precoverage, strong interaction between NO and CO results in "explosive" evolution of N_2 and CO_2 on planar Ir(210) (see Figure 18(B)). In addition, size effects in reduction of NO by CO on faceted Ir(210) for average facet size ranging from 5 to 14 nm are observed at high NO exposure and high CO exposure (1 ML CO) [11], which is evidenced by the pronounced change in spectral profile, peak position, and peak intensity of the TPD spectra of CO_2 and especially N_2 in response to the change in facet size (spectra not shown).

3.5.2 Reduction of NO by C_2H_2

Selective catalytic reduction (SCR) of NO by hydrocarbons is believed to be the most promising way to eliminate NO from automobile exhaust gases [142,240–245]. Studies show that among noble metals, the SCR of NO by hydrocarbons is most active on Pt- and Ir-based catalysts [241], and Ir is more active and selective than Pt at high temperatures [243]. Various hydrocarbons have been used for studying the SCR of NO on supported and unsupported catalysts, which include methane, acetylene, ethylene, propene, propane, and benzene. The reduction of NO by HCs has been studied on Pt [246], Ir [21], Pd [247], and Rh [248] single-crystal surfaces.

Reduction of NO by acetylene (C_2H_2) on planar Ir(210) and faceted Ir(210) with tunable facet sizes (5 and 14 nm) has been investigated by TPD [21]. Upon adsorption, C_2H_2 dissociates to form acetylide (CCH) and H species on the Ir surfaces at low C_2H_2 preoverage [16]. For adsorption of NO on C_2H_2-covered Ir, both planar and faceted Ir(210) exhibit high reactivity for reduction of NO with high selectivity to N_2 at low C_2H_2 precoverage, while the reaction is completely inhibited at high C_2H_2 precoverage.

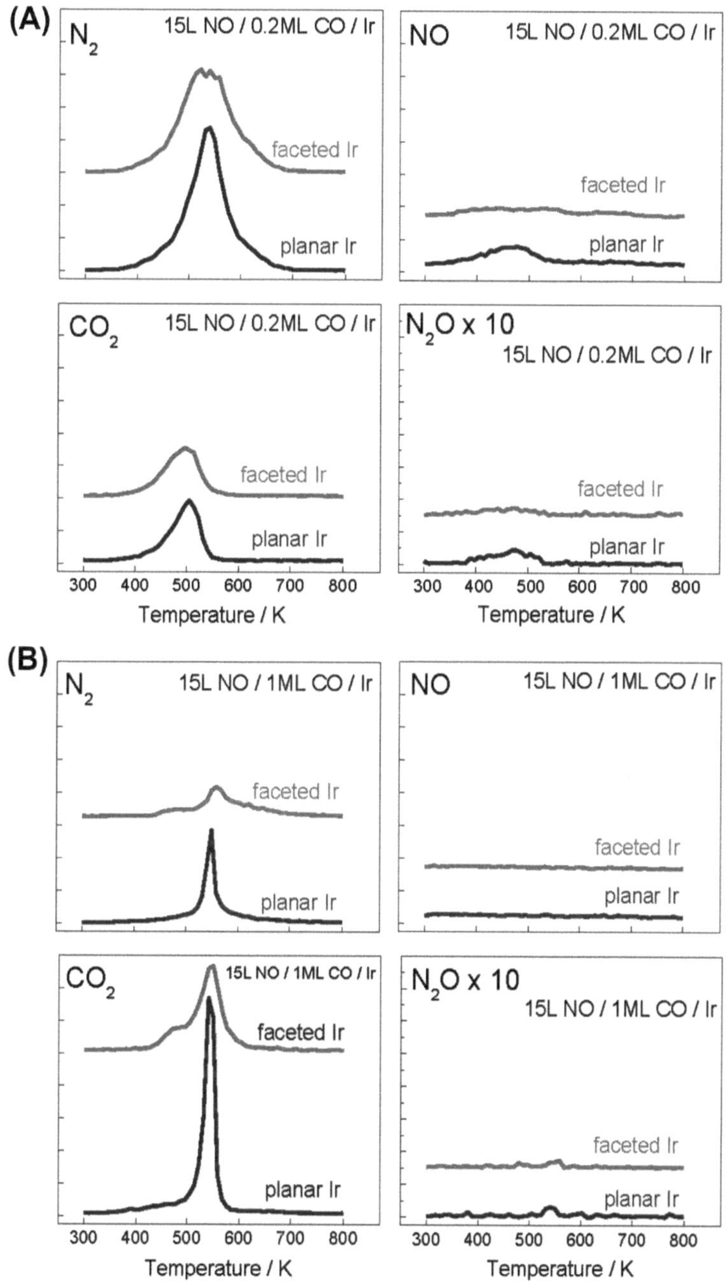

Figure 18 (A) Temperature programmed desorption (TPD) spectra of N_2, NO, N_2O and CO_2 from adsorption of 15 L NO on planar and faceted Ir(210) precovered by 0.2 ML CO. (B) TPD spectra of N_2, NO, N_2O and CO_2 from adsorption of 15 L NO on planar and faceted Ir(210) precovered by 1 ML CO. The faceted surface has an average facet size of 14 nm. *Reprinted with permission from Ref. [11]. Copyright (2010) Wiley-VCH Verlag GmbH & Co. KGaA, Weinheim.*

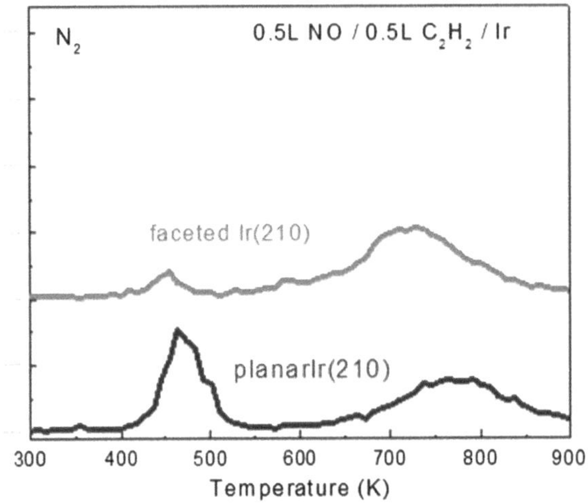

Figure 19 Temperature programmed desorption (TPD) spectra of N_2 from adsorption of 0.5 L NO on planar and faceted Ir(210) preexposed to 0.5 L C_2H_2 at 300 K. The faceted surface has an average facet size of 14 nm. *Reprinted with permission from Ref. [21]. Copyright (2013) American Chemical Society.*

Coadsorbed C_2H_2 significantly influences dissociation of NO. The N-, H-, and C-containing TPD products are dominated by N_2, H_2, CO, and CO_2 together with small amounts of H_2O. For adsorption of NO on C-covered Ir(210) at fractional C pre-coverage, formation of CO_2 is promoted while production of CO is reduced, indicating beneficial effects of surface carbon on the selectivity to C-containing products in reduction of NO. Reduction of NO by C_2H_2 is structure sensitive on faceted Ir(210) versus planar Ir(210) (Figure 19) as evidenced by the striking difference in TPD spectra of N_2 from reduction of NO on these two surfaces, but no evidence is found for size effects in the reduction of NO by C_2H_2 on faceted Ir(210) for average facet sizes of 5 and 14 nm.

3.6 Adsorption and Desorption of Hydrogen

The interaction of hydrogen (H_2) with metal surfaces has attracted considerable interest both as a model system for fundamental research [249–251] and as an energy-storage system for the hydrogen economy [252]. Hydrogen dissociatively adsorbs on many single crystal metal surfaces such as Ir [253,254], Ag [255], Pd [256–258], Ni [259,260], Pt [261], Rh [262–264], Mo [265], Co [266] and Ru [267–269], as confirmed by EELS measurements. H atoms generally prefer to adsorb on high coordination sites, e.g., a preference for N-fold ($N \geq 2$) sites has been identified for adsorption of H atoms on Pd [270,271], Pt [272], Rh [273] and Ir [22,274], with the exception of H atoms preferring to occupy atop sites on Ir(111) [275,276].

A combined TPD and DFT approach has been used to characterize adsorption and desorption of hydrogen on planar Ir(210) and faceted Ir(210) [22]. As shown in TPD spectra in Figure 20, only one H_2 peak is seen from planar Ir(210) at all coverages

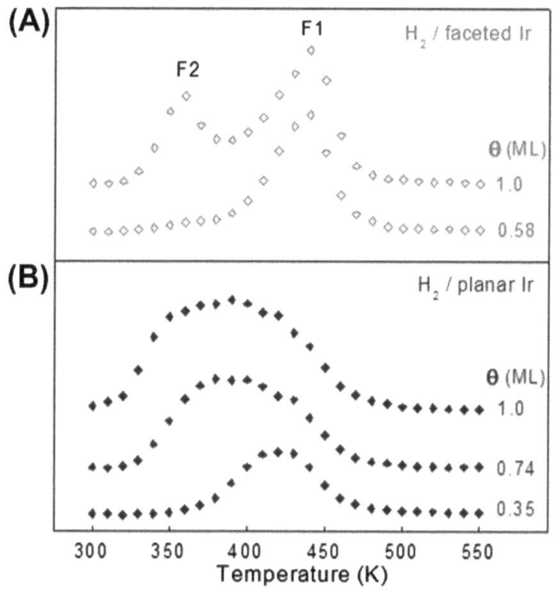

Figure 20 Temperature programmed desorption (TPD) spectra of H_2 from (A) faceted Ir(210) and (B) planar Ir(210) surfaces predosed with H_2 at 300 K. The faceted surface has an average facet size of 14 nm. *Reproduced from Ref. [22] with permission from the PCCP Owner Societies.*

whereas a single H_2 peak is observed at around 440 K at fractional ML coverage and an additional H_2 peak appears at around 360 K at 1 ML coverage from faceted Ir(210), implying structure sensitivity in recombination and desorption of hydrogen on faceted Ir(210) versus planar Ir(210). H atoms prefer to bind at the two-fold short-bridge sites on planar Ir(210) and faceted Ir(210). The surface phase diagrams of H/Ir in Figure 21 obtained from DFT calculations reveal the origin of the two TPD peaks of H_2 from faceted Ir(210): the peak at higher temperature is from desorption of H_2 on {311} facets while the peak at lower temperature is from desorption of H_2 on (110) facets.

4. APPLICATIONS OF FACETED METAL SURFACES

4.1 Growth of Metal Nanoclusters

Supported metal nanoclusters (nanoparticles) exhibit unique physical and chemical properties [277–279], which are different from their bulk materials. For example, Au is very active for CO oxidation when it is highly dispersed over oxide supports [278,280–282], whereas the surface of bulk Au is inactive for CO oxidation. Extensive studies have been performed to assemble metal nanoclusters on planar single-crystal surfaces due to advantage for fundamental study with single crystals while addressing supported catalysts, e.g., effects of particle size and role of the support material [224,277,283,284]. Deposition of metals onto planar oxide surfaces usually leads to heterogeneous

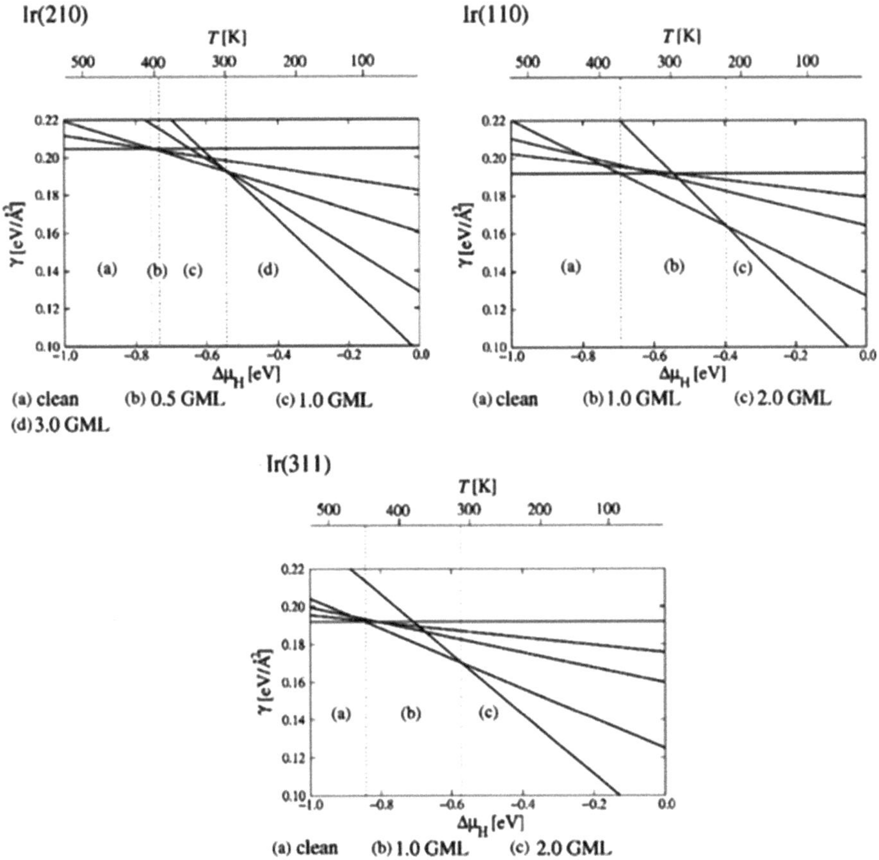

Figure 21 Surface phase diagrams for H adsorption on Ir surfaces showing the surface free energy as a function of the H chemical potential referenced to as $\Delta\mu_H = \mu_H - 1/2E_{H2}^{tot}$. *Reproduced from Ref. [22] with permission from the PCCP Owner Societies.*

nucleation and growth [279,285], in which metal nanoclusters preferentially form at defect sites of the surfaces (step and kink sites, oxygen-vacancy sites, etc.) rather than nucleate homogeneously and grow on terraces. In general, the size and shape distribution of the metal nanoclusters grown on the planar oxide surfaces are relatively broad.

Recently, a simple method that facilitates the growth of metal nanoclusters with narrow size distribution and regular spacings on the nanometer scale has been developed, which employs oxygen-covered faceted metal single-crystal surfaces as templates for growth via physical vapor deposition [38–40]. These templates contain uniform long ridges formed by facets and provide three types of nucleation sites for metal nanoclusters to grow: tops of ridges, valleys between ridges, and sides of ridges. On faceted O/Re$(12\bar{3}1)$ having two-sided ridges formed at low oxygen coverage [77], Wang et al. have successfully synthesized one dimensional arrays of Co

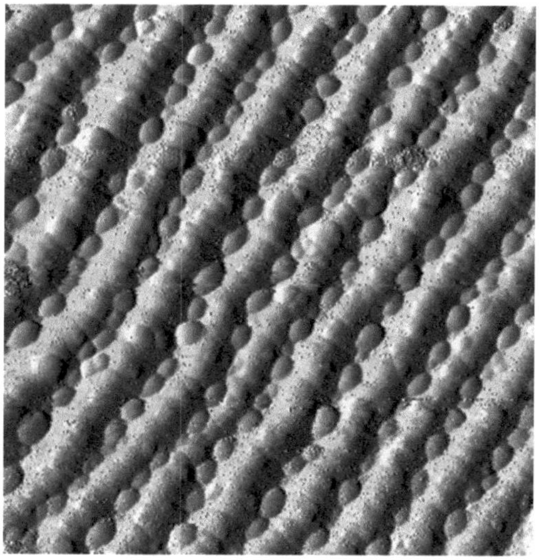

Figure 22 X-slope scanning tunneling microscopy image (50×50 nm) showing 1-D arrays of Au nanoclusters on a faceted O/Ru$(11\bar{2}0)$ surface after deposition of 0.3 ML gold at room temperature. *Reprinted from Ref. [40] with permission from Elsevier B.V.*

nanoclusters after deposition of Co on the faceted O/Re surface at 300 K followed by annealing in UHV at 800 K [38,39], which preferentially nucleate within valleys of the faceted surface. Shen et al. have used faceted O/Ru$(11\bar{2}0)$, composed of long uniform ridges [79], as a template to grow self-organized Au nanoclusters with narrow size distribution and regular spacings [40]. The growth of Au nanoclusters on faceted O/Ru is achieved after deposition of gold at room temperature, in contrast to the growth of Co nanoclusters on faceted O/Re which requires post annealing. The growth of Au nanoclusters is confined in the valleys between the ridges of faceted O/Ru, as shown in Figure 22. These studies demonstrate a novel method that enables the growth of metal nanoclusters on metal surfaces, where normally a layer-by-layer or 2D-like growth behavior would be expected, by lowering the support surface energy through oxygen chemisorption.

4.2 Synthesis of Nanoscale Model Electrocatalysts

Ultrathin metal films (monolayer, bilayers) show unusual reactivity in both heterogeneous catalysis [286–289] and electrocatalysis [289,290]. It is reported that Au bilayer exhibits exceptionally high catalytic activity for CO oxidation [286], which is comparable to that of Au nanoparticles [287].

The hydrogen evolution reaction (HER) is one of the most extensively investigated electrochemical reactions [291,292], in which hydrogen gas is produced at the cathode of an electrolytic cell by the reduction of hydrogen ions or the reduction of the water molecules of an aqueous solution. This reaction is of increasing interest as a route to

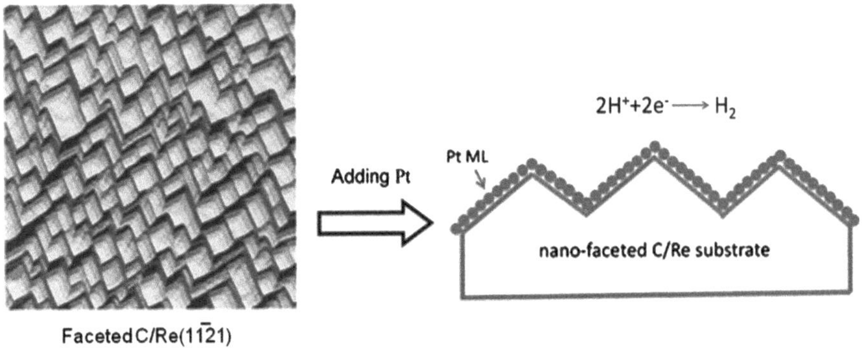

Figure 23 Schematic of the process to prepare nanostructured Pt ML electrocatalyst supported on carbon-covered faceted Re$(11\bar{2}1)$ at room temperature. *Reprinted with permission from Ref. [41]. Copyright (2012) American Chemical Society.*

carbon-free generation of H_2 and has great promise for use in the future for automobiles [290,293–298]. Pt is commonly used as an electrocatalyst for the HER. However, due to the high cost and limited supply of Pt, considerable effort has recently been devoted to the exploration of new HER catalysts that minimize the amount of Pt needed [290,296]. Of particular interest is the use of Pt ML catalysts for the HER in order to substantially reduce the Pt loading. For example, it was found that the activity of a Pt ML supported on planar tungsten carbide (WC) shows the same HER activity as a Pt foil electrode [290].

Recently, Yang et al. have used faceted C/Re$(11\bar{2}1)$ as a template for preparing nanostructured Pt ML electrocatalyst [41]. This was the first application of faceted metal single-crystal surface as template for synthesis of nanoscale model electrocatalyst with well-defined (facet) surface structure and controlled (facet) size on the nanometer scale. Faceted C/Re$(11\bar{2}1)$ consisting of three-sided pyramids exposing $(10\bar{1}1)$, $(01\bar{1}1)$ and $(11\bar{2}0)$ facets can stabilize Pt ML as schematically shown in Figure 23, which is prepared by physical vapor deposition of Pt at room temperature. After preparing a Pt ML supported on faceted C/Re$(11\bar{2}1)$, denoted as Pt ML/C/Re$(11\bar{2}1)$, its reactivity for HER was tested. Figure 24(A) shows polarization scans for the HER using Re$(11\bar{2}1)$, Pt(111), and Pt ML/C/Re$(11\bar{2}1)$ samples, which were performed in an Ar-purged 0.1 M $HClO_4$ solution at room temperature with a scan rate of $2\,mV\,s^{-1}$. Figure 24(B) gives plots of the HER overpotentials as a function of the logarithm of the exchange current density to produce the corresponding set of the Tafel curves. Clearly, the activity of Re$(11\bar{2}1)$ for the HER is much lower than Pt(111) as evidenced by the large overpotential. However, the Pt ML/C/Re$(11\bar{2}1)$ sample has a significantly improved activity with even better performance than Pt(111). The exchange current (j_0) values for Pt ML/C/Re$(11\bar{2}1)$ and Pt(111) are 6.30×10^{-4} and $3.98 \times 10^{-4}\,A\,cm^{-2}$, respectively. This work demonstrates the importance of nanoscale control in designing better electrocatalysts.

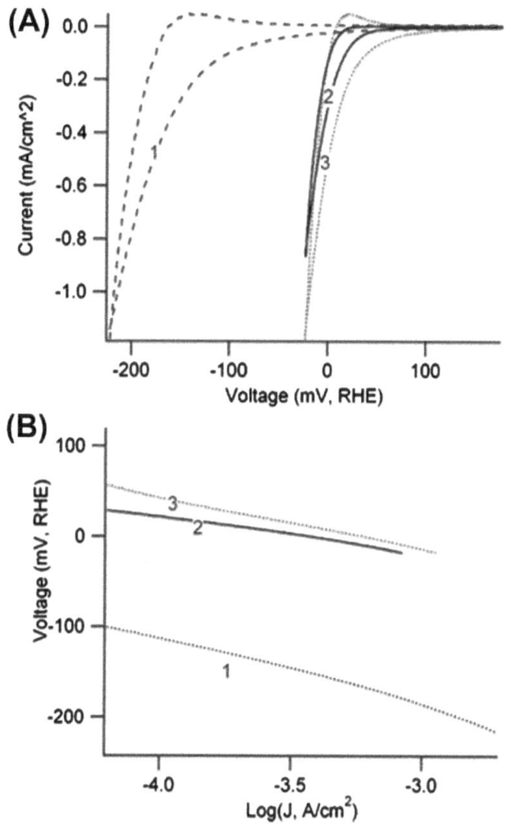

Figure 24 (A) Hydrogen evolution reaction (HER) polarization curves of (1) Re$(11\bar{2}1)$, (2) Pt(111) and (3) Pt ML supported on faceted C/Re$(11\bar{2}1)$ performed in Ar-purged 0.1 M $HClO_4$. (B) Corresponding Tafel plots for the data in (A). The scan rate in the polarization measurements was 2 mV s^{-1}. *Reprinted with permission from Ref. [41]. Copyright (2012) American Chemical Society.*

5. SUMMARY

We provide an overview of adsorbate-induced faceting of metal surfaces and facet identifications. A summary is given for a series of chemical reactions on clean and metal ML-covered faceted and planar surfaces of metal single crystals. The advantage of clean and metal ML-covered faceted surfaces for chemical reactions is to explore size effects in chemical reactions without change in facet structure when facet size varies as well as without switching from sample to sample. The utilization of adsorbate-covered faceted surfaces as nanotemplates for synthesis of metallic nanoclusters and nanoscale model electrocatalyst illustrates the importance of nanoscale control for achieving better catalysts and electrocatalysts.

ACKNOWLEDGMENT

We gratefully acknowledge support of this work by U.S. Department of Energy, Office of Basic Energy Sciences (Grant No. DE-FG02-93ER14331).

REFERENCES

[1] G.A. Somorjai, R.M. Rioux, Catal. Today 100 (2005) 201–215.
[2] G.A. Somorjai, J.Y. Park, Angew. Chem. Int. Ed. 47 (2008) 9212–9228.
[3] G.A. Somorjai, C.J. Kliewer, React. Kinet. Catal. Lett. 96 (2009) 191–208.
[4] A.R. Tao, S. Habas, P. Yang, Small 4 (2008) 310–325.
[5] Y. Xia, Y. Xiong, B. Lim, S.E. Skrabalak, Angew. Chem. Int. Ed. 48 (2008) 60–103.
[6] I. Lee, R. Morales, M.A. Albiter, F. Zaera, Proc. Natl. Acad. Sci. U.S.A. 105 (2008) 15241–15246.
[7] C.-Y. Chiu, Y. Li, L. Ruan, X. Ye, C.B. Murray, Y. Huang, Nat. Chem. 4 (2011) 393–399.
[8] K.M. Bratlie, H. Lee, K. Komvopoulos, P. Yang, G.A. Somorjai, Nano Lett. 7 (2007) 3097–3101.
[9] I. Lee, F. Delbecq, R. Morales, M.A. Albiter, F. Zaera, Nat. Mater. 8 (2009) 132–138.
[10] W. Chen, T.E. Madey, A.L. Stottlemyer, J.G. Chen, P. Kaghazchi, T. Jacob, J. Phys. Chem. C 112 (2008) 19113–19120.
[11] W. Chen, Q. Shen, R.A. Bartynski, P. Kaghazchi, T. Jacob, ChemPhysChem 11 (2010) 2515–2520.
[12] V.F. Puntes, K.M. Krishnan, A.P. Alivisatos, Science 291 (2001) 2115–2117.
[13] H. Lee, S.E. Habas, S. Kweskin, D. Butcher, G.A. Somorjai, P. Yang, Angew. Chem. Int. Ed. 45 (2006) 7988–7992.
[14] S.H. Joo, J.Y. Park, C.K. Tsung, Y. Yamada, P.D. Yang, G.A. Somorjai, Nat. Mater. 8 (2009) 126–131.
[15] R. Barnes, I.M. Abdelrehim, T.E. Madey, Top. Catal. 14 (2001) 53–61.
[16] W. Chen, I. Ermanoski, Q. Wu, T.E. Madey, H.H. Hwu, J.G. Chen, J. Phys. Chem. B 107 (2003) 5231–5242.
[17] W. Chen, I. Ermanoski, T.E. Madey, J. Am. Chem. Soc. 127 (2005) 5014–5015.
[18] W. Chen, I. Ermanoski, T. Jacob, T.E. Madey, Langmuir 22 (2006) 3166–3173.
[19] W. Chen, A.L. Stottlemyer, J.G. Chen, P. Kaghazchi, T. Jacob, T.E. Madey, R.A. Bartynski, Surf. Sci. 603 (2009) 3136–3144 and references therein.
[20] W. Chen, R.A. Bartynski, P. Kaghazchi, T. Jacob, J. Chem. Phys. 136 (2012) 224701–224710 and references therein.
[21] W. Chen, Q. Shen, R.A. Bartynski, P. Kaghazchi, T. Jacob, Langmuir 29 (2013) 1113–1121 and references therein.
[22] P. Kaghazchi, T. Jacob, W. Chen, R.A. Bartynski, Phys. Chem. Chem. Phys. 15 (2013) 12815–12820.
[23] D.P. Woodruff, J. Phys. Condens. Matter 6 (1994) 6067–6094.
[24] Q. Chen, N.V. Richardson, Prog. Surf. Sci. 73 (2003) 59–77.
[25] T.E. Madey, W. Chen, H. Wang, P. Kaghazchi, T. Jacob, Chem. Soc. Rev. 37 (2008) 2310–2327.
[26] G. Antczak, R. Błaszczyszyn, T.E. Madey, Prog. Surf. Sci. 74 (2003) 81–95.
[27] C.P. Norris, *Surface Science Research*, Nova Science Publishers. Inc., New York, 2005.
[28] A. Szczepkowicz, A. Ciszewski, R. Bryl, C. Oleksy, C.-H. Nien, Q. Wu, T.E. Madey, Surf. Sci. 599 (2005) 55–68.
[29] A. Szczepkowicz, R. Bryl, Phys. Rev. B 71 (2005) 113416.
[30] R. Bryl, T. Olewicz, T.V. de Bocarmé, N. Kruse, J. Phys. Chem. C 114 (2010) 2220–2226.
[31] R. Bryl, T. Olewicz, T.V. de Bocarmé, N. Kruse, J. Phys. Chem. C 115 (2011) 2761–2768.
[32] A. Szczepkowicz, Surf. Sci. 606 (2012) 202–208.
[33] R. Bryl, B. Lechowski, Vacuum 91 (2013) 41–46.
[34] P.L. Hansen, J.B. Wagner, S. Helveg, J.R. Rostrup-Nielsen, B.S. Clausen, H. Topsøe, Science 295 (2002) 2053–2055.
[35] P. Nolte, A. Stierle, N.Y. Jin-Phillipp, N. Kasper, T.U. Schulli, H. Dosch, Science 321 (2008) 1654–1658.
[36] M.A. Newton, C. Belver-Coldeira, A. Martínez-Arias, M. Fernández-García, Nat. Mater. 6 (2007) 528–532.
[37] P. Nolte, A. Stierle, N. Kasper, N.Y. Jin-Phillipp, N. Jeutter, H. Dosch, Nano Lett. 11 (2011) 4697–4700.
[38] H. Wang, Adsorbate-induced nanoscale faceting of rhenium surfaces (Ph.D. thesis), Rutgers University, 2008.
[39] M. Reyhan, H. Wang, T.E. Madey, Catal. Lett. 129 (2009) 46–49.
[40] Q. Shen, W. Chen, R.A. Bartynski, Surf. Sci. 605 (2011) 1457–1461.
[41] X. Yang, B.E. Koel, H. Wang, W. Chen, R.A. Bartynski, ACS Nano 6 (2012) 1404–1409.
[42] C. Herring, Phys. Rev. Lett. 82 (1951) 87–93.

[43] J.G. Che, C.T. Chan, Phys. Rev. B 57 (1998) 1875–1880.
[44] K.-J. Song, R.A. Demmin, C. Dong, E. Garfunkel, T.E. Madey, Surf. Sci. 227 (1990) L79–L85.
[45] K.-J. Song, T.E. Madey, in: S.Y. Tong, M.A. Van Hove, K. Takayanagi, X.D. Xie (Eds.), The Structure of Surfaces III, Springer-Verlag, Berlin, 1991, pp. 378–382.
[46] K.J. Song, C.Z. Dong, T.E. Madey, Langmuir 7 (1991) 3019–3026.
[47] T.E. Madey, J. Guan, C.-Z. Dong, S.M. Shivaprasad, Surf. Sci. 287–288 (1993) 826–830.
[48] J. Guan, Ultrathin metal films on tungsten(111) and molybdenum(111): surface reconstruction and morphology (Ph.D. thesis), Rutgers University, 1994.
[49] T.E. Madey, J. Guan, C.-H. Nien, C.-Z. Dong, H.-S. Tao, R.A. Campbell, Surf. Rev. Lett. 3 (1996) 1315–1328.
[50] C.-H. Nien, T.E. Madey, Surf. Sci. (1997) L527–L532.
[51] L. Schwenger, H.-J. Ernst, Surf. Sci. 347 (1996) 25–32.
[52] R. Szukiewicz, J. Kołaczkiewicz, Surf. Sci. 547 (2003) L837–L839.
[53] J. Kołaczkiewicz, M. Kuchowicz, R. Szukiewicz, Surf. Sci. 548 (2004) 246–258.
[54] C. Dong, L. Zhang, U. Diebold, T.E. Madey, Surf. Sci. 322 (1995) 221–229.
[55] J.J. Kolodziej, T.E. Madey, J.W. Keister, J.E. Rowe, Phys. Rev. B 65 (2002) 075413.
[56] J.C. Tracy, J.M. Blakely, Surf. Sci. 13 (1969) 313–336.
[57] F. Bonczek, T. Engel, E. Bauer, Surf. Sci. 97 (1980) 595–608.
[58] C. Egawa, S. Naito, K. Tamaru, Surf. Sci. 125 (1983) 605–612.
[59] C. Zhang, M.A. Van Hove, G.A. Somorjai, Surf. Sci. 149 (1985) 326–340.
[60] C. Zhang, A.J. Gellman, M.H. Farias, G.A. Somorjai, Mater. Res. Bull. 20 (1985) 1129–1135.
[61] D.R. Strongin, G.A. Somorjai, J. Catal. 118 (1989) 99–110.
[62] H. Cabibil, J.S. Lin, J.A. Kelber, Surf. Sci. 382 (1997) L645–L651.
[63] J.-S. Lin, H. Cabibil, J.A. Kelber, Surf. Sci. 395 (1998) 30–42.
[64] S. Reiter, E. Taglauer, Surf. Sci. 367 (1996) 33–39.
[65] N. Reinecke, E. Taglauer, Surf. Sci. 454–456 (2000) 94–100.
[66] S.M. Driver, D.P. Woodruff, Surf. Sci. 560 (2004) 35–44.
[67] I. Ermanoski, K. Pelhos, W. Chen, J.S. Quinton, T.E. Madey, Surf. Sci. 549 (2004) 1–23.
[68] C.W. Tucker, Acta Metall. 15 (1967) 1465–1474.
[69] Govind, W. Chen, H. Wang, T.E. Madey, Phys. Rev. B 81 (2010) 085415.
[70] J. Gustafson, A. Resta, A. Mikkelsen, R. Westerström, J.N. Andersen, E. Lundgren, J. Weissenrieder, M. Schmid, P. Varga, N. Kasper, X. Torrelles, S. Ferrer, F. Mittendorfer, G. Kresse, Phys. Rev. B 74 (2006) 035401.
[71] D.G. Castner, G.A. Somorjai, Surf. Sci. 83 (1979) 60–82.
[72] R.E. Kirby, C.S. McKee, M.W. Roberts, Surf. Sci. 55 (1976) 725–728.
[73] R.E. Kirby, C.S. McKee, L.V. Renny, Surf. Sci. 97 (1980) 457–477.
[74] H.V. Thapliyal, J.M. Blakely, J. Vac. Sci. Technol. 15 (1978) 600–605.
[75] M. Sander, R. Imbihl, R. Schuster, J.V. Barth, G. Ertl., Surf. Sci. 271 (1992) 159–169.
[76] E.R. Frank, R.J. Hamers, J. Catal. 172 (1997) 406–413.
[77] H. Wang, W. Chen, T.E. Madey, Phys. Rev. B 74 (2006) 205426.
[78] H. Wang, W. Chen, R.A. Bartynski, P. Kaghazchi, T. Jacob, J. Chem. Phys. 140 (2014) 024707.
[79] Q. Shen, W. Chen, H. Wang, Govind, T.E. Madey, R.A. Bartynski, Surf. Sci. 604 (2010) L12–L15.
[80] Q. Shen, W. Chen, H. Wang, R.A. Bartynski, J. Chem. Phys. 139 (2013) 084707.
[81] H. Wang, A.S.Y. Chan, W. Chen, P. Kaghazchi, T. Jacob, T.E. Madey, ACS Nano 1 (2007) 449–455.
[82] K. Pelhos, J.B. Hannon, G.L. Kellogg, T.E. Madey, Surf. Sci. 432 (1999) 115–124.
[83] I. Ermanoski, W. Swiech, T.E. Madey, Surf. Sci. 592 (2005) L299–L304.
[84] S.P. Chen, Surf. Sci. 274 (1992) L619–L626.
[85] C. Oleksy, Surf. Sci. 549 (2004) 246–254.
[86] J.G. Che, C.T. Chan, C.H. Kuo, T.C. Leung, Phys. Rev. Lett. 79 (1997) 4230–4233.
[87] J.G. Che, C.T. Chan, Surf. Sci. 401 (1998) L432–L436.
[88] J.G. Che, K.M. Zhang, X.D. Xie, Surf. Sci. 472 (2001) 179–186.
[89] P. Kaghazchi, T. Jacob, I. Ermanoski, W. Chen, T.E. Madey, ACS Nano 2 (2008) 1280–1288.
[90] P. Kaghazchi, T. Jacob, H. Wang, W. Chen, T.E. Madey, Phys. Rev. B 79 (2009) 132107.
[91] N.W. Ashcroft, N.D. Mermin, Solid State Physics, Holt, Rinehart and Winston, New York, 1976.

[92] P. Kaghazchi, T. Jacob, Phys. Rev. B 83 (2011) 035417.
[93] W.T. Tysoe, G.L. Nyberg, R.M. Lambert, J. Chem. Soc. Chem. Commun. (1983) 623–625.
[94] G. Pacchioni, R.M. Lambert, Surf. Sci. 304 (1994) 208–222.
[95] J.R. Lomas, C.J. Baddeley, M.S. Tikhov, R.M. Lambert, Langmuir 11 (1995) 3048–3053.
[96] N.R. Avery, J. Am, Chem. Soc. 107 (1985) 6711–6712.
[97] T.G. Rucker, M.A. Logan, T.M. Gentle, E.L. Muetterties, G.A. Somorjai, J. Phys. Chem. 90 (1986) 2703–2708.
[98] I.M. Abdelrehim, K. Pelhos, T.E. Madey, J. Eng Jr., J.G. Chen, J. Mol. Catal. A 131 (1998) 107–120.
[99] W.T. Tysoe, G.L. Nyberg, R.M. Lambert, Surf. Sci. 135 (1983) 128–146.
[100] J. Yoshinobu, T. Sekitani, M. Onchi, M. Nishijima, J. Phys. Chem. 94 (1990) 4269–4275.
[101] D.C. Papageorgopoulos, Q. Ge, S. Nimmo, D.A. King, J. Phys. Chem. B 101 (1997) 1999–2005.
[102] D. Stacchiola, H. Molero, W.T. Tysoe, Catal. Today 65 (2001) 3–11.
[103] W. Sesselmann, B. Woratschek, G. Ertl, J. Küppers, Surf. Sci. 130 (1983) 245–258.
[104] T.M. Gentle, E.L. Muetterties, J. Phys. Chem. 87 (1983) 2469–2472.
[105] B. Marchon, Surf. Sci. 162 (1985) 382–387.
[106] C.-H. Nien, Atomic resolution studies of overlayer-induced restructuring on tungsten (111) surfaces (Ph.D. thesis), Rutgers University, 1999.
[107] H. Persson, P. Thormählen, V.P. Zhdanov, B. Kasemo, J. Vac. Sci. Technol. A 17 (1999) 1721–1726.
[108] I.M. Abdelrehim, K. Pelhos, T.E. Madey, J. Phys. Chem. B 102 (1998) 9697–9707.
[109] B.E. Nieuwenhuys, D.I. Hagen, G. Rovida, G.A. Somorjai, Surf. Sci. 59 (1976) 155–176.
[110] T.S. Marinova, K.L. Kostov, Surf. Sci. 181 (1987) 573–585.
[111] C. Brucker, T. Rhodin, J. Catal. 47 (1977) 214–231.
[112] K. Yoshida, G.A. Somorjai, Surf. Sci. 75 (1978) 46–60.
[113] W.-H. Hung, S.L. Bernasek, Surf. Sci. 339 (1995) 272–290.
[114] D.G. Castner, B.A. Sexton, G.A. Somorjai, Surf. Sci. 71 (1978) 519–540.
[115] L.H. Dubois, D.G. Castner, G.A. Somorjai, J. Chem. Phys. 72 (1980) 5234–5238.
[116] D.L.S. Nieskens, F. Ample-Navarro, M.M.M. Jansen, D. Curulla-Ferré, J. Ricart, H. Niemantsverdriet, ChemPhysChem 7 (2006) 1068–1074.
[117] L.L. Kesmodel, L.H. Dubois, G.A. Somorjai, J. Chem. Phys. 70 (1979) 2180–2189.
[118] J.A. Stroscio, S.R. Bare, W. Ho, Surf. Sci. 148 (1984) 499–525.
[119] C. Egawa, S. Naito, K. Tamaru, J. Chem. Soc. Faraday Trans. 82 (1986) 3197–3204.
[120] O. Höfert, M.P.A. Lorenz, R. Streber, W. Zhao, A. Bayer, H.-P. Steinrück, C. Papp, J. Chem. Phys. 139 (2013) 164706.
[121] J.J. Eng, J.G. Chen, I.M. Abdelrehim, T.E. Madey, J. Phys. Chem. B 102 (1998) 9687–9696.
[122] C. Bianchini, K.G. Caulton, C. Chardon, M.-L. Doublet, O. Eisenstein, S.A. Jackson, T.J. Johnson, A. Meli, M. Peruzzini, Organometallics 13 (1994) 2010–2023.
[123] S.F. Yin, B.Q. Xu, X.P. Zhou, C.T. Au, Appl. Catal. A 277 (2004) 1–9.
[124] F. Schüth, R. Palkovits, R. Schlögl, D.S. Su, Energy Environ. Sci. 5 (2012) 6278–6289.
[125] D. Cheddie, in: D. Minic (Ed.), Hydrogen Energy – Challenges and Perspectives, InTech, 2012.
[126] C. Zamfirescu, I. Dincer, Fuel Process. Technol. 90 (2009) 729–737.
[127] A. Vāvere, R.S. Hansen, J. Catal. 69 (1981) 158–171.
[128] I.C. Bassignana, K. Wagemann, J. Küppers, G. Ertl, Surf. Sci. 175 (1986) 22–44.
[129] J.L. Gland, Surf. Sci. 71 (1978) 327–350.
[130] J.M. Gohndrone, C.W. Olsen, A.L. Backman, T.R. Gow, E. Yagasaki, R.I. Masel, J. Vac. Sci. Technol. A 7 (1989) 1986–1990.
[131] C. Egawa, T. Nishida, S. Naito, K. Tamaru, J. Chem. Soc. Faraday Trans. 1 80 (1984) 1595–1604.
[132] H. Dietrich, K. Jacobi, G. Ertl, Surf. Sci. 352–354 (1996) 138–141.
[133] K. Jacobi, Y. Wang, C.Y. Fan, H. Dietrich, J. Chem. Phys. 115 (2001) 4306–4313.
[134] Y. Wang, A. Lafosse, K. Jacobi, Surf. Sci. 507–510 (2002) 773–777.
[135] T.V. Choudhary, C. Sivadinarayana, D.W. Goodman, Catal. Lett. 72 (2001) 197–201.
[136] T.V. Choudhary, A.K. Santra, C. Sivadinarayana, B.K. Min, C.-W. Yi, K. Davis, D.W. Goodman, Catal. Lett. 77 (2001) 1–5.
[137] S.A.C. Carabineiro, B.E. Nieuwenhuys, Surf. Sci. 505 (2002) 163–170.
[138] S.A.C. Carabineiro, B.E. Nieuwenhuys, Surf. Sci. 532–535 (2003) 87–95.
[139] S.-F. Yin, Q.-H. Zhang, B.-Q. Xu, W.-X. Zhu, C.-F. Ng, C.-T. Au, J. Catal. 224 (2004) 384–396.

[140] A.M. Karim, V. Prasad, G. Mpourmpakis, W.W. Lonergan, A.I. Frenkel, J.G. Chen, D.G. Vlachos, J. Am, Chem. Soc. 131 (2009) 12230–12239.
[141] G. Papapolymerou, V. Bontozoglou, J. Mol. Catal. A 120 (1997) 165–171.
[142] V.I. Parvulescu, P. Grange, B. Delmon, Catal. Today 46 (1998) 233–316.
[143] F. Garin, Appl. Catal. A 222 (2001) 183–219.
[144] C.M. Comrie, W.H. Weinberg, R.M. Lambert, Surf. Sci. 57 (1976) 619–631.
[145] R.J. Gorte, L.D. Schmidt, Surf. Sci. 109 (1981) 367–380.
[146] J.M. Gohndrone, Y.O. Park, R.I. Masel, J. Catal. 95 (1985) 244–248.
[147] J.M. Gohndrone, R.I. Masel, Surf. Sci. 209 (1989) 44–56.
[148] S. Sugai, K. Takeuchi, T. Ban, H. Miki, K. Kawasaki, Surf. Sci. 282 (1993) 67–75.
[149] J. Kanski, T.N. Rhodin, Surf. Sci. 65 (1977) 63–76.
[150] J.C.L. Cornish, Surf. Sci. 235 (1990) 209–216.
[151] J.S. Villarrubia, W. Ho, J. Chem. Phys. 87 (1987) 750–765.
[152] M. Aryafar, F. Zaera, J. Catal. 175 (1998) 316–327.
[153] R.G. Sharpe, M. Bowker, Surf. Sci. 360 (1996) 21–30.
[154] I. Nakamura, T. Fujitani, H. Hamada, Surf. Sci. 514 (2002) 409–413.
[155] A.D. Karmazyn, V. Fiorin, D.A. King, Surf. Sci. 547 (2003) 184–192.
[156] C.P. Vinod, J.W.N. Hans, B.E. Nieuwenhuys, Appl. Catal. A 291 (2005) 93–97.
[157] Z. Wu, L. Xu, W. Zhang, Y. Ma, Q. Yuan, Y. Jin, J. Yang, W. Huang, J. Catal. 304 (2013) 112–122.
[158] N. Takehiro, F. Besenbacher, E. Lægsgaard, K. Tanaka, I. Stensgaard, Surf. Sci. 397 (1998) 145–153.
[159] S.S. Dhesi, S. Haq, S.D. Barrett, F.M. Leibsle, Surf. Sci. 365 (1996) 602–613.
[160] S.K. So, R. Franchy, W. Ho, J. Chem. Phys. 91 (1989) 5701–5706.
[161] W.A. Brown, P. Gardner, D.A. King, J. Phys. Chem. 99 (1995) 7065–7074.
[162] J.P. Fulmer, W.T. Tysoe, Surf. Sci. 233 (1990) 35–43.
[163] R. Klein, A. Shih, Surf. Sci. 69 (1977) 403–427.
[164] T. Zambelli, J. Wintterlin, J. Trost, G. Ertl, Science 273 (1996) 1688–1690.
[165] R. Ducros, M. Housley, G. Piquard, M. Alnot, Surf. Sci. 108 (1981) 235–252.
[166] B. Hammer, Phys. Rev. Lett. 83 (1999) 3681–3684.
[167] L.A. DeLouise, N. Winograd, Surf. Sci. 159 (1985) 199–213.
[168] O.R. Inderwildi, D. Lebiedz, O. Deutschmann, J. Warnatz, J. Chem. Phys. 122 (2005) 154702.
[169] O.R. Inderwildi, D. Lebiedz, O. Deutschmann, J. Warnatz, ChemPhysChem 6 (2005) 2513–2521.
[170] Q. Ge, M. Neurock, J. Am. Chem. Soc. 126 (2004) 1551–1559.
[171] R.J. Mukerji, A.S. Bolina, W.A. Brown, Surf. Sci. 547 (2003) 27–44.
[172] Z. Jiang, W. Huang, D. Tan, R. Zhai, X. Bao, Surf. Sci. 600 (2006) 4860–4869.
[173] M. Hirsimäki, S. Suhonen, J. Pere, M. Valden, M. Pessa, Surf. Sci. 402–404 (1998) 187–191.
[174] Z.-P. Liu, S.J. Jenkins, D.A. King, J. Am. Chem. Soc. 125 (2003) 14660–14661.
[175] C. Nyberg, P. Uvdal, Surf. Sci. 256 (1991) 42–49.
[176] A.D. Smeltz, R.B. Getman, W.F. Schneider, F.H. Ribeiro, Catal. Today 136 (2008) 84–92.
[177] W.D. Mieher, R.A. Pelak, W. Ho, Surf. Sci. 359 (1996) 23–36.
[178] M.E. Bartram, B.E. Koel, Surf. Sci. 219 (1989) 467–489.
[179] T.W. Root, L.D. Schmidt, Surf. Sci. 134 (1983) 30–45.
[180] H. Xu, K.Y.S. Ng, Surf. Sci. 365 (1996) 779–788.
[181] N.M.H. Janssen, A.R. Cholach, M. Ikai, K. Tanaka, B.E. Nieuwenhuys, Surf. Sci. 382 (1997) 201–213.
[182] Z.-P. Liu, S.J. Jenkins, D.A. King, J. Am, Chem. Soc. 126 (2004) 10746–10756.
[183] S. Khatua, Z.-P. Liu, D.A. King, Surf. Sci. 584 (2005) 214–224.
[184] F. Bozso, J. Arias, C.P. Hanrahan, J.T. Yates Jr., R.M. Martin, H. Metiu, Surf. Sci. 141 (1984) 591–603.
[185] J.G. Chen, W. Erley, H. Ibach, Surf. Sci. 224 (1989) 215–234.
[186] B. Chen, Y. Ma, L. Ding, L. Xu, Z. Wu, Q. Yuan, W. Huang, Chin. J. Catal. 34 (2013) 964–972.
[187] B.E. Hayden, K. Kretzschmar, A.M. Bradshaw, Surf. Sci. 125 (1983) 366–376.
[188] H. Conrad, R. Scala, W. Stenzel, R. Unwin, Surf. Sci. 145 (1984) 1–19.
[189] K.T. Queeney, C.M. Friend, J. Phys. Chem. B 102 (1998) 9251–9257.
[190] R.M. Heck, R.J. Farrauto, Appl. Catal. A 221 (2001) 443–457.
[191] U. Heiz, E.L. Bullock, J. Mater. Chem. 14 (2004) 564–577.
[192] H. Falsig, B. Hvolbæk, I.S. Kristensen, T. Jiang, T. Bligaard, C.H. Christensen, J.K. Nørskov, Angew. Chem. Int. Ed. 47 (2008) 4835–4839.

[193] L.M. Molina, B. Hammer, Phys. Rev. B 69 (2004) 155424.
[194] F.M. Leibsle, P.W. Murray, S.M. Francis, G. Thornton, M. Bowker, Nature 363 (1993) 706–709.
[195] J. Wintterlin, S. Völkening, T.V.W. Janssens, T. Zambelli, G. Ertl, Science 278 (1997) 1931–1934.
[196] J.A. Rodriguez, D.W. Goodman, Surf. Sci. Rep. 14 (1991) 1–107.
[197] Z.-P. Liu, P. Hu, Top. Catal. 28 (2004) 71–78.
[198] H. Over, Y.D. Kim, A.P. Seitsonen, S. Wendt, E. Lundgren, M. Schmid, P. Varga, A. Morgante, G. Ertl, Science 287 (2000) 1474–1476.
[199] S.M. McClure, D.W. Goodman, Chem. Phys. Lett. 469 (2009) 1–13.
[200] F. Gao, S.M. McClure, Y. Cai, K.K. Gath, Y. Wang, M.S. Chen, Q.L. Guo, D.W. Goodman, Surf. Sci. 603 (2009) 65–70.
[201] T. Jiang, D.J. Mowbray, S. Dobrin, H. Falsig, B. Hvolbæk, T. Bligaard, J.K. Nørskov, J. Phys. Chem. C 113 (2009) 10548–10553.
[202] T.J. Lerotholi, G. Held, D.A. King, Surf. Sci. 601 (2007) 1285–1295.
[203] R.B. Hoyle, A.T. Anghel, M.R.E. Proctor, D.A. King, Phys. Rev. Lett. 98 (2007) 226102.
[204] M. Kim, M. Bertram, M. Pollmann, A.v. Oertzen, A.S. Mikhailov, H.H. Rotermund, G. Ertl, Science 292 (2001) 1357–1360.
[205] X. Su, P.S. Cremer, Y.R. Shen, G.A. Somorjai, J. Am. Chem. Soc. 119 (1997) 3994–4000.
[206] M. Okumura, N. Masuyama, E. Konishi, S. Ichikawa, T. Akita, J. Catal. 208 (2002) 485–489.
[207] S. Lee, C. Fan, T. Wu, S.L. Anderson, J. Am. Chem. Soc. 126 (2004) 5682–5683.
[208] U. Heiz, A. Sanchez, S. Abbet, W.-D. Schneider, J. Am. Chem. Soc. 121 (1999) 3214–3217.
[209] R. Meyer, S.K. Shaikhutdinov, H.-J. Freund, Z. Phys. Chem. 218 (2004) 905–914.
[210] S.H. Joo, J.Y. Park, J.R. Renzas, D.R. Butcher, W. Huang, G.A. Somorjai, Nano Lett. 10 (2010) 2709–2713.
[211] M.J.P. Hopstaken, J.W. Niemantsverdriet, J. Chem. Phys. 113 (2000) 5457–5465.
[212] H. Uetsuka, K. Watanabe, H. Ohnuma, K. Kunimori, Surf. Rev. Lett. 4 (1997) 1359–1363.
[213] C.J. Zhang, P. Hu, J. Am. Chem. Soc. 123 (2001) 1166–1172.
[214] B.E. Nieuwenhuys, Adv. Catal. 44 (1999) 259–328.
[215] J. Yoshinobu, M. Kawai, J. Chem. Phys. 103 (1995) 3220–3230.
[216] Y. Yasui, M. Sawada, T. Aruga, N. Takagi, M. Nishijima, Surf. Sci. 397 (1998) 295–305.
[217] A. Alavi, P. Hu, T. Deutsch, P.L. Silvestrelli, J. Hutter, Phys. Rev. Lett. 80 (1998) 3650–3653.
[218] S.M. Vesecky, P. Chen, X. Xu, D.W. Goodman, J. Vac. Sci. Technol. A 13 (1995) 1539–1544.
[219] C.H.F. Peden, D.W. Goodman, D.S. Blair, P.J. Berlowitz, G.B. Fisher, S.H. Oh, J. Phys. Chem. 92 (1988) 1563–1567.
[220] L. Piccolo, C.R. Henry, J. Mol. Catal. A 167 (2001) 181–190.
[221] A.S. Wörz, K. Judai, S. Abbet, U. Heiz, J. Am. Chem. Soc. 125 (2003) 7964–7970.
[222] L. Piccolo, C.R. Henry, Appl. Surf. Sci. 162–163 (2000) 670–678.
[223] V.P. Zhdanov, B. Kasemo, Surf. Sci. Rep. 29 (1997) 31–90.
[224] C.R. Henry, Surf. Sci. Rep. 31 (1998) 231–325.
[225] J. Libuda, H.-J. Freund, Surf. Sci. Rep. 57 (2005) 157–298.
[226] M. Haneda, T. Yoshinari, K. Sato, Y. Kintaichi, H. Hamada, Chem. Commun. (2003) 2814–2815.
[227] A. Wang, L. Ma, Y. Cong, T. Zhang, D. Liang, Appl. Catal. B 40 (2003) 319–329.
[228] J. Shibata, H. Yoshida, A. Satsuma, T. Hattori, Chem. Lett. 33 (2004) 800–801.
[229] F. Solymosi, T. Bánsági, J. Catal. 202 (2001) 205–206.
[230] E. Ozensoy, C. Hess, D.W. Goodman, J. Am. Chem. Soc. 124 (2002) 8524–8525.
[231] T. Fujitani, I. Nakamura, A. Takahashi, M. Haneda, H. Hamada, J. Catal. 253 (2008) 139–147.
[232] I. Nakamura, K. Suzuki, A. Takahashi, M. Haneda, H. Hamada, T. Fujitani, J. Vac. Sci. Technol. A 25 (2007) 1143–1146.
[233] I. Nakamura, T. Fujitani, Catal. Surv. Asia 13 (2009) 22–29.
[234] T. Fink, J.-P. Dath, M.R. Bassett, R. Imbihl, G. Ertl, Surf. Sci. 245 (1991) 96–110.
[235] M.W. Lesley, L.D. Schmidt, Surf. Sci. 155 (1985) 215–240.
[236] Y.O. Park, W.F. Banholzer, R.I. Masel, Surf. Sci. 155 (1985) 341–365.
[237] A.V. Matveev, A.A. Sametova, V.V. Gorodetskii, Kinet. Catal. 45 (2004) 598–606.
[238] M. Daté, H. Okuyama, N. Takagi, M. Nishijima, T. Aruga, Surf. Sci. 341 (1995) L1096–L1100.
[239] A.V. Matveev, A.A. Sametova, N.N. Bulgakov, V.V. Gorodetskii, Chem. Sustainable Dev. 11 (2003) 181–188.

[240] X. Wang, in: G. Nikolic (Ed.), Fourier Transforms – New Analytical Approaches and FTIR Strategies, InTech, 2011.
[241] M.D. Amiridis, C. Mihut, M. Maciejewski, A. Baiker, Top. Catal. 28 (2004) 141–150.
[242] Y. Hu, K. Griffiths, P.R. Norton, Surf. Sci. 603 (2009) 1740–1750.
[243] R. Burch, J.P. Breen, F.C. Meunier, Appl. Catal. B 39 (2002) 283–303.
[244] M.D. Amiridis, T. Zhang, R.J. Farrauto, Appl. Catal. B 10 (1996) 203–227.
[245] R. Burch, P.J. Millington, Catal. Today 26 (1995) 185–206.
[246] Y. Hu, K. Griffiths, J. Phys. Chem. C 111 (2007) 9919–9926.
[247] S.M. Vesecky, J. Paul, D.W. Goodman, J. Phys. Chem. 100 (1996) 15242–15246.
[248] R.M. van Hardeveld, A.J.G.W. Schmidt, J.W. Niemantsverdriet, Catal. Lett. 41 (1996) 125–131.
[249] K. Christmann, Surf. Sci. Rep. 9 (1988) 1–163.
[250] B. Hammer, J.K. Nørskov, Surf. Sci. 343 (1995) 211–220.
[251] R. Löber, D. Hennig, Phys. Rev. B 55 (1997) 4761–4765.
[252] A.W.C. van den Berga, C.O. Areán, Chem. Commun. (2008) 668–681.
[253] D.V. Chakarov, T.S. Marinova, Surf. Sci. 204 (1988) 147–160.
[254] C.J. Hagedorn, M.J. Weiss, W.H. Weinberg, Phys. Rev. B (1999) R14016–R14018.
[255] G. Lee, P.T. Sprunger, M. Okada, D.B. Poker, D.M. Zehner, E.W. Plummer, J. Vac. Sci. Technol. A 12 (1994) 2119–2123.
[256] M. Jo, Y. Kuwahara, M. Onchi, M. Nishijima, Solid State Commun. 55 (1985) 639–642.
[257] U. Muschiol, P.K. Schmidt, K. Christmann, Surf. Sci. 395 (1998) 182–204.
[258] P. Schilbe, D. Farías, K.H. Rieder, Surf. Rev. Lett. 5 (1998) 473–478.
[259] P. Schilbe, S. Siebentritt, R. Pues, K.-H. Rieder, Surf. Sci. 360 (1996) 157–170.
[260] B. Voigtländer, S. Lehwald, H. Ibach, Surf. Sci. 208 (1989) 113–135.
[261] D.Y. Zemlyanov, M.Y. Smirnov, V.V. Gorodetskii, Catal. Lett. 43 (1997) 181–187.
[262] H. Yanagita, H. Fujioka, T. Aruga, N. Takagi, M. Nishijima, Surf. Sci. 441 (1999) 507–514.
[263] H.-J. Müssig, W. Stenzel, Y. Song, H. Conrad, Surf. Sci. 311 (1994) 295–307.
[264] D. Farías, S. Siebentritt, R. Apel, R. Pues, K.H. Rieder, J. Chem. Phys. 106 (1997) 8254–8261.
[265] M. Okada, A.P. Baddorf, D.M. Zehner, Surf. Sci. 373 (1997) 145–152.
[266] K.H. Ernst, E. Schwarz, K. Christmann, J. Chem. Phys. 101 (1994) 5388–5401.
[267] H. Shi, K. Jacobi, Surf. Sci. 313 (1994) 289–294.
[268] C.Y. Fan, K. Jacobi, Surf. Sci. 482–485 (2001) 21–25.
[269] K.L. Kostov, W. Widdra, D. Menzel, Surf. Sci. 560 (2004) 130–144.
[270] D. Tománek, Z. Sun, S.G. Louie, Phys. Rev. B 43 (1991) 4699–4713.
[271] W. Dong, J. Hafner, Phys. Rev. B 56 (1997) 15396–15403.
[272] D.C. Ford, Y. Xu, M. Mavrikakis, Surf. Sci. 587 (2005) 159–174.
[273] M. Mavrikakis, J. Rempel, J. Greeley, L.B. Hansen, J.K. Nørskov, J. Chem. Phys. 117 (2002) 6737–6744.
[274] D. Lerch, A. Klein, A. Schmidt, S. Müller, L. Hammer, K. Heinz, M. Weinert, Phys. Rev. B 73 (2006) 075430.
[275] W.P. Krekelberg, J. Greeley, M. Mavrikakis, J. Phys. Chem. B 108 (2004) 987–994.
[276] C.J. Hagedorn, M.J. Weiss, W.H. Weinberg, Phys. Rev. B 60 (1999) R14016–R14018.
[277] B.C. Gates, Chem. Rev. 95 (1995) 511–522.
[278] M. Valden, X. Lai, D.W. Goodman, Science 281 (1998) 1647–1650.
[279] C.C. Chusuei, X. Lai, K. Luo, D.W. Goodman, Top. Catal. 14 (2000) 71–83.
[280] S.D. Lin, M. Bollinger, M.A. Vannice, Catal. Lett. 17 (1993) 245–262.
[281] B.K. Min, C.M. Friend, Chem. Rev. 107 (2007) 2709–2724.
[282] A.A. Herzing, C.J. Kiely, A.F. Carley, P. Landon, G.J. Hutchings, Science 321 (2008) 1331–1335.
[283] D.R. Rainer, D.W. Goodman, J. Mol. Catal. A 131 (1998) 259–283.
[284] A.F. Carley, M.W. Roberts, G.J. Hutchings, M.S. Spencer, *Surface Chemistry and Catalysis*, Kluwer Academic/Plenum Publisher, New York, 2002.
[285] B.K. Min, W.T. Wallace, A.K. Santra, D.W. Goodman, J. Phys. Chem. B 108 (2004) 16339–16343.
[286] M.S. Chen, D.W. Goodman, Science 306 (2004) 252–255.
[287] M. Chen, D.W. Goodman, Acc. Chem. Res. 39 (2006) 739–746.
[288] J.G. Chen, C.A. Menning, M.B. Zellner, Surf. Sci. Rep. 63 (2008) 201–254.
[289] T.G. Kelly, J.G. Chen, Chem. Soc. Rev. 41 (2012) 8021–8034.

[290] D.V. Esposito, S.T. Hunt, A.L. Stottlemyer, K.D. Dobson, B.E. McCandless, R.W. Birkmire, J.G. Chen, Angew. Chem. Int. Ed. 49 (2010) 9859–9862.
[291] B. Hinnemann, P.G. Moses, J. Bonde, K.P. Jørgensen, J.H. Nielsen, S. Horch, I. Chorkendorff, J.K. Nørskov, J. Am. Chem. Soc. 127 (2005) 5308–5309.
[292] D.V. Esposito, S.T. Hunt, Y.C. Kimmel, J.G. Chen, J. Am. Chem. Soc. 134 (2012) 3025–3033.
[293] J. Greeley, J.K. Nørskov, L.A. Kibler, A.M. El-Aziz, D.M. Kolb, ChemPhysChem 7 (2006) 1032–1035.
[294] M.E. Björketun, A.S. Bondarenko, B.L. Abrams, I. Chorkendorff, J. Rossmeisl, Phys. Chem. Chem. Phys. 12 (2010) 10536–10541.
[295] J.K. Nørskov, T. Bligaard, A. Logadottir, J.R. Kitchin, J.G. Chen, S. Pandelov, U. Stimmingc, J. Electrochem. Soc. 152 (2005) J23–J26.
[296] I.E.L. Stephens, I. Chorkendorff, Angew. Chem. Int. Ed. 50 (2011) 1476–1477.
[297] T.F. Jaramillo, K.P. Jørgensen, J. Bonde, J.H. Nielsen, S. Horch, I. Chorkendorff, Science 317 (2007) 100–102.
[298] J. Greeley, T.F. Jaramillo, J. Bonde, I.B. Chorkendorff, J.K. Norskov, Nat. Mater. 5 (2006) 909–913.

CHAPTER 12

Nano-Array Catalysts for Energy and Environmental Catalysis

Yanbing Guo, Zheng Ren, Wen Xiao, Pu-Xian Gao*
Nanomaterials Science Laboratory, Department of Materials Science and Engineering & Institute of Materials Science, University of Connecticut, Storrs, CT, USA
*Corresponding author: E-mail: puxian.gao@ims.uconn.edu

1. INTRODUCTION

Heterogeneous catalysis is more widely used in industrial processes than homogeneous catalysis, because catalysts are much easier to isolate and separate [1]. Heterogeneous catalysts are essential to many chemical processes for new chemical bond formation and old bond breaking in a wide array of industrial applications including exhaust after-treatment, fine chemical processing, fuel production, and pharmaceutical industry. For decades, heterogeneous catalysis has been intensively studied, and catalytic materials such as precious metals have been developed to meet various industrial needs [2]. High surface-to-volume ratio is a key characteristic in catalysts, particularly in nanosized catalysts, which exhibit improved catalytic performance compared to the catalysts in bulk form.

Many metal catalysts are in the nanometer size; therefore the relevant catalysis research was referred to as 'nanoscience' a few decades before the term was popularized [3]. The rapid development of nanochemistry in the past two decades has greatly boosted the nanocatalysts' development through rational manipulations over the nanostructures' geometry, composition, and surface properties toward tunable catalytic activity. Besides the size effect, the shape and morphology control of nanocatalyts is generally believed to be another important strategy for catalyst design with different exposed crystal facets usually displaying different physical and chemical characteristics, which could influence the adsorption and desorption behaviors of molecules on nanocatalyst surfaces.

As an important class of nanomaterials, one-dimensional (1D) nanostructures such as nanorods, nanowires, nanotubes, and nanobelts can exhibit enhanced catalytic activity derived from their anisotropic structures, high surface-to-volume ratios, and preferential exposures of active crystal planes in photocatalysis, electrocatalysis, and gas phase catalysis. When assembled from the free-standing 1D nanostructures onto solid substrates, the nanostructure array (nano-array)-based catalysts emerged recently as a promising class of hierarchical catalytic materials due to its merits in: (1) mechanical stability: improved adhesion between direct grown nano-array catalysts and substrate; (2) thermal stability: reduced materials aggregation and surface area loss induced by high temperature

Catalysis by Materials with Well-Defined Structures
http://dx.doi.org/10.1016/B978-0-12-801217-8.00012-8

sintering; (3) recyclability: ease to collect and recycle due to its attachment with substrate; and (4) structural and functional efficiency: enhanced materials dispersion and usage, as well as catalytic performance [4].

In this chapter, we intend to provide a comprehensive review of the latest progress in the fabrication, characterization, and catalytic performance of various nano-array catalysts. The target applications discussed here mainly include gas phase catalytic reactions such as oxidations of CO, NO, and hydrocarbon for automotive emission control, electrocatalysis, and photocatalysis for environmental remediation as well as energy and fuel production.

2. NANO-ARRAY CATALYSTS FOR HETEROGENEOUS CATALYSIS

2.1 Nano-Array Catalysts for CO and NO Oxidation

Carbon monoxide (CO) and nitrogen monoxide (NO) are two major toxic air pollutants, which contribute to some serious environmental problems such as acidic rain and smog. NO is also a greenhouse gas that could contribute to ozone layer depletion. Both CO and NO are generated mainly from incomplete or high temperature fossil coal or fuel combustions. Heterogeneous catalysts have been generally utilized to reduce the emissions of CO and NO. A typical example goes to the precious metals such as Pt and Pd, which have been widely used in traditional three-way catalysts for automobiles. It has been a challenging and sustained topic on how to eliminate or reduce the usage of precious metals to lower the cost of industrial catalysts. Despite the reported issues on sintering and chemical poisoning resistance back to 1970s [5,6]. Mixed metal oxides and perovskite-type metal oxides (e.g., $La_xSr_{1-x}CoO_3$(LSCO), $La_xSr_{1-x}MnO_3$(LSMO)) have been extensively studied and found to be good candidates for catalytic exhaust oxidation reactions [7–9]. Gao and coworkers fabricated TiO_2/LSMO nanorod arrays on planar substrates (such as Si and glass) via a combination of hydrothermal synthesis and radio frequency (RF) magnetron sputtering methods [10] a significant enhancement of catalytic performance has been achieved through incorporation of thin LSMO layer on hierarchical TiO_2 nanorods. As shown in Figure 1(A), aligned and densely packed TiO_2/LSMO nanorods vertically grew on the substrate. The cross-section view of nanorod array in Figure 1(B) clearly showed that the length of the composite nanorods was about 1 μm and the diameter ranged from 30 to 100 nm. The as-prepared composite nanorods were composed of crystalline TiO_2 nanorod core with amorphous LSMO shell, while the LSMO shell turned into crystal after annealing in 800 °C for 3 h. Figure 1(C) showed the typical single TiO_2/LSMO core-shell nanorod after annealing. A ridged shell of LSMO could be clearly observed. High-resolution transmission electron microscopy was also applied to characterize the detailed crystal structure. The lattice spacing of 0.38 nm (Figure 1(D)) revealed that the LSMO nanoshell grew along the (110) LSMO atomic planes. The energy dispersive X-ray spectrum (EDS) and X-ray diffraction (XRD) pattern were also collected from the samples and further confirmed the composition and

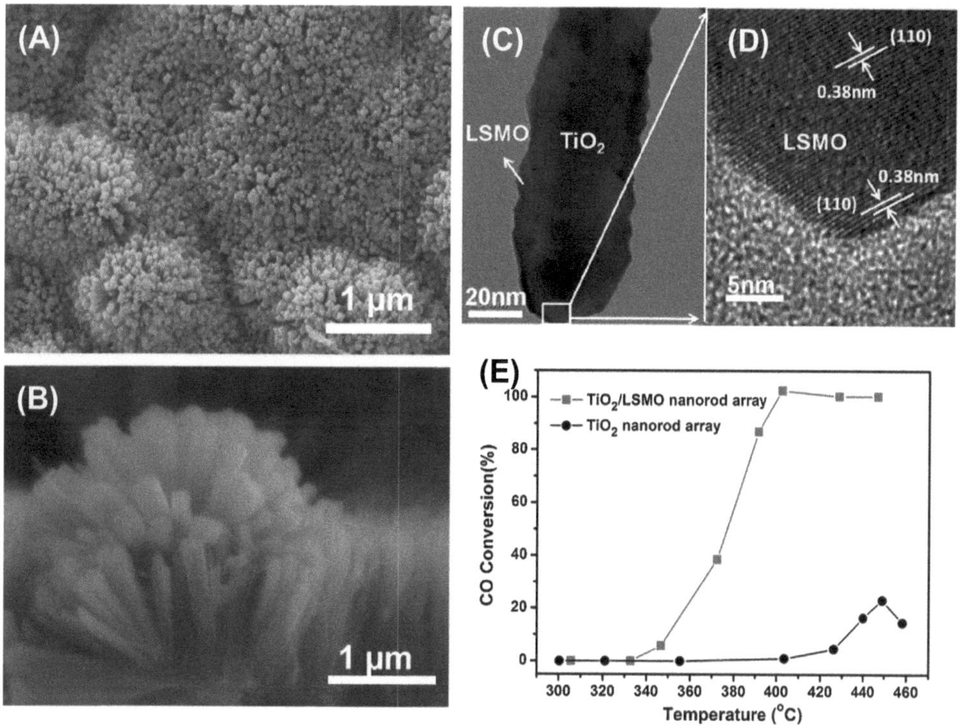

Figure 1 (A) Top view SEM images of TiO_2/LSMO nanorod arrays on Si substrates; (B) a cross-sectional view of TiO_2/LSMO nanorod array; (C) TEM of a typical TiO_2/LSMO nanorod after 800 °C and 3 h annealing; (D) lattice image of LSMO shell; (E) CO oxidation conversion efficiency as a function of reaction temperature over the 800 °C annealed TiO_2/LSMO composite nanorod array and TiO_2 nanorod array. Reaction conditions: $0.9 \times 0.45\ cm^2$ substrate, 1% CO/10% O_2/N_2, 30 ml/min. *Reproduced with permission from [10].*

crystal structure of TiO_2/LSMO core-shell nanorod arrays. Compared with pure TiO_2 nanorod array, BET (Brunauer, Emmett, and Teller) surface area measurement suggested that TiO_2/LSMO core-shell nanorod array had a smaller surface area. CO oxidation reaction was selected as a probe reaction to investigate the performance of the 800 °C annealed TiO_2/LSMO composite nanorod array and TiO_2 nanorod array on Si substrate separately. 1% CO and 10% O_2/N_2 balanced by Ar was introduced at a flow rate of 30 ml/min. TiO_2/LSMO composite nanorod array showed a better performance than TiO_2 nanorod array with 90 °C lower onset temperature. At 400 °C, TiO_2/LSMO core-shell nanorod array achieved 100% CO conversion efficiency, while pure TiO_2 nanorod array just showed ca. 20% CO conversion efficiency. And the 100% conversion temperature was also lower than the reported powder form LSMO catalysts. Oxygen vacancy in LSMO film is suggested to contribute to the improved performance.

Three-dimensional (3D) monolithic catalytic devices, coated with catalytically active materials on the parallel honeycomb channels of micrometer to millimeter in diameter, are a family of very important heterogeneous catalysts in practical industrial applications. To extend the advantages enabled by nanoscale dimension into 3D space, direct assemblies of metal oxide nano-array on 3D monoliths (made of cordierite or stainless steel) were reported by Gao and coworkers [4]. A wide spectrum of metal oxide nano-arrays was in situ grown on the cordierite and stainless steel monoliths via a green, low temperature hydrothermal process (Figure 2(A)). The 1-inch diameter monolithic catalysts shown in Figure 2(B) were ZnO nanorod arrays on stainless steel and TiO_2 nanorod arrays, CeO_2 nanotube arrays (NTAs), and Co_3O_4 porous nanowire arrays on cordierite, respectively. SEM images in Figure 2(C) and (D) depicted ZnO nanorod array uniformly coated on the channel surfaces of cordierite substrates. The EDS mapping inset in Figure 2(C) showed the Zn element distribution in a typical ZnO nano-array rooted channel surface, which further confirmed the excellent uniformity of ZnO nanorod arrays. Figure 2(E) demonstrated the morphology of Co_3O_4 porous nanowire array grown in the corner of a cordierite monolith channel. The cross-section SEM images and corresponding transmission electron microscope (TEM) images of ZnO nanorod array, TiO_2 nanorod array, CeO_2 NTA, and Co_3O_4 porous nanowire array were illustrated in Figure 2(F)–(M). Single crystal ZnO nanowire and TiO_2 nanorod with well-defined surface grew along [0001] and [001] direction, respectively. Polycrystalline CeO_2 NTA, which was prepared by using ZnO nanorods as template, showed uniform diameter distribution with an average diameter of 80 nm and a typical wall thickness of 20 nm. Co_3O_4 porous nanowire array could also be assembled on the monolith channels with a length up to 12 μm and a diameter ranging from 100 to 200 nm. The pore distribution of Co_3O_4 porous nanowire array were also characterized and showed an average pore size of 20 nm. As critical requirements of automotive catalysts, the thermal and mechanical stability of as-prepared nano-array monolithic catalysts were investigated. ZnO, TiO_2, and Co_3O_4 nano-array showed exceptional thermal stabilities with no obvious morphology and phase change after 800 °C annealing temperature for 100 h. The decrease of surface area was observed for the initial 24 h and became stable in the following period of time. The monolithic nano-array catalysts also showed excellent mechanical properties with well-retained morphology under both static (50 L min^{-1}) and pulsatile (60 L min^{-1}) air flow. On the other hand, the thermal and mechanical stability of two-step template method prepared CeO_2 nanotube were not as good as the one-step hydrothermal derived oxides. The morphology change and weight loss of CeO_2 NTA were observed after 800 °C thermal annealing and the long time air flushing. The relatively poor thermal and mechanical stabilities might be attributed to the polycrystalline nature of the template-derived nanotubes, their nonuniform tube wall thickness along the nanotube length, and the weak adhesion between nano-array and substrate as a result of the template method.

Metal oxide nano-array supported Pt catalysts were prepared by depositing 2 nm Pt nanoparticle on metal oxide nano-array through a colloidal deposition process. TEM

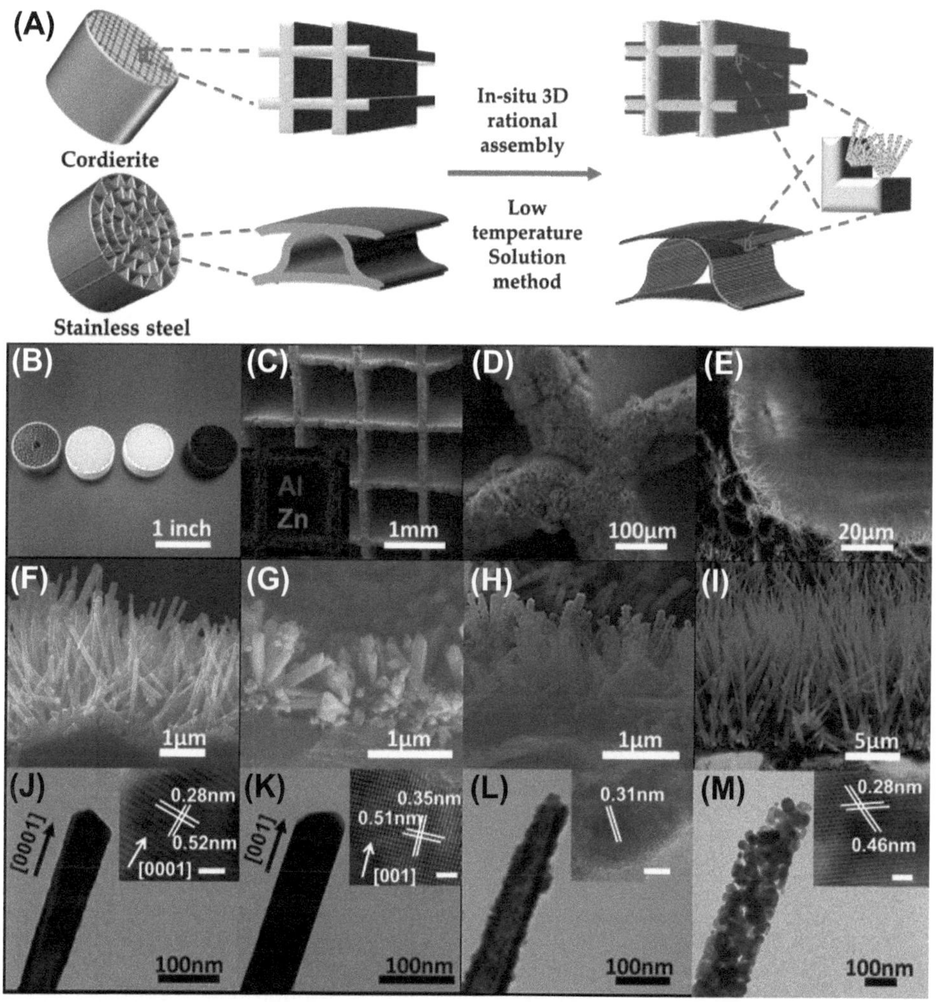

Figure 2 (A) Schematic nano-array assembly process on typical cordierite and stainless steel monoliths. (B) Selected nano-array rooted monoliths. SEM images of metal oxide nano-array on monoliths: (C) Low magnification view of cordierite monolithic channels grown with ZnO nanorod arrays; inset: Zn and Al element map of a single monolith channel from EDXS. (D), (E) Top and cross-sectional views of Co_3O_4 porous nanowire arrays on cordierite monolith. (F) and (J), (G) and (K), (H) and (L), and (I) and (M), are, respectively, ZnO nanorod arrays, TiO_2 nanorod arrays, CeO_2 nanotube arrays, Co_3O_4 porous nanowire arrays on cordierite monoliths, and their corresponding TEM images. Scale bars in all the insets are 2 nm. *Reproduced with permission from [4].*

images of a typical metal oxide nano-structure supported Pt depicted in Figures 3(A)–(D) demonstrated the uniform distribution of Pt nanoparticles all over the nanostructure, and the deposited Pt nanoparticles had a narrow size distribution. CO oxidation was performed as a probe reaction to evaluate the activity of nano-array supported catalysts. As shown in Figure 3(E), the light-off temperatures (defined as 50% CO conversion temperature) of Pt loaded on ZnO, TiO_2, CeO_2, and Co_3O_4 were 260 °C, 258 °C, 193 °C, and

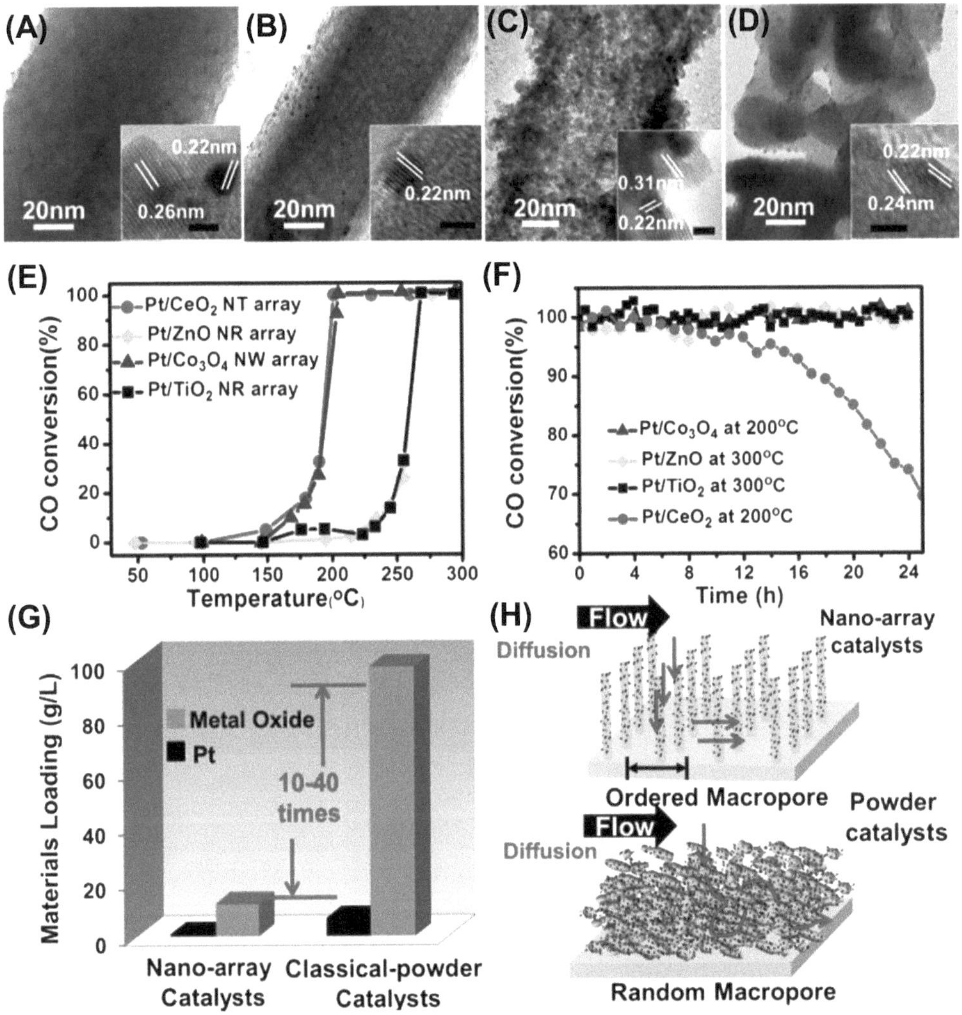

Figure 3 TEM images of individual Pt nanoparticle-loaded metal oxide nanostructure: (A) ZnO nanorod; (B) TiO_2 nanorod; (C) CeO_2 nanotube; and (D) Co_3O_4 nanowire. Scale bars in all the insets are 2 nm. (E) CO oxidation conversion as a function of temperature over and (F) catalytic stability of Pt-loaded (CeO_2, ZnO, Co_3O_4, and TiO_2) nano-array monolithic catalysts; (G) materials utilization efficiency of nano-array catalysts and classical-powder washcoated catalysts; (H) A schematic transport model of nano-array catalysts and traditional nano-powder catalysts. *Reproduced with permission from [4].*

195 °C, respectively. The catalytic stabilities of as-prepared Pt-loaded metal oxide nano-arrays were shown in Figure 3(F), in which nano-array catalysts were applied at the 100% conversion temperature for 24 h. While Pt/CeO_2 nanotube catalysts sustained the 100% conversion efficiency at the initial 6 h and decreased to 70% conversion efficiency in 24 h, it is observed that Pt-loaded ZnO, Co_3O_4, and TiO_2 nanorod array exhibited very stable catalytic activity with no decrease of efficiency. The decrease of

catalytic activity of Pt/CeO_2 NTA catalysts may be due to the poor thermal stability of polycrystalline nanotube structure. Compared with the literature reported powder form Pt loaded catalysts of similar catalytic performance, the Pt loading on the metal oxide nano-array catalysts was significantly lower (around 10–40 times lower) (Figure 3(G)). As shown in Figure 3(H), the improved catalytic activity of nano-array catalysts was suggested to be due to the intrinsic merits of nano-array structure with ordered macropore channels between the adjacent nanostructures. The ordered macropore channels could not only help the catalysts preparation for an optimized metal loading and dispersion but also promote the gas phase reaction by enabling short gas diffusion pathway.

Similar results were also obtained by Wang and coworkers on CeO_2 NTA-based catalysts [11]. The start-up temperature of CeO_2 NTA on cordierite was lower than the CeO_2 nanoparticle wash-coated cordierite monolithic catalysts. In CO oxidation, the light-off temperature of CeO_2 NTA was around 260 °C, about 40 °C lower than CeO_2 nanoparticle-coated monolith.

Zheng and coworkers [12] reported the improvement of catalytic activity of CuO nanowire array with nonequilibrium RF plasma treatment. CuO nanowire array with nanowire diameters of 100–200 nm and lengths of ~50 μm were prepared on Cu substrate by a thermal annealing method. The as-prepared CuO nanowire array was treated by nonequilibrium RF plasma with either Ar or H_2 flow. CO oxidation was conducted on CuO nanowire array (as prepared CuO nanowire array), CuO_Ar_RF nanowire array (Ar treated), and CuO_H_2_RF nanowire array (H_2 treated) to investigate the plasma effect on CuO catalytic activity at 140 °C with flow rate of 300sccm, respectively. For both fuel-rich (2.0 vol% CO and 0.7 vol% O_2 in He) and fuel-lean (2.0 vol% CO, 3.0 vol% O_2 in He) conditions, CuO_H_2_RF nanowire array had the highest activity. While CuO nanowire array showed the lowest activity, CuO_Ar_RF nanowires had the medium performance. After further studying surface chemistry, the kinetic as well as the bulk oxygen diffusion, the enhancement from the plasma treatment is ascribed to the creation of grain boundaries and the reduction of Cu(II) to the more active Cu(I), which had higher diffusion rate of bulk oxygen to surface.

Contreras and coworkers [13] investigated the CO oxidation on model Pt nanowire array catalysts made by lithography method. Dense arrays of platinum (Pt) nanowires were deposited on ZrO_2, SiO_2, Al_2O_3, and CeO_2 thin film, respectively. These unique 2D nano-arrays could serve as model catalysts that offer possibilities to investigate the metal–support interaction between Pt and metal oxides. CO oxidation was carried out in the 513–613 K temperature range, where the reaction kinetics was mass-transport limited. ZrO_2 was found to be able to stabilize Pt nanowires, and the stable Pt led to high turnover frequency (TOF) for ZrO_2/Pt catalysts. While CeO_2 supported nanowire array catalysts, very strong metal-support interaction between CeO_2 and Pt contributed to the high activity rather than Pt. SiO_2- and Al_2O_3-supported Pt catalysts showed lower TOF, which may be caused by the migration of Pt into supports and thus low population of Pt exposed. Further structure-engineered nano-array catalysts could help advance the understanding of CO oxidation mechanism and thus contribute to new efficient catalysts development.

Nitrogen oxide (NO_x) is one of the major pollutant gases, which not only contribute to the greenhouse effect but also is a main cause of acid rain and smog. The state-of-the-art catalytic technologies for NO_x removal are selective catalytic reduction and NO_x storage and reduction (NSR). Compared with NO, NO_2 is easier to be reduced into nitrogen. Therefore, for both strategies, NO oxidation is the critical step that determines the efficiency of NO_x removal. Nano-array catalysts based on some catalytically active metal oxides have been demonstrated to be effective for NO oxidation.

For instance, Gao and coworkers [14] reported the in situ fabrication of Co_3O_4 nanowire array on 3D monolith substrates for highly efficient CO oxidation catalysts. Moreover, the NO oxidation performance of Co_3O_4 nano-array monolithic catalysts was also investigated and exceptional catalytic NO to NO_2 conversion had been achieved at low temperature. As shown in Figure 4(A) and (B) were the SEM images of Co_3O_4 nanowire arrays

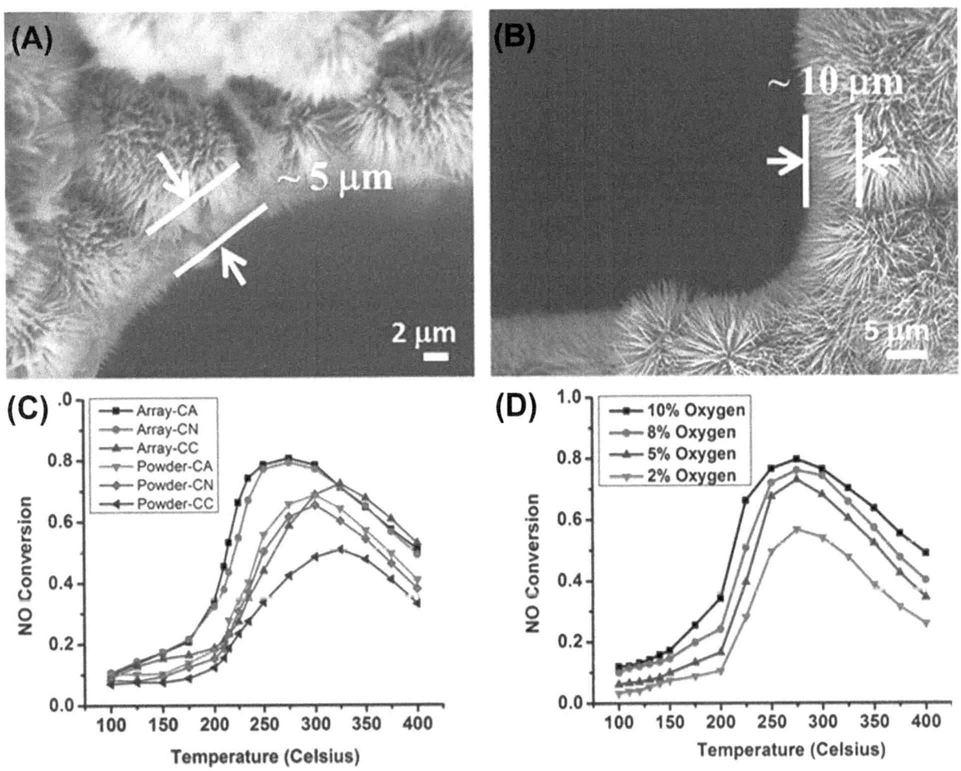

Figure 4 (A), (B) SEM images of Co_3O_4 nano-arrays grown on the corner of an individual monolith channel prepared by cobalt nitrate and cobalt chloride, respectively; (C) NO oxidation performance of catalytic Co_3O_4 nano-arrays and powders prepared from different precursors; (D) catalytic NO conversion performance of CA with varied oxygen concentration. *Reproduced with permission from [14].*

grown on the corner of a monolith channel derived from different Co precursors. Co_3O_4 nanowire array derived from cobalt nitrate (Figure 3(A)) was about 5 μm in length and 50 nm in diameter, while Co_3O_4 nanowire array prepared from cobalt chloride was about 10 μm in length and 100 nm in diameter given the same amount of growth time. XRD confirmed the same phase of all derived Co_3O_4 nanowire arrays.

The temperature dependence of NO conversions was studied by recording the NO concentration from room temperature to 400 °C at each individual temperature by FT-IR spectroscopy. For the experiment, the gas mixture of 500 ppm NO and 10% oxygen balanced by nitrogen were introduced into the reaction system with flow rate set as 200 mL per minute to achieve a space velocity of ~50,000 h^{-1}. As shown in Figure 4(C), compared with powder form counterparts prepared from the same hydrothermal method, Co_3O_4 nanowire array derived from cobalt chloride (CC), cobalt nitrate (CN), and cobalt acetate (CA) achieved maximum NO conversion at ~25 °C lower temperature. The maximum NO conversion efficiencies on Co_3O_4 nano-arrays were also 15–20% higher than the counterpart powders. The NO light-off curves of Co_3O_4 nano-arrays from CA and CN were quite similar with the drastically increase of conversion at ~150 °C and maximum conversion (80%) at ~275 °C. The Co_3O_4 nano-arrays derived from CC start the reaction quickly at ~200 °C and reaches its maximum conversion (75%) at ~325 °C. In the typical test, 10% Oxygen in the gas mixture is far more than 500 ppm NO so that it was reasonable to treat the reversible NO oxidation into an irreversible process. In the real application, both lean (oxygen rich atmosphere) and rich (fuel rich atmosphere) conditions existed. To evaluate the catalysts performance under different oxygen ratio condition, NO oxidation under different ratio of oxygen were performed on Co_3O_4 nano-array synthesized from CA (Figure 4(D)). With the decreased oxygen concentration, the NO conversion efficiency also decreased. A maximum conversion of 55% was achieved in 2% oxygen atmosphere, which indicated that the catalysts were reliable under both rich and lean conditions. The high efficiency of Co_3O_4 nano-array catalysts was attributed to the well-defined array structure, which had an increased surface area and a porous accompanied with the porous nanowire array.

2.2 Nano-Array Catalysts for Hydrocarbon Combustion

Hydrocarbon represents an important and widely used chemical fuel for transportation and electric power generation. As a typical example of hydrocarbon, methane represents a promising chemical fuel for the future not only owing to its abundance in the minerals like shale and natural gas hydrate in the deep sea but also its large energy capacity per volume. The combustion reaction of hydrocarbon, however, requires high temperature, which is usually above 1000 °C. Such high temperature reaction is not favored in terms of energy consumption. In the meantime, nitrogen and oxygen are going to react at temperature around 1500 °C to form NO_x, a notorious pollutant leading to severe environmental problems such as acidic rain and smog. It is thus of significance to develop efficient catalyst for low temperature catalytic combustion in terms of both energy efficiency and environmental friendliness. In this section we will first review the catalytic materials that have been studied for hydrocarbon oxidation. Then we will focus on the monolithic integration of these materials for nano-array catalysts fabrication.

Traditional low temperature hydrocarbon combustion catalysts are usually noble metal nanoparticles. For example, palladium (Pd) nanoparticles have been found to demonstrate high activity toward methane oxidation at low temperature [15–18] although whether metallic Pd or surface oxidized palladium (PdO_x) functions as the active site is still under debate. Pt has also been investigated. Generally Pd is a better catalyst for hydrocarbon oxidation but there are some exceptions where Pt gives better performance for lower alkane combustion. The volatility at high temperature precludes other noble metals for the application in the catalytic hydrocarbon combustion.

Although noble metals are the prevailing catalysts for emission after-treatment and clean energy low temperature combustion, much effort has been devoted to lower the cost by looking for catalyst of earth abundant elements or promote the utilization efficiency of noble metals. Specifically, for catalytic hydrocarbon combustion transition metal oxides as well as perovskites have been studied. Paredes and coworkers investigated catalytic activity of a series of bulk transition metal oxide for methane combustion [19]. The transition metal oxides include Cr_2O_3, NiO, Mn_2O_3, Co_3O_4, and CuO. Catalysts activity was studied by recording light-off curves for the combustion of 2000 ppm of methane in air using 2 g of catalyst, at temperatures from 250 to 650 °C. Figure 5 displayed the typical light-off curves for catalytic methane combustion. It was observed that Co_3O_4 exhibited the best conversion performance while NiO is the worst. The catalytic activity decreased as the

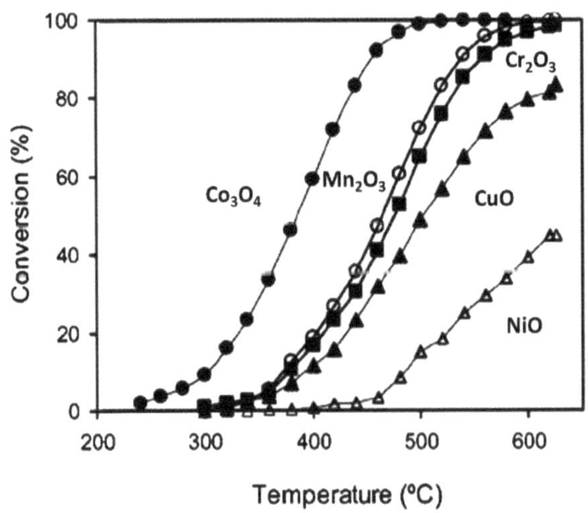

Figure 5 Light-off curves for the combustion of 2000 ppm methane over various metal oxide catalysts. *Reprinted with permission from [19].*

sequence of $Co_3O_4 > Mn_2O_3 > Cr_2O_3 > CuO > NiO$. The normalized methane conversion activity revealed that Co_3O_4 had far better activity than other oxides in the low temperature region. At high temperature Mn_2O_3 and CuO were slightly better.

In addition to the bulk oxide, metal oxide nanostructures had demonstrated promising catalytic activity for low temperature hydrocarbon combustion [20,21]. Taking Fe_2O_3 and Co_3O_4 nanostructures with controlled morphology as an example, Fe_2O_3 nanocrystals can be tuned to form nanorods, nanotubes, and nanocubes. Co_3O_4 nanostructure was controlled to be nanosheets, nanobelts, and nanocubes [20]. The Fe_2O_3 nanorods displayed the best performance of catalytic methane combustion [21]. The surface density of iron atoms, the catalytic active sites, was the largest on $\{110\}$ planes. The Fe_2O_3 nanorods were found to be enclosed by six equivalent crystal planes with $\{110\}$ and two equivalent planes with $\{001\}$. The nanotubes were enclosed by two preferentially exposed planes with $\{010\}$ and two planes with $\{001\}$ and the nanocubes were enclosed by planes with $\{012\}$, $\{102\}$, and $\{\bar{1}1\bar{2}\}$. Therefore, the nanorods exhibited the highest catalytic activity with large populated catalytic active $\{110\}$ planes. Similar phenomena were observed in the Co_3O_4 systems where nanosheets exhibited the best catalytic performance. It can be seen from these two examples that the catalytic activity can be manipulated through controlled crystal growth by solution chemistry. The tunable catalytic activity based on the correlation between crystallography and active site distribution had opened up new opportunity for rational nanocatalyst design via controlled nanostructure self-assembly. This advantage had inspired much research effort to artificially integrate nanostructures into well-defined geometry and orientation to tailor the catalytic activity and promote the materials utilization efficiency.

As a typical configuration, monolithic nano-array catalysts had been reported for the catalytic hydrocarbon combustion. The major advantages of nano-array geometry include free of binder for the catalyst preparation, enhanced internal diffusion by ordered orientation, increased surface area of nanoscale dimension, better thermo-mechanical property, and facilitated gas-surface interaction. Zheng and coworkers prepared CuO nanowire arrays as monolithic catalysts by directly annealing a piece of copper mesh at 520 °C for 2 days [22]. After the annealing, the color of the copper mesh changed from bright yellow to black, indicating that Cu was oxidized to CuO. The morphology, crystal structure, composition, and oxidation state of the as-grown CuO nanowires were further examined by electron microscopy. Figure 6 showed that the CuO nanowires were formed perpendicularly to the copper mesh with an average diameter of 100 nm and length of 50 μm. The coverage over the large copper mesh was very uniform with nanowires successfully growing everywhere. The catalytic activity of the CuO NWs for CH_4 oxidation was measured in a flow reactor, which consisted of a quartz tube (90 cm long, 2 cm inner diameter) housed in a tube furnace. The inlet gas flow was a mixture of 1.5 vol% CH_4 and 6 vol% O_2 balanced by helium; the total flow rate was set to be 100 sccm.

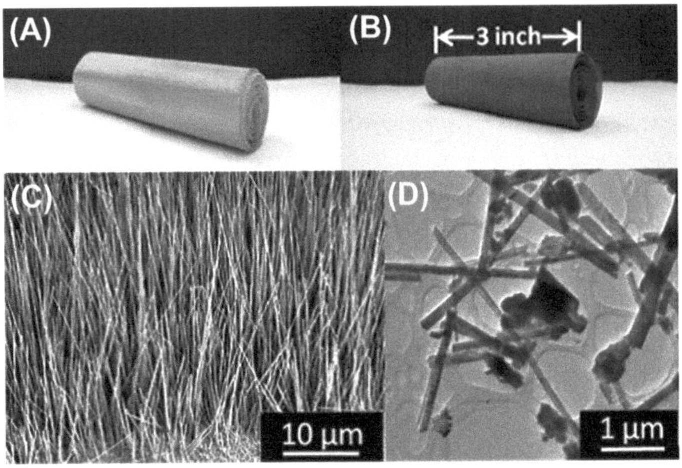

Figure 6 Photographs of (A) the copper mesh and (B) the annealed copper mesh (520 °C, 2 days) with the CuO nanowires. (C) SEM image and (D) TEM image of the as-grown CuO nanowires. *Reprinted with permission from [22].*

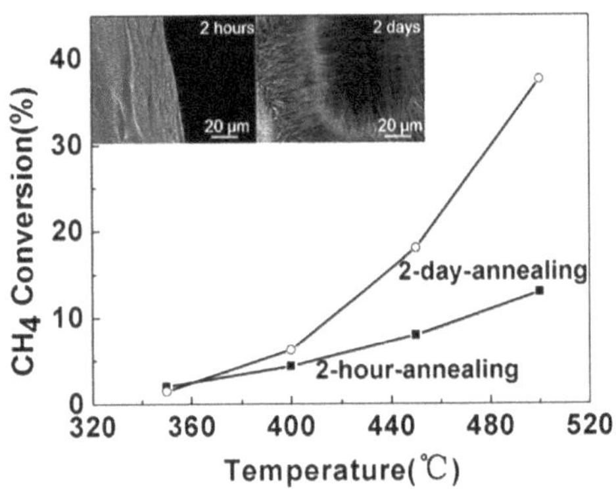

Figure 7 Catalytic methane combustion performance of CuO nanowires array catalyst achieved by thermal annealing for 2 h and 2 days respectively. The insets are the typical SEM characterizations of the CuO nanowires arrays grown from copper mesh by different annealing time. *Reprinted with permission from [22].*

The CuO nanowire arrays obtained after thermal annealing for 2 h and 2 days were selected to study the annealing effects on the nanowires growth and the catalytic methane combustion performance. In Figure 7, the SEM characterizations of both copper meshes annealed for 2 h and 2 days were displayed. The longer annealing time was found to promote the nanowire growth, which can be concluded from increased length of nanowires by 2 days annealing compared with 2 h annealing. Moreover, the coverage density also increased with prolonged annealing time. As shown in Figure 7, the nanowires were much more densely packed with better uniformity when the copper mesh is annealed for 2 days. With longer nanowires grown from the substrate, the 2 days annealed CuO nanowire arrays exhibited better catalytic performance of methane combustion. At 500 °C, the methane conversion was close to 40%, which is 25% higher than that of 2 h annealed nanowire arrays.

As discussed previously, Co_3O_4 showed the highest catalytic activity for the hydrocarbon combustion. It would be beneficial if monolithic Co_3O_4 nanowire arrays can be achieved. Recently, we had successfully proposed a hydrothermal method to directly integrate 3D hierarchical Co_3O_4 nanowire arrays onto the commercial monolithic ceramic honeycombs substrate [14,23]. Similar strategy had been reported by Li et al. and Feng et al. by using ammonia as the precipitator. Zheng and coworkers directly grew Co_3O_4 nanowire arrays on stainless steel mesh substrate by ammonia-evaporation-induced growth [24]. During the synthesis, the addition of Cu^{2+} into the precursor solution was found to promote the nanowire arrays growth on the stainless steel substrate. Figure 8 showed the as-prepared Co_3O_4 nanowire arrays supported on stainless steel mesh with different amount of Cu^{2+} being introduced. As can be seen from Figure 8, only large bundles of 1D Co_3O_4 can be achieved without the existence of Cu^{2+}. The microstructures were pretty long and aggregated with each other. With the increased concentration of Cu^{2+}, much more ordered nanoarrays geometry started to form and the coverage density was increased, which led to uniform distribution of Co_3O_4 nanowires. It was worth noticing that further increasing the Cu^{2+} concentration will give rise to the deteriorated nanoarrays morphology. Instead of forming ordered and well-distributed nanowire arrays, the nanowires became shortened and the aggregation took place making the wires thick.

Figure 9 demonstrated the catalytic methane combustion performance of Co_3O_4 nanowire arrays with different amount of Cu^{2+} used in the precursor solution. Compared with original Co_3O_4 without Cu^{2+} addition, the catalytic activity of methane combustion had been promoted by Cu^{2+} incorporation. Within an appropriate concentration region, the introduced Cu^{2+} enhanced the mass loading and the uniformity of the Co_3O_4 nanowire arrays, which thus promoted the catalytic reaction performance. However, further increasing the Cu^{2+} will deteriorate the catalytic activity. Based on the previous SEM characterization, the increased Cu^{2+} concentration led to shortened nanowires and aggregation, which might cause the surface area loss and thus gave rise to worse catalytic performance.

Another group of promising and cost-effective material candidate for hydrocarbon treatment is perovskite-type metal oxides (ABO_3), which are complex oxides with multiple structural and functional properties. In addition to the functional properties that find them applications for electronics, dielectrics, energy storage, and conversion, the catalytic property of perovskite is also intriguing. The good redox property had made them an important group of materials as oxidation catalysts. The catalytic property of perovskites can be deliberately tailored by different combination of metal elements on both A and B

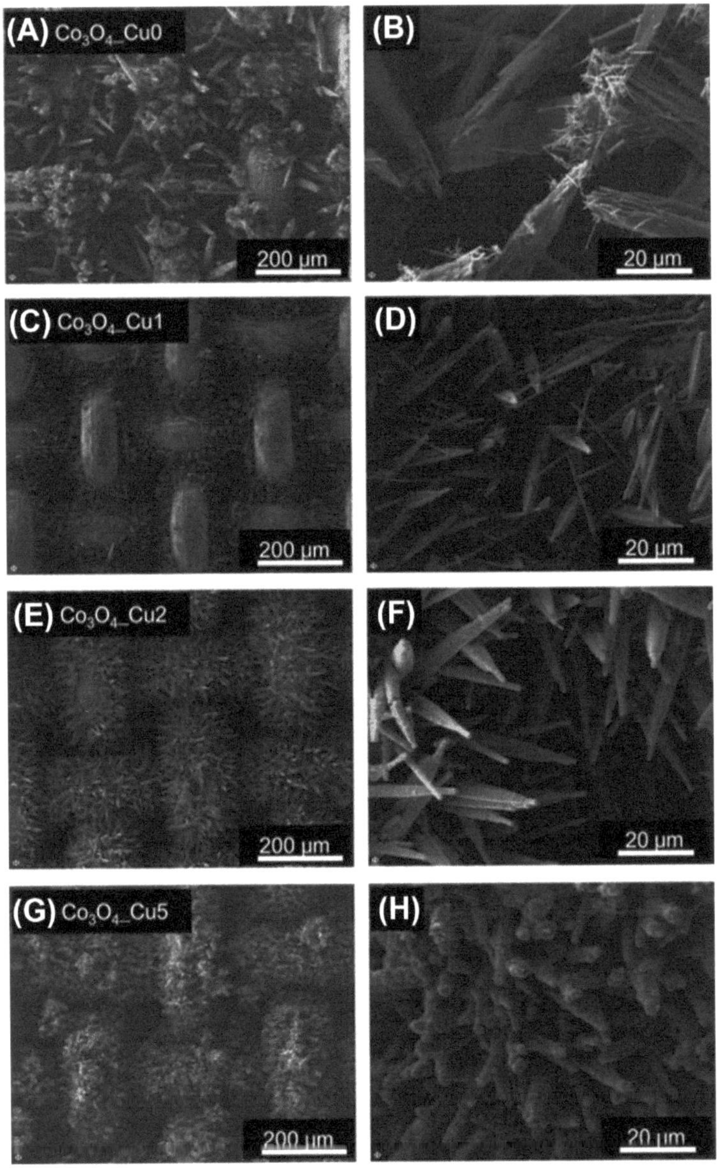

Figure 8 SEM characterizations of Co_3O_4 nanowires arrays supported on stainless steel substrates with different morphologies achieved by adding different amount of Cu^{2+} in the precursor solution for the synthesis. *Reprinted with permission from [24].*

sites. A vast amount of literature had demonstrated the possibility to employ different elements in different permutations and combinations, resulting in robust perovskite structures. Besides numerous options of A site and B site elements, partial substitution of A or B site cations was able to result in profound modification upon the catalytic activity.

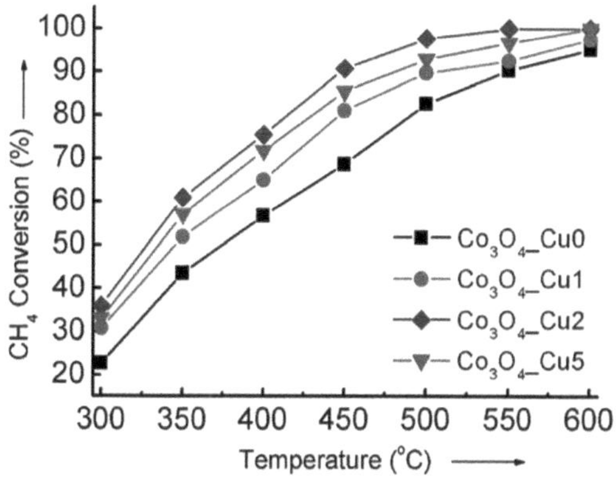

Figure 9 Catalytic methane combustion performance of monolithic Co_3O_4 nanowire arrays supported on stainless steel mesh with different Cu^{2+} concentration in the precursor solution. *Reprinted with permission from [24].*

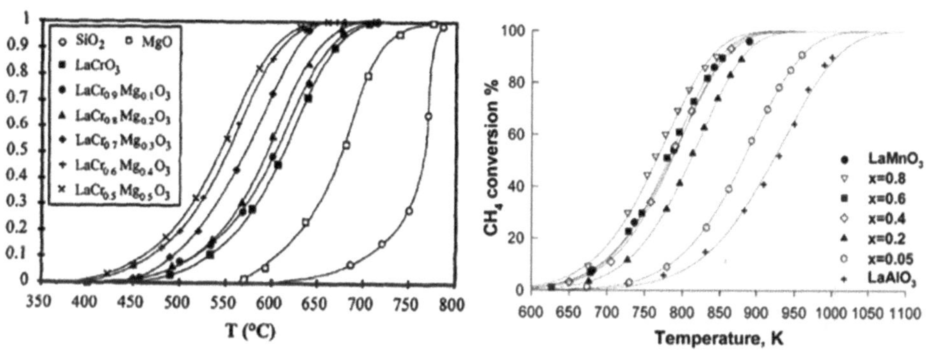

Figure 10 Adjusted catalytic activity of methane combustion by changing the dopant species and their concentrations in the B sites of perovskites. *Reprinted with permission from [25–26].*

Herein we will introduce two examples of how the cations in the perovskite structure influence the catalytic activity toward hydrocarbon combustion. Figure 10 illustrated the B site effects on the catalytic methane combustion [25]. For the $LaCrO_3$ perovskite, La occupies A sites while Cr sits in B sites. Typically, the B site element was usually the active site and was thus more closely related to the catalytic activity. By partially replacing Cr in the B site of perovskite with Mg, enhanced catalytic performance of methane combustion was observed. When the Mg doping concentration increased (Mg/Cr ratio increases from 1/9 to 1/1), the temperature at which full conversion of methane takes place was lowered from 700 °C to 600 °C. Similarly, for $LaAlO_3$ perovskite, replacing Al with more active Mn in the B sites led to much enhanced catalytic performance [26]. The temperature for full conversion of methane underwent a 200 K decrease. The catalytic activity can also be adjusted by introducing different dopants in A sites of perovskite.

To combine the benefit enabled by nanowire arrays and the numerous possibilities toward adjusted catalytic activity of perovskite through doping, the challenge was mainly the scale up of perovskite nanostructures. The synthesis of perovskite nanostructures with controlled morphology and orientation was difficult due to the complex nature of the compound itself and the lacking of understanding of the pertinent solution chemistry. It was interesting to note, however, the fabrication of perovskite thin film had been well developed in materials physics study. Besides the expensive vapor deposition strategy sol-gel processing as well as spin coating had been proved to prepare polycrystalline thin film with good quality. Therefore, an indirect growth technique for perovskite nanoarrays structures had been proposed by our group recently. The fabrication process starts with the preparation of monolithic metal oxide nanowire arrays such as ZnO and TiO_2 nanoarrays. Using these nanoarrays as template, the following sol-gel process was able to successfully coat a thin layer of perovskite onto the nanoarrays. The composite structures had been discovered to serve as an efficient oxidation catalyst such as catalytic CO oxidation and photocatalytic organic dye decomposition. We believed the same composite nanoarrays structure will be able to function as an effective catalyst for hydrocarbon combustion as well.

2.3 Nano-Array Catalysts for Electrocatalysis

Being the world's largest form of energy consumption by end-user, electricity serves as a necessary resource in people's daily life routine [27]. The demand for electricity is pushed by the growth of population as well as the boom of electronic industry worth billions of US dollars. The growth of electricity demand lasts for years and the total electricity demand in United States is expected to grow by 28% in 2040 [28]. The above trend poses an urgent demand on the development of green energy technologies for clean energy production, storage, and conversion. Among all the possible solutions so far, electrochemical technologies such as hydrogen production and fuel cells are particularly interesting owing to their ability to produce highly dense but pollution-free energy with good efficiencies [27]. As a result, research in electrocatalysis and interfacial electrochemistry to accelerate electrochemical process has attracted tremendous attention in recent years [29].

Electrocatalysis is known as the interfacial catalytic phenomenon that promotes electrochemical reaction by reducing its required overpotential. The process involves transfer or accumulation of charges at interface between electrode and electrolyte. As a result, such heterogeneous catalytic system is closely related to the electrochemical parameters of the interfacial system. Electrocatalytic behavior is not only dependent on the nature of involved materials (electrode, electrolyte, or catalyst) but also controlled by the formed interfaces in the system. In recent years, electrocatalysis in hydrogen evolution reaction and fuels cells has been extensively studied including, oxygen reduction reaction (ORR), oxygen evolution reaction (OER), oxidation of methanol/ethanol, and so on.

Design and discovery of electrocatalyst has been one of the mainstreams of research on electrocatalysis, aiming to achieve high-rate and high-efficiency electrochemical reactions. Various types of catalysts or catalytic electrodes have been designed to accelerate particular electrochemical reactions. The morphologies and crystal structure of catalysts are proven to play an important role; for instance, nanoparticles, nanorod, thin-film, and bulk of same catalytic material demonstrate distinct catalytic behaviors [30]. Among them, 1D nano-array catalysts attract great attention owing to large surface area, directional charge transport, long-term and high-temperature stability, as well as excellent and tunable catalytic performance [4]. In this section, three types of nano-array catalysts for electrocatalysis are highlighted: carbon-based nano-arrays, metal-oxide-based nano-arrays, and metal/composite nano-arrays.

2.3.1 Carbon-Based Nano-Arrays

Currently, one of the most popular carbon-based nano-arrays is carbon nanotubes (CNTs). Numerous studies on CNTs have been done with their implications in heterogeneous catalysis [31,32]. Multiwalled CNTs (MWCNTs) are shown to be excellent support materials for noble-metal catalysts such as Pt and Pd, as a result of their large surface area, high electrical conductivity, and robust chemical stability [33,34]. Nevertheless, the overall catalytic activity not only depends on the attached metallic catalysts but also is strongly affected by the intrinsic property of MWCNTs and the interaction between the anchored catalysts and the surface of MWCNTs [35]. Due to the fact that as-prepared MWCNTs are usually chemically inert, proper functionalization of MWCNTs is required to activate the surface in order to achieve efficient deposition of catalytic nanoparticles [36]. Typical functionalization methods include harsh oxidative processes [37], covalent and noncovalent functionalization [38,39], and polyelectrolyte functionalization method [40].

A typical example of MWCNT-supported electrocatalysts was demonstrated by Chakraborty and co-workers on MWCNT-supported Pd particles [41]. MWCNTs were prepared and further wrapped by polyelectrolyte poly(diallyldimethylammonium chloride) (PDDA). Consequently, rod-like Pd nanostructures (nPds) of average size ~8 nm were stably deposited onto the walls of PDDA-wrapped MWCNTs. SEM and TEM images were shown in Figure 11(A) and (B), respectively.

Cyclic voltammogram of the synthesized nPd-MWCNT electrode for oxidation of formic acid was measured as displayed in Figure 11(C). Only one peak for the oxidation of formic acid was observed at 0.64 V, which was attributed to high electronic and ionic transport capacity of MWCNT film with enhanced electron transfer kinetics for the electrode reactions. Electrocatalytic performance of nPd was largely affected by particle coverage on the MWCNTs as seen in Figure 11(D). High coverage enhanced the catalytic activity as a result of ensemble behavior of the electrode and the high catalytic activity of the nPds. Apart from oxidation of formic acid, electroreduction of oxygen by nPd-MWCNTs was also investigated. As seen in Figure 11(E), the voltammetric response for nPd-MWCNTs indicated that reduction of oxygen

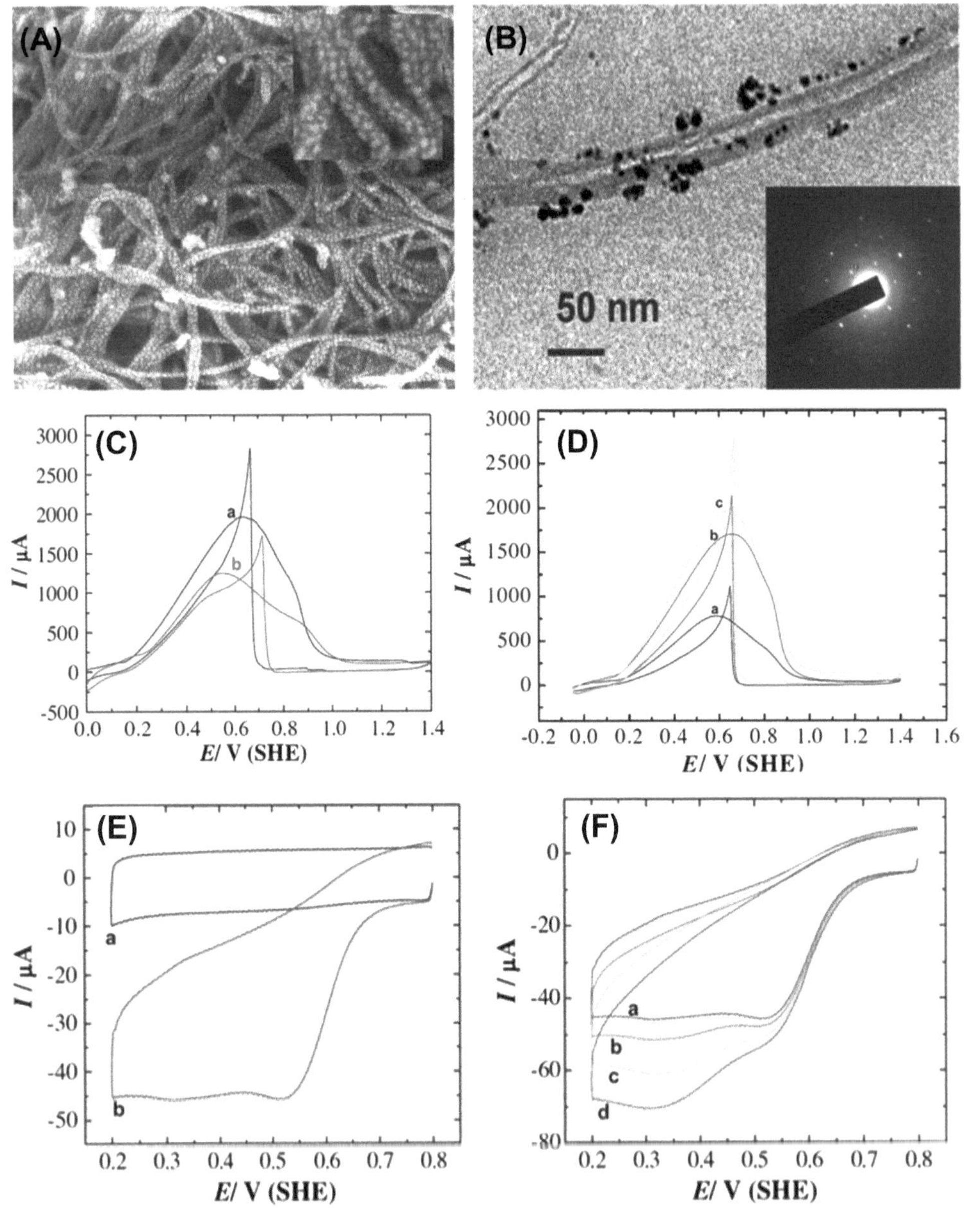

Figure 11 (A) SEM and (B) TEM images of Pd nanoparticles on the PDDA-wrapped MWCNTs-modified electrode. (C) Cyclic voltammograms of the Pd nanoparticles on the MWCNT electrode toward oxidation of formic acid in 0.1 M $HClO_4$ (black) and H_2SO_4 (red (dark gray in print version)); (D) cyclic voltammograms of the Pd nanoparticles toward oxidation of formic acid in 0.1 M $HClO_4$ for different particle coverage 7.94 (black), 11.8 (red (dark gray in print version)), and 16.7 (green (light gray in print version)). (E) Cyclic voltammograms of nPd-MWCNTs electrode in the absence (black) and presence (red (dark gray in print version)) of oxygen and (F) in presence of oxygen at different H_2O_2 concentrations (a = 0; b = 0.2; c = 0.6; d = 1 mM) in 0.1 M $HClO_4$. *Reproduced with permission from [41].*

followed a two-step process: peak at 0.53V corresponded to the reduction of O_2 to H_2O_2 and the small hump at 0.31V corresponded to the reduction of H_2O_2 to H_2O. The addition of H_2O_2 in solution gradually enhanced the peak at 0.31V, as seen in Figure 11(F), which further confirms that the peak resulted from the reduction of H_2O_2.

The usage of noble-metal catalysts such as Pt and Pd is greatly restricted by their high cost due to their limited natural resources. Therefore, the development of highly efficient noble-metal-free electrocatalysts is of great significance. Recently, Gong and co-workers reported that nitrogen-doped carbon NTAs were able to act as an effective noble metal-free electrocatalyst for oxygen reduction reaction with higher electro-catalytic activity, lower overpotential, and better long-term stability [42]. In their work, vertically aligned nitrogen-doped carbon nanotubes (VA-NCNTs) were synthesized by pyrolysis of iron (II) phthalocyanine. Voltammetric responses of unpurified and purified VA-NCNTs were investigated separately. Two well-defined peaks in unpurified voltammetric responses in absence of oxygen (upper, dotted curve) presented at −1.0V and −0.65V, which were ascribed to redox reactions induced by Fe catalyst residues. After being purified by electrochemical oxidation, these two peaks disappeared from the voltammetric responses as expected. The oxygen reduction voltammetric response of the purified VA-NCNTs was further studied, and high reduction potential at −0.15V was observed indicating lower overpotential. Furthermore, quantum mechanics calculations demonstrated that the nitrogen dopant induced charge delocalization among surrounding carbon atoms. The charge delocalization changed the chemisorption mode of O_2 from end-on adsorptions to side-on diatomic adsorptions, which weakened O–O bonds and hence gave rise to enhanced oxygen reduction reaction rate at nitrogen-doped CNT electrode.

2.3.2 Metal-Oxide-Based Nano-Arrays

Apart from carbon-based materials, metal-oxide-based materials such as RuO_2, IrO_2, and Co_3O_4 represent another major group of electrocatalysts. Compared to carbon-based electrocatalysts in ORR, metal-oxide-based catalysts show outstanding electro-catalytic activity for OER, which is important for application in water splitting and hydrogen production [43]. In addition, different from carbon-based electrocatalysts that store electrostatic charge and favor high-power delivery, metal-oxide-based electrocatalysts store charge through a redox process and thus result in high energy density [44]. This enables metal-oxides-based catalysts to be promising candidates as electrodes in supercapacitors. Moreover, metal-oxides are good in terms of mechanical and structural stabilities, which enable prolonged lifetime. In particular, metal-oxide nano-arrays allow efficient mass transfer due to the open space between nanowires in conjunction with the large surface area.

Li and co-workers successfully synthesized pure and Ni-doped Co_3O_4 ($Ni_xCo_{3-x}O_4$) mesoporous nanowire arrays on conductively Ti foil substrate [43]. Surface of the nanowires was enriched in Ni based on XPS study. The synthesized $Ni_xCo_{3-x}O_4$ nanowire showed significantly enhanced surface area and roughness compared to pure Co_3O_4 nanowires. Moreover, the Ni doping largely improved electrical conductivity of pure Co_3O_4 nanowire by four orders of magnitude. The polarization curves and cyclic voltammograms for oxygen evolution reaction of pure and Ni-doped Co_3O_4 nanowires demonstrated that Ni-doped Co_3O_4 nanowires showed enhanced electrocatalytic activity with higher polarization current density at the same overpotential. The authors attributed the enhanced electro catalytic activity to the enlarged surface area and improved electrical conductivity of Ni-doped Co_3O_4 nanowires. Moreover, they demonstrated that the crushed sample had obviously lower catalytic activity than nano-arrays, confirming the crucial role of nano-arrays in catalytic property enhancement.

In addition to oxygen evolution reaction, application of metal-oxide-based nano-arrays in electrochemical capacitors (ECs) were also demonstrated by Liu and coworkers [44]. In their work, Co_3O_4@MnO_2 core/shell hybrid nanowire array were studied as hybrid electrode architecture in EC applications. SEM images of as-prepared hybrid nanowire array on stainless steel substrate were shown in Figure 12(A) and (B), respectively. Combination of Co_3O_4 and MnO_2 induced great electrochemical performance much better than that of each individual material. It could be seen from Figure 12(C) that the presence of MnO_2 expanded the cyclic voltammograms curve, resulting in a large increase in the integrated area. The authors ascribed the enhancement to cations (Li^+) adsorption on MnO_2 surface from electrolyte (LiOH): $(MnO_2)_{surface} + Li^+ + e^- \rightarrow (MnO_2^-Li^+)_{surface}$. The enhanced CV integrated area further resulted in high pseudo-capacitance as confirmed by capacitance measurement shown in Figure 12(D). Other research on metal-oxide nanoarrays electrocatalysis for high-performance supercapacitor applications include Xu and coworkers' work on Co_9S_8 nanorod//Co_3O_4@RuO_2 nanosheet arrays [45] and Xia and coworkers' work on Co_3O_4@Pt@MnO_2 nanowire arrays [46].

2.3.3 Metal/Composite Nano-Arrays

Although great progress has been made on carbon-based and metal-oxide-based nano-array catalysts, noble-metal (Pt, Pd, and so on)-based catalysts still show higher catalytic activity for both oxygen reduction and methanol oxidation reactions [47,48]. As discussed in previous section, Pt or Pd nanoparticle catalysts can be used in conjunction with carbon- or metal-oxide-based nanoarrays; however, 0D structured nanoparticles give rise to a large number of defects and low-coordination surface atoms, further limiting catalytic activity for oxygen reduction reaction [49]. To address this issue, 1D metallic nanostructures are developed with the highly anisotropy structure with highly active surface, while maintaining good surface structure with relatively fewer defects [50]. As an

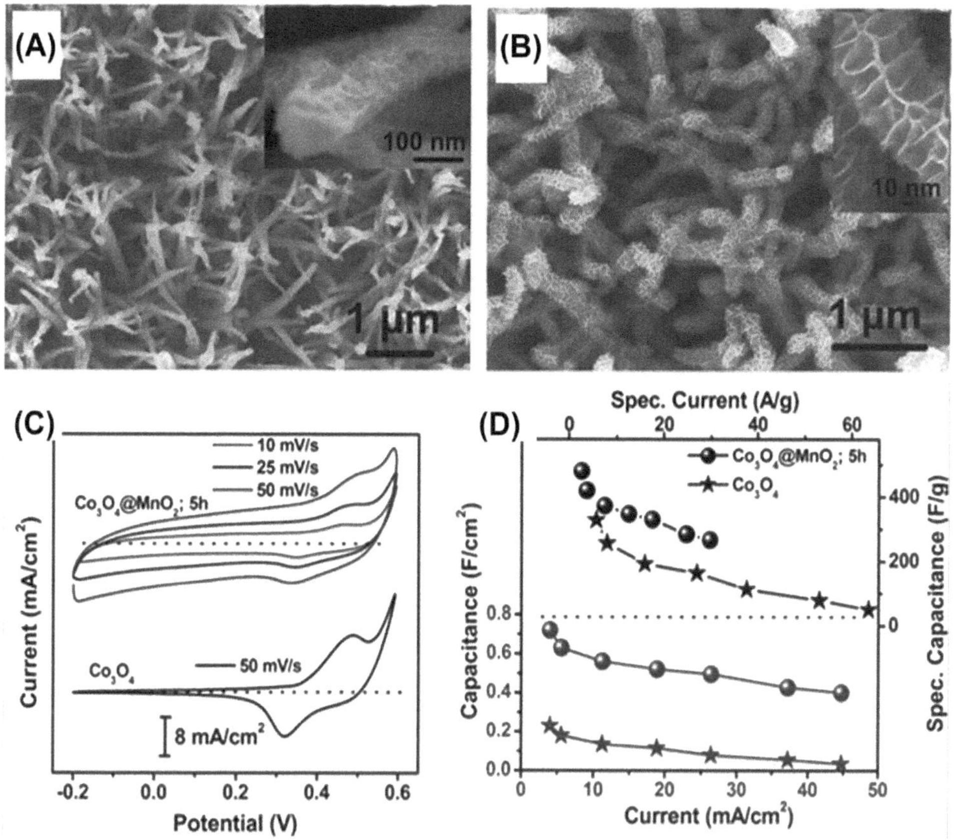

Figure 12 (A) and (B) SEM images of Co_3O_4@MnO_2 core/shell hybrid nanowire arrays after hydrothermal growth for 1 and 5 h, respectively. (C) Cyclic voltammograms of the Co_3O_4@MnO_2 core/shell hybrid nanowire arrays (upper) after 5 h of growth and the bare Co_3O_4 nanowire arrays (down). (D) Areal capacitance (gray curve) and specific capacitance (black curve) of the both hybrid and Co_3O_4 nanowire arrays as a function of current density. *Reproduced with permission from [44].*

example, Koenigsmmann and coworkers had studied the high-quality and single crystalline Pd and Pd-Pt core-shell nanowires.

In their work, arrays of single crystalline Pd nanowires with diameters of 270, 45, and 2 nm were synthesized through a template-assisted U-tube synthetic method with controllable dimensions. SEM and TEM images of the 45 nm Pd nanowire arrays were shown in Figure 13. Polarization curves were obtained to test their activity toward oxygen reduction reaction as seen in Figure 13(D). The measured specific activity toward ORR for 2-nm diameter Pd nanowire was 3.62 mA cm^{-2}, which was double the value of 1.84 mA cm^{-2} for 270-nm diameter ones. It was notable that the trend was just the reverse of Pd nanoparticles, for which smaller size results in lower specific activity due to increased defects density [51]. The authors attributed the size-dependent activity

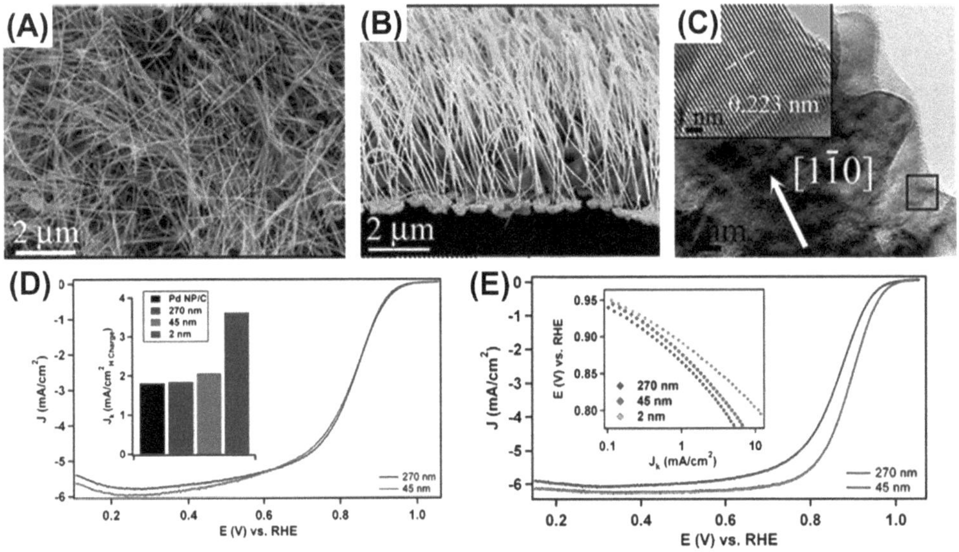

Figure 13 (A) and (B) SEM images of isolated and free-standing Pd nanowire arrays. (C) TEM image of a typical Pd nanowire. Inset: high-resolution TEM image of selecting area (black box) showing well-defined (111) lattice planes. (D) and (E) Polarization curves of 45- and 270-nm diameter Pd (d) and Pd–Pt core-shell (E) nanowires in oxygen-saturated 0.1 M $HClO_4$ solution at 20 °C. *Reproduced with permission from [45].*

enhancement to significant enlarged active surface for small-size 1D nano-array catalysts. The active surface resulted from lattice contraction of the surface when diameter of the Pd nanowires decreased to ~2 nm. Moreover, the as-synthesized Pd nanowires were further coated with a monolayer of Pt. The electrochemical polarization curves in an oxygen-saturated solution were shown in Figure 13(E). The overall ORR activity of the Pd–Pt core-shell structure was largely enhanced as compared with bare Pd nanowires. The performance enhancement was believed to arise from contraction of Pd nanowire surface, strain-induced contraction of Pt monolayer, and electronic 'ligand effect' [52].

Although noble-metal catalysts show high efficiency and activity toward oxygen reduction reaction, major drawbacks of such catalysts include high cost and low CO tolerance. As Pt catalyst can be easily poisoned by CO adsorption onto catalyst surface [41], it suffers from low efficiency for oxidation of organic molecules, for instance, in methanol oxidation reaction for direct methanol fuel cells. To address the issue, Pt-based composite catalyst was developed by Ding and coworkers [30], which on one hand decreased Pt consumption, while on the other hand increased CO tolerance. In their work, highly ordered Pt–Ni–P 1D NTAs were synthesized by ZnO template-assisted electrodeposition. The wall of each NTA was porous and piled by nanocrystals of 3–4 nm. The inner diameter and wall thickness of NTAs were ~400 and ~70 nm, respectively. Cyclic voltammograms for Pt–Ni–P and Pt–Ni NTAs toward methanol oxidation reaction were obtained under condition of 0.5 M CH_3OH and 0.5 M H_2SO_4. It can be seen that Pt–Ni–P NTAs demonstrated much higher catalytic activity, whose specific peak current density was 4 times higher than that of Pt–Ni NTAs. The results indicate that phosphorus P plays a vital role in promoting methanol oxidation reaction. Furthermore, CO stripping measurements of Pt–Ni–P and Pt–Ni NTAs were conducted. On-site potential of CO oxidation on Pt–Ni–P was determined to be 0.28 V, which was significantly less than that of 0.36 V on Pt–Ni NTAs, indicating easier removal of CO from Pt–Ni–P surface. The authors suggested that P could lower the adsorption strength of CO on surface of Pt–Ni–P NTAs and thus enhance CO tolerance of the metal composite system.

In summary, three categories of nano-arrays in electrocatalysis are highlighted in this section, including carbon-based nano-arrays, metal oxide-based nano-arrays and metal/composite nano-arrays. Their electrocatalytic behaviors have been discussed in chemical energy transformation and fuel production through oxygen reduction reaction, oxygen evolution reaction, methanol oxidation reaction. Carbon-based nano-arrays such as carbon nanotubes-based catalysts are widely used to accelerate oxygen reduction reaction. Differently, metal oxide-based nano-arrays are generally utilized to promote oxygen evolution reaction or achieve high-storage ECs. Moreover, metallic nano-array catalysts still have the highest catalytic activity toward ORR. CO tolerance of metallic nano-array catalysts remains an issue and can be mitigated by proper design of metal composites. The development of nano-array catalysts for electrocatalysis paves the way toward rational design of high-efficiency fuel cells and high-performance supercapacitors for chemical energy conversion and fuel production.

2.4 Nano-Array Catalysts for Photocatalysis

2.4.1 Nano-Array Catalyst for Photocatalytic Degradation of Organic Dyes

Organic dyes are one of the largest pollutants in water. Photocatalytic degradation of organic dyes is reported to be one of the potential methods to effectively remove the pollutants. Semiconductor nanostructured materials are among the most efficient photocatalysts. ZnO nanowires have been reported to be able to decompose methylene blue [53]. The recombination of photo-generated electron and holes is considered to be the major hurdle limiting the efficiency of photocatalysts. One of the possible solutions is to decorate precious metals on the metal oxide nanostructures, which will form Schottky barrier on the interface and help transfer electron into precious metals. Ag/ZnO nanorod array with different Ag loading amount were systematically investigated to achieve the highest photocatalytic degradation ratio, and the molar content of 0.37% was found to be the best dosage [54].

Gao and coworkers [55] reported the photocatalytic performance of ZnO/LSCO composite nanorod array on decomposing of methyl orange in water. ZnO/LSCO composite nanorod array were grown on Si substrate by two-step method. Firstly, ZnO nanorod

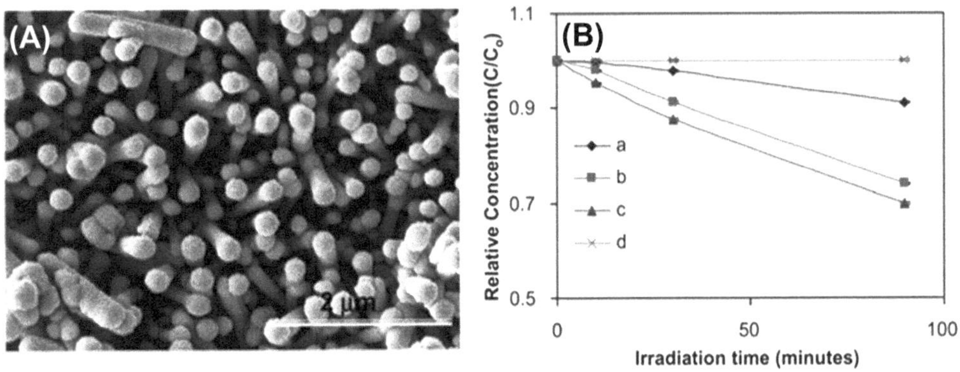

Figure 14 (A) Top view SEM images of ZnO/LSCO composite nanorod arrays. (B) Photodegradation of methyl orange in the presence of different nanorod arrays: a, ZnO nanorod arrays; b, LSCO thin film; c, ZnO/LSCO composite nanorod arrays; and d, blank control. *Reproduced with permission from [55].*

arrays were grown on Si substrate by hydrothermal method. Then LSCO nanoshell was either coated on the ZnO nanorod by solution dip coating method or sputtering method. Shown in Figure 14(A) was the top view SEM image of ZnO/LSCO nanorod array by sputtering method. The rough grainy and porous surface could be observed from the SEM. The diameter of the ZnO/LSCO nanorod was around 100 nm, and the average length was about 1 μm. Photocatalytic decomposition of methyl orange was used as a probe reaction to investigate the photocatalytic activity of ZnO/LSCO-based nano-array photocatalysts. In the photodegradation test, the concentration of methyl orange was controlled as $10\,mg\,L^{-1}$. Prior to the irradiation, adsorption/degradation equilibrium was established by sonicating the dye solution with the immersed ZnO/LSCO nano-array substrate in the dark for 30 min. The solutions with the substrate were then irradiated by UV light. The concentrations of methyl orange were monitored and analyzed by measuring the absorbance at 464 nm wavelength with UV-Vis absorption spectroscopy. Time-dependent concentration change was depicted in Figure 14(B). Three different control samples such as ZnO nanorod array on Si, LSCO thin film on Si, and pure Si substrate were used in the test. Among the tested samples, ZnO/LSCO composite nanorod array showed the best performance with the highest reaction constant around $3.18 \times 10^{-3}\,min^{-1}$. The enhanced photocatalytic activity may originate from the increased surface area, transition metal sites from the LSCO shell, and the interaction between ZnO and LSCO.

Photo active nanostructure array can also be grown on 3D surfaces to make a new type of 3D nano-array photocatalysts. Gao and coworkers [56] fabricated a recyclable photocatalysts by directly growing ZnO nanorod array on a magnetic core-shell particles ($Fe_xO_y@SiO_2$). Figure 15(A) showed the low magnification SEM images of a bundle of as-prepared $Fe_xO_y@SiO_2$-ZnO koosh ball structures. The detailed structure information can be collected from the single koosh ball in Figure 15(B). The diameter

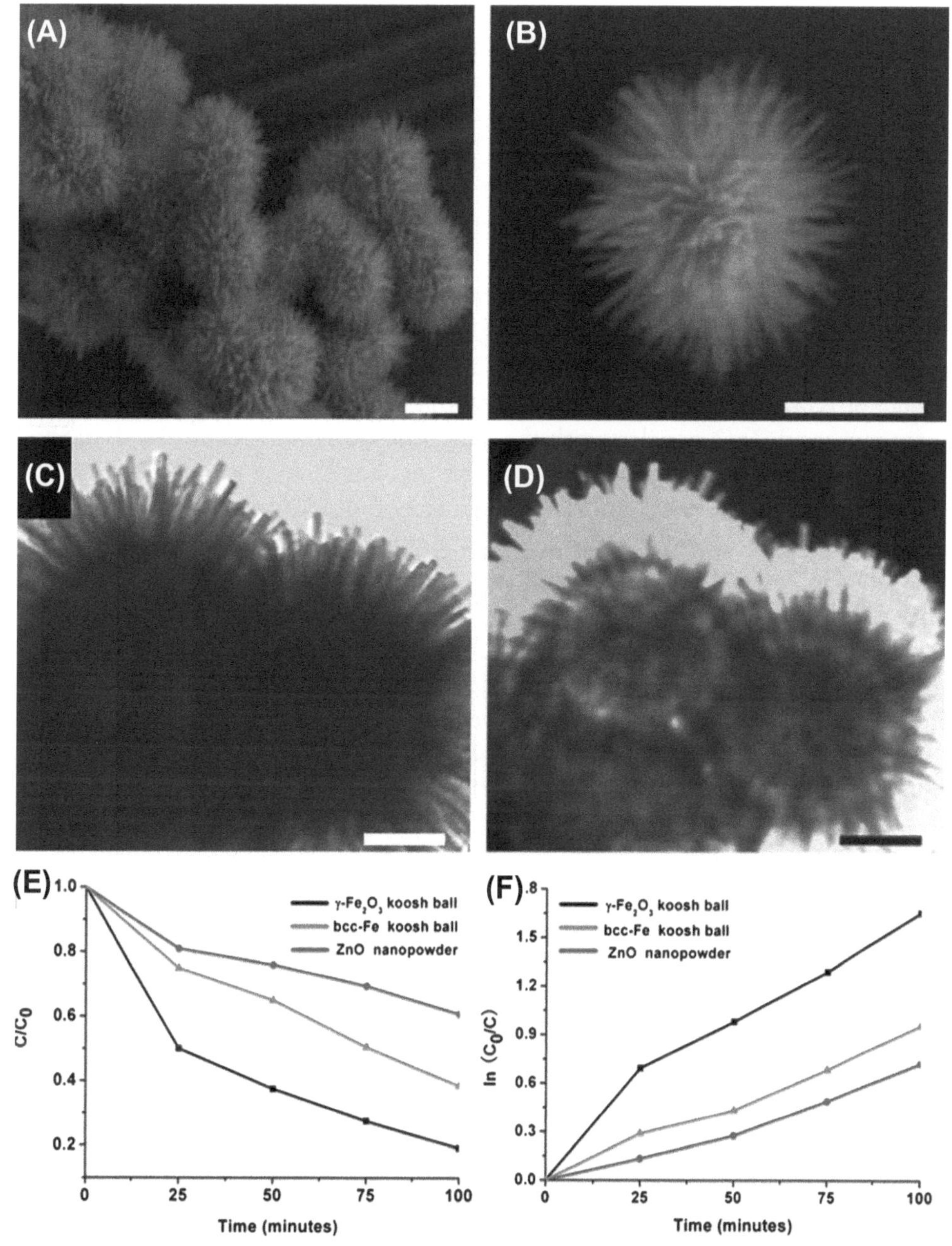

Figure 15 (A) A low magnification SEM image of typical koosh ball architectures. (B) A high magnification SEM image of a typical koosh ball. (C) and (D) are bright-field and dark-field TEM image of koosh ball architectures. (E) Photodegradation of RhB by 2.5 mg commercial ZnO nanopowders; bcc-Fe@SiO_2–ZnO koosh balls and γ-Fe_2O_3@SiO_2–ZnO koosh balls. (F) Kinetic analysis for photocatalytic reactions. Scale bar: (A) 1 μm; (B–D) 500 nm. *Reproduced with permission from reference [56].*

of the koosh ball was around 1 μm with ZnO (50 nm in diameter and 200 nm in length) nanorod array uniformly grown out of the Fe_xO_y@SiO_2 core-shell surface. The bright and dark field TEM images of as-prepared koosh ball were shown in Figure 15(C) and Figure 15(D), respectively, from which we can clearly see the layered geometry with Fe_xO_y@SiO_2 core-shell structure as the support for nanorod arrays. More TEM and SEM analysis indicated that the diameters of Fe_xO_y core and SiO_2 shell were around 700 and 50 nm, respectively. It was worth noticing that the phase of Fe_xO_y core could be tuned from bcc-Fe, γ-Fe_2O_3, α-Fe_2O_3 to Fe_3O_4 in H_2 reduction treatment, which expanded the application of as-prepared koosh ball structure with varied magnetic property. Moreover, the photo-degradation of Rhodamine B (RhB) was investigated on the as-prepared koosh balls. As shown in Figure 15(E) were the time-dependent concentrate change of RhB for γ-Fe_2O_3@SiO_2–ZnO koosh ball, bcc-Fe@SiO_2–ZnO koosh ball, and commercial ZnO powders with the same ZnO loading amount. Bcc-Fe@SiO_2–ZnO-controlled sample was designed and prepared by reducing γ-Fe_2O_3 in H_2 atomsphere to help probe the controversial arguments on ZnO nanowire photocatalytic mechanism: the removal of ligand impurity or oxygen defects. From Figure 15(E), γ-Fe_2O_3@SiO_2–ZnO koosh ball had better conversion efficiency than bcc-Fe@SiO_2–ZnO koosh ball, and commercial ZnO powders showed the lowest efficiency. It was proved that oxygen defects on the ZnO nanowire surface, which served as trapping sites for photo generated electron and holes, played a dominant role for the photocatalytic process of koosh ball. The photogenerated electron and holes then interacted with the surface absorbed RhB molecule to degrade the dye molecules. The corresponding kinetic analysis were plotted in Figure 15(F), which indicated that the linear relation between ln(Co/C) and reaction time matched with the quasi-first order reaction nature of RhB photodegradation. The enhanced photocatalytic activity on koosh ball structure were mainly attributed to the enhanced surface area and promoted surface oxygen defects. The Fe_xO_y core and SiO_2 shell may contribute to the photodegradation by supplying a co-catalyst site and prohibit the recombination of electron-hole on the interface of ZnO and Fe_xO_y separately.

TiO_2-based nanocatalysts are another group of widely investigated photocatalysts for organic dye pollutant degradation. However, the problem associated with TiO_2 nanocatalysts is that it only works efficiently in the ultraviolet wavelength region. A great amount of effort had been put to improve the catalytic performance of TiO_2-based catalysts in the visible light region. One approach is to dope TiO_2 with other element to tune its band gap into visible region. Nitrogen-doped TiO_2 nanorod and NTAs were fabricated by He and coworkers [57] with a hydrothermal method. And their corresponding photocatalytic performances were evaluated under different light sources. The photodegradation of methyl orange were investigated on commercial TiO_2 nanopowders, N-doped TiO_2 nanorod array, and NTA. N-doped TiO_2 NTs showed the fastest degradation rate due to their ~8 times higher surface area than that of TiO_2 nanorod arrays. Precious metal Ag nanoparticles loaded Nitrogen-doped TiO_2 nanopillar arrays were also prepared to help reducing the electron-hole recombination and improving the photodegradation rate by Barber and coworkers [58]. Nitrogen-doped TiO_2 after 550 °C annealing incorporation with Ag nanoparticles showed the highest photodegradation rate of RhB. TiO_2-based heterostructure nano-arrays were also reported to show excellent photocatalytic and photoelectrochemical properties. Cr-doped $SrTiO_3/TiO_2$ heterostructure NTAs were fabricated and studied for dye degradation by Bi and coworkers [59]. The as-prepared heterostructure NTA displayed much higher photocatalytic activity than both pure TiO_2 NTAs and $SrTiO_3/TiO_2$ NTAs under visible light irradiation.

GaN nano-arrays were also proposed to be good photocatalysts for dye degradation especially in the case of harsh environment under a wide range of pHs. GaN nano-array photocatalysts were prepared by Yi and coworkers via a metal oxide chemical vapor deposition method [60]. Compared with the well-known ZnO and TiO_2 photocatalysts, GaN nano-array showed much better chemical stability under strong acidic or basic conditions, which made it very useful for many industrial processes. The photocatalytic activity of GaN nano-array was enhanced in the acid pH region between pH 2 and 5, and the photocatalytic activity of GaN nano-array was much better than ZnO and TiO_2 in the lower pH range.

Other semiconductor nanostructure array such as CdSe NTA [61], Si nanowire array [62], and Sn-filled SnO_2 NTA [63] were also fabricated and showed the photocatalytic activity on organic dye degradation under visible light and UV light.

2.4.2 Nano-Array Catalysts for Photocatalytic Water Splitting

Hydrogen has been widely recognized as a clean and more efficient fuel due to its advantages such as no greenhouse gas emission and high energy conversion to electricity through fuel cells. Even though the majority of H_2 are produced from fossil fuel, H_2 production from renewable photon energy is more attractive. TiO_2 nanocatalysts were widely used as photocatalysts for waste water treatment. One-dimensional TiO_2 nanostructure (nanowire, nanorod, and nanotube) showed better photocatalytic performance due to its direct electron transport pathway, lower electron-hole recombination rate. Reduced graphene oxide (RGO) was reported to be able to promote the photocatalytic performance of TiO_2 nanoparticles because of its high surface area and excellent electron conductivity. Tian and coworkers synthesized well-defined RGO-TiO_2 composite by in situ growing TiO_2 nanorod array on the RGO sheet and investigated its photocatalytic production of H_2 [64]. The hierarchical composites of RGO–TiO_2 nanowire arrays facilitate the improved interface adhesion and stronger interaction, which promotes interfacial charge transfer and reduced self-agglomeration. Figure 16(A) shows the SEM images of as-prepared 3% RGO–TiO_2 nanowire arrays. Anatase TiO_2 nanorod arrays (proved by XRD analysis) uniformly coated on the surface of graphene sheet. Photocatalytic H_2 evolution rate of RGO–TiO_2 with different mass loadings 3% mechanical mixing RGO–TiO_2 and pure TiO_2 under UV light irradiation were depicted in Figure 16(B). Compared with pure

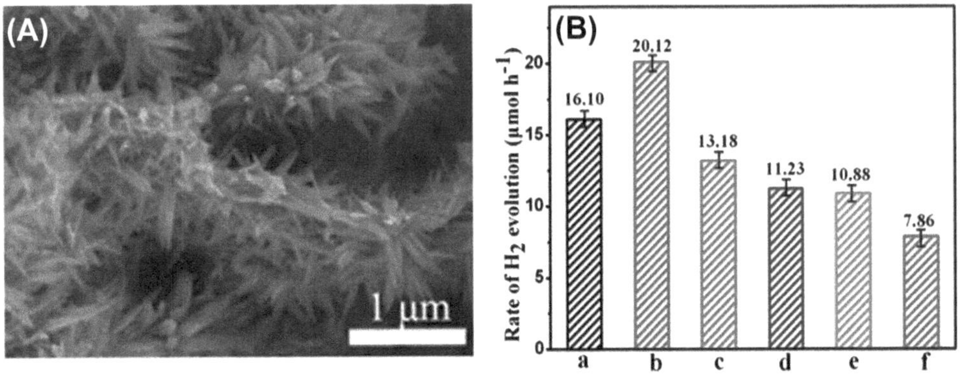

Figure 16 (A) SEM image of 3 wt% TiO_2 nanorod array on reduced graphene oxide (RGO-TiO_2); (B) photocatalytic H_2 evolution rates under UV light illumination: 2 wt% RGO–TiO_2 (a), 3 wt% RGO–TiO_2 (b), 4 wt% RGO–TiO_2 (c), 5 wt% RGO–TiO_2 (d), 3 wt% RGO–TiO_2 generated by mechanical mixing (e) and pure TiO_2 (f). Catalyst amount is 0.07 g. *Reproduced with permission from reference [64].*

TiO_2, the presence of RGO enhanced the photocatalytic performance of all the RGO–TiO_2 samples, which confirms the promotion effect of RGO upon TiO_2 due to its enlarged surface area and electron conductivity. In situ grown 3% RGO–TiO_2 nano-array showed doubled H_2 evolution rate than the mechanical mixing 3% RGO–TiO_2. It suggested that the synergetic catalytic effect and strong interaction between RGO and anatase TiO_2 in the in situ grown hierarchical array can provide improved electron-hole pair separation and boosted catalytic active sites. Interestingly, among the different mass loading RGO–TiO_2 samples, 3% RGO–TiO_2 showed the maximum H_2 evolution rate than 2%, 4%, and 5% samples, which could be attributed to the optimized balance of exposed TiO_2 active sites and TiO_2/RGO interface.

The excellent photocatalytic H_2 production performance exhibited from in situ grown RGO–TiO_2 nano-array demonstrated the synergetic interface effect and provided a facile strategy to prepare photocatalytically active nano-array catalysts.

GaN nanowire array and Rh/Cr_2O_3 core-shell nanostructures loaded GaN nanowire array were also reported to be a stable and effective photocatalyst. Mi and coworkers [65] synthesized the wafer-level GaN nanowire array by molecular beam epitaxy method. The photocatalytic tests were performed by using a 300 W xenon lamp as an outer irradiation source and gas chromatography as a detector. Pristine GaN nanowire array showed good H_2 generation and O_2 generation properties in the half reaction tests but lower activity of the theoretical prediction. It is because of the presence of dangling bonds and/or surface defects in practical GaN nanowire surface. To further improve the photocatalytic activity, Rh/Cr_2O_3 core-shell nanostructure cocatalysts were loaded on GaN nanowire array through photodeposition method. And Cr_2O_3 shell was introduced to prevent the back formation of H_2O from H_2 and O_2 on the precious metal surface, which was typical in most of the precious metals. The overall water splitting experiments were performed on the Rh/Cr_2O_3 core-shell loaded GaN nanowire array. And the turnover number of as-prepared composite nanowire array per unit time was dramatically higher than the literature reported powder samples. Further improvements were also possible by tuning the size, morphology, as well as band gap of GaN nanowires.

Si/Cu_2O core-shell nanowire arrays were also reported as a good candidate for photocatalytic water splitting by Zheng and coworkers [66]. Cu_2O nanocrystals were electroless deposited on Si nanowire array (80–200 nm in diameter and 10 μm in length) and the thickness of Cu_2O nanocrystal shell could be controlled by deposition time. The photocatalytic water splitting tests were carried out on pure Si nanowire arrays, planar Si/Cu_2O structure, and Si NW/Cu_2O core-shell nanowire arrays under simulated sunlight irradiation. Si NW/Cu_2O core-shell nanowire arrays showed the remarkable performance due to a light trapping effect of NW array structure, the formation of heterojunctions, and the increased effective surface area. Pt loading could improve the performance even more. More and more high-efficient composite catalysts will be developed in the near future.

3. CONCLUSION

The nano-array based catalysts with well-defined structure, ultraefficient, robust, and multifunctional performance represented a new class of nanocatalysts, which exhibited superior performance to the reported counterparts in powder form in a broad array of applications including exhaust aftertreatment, catalytic combustion, electrocatalysis, and photocatalysis. The modeled nano-array-based catalysts could not only serve as an ideal platform of exploring the origin of catalytic activity, mechanism of nanoscale diffusion, but also hold the potential to be scaled up for industrially relevant level by integrating nanostructures directly with the commercial 3D substrates. Essentially, the nano-array catalysts could bridge the gap between nano-catalysis science and practical industrial applications. Considering a great deal of effort involved in this area, more efficient nano-array catalysts are expected to be demonstrated in the next few years. The nano-array catalysts hold the potential to be the next generation automotive and industrial catalysts.

ACKNOWLEDGMENT

The authors are grateful for the generous financial support from the US Department of Energy, the US National Science Foundation, and the Alexander von Humboldt Foundation. Z. R. acknowledges the General Electric Graduate Fellowship for Innovation.

REFERENCES

[1] V. Polshettiwar, R. Luque, A. Fihri, H. Zhu, M. Bouhrara, J.-M. Basset, Magnetically recoverable nanocatalysts, Chem. Rev. 111 (2011) 3036–3075.
[2] S.M. George, Introduction: heterogeneous catalysis, Chem. Rev. 95 (1995) 475–476.

[3] A.J. Gellman, N. Shukla, Nanocatalysis: more than speed, Nat. Mater. 8 (2009) 87–88.

[4] Y. Guo, Z. Ren, W. Xiao, C. Liu, H. Sharma, H. Gao, A. Mhadeshwar, P.-X. Gao, Robust 3-D configurated metal oxide nano-array based monolithic catalysts with ultrahigh materials usage efficiency and catalytic performance tunability, Nano Energy 2 (2013) 873–881.

[5] R.J.H. Voorhoeve, D.W. Johnson Jr, J.P. Remeika, P.K. Gallagher, Perovskite oxides: materials science in catalysis, Science 195 (1977) 827–833.

[6] R.J.H. Voorhoeve, J.P. Remeika, P.E. Freeland, B.T. Matthias, Rare-earth oxides of manganese and cobalt rival platinum for the treatment of carbon monoxide in auto exhaust, Science 177 (1972) 353–354.

[7] J.E. Parks, Less costly catalysts for controlling engine emissions, Science 327 (2010) 1584–1585.

[8] W. Wang, G. McCool, N. Kapur, G. Yuan, B. Shan, M. Nguyen, U.M. Graham, B.H. Davis, G. Jacobs, K. Cho, X. Hao, Mixed-phase oxide catalyst based on Mn-Mullite (Sm, Gd)Mn_2O_5 for no oxidation in diesel exhaust, Science 337 (2012) 832–835.

[9] C.H. Kim, G. Qi, K. Dahlberg, W. Li, Strontium-doped perovskites rival platinum catalysts for treating NOx in simulated diesel exhaust, Science 327 (2010) 1624–1627.

[10] Y. Guo, Z. Zhang, Z. Ren, H. Gao, P.-X. Gao, Synthesis, characterization and CO oxidation of TiO2/(La,Sr)MnO3 composite nanorod array., Catal. Today. 184 (2012) 178–183.

[11] Y-j. Feng, L-l. Liu, X-d Wang, Hydrothermal synthesis and automotive exhaust catalytic performance of CeO_2 nanotube arrays, J. Mater. Chem. 21 (2011) 15442–15448.

[12] Y. Feng, X. Zheng, Plasma-enhanced catalytic CuO nanowires for CO oxidation, Nano Lett. 10 (2010) 4762–4766.

[13] A.M. Contreras, X.M. Yan, S. Kwon, J. Bokor, G.A. Somorjai, Catalytic CO oxidation reaction studies on lithographically fabricated platinum nanowire arrays with different oxide supports, Catal. Lett. 111 (2006) 5–13.

[14] Z. Ren, Y. Guo, Z. Zhang, C. Liu, P.-X. Gao, Nonprecious catalytic honeycombs structured with three dimensional hierarchical Co_3O_4 nano-arrays for high performance nitric oxide oxidation, J. Mater. Chem. A 1 (2013) 9897–9906.

[15] R.F. Hicks, H. Qi, M.L. Young, R.G. Lee, Structure sensitivity of methane oxidation over platinum and palladium, J. Catal. 122 (1990) 280–294.

[16] T.R. Baldwin, R. Burch*, Catalytic combustion of methane over supported palladium catalysts: I. Alumina supported catalysts, Appl. Catal. 66 (1990) 337–358.

[17] L.-h. Xiao, K.-p. Sun, X.-l. Xu, X.-n. Li, Low-temperature catalytic combustion of methane over Pd/CeO_2 prepared by deposition–precipitation method, Catal. Comm. 6 (2005) 796–801.

[18] A.D. Mayernick, M.J. Janik, Methane oxidation on Pd–Ceria: a DFT study of the mechanism over $Pd_xCe_{1-x}O_2$, Pd, and PdO, J. Catal. 278 (2011) 16–25.

[19] J.R. Paredes, E. Díaz, F.V. Díez, S. Ordóñez, Combustion of methane in lean mixtures over bulk transition-metal oxides: evaluation of the activity and self-deactivation, Energy Fuels 23 (2008) 86–93.

[20] L. Hu, Q. Peng, Y. Li, Selective synthesis of Co_3O_4 nanocrystal with different shape and Crystal plane effect on catalytic property for methane combustion, J. Am. Chem. Soc. 130 (2008) 16136–16137.

[21] Q.-X. Gao, X.-F. Wang, J.-L. Di, X.-C. Wu, Y.-R. Tao, Enhanced catalytic activity of [small alpha]-Fe_2O_3 nanorods enclosed with {110} and {001} planes for methane combustion and CO oxidation, Catal. Sci. Tech. 1 (2011) 574–577.

[22] Y. Feng, P.M. Rao, D.R. Kim, X. Zheng, Methane oxidation over catalytic copper oxides nanowires, Proc. Combust. Inst. 33 (2011) 3169–3175.

[23] J. Mei, Y. Diao, A.L. Appleton, L. Fang, Z. Bao, Integrated materials design of organic semiconductors for field-effect transistors, J. Am. Chem. Soc. 135 (2013) 6724–6746.

[24] Y. Feng, X. Zheng, Copper ion enhanced synthesis of nanostructured cobalt oxide catalyst for oxidation of methane, ChemCatChem 4 (2012) 1551–1554.

[25] G. Saracco, G. Scibilia, A. Iannibello, G. Baldi, Methane combustion on Mg-doped $LaCrO_3$ perovskite catalysts, Appl. Catal., B 8 (1996) 229–244.

[26] S. Cimino, L. Lisi, S. De Rossi, M. Faticanti, P. Porta, Methane combustion and CO oxidation on $LaAl_{1-x}MnxO_3$ perovskite-type oxide solid solutions, Appl. Catal., B 43 (2003) 397–406.

[27] N. Chouhan, R.-S. Liu, in: Electrochemical Technologies for Energy Storage and Conversion, Wiley-VCH Verlag GmbH & Co. KGaA, 2011, pp. 1–43.

[28] U.S. Energy Information Administration, Washington, DC 2013, 71.
[29] N.M. Markovic, Electrocatalysis: Interfacing electrochemistry, Nat. Mater. 12 (2013) 101–102.
[30] L.-X. Ding, A.-L. Wang, G.-R. Li, Z.-Q. Liu, W.-X. Zhao, C.-Y. Su, Y.-X. Tong, Porous Pt–Ni–P composite nanotube arrays: highly electroactive and durable catalysts for methanol electrooxidation, J. Am. Chem. Soc. 134 (2012) 5730–5733.
[31] R.Z. Ma, J. Liang, B.Q. Wei, B. Zhang, C.L. Xu, D.H. Wu, Study of electrochemical capacitors utilizing carbon nanotube electrodes, J. Power Sources 84 (1999) 126–129.
[32] C. Niu, E.K. Sichel, R. Hoch, D. Moy, H. Tennent, High power electrochemical capacitors based on carbon nanotube electrodes, Appl. Phys. Lett. 70 (1997) 1480–1482.
[33] T. Matsumoto, T. Komatsu, K. Arai, T. Yamazaki, M. Kijima, H. Shimizu, Y. Takasawa, J. Nakamura, Reduction of Pt usage in fuel cell electrocatalysts with carbon nanotube electrodes, Chem. Comm. (2004) 840–841.
[34] Z. Liu, X. Lin, J.Y. Lee, W. Zhang, M. Han, L.M. Gan, Preparation and characterization of platinum-based electrocatalysts on multiwalled carbon nanotubes for proton exchange membrane fuel cells, Langmuir 18 (2002) 4054–4060.
[35] Z.Q. Tian, S.P. Jiang, Y.M. Liang, P.K. Shen, Synthesis and characterization of platinum catalysts on multiwalled carbon nanotubes by intermittent microwave irradiation for fuel cell applications, J. Phys. Chem. B 110 (2006) 5343–5350.
[36] K. Balasubramanian, M. Burghard, Chemically functionalized carbon nanotubes, Small 1 (2005) 180–192.
[37] R. Yu, L. Chen, Q. Liu, J. Lin, K.-L. Tan, S.C. Ng, H.S.O. Chan, G.-Q. Xu, T.S.A. Hor, Platinum deposition on carbon nanotubes via chemical modification, Chem. Mater. 10 (1998) 718–722.
[38] Y.L. Hsin, K.C. Hwang, C.-T. Yeh, Poly(vinylpyrrolidone)-Modified graphite carbon nanofibers as promising supports for PtRu catalysts in direct methanol fuel cells, J. Am. Chem. Soc. 129 (2007) 9999–10010.
[39] X.L. Li, Y.Q. Liu, L. Fu, L.C. Cao, D.C. Wei, Y. Wang, Efficient synthesis of carbon nanotube–nanoparticle hybrids, Adv. Funct. Mater. 16 (2006) 2431–2437.
[40] W. Shuangyin, J. San Ping, W. Xin, Polyelectrolyte functionalized carbon nanotubes as a support for noble metal electrocatalysts and their activity for methanol oxidation, Nanotechnology 19 (2008) 265601.
[41] S. Chakraborty, C. Retna Raj, Electrocatalytic performance of carbon nanotube-supported palladium particles in the oxidation of formic acid and the reduction of oxygen, Carbon 48 (2010) 3242–3249.
[42] K. Gong, F. Du, Z. Xia, M. Durstock, L. Dai, Nitrogen-doped carbon nanotube arrays with high electrocatalytic activity for oxygen reduction, Science 323 (2009) 760–764.
[43] R. Habibi, A.M. Rashidi, J.T. Daryan, A.M.a. zadeh, Study of the rod-like and spherical nano-ZnO morphology on H_2S removal from natural gas, Appl. Surf. Sci. 257 (2010) 434–439.
[44] J. Liu, J. Jiang, C. Cheng, H. Li, J. Zhang, H. Gong, H.J. Fan, Co_3O_4 Nanowire@MnO_2 ultrathin nanosheet Core/Shell arrays: a new class of high-performance pseudocapacitive materials, Adv. Mater. 23 (2011) 2076–2081.
[45] J. Xu, Q. Wang, X. Wang, Q. Xiang, B. Liang, D. Chen, G. Shen, Flexible asymmetric supercapacitors based upon Co_9S_8 Nanorod//Co_3O_4@RuO_2 nanosheet arrays on carbon cloth, ACS Nano 7 (2013) 5453–5462.
[46] H. Xia, D. Zhu, Z. Luo, Y. Yu, X. Shi, G. Yuan, J. Xie, Hierarchically structured Co_3O_4@Pt@MnO_2 nanowire arrays for high-performance supercapacitors, Sci. Rep. 3 (2013) 2978.
[47] C. Koenigsmann, S.S. Wong, One-dimensional noble metal electrocatalysts: a promising structural paradigm for direct methanol fuel cells, Energy Environ. Sci. 4 (2011) 1161–1176.
[48] A. Morozan, B. Jousselme, S. Palacin, Low-platinum and platinum-free catalysts for the oxygen reduction reaction at fuel cell cathodes, Energy Environ. Sci. 4 (2011) 1238–1254.
[49] N.M. Marković, T.J. Schmidt, V. Stamenković, P.N. Ross, Oxygen reduction reaction on Pt and Pt bimetallic surfaces: a selective review, Fuel Cells 1 (2001) 105–116.
[50] L. Cademartiri, G.A. Ozin, Ultrathin Nanowires—A materials chemistry perspective, Adv. Mater. 21 (2009) 1013–1020.
[51] Y. Xu, M. Shao, M. Mavrikakis, R.R. Adzic, in: Fuel Cell Catalysis, John Wiley & Sons, Inc., 2008, pp. 271–315.

[52] J. Zhang, M.B. Vukmirovic, Y. Xu, M. Mavrikakis, R.R. Adzic, Controlling the catalytic activity of platinum-monolayer electrocatalysts for oxygen reduction with different substrates, Angew. Chem., Int. Ed. 44 (2005) 2132–2135.
[53] C. Ma, Z. Zhou, H. Wei, Z. Yang, Z. Wang, Y. Zhang, Rapid large-scale preparation of ZnO nanowires for photocatalytic application, Nanoscale Res. Lett. 6 (2011) 1–5.
[54] C. Ren, B. Yang, M. Wu, J. Xu, Z. Fu, Y. lv, T. Guo, Y. Zhao, C. Zhu, Synthesis of Ag/ZnO nanorods array with enhanced photocatalytic performance, J. Hazard. Mater. 182 (2010) 123–129.
[55] D. Jian, P.-X. Gao, W. Cai, B.S. Allimi, S. Pamir Alpay, Y. Ding, Z.L. Wang, C. Brooks, Synthesis, characterization, and photocatalytic properties of ZnO/(La,Sr)CoO_3 composite nanorod arrays, J. Mater. Chem. 19 (2009) 970–975.
[56] Z. Ren, Y. Guo, G. Wrobel, D.A. Knecht, Z. Zhang, H. Gao, P.-X. Gao, Three dimensional koosh ball nanoarchitecture with a tunable magnetic core, fluorescent nanowire shell and enhanced photocatalytic property, J. Mater. Chem. 22 (2012) 6862–6868.
[57] Z. He, H.Y. He, Synthesis and photocatalytic property of N-doped TiO_2 nanorods and nanotubes with high nitrogen content, Appl. Surf. Sci. 258 (2011) 972–976.
[58] W. Jiang, N. Ullah, G. Divitini, C. Ducati, R.V. Kumar, Y. Ding, Z.H. Barber, Vertically oriented TiO_xN_y nanopillar arrays with embedded Ag nanoparticles for visible-light photocatalysis, Langmuir 28 (2012) 5427–5431.
[59] Z. Jiao, T. Chen, J. Xiong, T. Wang, G. Lu, J. Ye, Y. Bi, Visible-light-driven photoelectrochemical and photocatalytic performances of Cr-doped $SrTiO_3/TiO_2$ heterostructured nanotube arrays, Sci. Rep. 3 (2013) 2720.
[60] H.S. Jung, Y.J. Hong, Y. Li, J. Cho, Y.-J. Kim, G.-C. Yi, Photocatalysis using GaN nanowires, ACS Nano 2 (2008) 637–642.
[61] H. Zhu, Q. Li, Visible light-driven CdSe nanotube array photocatalyst, Nanoscale Res. Lett. 8 (2013) 1–6.
[62] Y. Liu, G. Ji, J. Wang, X. Liang, Z. Zuo, Y. Shi, Fabrication and photocatalytic properties of silicon nanowires by metal-assisted chemical etching: effect of H_2O_2 concentration, Nanoscale Res. Lett. 7 (2012) 1–9.
[63] H. Wang, F. Sun, Y. Zhang, K. Gu, W. Chen, W. Li, Photochemical construction of free-standing Sn-filled SnO_2 nanotube array on a solution surface for flexible use in photocatalysis, J. Mater. Chem. 21 (2011) 12407–12413.
[64] X. Cao, G. Tian, Y. Chen, J. Zhou, W. Zhou, C. Tian, H. Fu, Hierarchical composites of TiO_2 nanowire arrays on reduced graphene oxide nanosheets with enhanced photocatalytic hydrogen evolution performance, J. Mater. Chem. A 2 (2014) 4366–4374.
[65] D. Wang, A. Pierre, M.G. Kibria, K. Cui, X. Han, K.H. Bevan, H. Guo, S. Paradis, A.-R. Hakima, Z. Mi, Wafer-level photocatalytic water splitting on GaN nanowire arrays grown by molecular beam epitaxy, Nano Lett. 11 (2011) 2353–2357.
[66] X. Zuzhou, Z. Maojun, L. Sida, M. Li, S. Wenzhong, Silicon nanowire array/Cu_2O crystalline core–shell nanosystem for solar-driven photocatalytic water splitting, Nanotechnology 24 (2013) 265402.

INDEX

Note: Page numbers followed by "f" and "t" indicate figures and tables respectively.

U

V

W

X

Z

FSC
www.fsc.org
MIX
Papier | Fördert
gute Waldnutzung
FSC® C083411

Kolibri 360 GmbH
Ferdinand-Jühlke-Straße 7
99095 Erfurt, Deutschland
produktsicherheit@kolibri360.de